IMAGE-GUIDED HYPOFRACTIONATED STEREOTACTIC RADIOSURGERY

IMAGE-GUIDED HYPOFRACTIONATED STEREOTACTIC RADIOSURGERY

A Practical Approach to Guide Treatment of Brain and Spine Tumors

edited by

Arjun Sahgal, MD
Simon S. Lo, MD
Lijun Ma, PhD
Jason P. Sheehan, MD

CRC Press
Taylor & Francis Group
Boca Raton London New York

CRC Press is an imprint of the
Taylor & Francis Group, an **informa** business

CRC Press
Taylor & Francis Group
6000 Broken Sound Parkway NW, Suite 300
Boca Raton, FL 33487-2742

Printed and bound in India by Replika Press Pvt. Ltd.

Printed on acid-free paper
Version Date: 20160201

International Standard Book Number-13: 978-1-4987-2283-4 (Hardback)

Visit the Taylor & Francis Web site at
http://www.taylorandfrancis.com

and the CRC Press Web site at
http://www.crcpress.com

We are grateful to our families and friends for their support. We are especially grateful to our patients who endure the cancer journey and trust us with their lives.

Contents

Preface

There is no doubt that single-fraction brain radiosurgery has transformed the management of brain metastases. Patients are no longer reflexively treated with whole brain radiation therapy, and it has been recently proven that radiosurgery alone enables many patients to maximize both quality of life and neurocognitive function. At the time of its development, brain radiosurgery was limited to a few sites and technologies, and required an invasive head frame. With improvements and increasing radiosurgery demand, the technology has evolved such that we can deliver radiosurgery within both academic and community practices. Technological advances such as image guidance, micro-multileaf collimators, intensity-modulated radiotherapy, robotic technology, and frameless stereotaxy are quickly becoming standard features on modern linear accelerators, thereby enabling patient access to what is now considered a standard practice for patients with limited brain metastases.

In this book, we provide detailed chapters on the technology, as it is imperative that we understand the capabilities of the technology to maximize efficacy. One such advance is to hypofractionate brain metastases. It has been shown that local control decreases with increasing tumor volume and, by taking advantage of a few fraction approaches, we can dose escalate while maintaining acceptable risks of radiation necrosis. We provide detailed chapters specific to the rationale and clinical outcomes with hypofractionated brain metastases, benign brain tumors, gliomas, and surgical metastases cavities. This trend to hypofractionate will fast become a standard approach, and we are only at the beginning of a transformative phase in the optimization of radiosurgical management of brain tumors.

With radiosurgery having a firm role in the brain, and advances in technology permitting high-precision conformal radiation to body sites, it was only natural for the field to develop stereotactic body radiotherapy (SBRT). One of its applications was in the spine. Similar to the brain, the idea was to maximize local tumor control and pain control, and prevent neurologic catastrophes (malignant epidural spinal cord compression). SBRT has been applied to intact spinal metastases, previously radiated metastases and, increasingly, residual tumors in the postoperative patient. In fact, the trajectory is similar to that of brain indications such that benign spinal tumors are also treated with spine SBRT. What is distinct from the evolution of brain radiosurgery is the ability at the forefront to deliver radiation in a hypofractionated approach rather than just a single-fraction one. There are advantages and disadvantages to either approach, and we provide chapters on the rationale and clinical experience with both fractionation schemes specific to spinal tumors.

We have learned a tremendous amount in the last five years, and the current work includes expert perspectives that summarize contemporary philosophies on SBRT and hypofractionation. We also have a dedicated chapter focused on the vascular effect with hypofractionation, as we have increasingly learned that these pathways may explain the often dramatic responses seldom seen with low-dose standard fractionation schemes. Last, as imaging is crucial to our field, we have a dedicated chapter to advanced magnetic resonance (MR) imaging of the brain, and we expect in the future that spinal imaging will advance as well with functional applications.

It is our honor to provide readers with this compilation specific to brain and spinal indications with hypofractionation, and we include a checklist with each clinical chapter for implementing these approaches into your practices.

Editors

Arjun Sahgal, MD (chief editor), is a leader in the field of high-precision stereotactic radiation to the brain and spine. After training at the University of Toronto, Ontario, Canada, in radiation oncology, he completed a fellowship at the University of California, San Francisco, in brain and spine radiosurgery with Dr. David Larson. Since then he has been recognized as a national and international clinical expert and research leader in radiosurgery. His main focus is on developing spine stereotactic body radiotherapy as an effective therapy for patients with spinal tumors. He has published numerous book chapters on the subject and more than 200 peer-reviewed papers in high-impact journals, including *Journal of Clinical Oncology* and *The Lancet Oncology*. He has edited or written several books specific to research on brain and bone metastases and is an editorial board member for several journals. He was chairman of the International Stereotactic Radiosurgery Society meeting (June 2013) and was a board member for the Brain Tumour Foundation of Canada and the International Stereotactic Radiosurgery Society. He has been invited to speak at several international meetings, has been a visiting professor at various universities, and leads several research groups. His further research activities involve integrating MRI into radiotherapy delivery, combining novel pharmacologic therapies with radiosurgery, and MRI-guided focused ultrasound.

Simon S. Lo, MD, is professor of radiation oncology at Case Western Reserve University, Cleveland, Ohio, and director of radiosurgery services and neurologic radiation oncology at University Hospitals Seidman Cancer Center, Case Comprehensive Cancer Center, Cleveland, Ohio. Dr. Lo graduated from the Faculty of Medicine of The Chinese University of Hong Kong and did his residency in clinical oncology (Royal College of Radiologists, UK curriculum) at Queen Elizabeth Hospital, Hong Kong. He subsequently completed a residency in radiation oncology at the University of Minnesota, Minneapolis, and also received a grant from the American College of Radiation Oncology for a gastrointestinal radiation oncology fellowship at the Mayo Clinic (Minnesota). He was a visiting resident at Princess Margaret Hospital, University of Toronto, Ontario, Canada. He is currently chair of the American College of Radiology Appropriateness Criteria Expert Panel in Bone Metastasis and is the radiation oncology track co-chair for a Radiological Society of North America (RSNA) refresher course. He is an expert in brain and spinal tumors, stereotactic radiosurgery, and stereotactic body radiotherapy (SBRT). He has published more than 135 peer-reviewed papers, more than 50 book chapters, and three textbooks, including a comprehensive textbook in SBRT (27,000 downloads in 32 months). He has given lectures on SBRT to the American Society for Radiation Oncology (ASTRO), RSNA, the Radiosurgery Society, the International Stereotactic Radiosurgery Society, and the American Thoracic Society conferences and in multiple U.S. and international academic centers. He was also a member of both the ASTRO bone and brain metastases taskforces and contributed to the ASTRO guidelines for bone and brain metastases. He is on the editorial boards of multiple oncology journals and is a reviewer for *The Lancet*, *The Lancet Oncology*, *Nature Reviews Clinical Oncology*, *Journal of Clinical Oncology*, *Radiotherapy & Oncology*, and *International Journal of Radiation Oncology: Biology and Physics*. His areas of research are in brain tumors, stereotactic radiosurgery, radiobiological modeling for ablative radiotherapy, SBRT for lung, liver, and spinal tumors, and toxicities associated with SBRT.

Lijun Ma, PhD, is professor in residence of radiation oncology physics and director of the Physics Residency Program at the University of California, San Francisco. Dr. Ma has served in American Association of Physicists in Medicine on multiple task groups and working groups. He currently co-chairs the normal tissue complication probability spine subcommittee and serves on the editorial board of *Medical Physics*. He is board certified by the American Board of Medical Physics and is a member of the American College of Radiology. He has been active professionally in the International Society of Stereotactic Radiosurgery and has served on its executive board. Dr. Ma has published more than 100 papers and more than 20 book chapters, and is a holder of three international patents.

Jason P. Sheehan, MD, graduated with highest distinction in bachelors of chemical engineering at the University of Virginia, Charlottesville, Virginia, where he subsequently earned a master of science in biomedical engineering and a doctorate in biological physics. He earned his medical degree from the University of Virginia and completed his neurosurgical residency at the University of Virginia along with fellowships in stereotactic and functional neurosurgery at the University of Pittsburgh and microsurgery at the Auckland Medical Center in New Zealand.

After his neurosurgical training, he joined the faculty of the University of Virginia's Department of Neurological Surgery. He currently serves as the Harrison Distinguished Professor of Neurological Surgery. He is also the vice chairman of academic affairs, associate director of the residency program, and director of stereotactic radiosurgery.

Dr. Sheehan's research effort focuses on translational and clinical studies for minimally invasive intracranial and spinal surgery. He has published more than 300 papers and has served as the editor for several books. He has received the National Brain Tumor Foundation's Translational Research Award, the Young Neurosurgeon Award from the World Federation of Neurological Surgeons, the Integra Award, the Synthes Skull Base Award, and the Crutchfield Gage Research Award. He serves on the editorial boards of *Neurosurgery, Journal of Neurosurgery, Journal of Neuro-Oncology,* and the *Journal of Radiosurgery and SBRT.* He is a member of the American Association of Neurological Surgeons (AANS), the Congress of Neurological Surgeons (CNS), the Society for Neuro-Oncology, the Society of Pituitary Surgeons, the American Society of Therapeutic Radiology and Oncology, the International Stereotactic Radiosurgery Society, and the Neurosurgical Society of the Virginias. He serves on the executive committee for the AANS/CNS section on tumors and is chair of the radiosurgery committee for the AANS/CNS section on tumors. He is listed in Best Doctors of America.

Contributors

Justus Adamson
Department of Radiation Oncology
Duke University
Durham, North Carolina

Paula Alcaide Leon
Department of Medical Imaging
St. Michael's Hospital
University of Toronto
Toronto, Ontario, Canada

Lilyana Angelov
Department of Neurosurgery
and
Rose Ella Burkhardt Brain Tumor and
    Neuro-Oncology Center
Cleveland Clinic
Cleveland, Ohio

Steven Babic
Department of Medical Physics
Odette Cancer Centre
Sunnybrook Health Sciences Centre
and
Department of Radiation Oncology
University of Toronto
Toronto, Ontario, Canada

Ehsan H. Balagamwala
Department of Radiation Oncology
Taussig Cancer Institute
Cleveland Clinic
Cleveland, Ohio

Igor Barani
Department of Radiation Oncology
and
Department of Neurological Surgery
University of California, San Francisco
San Francisco, California

Sukhjeet S. Batth
Department of Radiation Oncology
Keck School of Medicine
University of Southern California
Los Angeles, California

Kathryn Beal
Department of Radiation Oncology
Memorial Sloan Kettering Cancer Center
New York, New York

John M. Boyle
Department of Radiation Oncology
Duke University
Durham, North Carolina

Martin Brown
Department of Radiation Oncology
Stanford University
Stanford, California

Paul D. Brown
Department of Radiation Oncology
MD Anderson Cancer Center
The University of Texas
Houston, Texas

Michael Chan
Department of Medical Imaging
University of Toronto
Toronto, Ontario, Canada

Eric L. Chang
Department of Radiation Oncology
Keck School of Medicine
and
Norris Cancer Center
University of Southern California
Los Angeles, California

Samuel T. Chao
Department of Radiation Oncology
Taussig Cancer Institute
and
Rose Ella Burkhardt Brain Tumor and
    Neuro-Oncology Center
Cleveland Clinic
Cleveland, Ohio

Steven J. Chmura
Department of Radiation and Cellular Oncology
The University of Chicago
Chicago, Illinois

Or Cohen-Inbar
Department of Neurological Surgery
University of Virginia
Charlottesville, Virginia

John Cuaron
Department of Radiation Oncology
Memorial Sloan Kettering Cancer Center
New York, New York

Gregory J. Czarnota
Department of Radiation Oncology
and
Imaging Research and Physical Sciences
Sunnybrook Health Sciences Centre
and
Department of Medical Biophysics
University of Toronto
Toronto, Ontario, Canada

Martina Descovich
Department of Radiation Oncology
University of California, San Francisco
San Francisco, California

Ahmed El Kaffas
Department of Radiation Oncology
and
Imaging Research and Physical Sciences
Sunnybrook Health Sciences Centre
Toronto, Ontario, Canada
and
Department of Radiology
Stanford University
Stanford, California

Susannah Ellsworth
Department of Radiation Oncology
Indiana University
Indianapolis, Indiana

John C. Flickinger
Department of Radiation Oncology
and
Department of Neurological Surgery
University of Pittsburgh Medical Center
Pittsburgh, Pennsylvania

Chris Heyn
Department of Medical Imaging
Sunnybrook Health Sciences Centre
University of Toronto
Toronto, Ontario, Canada

Peter C. Gerszten
Department of Neurological Surgery
and
Department of Radiation Oncology
University of Pittsburgh Medical Center
Pittsburgh, Pennsylvania

Stefan Glatz
Department of Radiation Oncology
University Hospital Zurich
Zurich, Switzerland

Samuel Bergeron Gravel
Departement of Radiation Oncology
Centre Hospitalier Universitaire de Québec
Université Laval
Québec City, Québec, Canada

Matthias Guckenberger
Department of Radiation Oncology
University Hospital Zurich
Zurich, Switzerland

Ahmed Hashmi
Department of Radiation Oncology
Odette Cancer Centre
Sunnybrook Health Sciences Centre
University of Toronto
Toronto, Ontario, Canada

Zhibin Huang
Department of Radiation Oncology
East Carolina University
Greenville, North Carolina

Brian D. Kavanagh
Department of Radiation Oncology
Anschutz Medical Campus
University of Colorado
Denver, Colorado

Kevin D. Kelley
Northwell Health
Lake Success, New York
and
Department of Radiation Medicine
Center for Advanced Medicine
Hofstra University
Hempstead, New York

John P. Kirkpatrick
Department of Radiation Oncology
Duke University
Durham, North Carolina

Jonathan P.S. Knisely
Northwell Health
Lake Success, New York
and
Department of Radiation Medicine
Center for Advanced Medicine
Hofstra University
Hempstead, New York

David Larson
Department of Radiation Oncology
University of California, San Francisco
San Francisco, California

Young Lee
Department of Medical Physics
Odette Cancer Centre
Sunnybrook Health Sciences Centre
and
Department of Radiation Oncology
University of Toronto
Toronto, Ontario, Canada

Simon S. Lo
Department of Radiation Oncology
Case Comprehensive Cancer Center
Cleveland, Ohio

Lijun Ma
Department of Radiation Oncology
University of California, San Francisco
San Francisco, California

Ariel E. Marciscano
Department of Radiation Oncology and Molecular
    Radiation Sciences
Johns Hopkins University
Baltimore, Maryland

Mihaela Marrero
Northwell Health
Lake Success, New York
and
Department of Radiation Medicine
Center for Advanced Medicine
Hofstra University
Hempstead, New York

Nina A. Mayr
Department of Radiation Oncology
University of Washington
Seattle, Washington

Mary Frances McAleer
Department of Radiation Oncology
MD Anderson Cancer Center
The University of Texas
Houston, Texas

Christopher McGuinness
Department of Radiation Oncology
University of California, San Francisco
San Francisco, California

David Mercier
Departement of Neurosurgery
Centre Hospitalier Universitaire de Québec
Université Laval
Québec City, Québec, Canada

Michael T. Milano
Department of Radiation Oncology
School of Medicine and Dentistry
University of Rochester Medical Center
Rochester, New York

Jacob Miller
Center for Spine Health
Lerner College of Medicine
Cleveland Clinic
Cleveland, Ohio

Sten Myrehaug
Department of Radiation Oncology
Odette Cancer Center
Sunnybrook Hospital
Toronto, Ontario, Canada

Alan Nichol
BC Cancer Agency
Vancouver, British Columbia, Canada

Richard Popple
Department of Radiation Oncology
University of Alabama at Birmingham
Birmingham, Alabama

Kristin J. Redmond
Department of Radiation Oncology and
    Molecular Radiation Sciences
Johns Hopkins University
Baltimore, Maryland

Johannes Roesch
Department of Radiation Oncology
University Hospital Zurich
Zurich, Switzerland

Arjun Sahgal
Department of Radiation Oncology
Odette Cancer Center
Sunnybrook Hospital
Toronto, Ontario, Canada

Joseph K. Salama
Department of Radiation Oncology
Duke University
Durham, North Carolina

David Schlesinger
Department of Radiation Oncology
and
Department of Neurological Surgery
University of Virginia Health System
Charlottesville, Virginia

Jason P. Sheehan
Department of Neurological Surgery
and
Department of Radiation Oncology
and
Department of Neuroscience
University of Virginia Health System
Charlottesville, Virginia

Hany Soliman
Department of Radiation Oncology
Odette Cancer Centre
Sunnybrook Health Sciences Centre
University of Toronto
Toronto, Ontario, Canada

Paul W. Sperduto
Minneapolis Radiation Oncology
University of Minnesota Gamma Knife
Minneapolis, Minnesota

John H. Suh
Department of Radiation Oncology
Taussig Cancer Institute
and
Rose Ella Burkhardt Brain Tumor and Neuro-
    Oncology Center
Cleveland Clinic
Cleveland, Ohio

Hiroshi Tanaka
Division of Radiation Oncology
Tokyo Metropolitan Cancer and
    Infectious Diseases Center
Komagome Hospital
Tokyo, Japan

Bin S. Teh
Department of Radiation Oncology
Weill Cornell Medical College
Houston Methodist Hospital
Houston, Texas

Isabelle Thibault
Departement of Radiation Oncology
Centre Hospitalier Universitaire de Québec
Université Laval
Québec City, Québec, Canada

Nicholas Trakul
Department of Radiation Oncology
Keck School of Medicine
University of Southern California
Los Angeles, California

Daniel M. Trifiletti
Department of Radiation Oncology
University of Virginia Health System
Charlottesville, Virginia

Chia-Lin Tseng
Department of Radiation Oncology
Odette Cancer Centre
Sunnybrook Health Sciences Centre
University of Toronto
Toronto, Ontario, Canada

Kenneth Y. Usuki
Department of Radiation Oncology
School of Medicine and Dentistry
University of Rochester Medical Center
Rochester, New York

Cari Whyne
Orthopaedic Biomechanics Laboratory
Department of Surgery
Sunnybrook Research Institute
University of Toronto
Toronto, Ontario, Canada

Shun Wong
Division of Radiation Oncology
Tokyo Metropolitan Cancer and Infectious
    Diseases Center
Komagome Hospital
Tokyo, Japan

Yoshiya Yamada
Department of Radiation Oncology
Memorial Sloan Kettering Cancer Center
New York, New York

William T. Yuh
Department of Radiology
University of Washington
Seattle, Washington

1

Invited perspectives on hypofractionated stereotactic radiosurgery*

Contents

1.1 RATIONALE FOR SPINE STEREOTACTIC RADIOSURGERY

Yoshiya Yamada

There is mounting evidence that stereotactic radiosurgery (SRS), defined as highly conformal, precision high-dose single-fraction irradiation, is an effective and safe treatment for even the most radioresistant tumors of the spine. A growing body of data suggests that stereotactic spine radiosurgery (SSRS) differs from conventional palliative radiation of solid tumor spine metastases in several important domains. Clinically, investigators have noted that response to conventional radiation differs according to tumor histology. It is well accepted that lymphomas and myeloma, for example, are very radioresponsive phenotypes, and breast cancer and prostate cancer are also relatively favorable response tumors. In contrast, melanomas, sarcomas, renal cell carcinomas, gastrointestinal tumors, and non-small-cell lung cancer (NSCLC) are classified as less responsive or poor responders to conventional palliative radiotherapy. SSRS appears to provide a much higher level of tumor control, which is independent of histology and is also much more durable than conventional palliative radiation. Both prospective and retrospective data suggest that conventional radiation results in a median benefit of less than 4 months in the case of radioresistant disease, while SRS series report local control in excess of 85% with median follow-up that exceeds 1 year. In an era of improving survival in many with stage IV disease, those with spine metastases treated with conventional palliative radiation therapy are surviving beyond the median duration of benefit of 3–4 months. Thus an increasing number of patients are experiencing local relapse of spinal tumors, resulting in significant pain and potentially devastating neurologic compromise. Furthermore, it is not uncommon for patients who have metastases in soft tissue sites that are controlled with systemic agents but are less effective in osseous sites. These can be exceedingly vexing clinical problems in which spinal recurrence threatens performance status and with few options for salvage—that is, until the advent of spine radiosurgery.

It is becoming increasingly apparent that the linear quadratic model does not accurately predict the better-than-expected levels of tumor control that can be achieved with radiosurgery. Many radiobiologists agree that the assumptions upon which the standard linear quadratic model is based are not valid in doses of radiation that exceed 6 Gy/fraction. For example, brain metastases can be effectively controlled at doses of 18–21 Gy in a single fraction, whereas doses in excess of 25 Gy would be predicted to be necessary for similar rates of tumor control. These observations suggest that different mechanisms of response are at play in the setting of high-dose-per-fraction radiotherapy.

Preclinical work from Fuks and Kolesnick at Memorial Sloan Kettering suggests that endothelial apoptosis mediated by ceramide in the plasma membrane of tumor endothelial cells may be an important

---

* This chapter consists of three expert perspective articles invited by the editors. The reader may notice some differences in viewpoints between the authors, reflecting that the field remains in active development without a final, uniform consensus on certain key issues.

mechanism of response in the setting of radiosurgery. This phenomenon appears to a threshold response only observed in doses that exceed 8–10 Gy/fraction and hence are not relevant in conventional palliative radiotherapy. When mice that have been engineered to be unable to mount a ceramide response, implanted tumors which respond briskly to radiosurgerical doses of radiation in wild-type mice demonstrate tepid or no response. Moreover, the addition of VEGF inhibitors can render a highly resistant tumor very radioresponsive. For example, the use of axitinib, a potent VEGF inhibitor, appears to significantly increase the radioeffectiveness of radiosurgery. Dynamic contrast-enhanced MRI, a technique that can accurately measure blood volume in tumor microvasculature, has demonstrated significant reductions in perfusion after spine radiosurgery.

Immune response may also be important in tumor responses to high-dose-per-fraction radiation, in which tumors irradiated with high doses become more immunogenic. Preclinical models also suggest that immune response mediated by CD8+ lymphocytes may also play an important role in the setting of high-dose radiation. It has been noted that tumors implanted in mice without T cells exhibit minimal response to radiosurgical doses of radiation, whereas the same tumors implanted in wild-type mice with competent CD8+ T cells to the same dose demonstrate complete responses. When T cells are suppressed in wild-type mice, the same pattern of radioresistance appears, suggesting that T cells are an important factor in response to high-dose radiation. Abscopal effects may be immune mediated and appear to be most likely in the setting of high-dose-per-fraction radiotherapy with the presence of immune modulators that enhance the immune effect.

Despite the very high doses of radiation utilized in radiosurgery, toxicity is limited. Spinal cord injury is less than 1% of cases. Esophageal toxicity is 6% or less, and grade three complications are extremely rare. Although these observations need more rigorous scrutiny, the lack of severe toxicity may be due to the extremely small normal tissue margins that are used for radiosurgery. Thus, the high-dose regions are extremely tightly controlled, and every effort is made to constrain the dose to within the tumor and minimize the amount of radiation given to adjacent normal tissues. The exception is the risk of vertebral body fracture, because in cases of vertebral body involvement, the target volume typically includes the entire vertebral body and sparing the vertebral body is not possible.

In summary, there are compelling reasons to consider radiosurgery for the management of spine metastases. In particular, tumors that are less responsive to conventional radiotherapy and those with expected survivals beyond 6 months are most likely to benefit from more aggressive treatment. Investigators have described unique mechanisms of response, such as endothelial apoptosis and immune responses, that go beyond the traditional radiobiology that underlies conventional radiation treatment. Image-guided technology has enabled the use of extremely small normal tissue margins, coupled with highly conformal delivery, minimized the dose and the volume of nearby critical structures that receive high doses of radiation, and is an important reason why severe toxicity is extremely rare.

BIBLIOGRAPHY

(August 1999) 8 Gy single fraction radiotherapy for the treatment of metastatic skeletal pain: Randomised comparison with a multifraction schedule over 12 months of patient follow-up. Bone Pain Trial Working Party. *Radiother Oncol* 52(2):111–121.

Al-Omair A, Smith R, Kiehl TR, Lao L, Yu E, Massicotte EM, Keith J, Fehlings MG, Sahgal A (May 2013) Radiation-induced vertebral compression fracture following spine stereotactic radiosurgery: Clinicopathological correlation. *J Neurosurg Spine* 18(5):430–435.

Bilsky MH, Laufer I, Burch S (October 15, 2009) Shifting paradigms in the treatment of metastatic spine disease. *Spine (Phila Pa 1976)* 34(22 Suppl):S101–S107.

Chang EL, Shiu AS, Mendel E, Mathews LA, Mahajan A, Allen PK, Weinberg JS et al. (August 2007) Phase I/II study of stereotactic body radiotherapy for spinal metastasis and its pattern of failure. [In Eng]. *J Neurosurg Spine* 7(2):151–160.

Chu S, Karimi S, Peck KK, Yamada Y, Lis E, Lyo J, Bilsky M, Holodny AI (October 15, 2013) Measurement of blood perfusion in spinal metastases with dynamic contrast-enhanced magnetic resonance imaging: Evaluation of tumor response to radiation therapy. *Spine (Phila Pa 1976)* 38(22):E1418–E1424.

Cox BW, Jackson A, Hunt M, Bilsky M, Yamada Y (August 1, 2012) Esophageal toxicity from high-dose, single-fraction paraspinal stereotactic radiosurgery. [In Eng]. *Int J Radiat Oncol Biol Phys* 83(5):e661–e667.

Garcia-Barros M, Paris F, Cordon-Cardo C, Lyden D, Rafii S, Haimovitz-Friedman A, Fuks Z, Kolesnick R (May 16, 2003) Tumor response to radiotherapy regulated by endothelial cell apoptosis. *Science* 300(5622):1155–1159.

Gerszten PC, Burton SA, Quinn AE, Agarwala SS, Kirkwood JM (2005) Radiosurgery for the treatment of spinal melanoma metastases. [In Eng]. *Stereotact Funct Neurosurg* 83(5–6):213–221.

Gerszten PC, Mendel E, Yamada Y (October 15, 2009) Radiotherapy and radiosurgery for metastatic spine disease: What are the options, indications, and outcomes? [In Eng]. *Spine (Phila Pa 1976)* 34(22 Suppl):S78–S92.

Gibbs IC, Patil C, Gerszten PC, Adler JR, Jr., Burton SA (February 2009) Delayed radiation-induced myelopathy after spinal radiosurgery. [In Eng]. *Neurosurgery* 64(2 Suppl):A67–A72.

Guckenberger M, Sweeney RA, Flickinger JC, Gerszten PC, Kersh R, Sheehan J, Sahgal A (2011) Clinical practice of image-guided spine radiosurgery—Results from an international research consortium. *Radiat Oncol* 6:172.

Guerrero M, Li XA (October 21, 2004) Extending the linear-quadratic model for large fraction doses pertinent to stereotactic radiotherapy. *Phys Med Biol* 49(20):4825–4835.

Katagiri H, Takahashi M, Inagaki J, Kobayashi H, Sugiura H, Yamamura S, Iwata H (December 1, 1998) Clinical results of nonsurgical treatment for spinal metastases. *Int J Radiat Oncol Biol Phys* 42(5):1127–1132.

Kirkpatrick JP, Meyer JJ, Marks LB (October 2008) The linear-quadratic model is inappropriate to model high dose per fraction effects in radiosurgery. *Semin Radiat Oncol* 18(4):240–243.

Kolesnick R, Fuks Z (September 1, 2003) Radiation and ceramide-induced apoptosis. *Oncogene* 22(37):5897–5906.

Lee Y, Auh SL, Wang Y, Burnette B, Wang Y, Meng Y, Beckett M et al. (July 16, 2009) Therapeutic effects of ablative radiation on local tumor require CD8+ T cells: Changing strategies for cancer treatment. *Blood* 114(3):589–595.

Maranzano E, Bellavita R, Rossi R, De Angelis V, Frattegiani A, Bagnoli R, Mignogna M et al. (May 20, 2005) Short-course versus split-course radiotherapy in metastatic spinal cord compression: Results of a phase III, randomized, multicenter trial. *J Clin Oncol* 23(15):3358–3365.

Rao SS, Thompson C, Cheng J, Haimovitz-Friedman A, Powell SN, Fuks Z, Kolesnick RN (April 2014) Axitinib sensitization of high single dose radiotherapy. *Radiother Oncol* 111(1):88–93.

Rose PS, Laufer I, Boland PJ, Hanover A, Bilsky MH, Yamada J, Lis E (October 20, 2009) Risk of fracture after single fraction image-guided intensity-modulated radiation therapy to spinal metastases. [In Eng]. *J Clin Oncol* 27(30):5075–5079.

Sahgal A, Ma L, Gibbs I, Gerszten PC, Ryu S, Soltys S, Weinberg V et al. (September 16, 2009) Spinal cord tolerance for stereotactic body radiotherapy. [In Eng]. *Int J Radiat Oncol Biol Phys* 77(2):548–553.

Seung SK, Curti BD, Crittenden M, Walker E, Coffey T, Siebert JC, Miller W et al. (June 6, 2012) Phase 1 study of stereotactic body radiotherapy and interleukin-2—Tumor and immunological responses. *Sci Transl Med* 4(137):137ra74.

Yamada Y, Bilsky MH, Lovelock DM, Venkatraman ES, Toner S, Johnson J, Zatcky J, Zelefsky MJ, Fuks Z (January 28, 2008) High-dose, single-fraction image-guided intensity-modulated radiotherapy for metastatic spinal lesions. [In Eng]. *Int J Radiat Oncol Biol Phys* 71(2):484–490.

1.2 HYPOFRACTIONATION RATIONALE AND EVIDENCE

Martin Brown

There is a continuing controversy as to whether there is an increased efficacy of high-dose-per-fraction radiotherapy. This has come about from suggestions from preclinical studies that vascular damage by high radiation doses might secondarily lead to increased tumor cell killing (Garcia-Barros et al., 2003; Park et al., 2012), and/or from preclinical and clinical data that high-dose-per-fraction radiation could enhance antitumor immunity (Dewan et al., 2009; Postow et al., 2012). The notion of extra efficacy of high dose per fraction has also emerged from clinical data showing that stereotactic body radiotherapy (SBRT) is producing remarkably high local control in various settings (Timmerman et al., 2010; Rubio et al., 2013). This controversy is directly relevant to the issue of single-dose versus hypofractionated irradiation because it could be argued by the proponents of the extra efficacy of high-dose irradiation that if the individual doses in the hypofractionated regime fall below what is considered to be the cutoff for the vascular damage component (8–10 Gy/fraction), then the hypofractionated regime would be less efficacious than single fractions. However, against these data for extra efficacy of high-dose-per-fraction irradiation, which are based on a stromal component to the tumor response, are preclinical data that argue that it is tumor cell radiosensitivity not stromal damage that governs tumor response (Budach et al., 1993; Li et al., 2014). Further, at least for NSCLC, we have demonstrated that it is the high local tumor doses, not any special efficacy (or "new biology") of high single or fractionated doses, that account for the high local control rates of SBRT (Brown et al., 2013). Further, we have recently reviewed the preclinical and clinical data and have concluded that the response of tumors at high doses is predicted from the classical radiobiology of the 5Rs and that there is no evidence as yet for any special efficacy of high tumor doses beyond that expected from the high cell killing of these doses (Brown et al., 2014). This conclusion is an important prerequisite for any consideration of single dose versus hypofractionated SRS or SBRT. It means that the two modes can be compared using standard radiobiology considerations without invoking any "new biology" at high single doses that might not occur with the hypofractionated doses.

Having established that there is, in most situations, no extra efficacy of high single doses, the question of whether single doses or hypofractionated irradiation is preferable can be addressed using standard radiobiological modeling. The considerations and conclusions from such modeling are as follows:

- Comparisons between different radiation schedules have to be done at equal _biologically effective doses_ (_BEDs_). The conversion formula using the linear quadratic model is $BED = n * d * (1 + d/(\alpha/\beta))$, where n is the number of fractions, d is the dose/fraction, and α/β is a tissue-specific constant, which for the brain is usually assumed to be 3 Gy. This assures that normal tissue damage is the same for each regimen.
- Tumor hypoxia is the main issue that affects the comparison between single and hypofractionated courses. With hypofractionation, there is the opportunity for the hypoxic areas of the tumor to reoxygenate between fractions, thereby increasing tumor cell killing. The results of a modeling study comparing different fractionation patterns is shown in Figure 1.1.
- The modeling in Figure 1.1 shows that there is a definite loss of tumor cell killing in moving from 5 fractions to 1 fraction for the same normal tissue injury (same BED). _So hypofractionation is predicted to be superior to single doses, producing more tumor cell killing for the same BED._
- Is there clinical evidence that stereotactic hypofractionation is superior to single doses for brain tumors? We have recently examined this from an analysis of 2965 patients who were treated with single doses or hypofractionated regimens for NSCLC or brain metastases. _We found that there is a significant lack of efficacy of single doses compared to hypofractionated irradiation for brain metastases for the same BED_ (Shuryak et al., 2015). This is consistent with the modeling study in Figure 1.1.

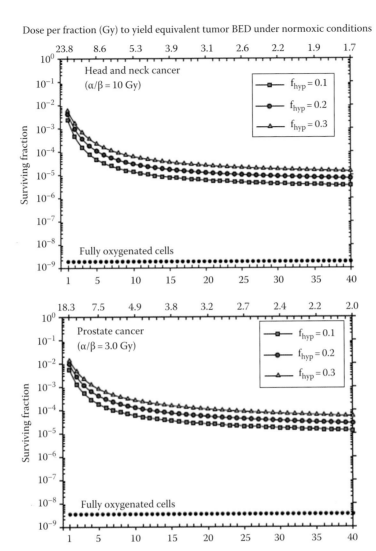

Figure 1.1 Tumor cell survival as a function of dose fraction for equal biologically effective doses assuming daily fractionation and full reoxygenation between doses. Two different α-/β-values are shown with three different values of the hypoxic fraction. (From Carlson, D.J. et al., *Int. J. Radiat. Oncol. Biol. Phys.*, 79, 1188, 2011. With permission.)

REFERENCES

Brown JM, Brenner DJ, Carlson DJ (2013) Dose escalation, not "new biology," can account for the efficacy of SBRT with NSCLC. *Int J Radiat Oncol Biol Phys* 85(5):1159–1160.

Brown JM, Carlson DJ, Brenner DJ (2014) The tumor radiobiology of SRS and SBRT: Are more than the 5 Rs involved? *Int J Radiat Oncol Biol Phys* 88(2):254–262.

Budach W, Taghian A, Freeman J, Gioioso D, Suit HD (1993) Impact of stromal sensitivity on radiation response of tumors. *J Natl Cancer Inst* 85(12):988–993.

Carlson DJ, Keall PJ, Loo BW, Jr., Chen ZJ, Brown JM (2011) Hypofractionation results in reduced tumor cell kill compared to conventional fractionation for tumors with regions of hypoxia. *Int J Radiat Oncol Biol Phys* 79:1188–1195.

Dewan MZ, Galloway AE, Kawashima N, Dewyngaert JK, Babb JS, Formenti SC, Demaria S (2009) Fractionated but not single-dose radiotherapy induces an immune-mediated abscopal effect when combined with anti-CTLA-4 antibody. *Clin Cancer Res* 15(17):5379–5388.

Garcia-Barros M, Paris F, Cordon-Cardo C, Lyden D, Rafii S, Haimovitz-Friedman A, Fuks Z, Kolesnick R (2003) Tumor response to radiotherapy regulated by endothelial cell apoptosis. *Science* 300(5622):1155–1159.

Li W, Huang P, Chen DJ, Gerweck LE (2014) Determinates of tumor response to radiation: Tumor cells, tumor stroma and permanent local control. *Radiother Oncol* 113(1):146–149.

Park HJ, Griffin RJ, Hui S, Levitt SH, Song CW (2012) Radiation-induced vascular damage in tumors: Implications of vascular damage in ablative hypofractionated radiotherapy (SBRT and SRS). *Radiat Res* 177(3):311–327.

Postow MA, Callahan MK, Barker CA, Yamada Y, Yuan J, Kitano S, Mu Z et al. (2012) Immunologic correlates of the abscopal effect in a patient with melanoma. *N Engl J Med* 366(10):925–931.

Rubio C, Morera R, Hernando O, Leroy T, Lartigau SE (2013) Extracranial stereotactic body radiotherapy. Review of main SBRT features and indications in primary tumors. *Rep Pract Oncol Radiother* 18(6):387–396.

Shuryak I, Carlson DJ, Brown JM, Brennan DJ (2015) High-dose and fractionation effects in stereotactic radiation therapy: Analysis of tumor control data from 2,965 patients. *Radiat Oncol* 115(3):327–334.

Timmerman R, Paulus R, Galvin J, Michalski J, Straube W, Bradley J, Fakiris A et al. (2010) Stereotactic body radiation therapy for inoperable early stage lung cancer. *JAMA* 303(11):1070–1076.

1.3 THE FUTURE OF HIGH-DOSE-PER-FRACTION RADIATION

Brian D. Kavanagh

Imagining the future of anything involves one of three fundamental philosophical approaches. First, it could be assumed that the trajectory is precisely calculable, that influencing forces will remain constant, and that a flight path may thus be determined: NASA scientists launch rockets to other planets based on a confident assessment of predictable gravitational conditions. Alternatively, the probability of independent random events that impact a net result can be estimated, and the outlook is then rendered as a Gaussian distribution of possibilities: investors speculate crop prices based on imprecise climate trends and recent history-based population growth expectations. Finally, there is the element of surprise that no one can easily anticipate: here, we enter the domain of psychic fortune tellers to argue that a man with blue eyes will meet a girl with red hair on the night of the next full moon and fall in love and live happily ever after. And so, let us consider the future of high-dose-per-fraction irradiation according to each of these guiding principles and venture a vision of what lies ahead.

If today's momentum is an indicator of things to come, then there will continue to be energetic development of patient management strategies that involve efficient and effective radiation treatment regimens across the full spectrum of patients for whom radiotherapy plays a role in treating a benign or malignant condition. Economic forces, which include forthcoming alternative payment models incentivizing shorter courses of treatment, and clinical forces, in the form of more and more data showing various advantages to patients (convenience, efficacy, lower toxicity, etc.), are aligned in pushing the field toward increasing utilization of short-course, high-dose-per-fraction radiation in the place of lengthier, more traditional course of treatment involving low daily doses of radiotherapy stretching over many weeks.

Mitigating the unbridled enthusiasm for unlimited expansion of ultraconsolidated radiotherapy treatment regimens are the obvious caveats: we need long-term follow-up data for some common cancers before we know the ideal patient selection criteria for high-dose-per-fraction treatment, and there will likely always be anatomic limitations in the form of sensitive normal tissues that cannot so easily tolerate shortened treatment courses. On the other hand, already in the clinic are devices and materials that might enable hypofractionation to be accomplished more easily. For example, the full potential of particle therapy and its steep dose gradients away from tumor-bearing areas has only just recently begun to be explored as a means of shortening treatment courses, and the moment of this writing coincides with the dawn of the era when injectable gels that can separate the tumor from adjacent normal tissues (e.g., prostate from rectum) and thus facilitate hypofractionation enter routine clinical use. The net result of these various clinical, biological, and technical inputs would still lead even the most pessimistic meteorologist to forecast at worst partly cloudy skies but overall very little chance of enough rain to dampen the summertime picnic celebration for those who find these current shifts toward the more frequent use of tightly targeted, potent, tumor-ablative radiation therapy doses to be a reason to rejoice.

Enter, stage left, the *deus ex machina* that nobody sees coming, the disruptive biological or technical innovation that blows everything up. The trendy guess in 2015 would be that some form or forms of an immune system–based intervention will suddenly yield breakthrough results across the board, turning cancer cells into little more than temporarily pesky annoyances that get mopped up by the patient's native T cells and macrophages. Even better, there is preclinical evidence and even a smattering of provocative clinical observations that in certain settings there might be a way to exploit the immunological reaction to high-dose radiation in a way that enhances or otherwise improves the antitumoral effects of immune system–directed therapy.

However, while several forms of immunological therapy have already proven genuinely helpful for an assortment of tumor types, it should be recalled that at one time not so long ago, antiangiogenic agents were expected to revolutionize cancer treatment. True, a few modest gains have been made here and there, but blocking angiogenesis has not proven to be the panacea it once appeared to be. Likewise, immunotherapy in all its protean forms is still not likely to solve the entire cancer problem, despite remarkable gains in selected circumstances.

But what if the next great breakthrough is an agent that reverses everything we now think about high-dose-per-fraction radiation? What if someone develops a compound that can be injected, ingested, or inhaled and that is best combined with low-dose radiation to a large volume rather than high-dose radiation to a small volume? If that were the case, the big question is whether the future radiation medicine community will be intellectually agile enough to shed its biases and consider an open-minded response to such a plot twist, in the same way that open-mindedness has been necessary as current investigators forge ahead into previously uncharted, dogma-shattering domains of high-dose-per-fraction radiation therapy. The challenge for future investigators will be to maintain curiosity, avoid complacency, and uphold integrity so that current high-dose-per-fraction irradiation strategies will become more refined, better selected, and optimally utilized for patients in the years to come.

2

Tumor vascular conundrum: Hypoxia, ceramide, and biomechanical targeting of tumor vasculature

Ahmed El Kaffas and Gregory J. Czarnota

Contents

2.1 TUMOR VASCULAR INFLUENCE ON RADIATION RESPONSE

Tumor blood vessels are the lifeline of cancer; without vascular development, a cancer cannot progress or metastasize. Angiogenesis is considered to be the prevalent method by which vascular development takes place in tumors (Chen et al., 2010). Unlike in normal tissue, tumor angiogenesis is minimally regulated and thought to be continuously active in order to support rapid tumor cell proliferation and a growing tumor mass (Dvorak, 1986). Tumor angiogenesis occurs mainly through endothelial sprouting and involves the upregulation of vascular endothelial growth factor (VEGF), interleukin (IL-8), epithelial growth factor, platelet-derived growth factor (PDGF), and basic fibroblast growth factor (bFGF). More specifically, the process starts with a degradation of the extracellular matrix surrounding endothelial cells to promote endothelial migration/sprouting. This also includes the detachment of vascular pericytes, henceforth increasing vessel porosity and affecting existing blood flow and vessel pressure. An endothelial "tip" cell leads the sprouting of new vessels (toward a VEGF signal source), while a "stalk" cell follows and proliferates to form a lumen.

Functionally, VEGF-based vessel formation is regulated by the NOTCH-1 pathway and expression of delta-like ligand 4 (Dll4) in tip cells. During the final stages of vessel formation, endothelial cells produce PDGF to recruit pericytes, leading to vessel stabilization. Tumor blood vessels tend to differ significantly from normal tissue blood vessels. Structurally, tumor vessels are tortuous, dilated, and leaky; lack pericyte coverage; and have abnormal basement membranes. Moreover, vascular density and vessel diameters are physiologically different in tumors compared to blood vessels in normal tissue. Such structural characteristics are associated with aberrations in local tumor blood flow and unusual fluid dynamics (Goel et al., 2012). Other reported modes of tumor vascularization include "co-option" of neighboring persisting vessels, intussusceptions (splitting of existing vessels into two or more vessels), and recruitment of circulating endothelial progenitor cells from the bone marrow. Reports have also suggested that tumor cells themselves can act as blood vessels in a process called vascular mimicry (Mahadevan and Hart, 1990; Folkman, 1996; Folberg et al., 2000; Kerbel, 2000; Fukumura and Jain, 2007, 2008; Gordon et al., 2010; Fan et al., 2012; Tahergorabi and Khazaei, 2012).

Angiogenesis can be triggered by a combination of factors including (1) hypoxia, (2) oncogene-mediated growth signals, (3) metabolic and/or mechanical stress, (4) genetic mutations, and/or (5) hormones and cytokines. Not surprisingly, each of these is a hallmark of the tumor microenvironment (Hanahan and Weinberg, 2000). Tumor vascular development is also dependent on tumor type, site, cell proliferation rate, and stage of disease. Common tumor functional/physiological abnormalities result in, or are enhanced by, the formation of abnormal tumor blood vessels. These include increased interstitial fluid pressure, decreased tumor oxygenation, and increased intratumoral metabolites contributing to the hypoxic and physiologically complex microenvironment characteristic of many tumor lines (Siemann, 2006). The tumor vascular network is constantly changing leading to large fluctuations in blood flow and, at times, a reversal in flow direction. Irregular distributions of vessel density throughout the tumor microenvironment contributes to acute, chronic, and cyclic hypoxia, which in turn is a major cause of conventional radiation therapy failure and can promote metastasis (Dewhirst et al., 2008).

Presently, the clinically established approach to delivering radiation to a cancer patient is through fractionated radiotherapy. This comprises single-dose fractions over a period of time to attain a set total dose. In principle, this permits healthy tissue recovery. The size of a single-dose fraction typically ranges between 1.8 and 4 Gy and is set mostly based on accumulated clinical empirical experience over the past century. On a molecular level, radiation is known to damage DNA, thus hindering clonogenic cell proliferation and inducing cell death (Lehnert, 2007). Nonetheless, it has been demonstrated that tumor-derived cell lines respond to radiation differently *in vitro* than *in vivo* (Prise et al., 2005; Lehnert, 2007). This phenomenon is potentially attributed to the temporal and spatial complexity of the tumor microenvironment, as described earlier. Yet many clinicians continue to prescribe treatments based on dose curves obtained *in vitro*. Moreover, low-dose fractions of radiation may promote radioresistance by causing the secretion of cytokines such as VEGF and bFGF that act to protect cells (in particular endothelial cells) from radiation effects and augment the complexity of the tumor microenvironment (Gorski et al., 1999). Radioresistance has been associated with poor prognosis and treatment response (Garcia-Barros et al., 2003; Fuks and Kolesnick, 2005). Given that radiation therapy is prescribed to more than 50% of patients diagnosed with solid masses worldwide, a deep understanding of tumor tissue radiobiology considering the tumor as a whole rather than simply cellular-based tumor radiobiology is imperative. New developments in endothelial radiobiology have suggested that blood vessels regulate tumor response to radiation therapy and have led clinicians and researchers to further study the role of stroma in radiation response as described further in the succeeding text.

2.2 HIGH-DOSE RADIATION EFFECTS ON TUMOR ENDOTHELIAL CELLS

New technologies in radiation oncology such as intensity-modulated radiation have enhanced how patients are treated with ionizing radiation. These enable delivering specific doses of radiation, precisely targeted to anatomical sites, thus sparing normal tissue and maximizing gross tumor volume dose. In essence, the concept of many weeks of fractionated therapy could be abandoned in the near future as a greater emphasis is put on delivering single or a few large doses. It is however imperative that before doing so, a high-dose radiobiology is developed as suggested by numerous investigators (Kirkpatrick et al., 2008; Brown et al., 2014; Song et al., 2014). Although recognized to regulate the tumor microenvironment, the role of blood vessels in tumor therapy response has been an important topic of scientific investigation in recent years.

At conventional radiation doses (1.8–2 Gy), endothelial cells are reported to mainly respond via a molecular DNA damage mechanism reminiscent of that in tumor cells. However, at larger doses (>8–10 Gy), endothelial cells have been demonstrated to activate cell membrane–derived ceramide-dependent apoptotic pathways, independent of DNA damage. Ceramide is a sphingolipid-derived molecule that is cytotoxic to endothelial cells. This suggests a secondary mechanism of tumor damage at high radiation doses. Indeed, substantive evidence now suggests that tumor blood vessels act to regulate tumor response to radiotherapy (Garcia-Barros et al., 2003, 2004; Kolesnick, 2003; Fuks and Kolesnick, 2005; Carpinteiro et al., 2008; Chometon and Jendrossek, 2009; García-Barros et al., 2010), challenging the canonical notion that tumor response is primarily dependent on an inherent radiosensitivity of clonogenic tumor cancer cells.

Paris et al. (2001) first suggested that endothelial cells are the primary lesion during gastrointestinal (GI) irradiation at doses between 8 and 16 Gy. They demonstrated that this leads to secondary viable cell damage and radiation-induced GI syndrome, resulting in intestinal crypt stem cell lethality via reproductive cell death. Upregulation of acid sphingomyelinase (ASMase) involved in ceramide formation and subsequent crypt vessel network apoptosis was deemed responsible for observations (Paris et al., 2001; Maj et al., 2003; Gaugler et al., 2007; Rotolo et al., 2008, 2009, 2010). Preirradiation administration of bFGF countered the ASMase-dependent apoptosis, protecting gut endothelial and epithelial cells (Paris et al., 2001; Maj et al., 2003) from radiation-induced cell death. Other investigators challenged these findings, suggesting that p53-dependent apoptosis regulates GI syndrome resulting from radiation, independent of endothelial apoptosis (Kirsch et al., 2010). Although conflicting results were reported, these studies stirred a series of studies that would aim to understand if and how tumor endothelial cells play a role in tumor response to radiation therapy. Studies to date have ultimately demonstrated an important role for blood vessels in radiation responses.

Stimulated by these findings, Folkman and Camphausen (2001) speculated on the question: "What does radiotherapy do to endothelial cells?" Concurrently, researchers were prompted to query radiation-induced effects on endothelial cells and whether vascular dysfunction could be targeted to regulate tumor response to radiation therapy at high doses. It was suggested that if the microvasculature is the primary target of radiation in the intestine, as posited by Paris, Fuks and Kolesnick and colleagues. It was posited that if damage to the epithelial tumor stem cell is a secondary event, then such a relationship may also hold even in tumors where endothelial cells also support surrounding tumor cells. Reports of differing tumor radiosensitivities *in vivo* and *in vitro* could be therefore explained by the presence of such host-derived supporting cells (i.e., endothelial cells). It was also surmised that one could also take advantage of increased tumor endothelial cell radiosensitivity (compared to normal endothelial cells) to better target tumors. In 2003, Garcia-Barros et al. (2003) demonstrated results from *asmase* knockout animals supporting high-dose (>8–10 Gy) radiation targeting of endothelial cells as regulators of tumor response to radiotherapy. A series of scientific studies by various groups (particularly those of Kolesnick and Fuks at the Memorial Sloan-Ketterling Cancer Center in New York City) have since repeatedly confirmed these findings and advanced knowledge on endothelial radiobiology, particularly at high radiation doses.

The collective evidence gathered to date suggests that upregulation of ASMase occurs in response to high radiation doses (>8–10 Gy), resulting in sphingomyelin hydrolyzation producing the apoptosis second messenger ceramide within the first 6 hours following irradiation. The breakdown of sphingomyelin is suggested to take place in the membrane of endothelial cells, resulting in subsequent rapid endothelial cell death. Lower and more commonly utilized single 1.8–4 Gy doses of radiation do not result in sufficient ceramide production to activate endothelial apoptosis. Thus, at such doses, radiation is thought to mainly induce DNA-based damage in cells. In fact, these doses have been found to be minimally effective due to hypoxia, reperfusion, and reactive oxygen species formation (Fuks and Kolesnick, 2005; Moeller et al., 2005; Moeller and Dewhirst, 2006). The presence of a 20-fold enrichment of a nonlysosomal secretory form of the ASMase enzyme in the membrane of endothelial cells (due to involvement in membrane remodeling in response to flow shear stress and various endothelial mechanotransduction activities) is believed to make these cells particularly susceptible to high-dose radiation damage. It was also demonstrated that bFGF or sphingosine-1-phosphate (S1P) could prevent upregulation of ASMase and avert ceramide-based apoptosis. Such agents also have effects on tumor blood flow (Figure 2.1). Nevertheless, the exact mechanism of the endothelial–tumor cell linkage remains poorly understood.

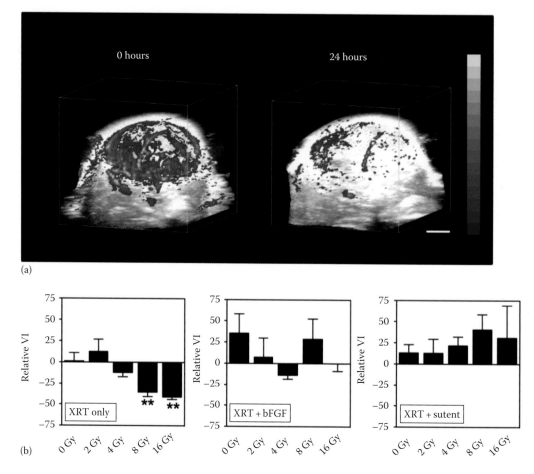

Figure 2.1 (a) Representative maximum intensity projections of a tumor volume at 0 and 24 hours after treatment with 16 Gy radiation. Data were obtained with 3D high-frequency ultrasound. A decrease in total power Doppler flow signal (blood volume detected) can be observed 24 hours after treatment. Red color represents the lowest power Doppler intensity (15 dB), while yellow represents the highest power Doppler intensity (40 dB). The scale bar represents 2 mm. (b) Relative change in VI at 24 hours for a range of radiation doses (0, 2, 4, 8, and 16 Gy) alone, in combination with bFGF or Sunitinib. Results indicate that there is a significant decrease in power Doppler flow signal only when tumors are treated with single doses of 8 and 16 Gy. Tumors pretreated with bFGF appeared to be unaffected by radiation doses. Pretreating with Sunitinib also appeared to cause no significant change in VI at doses lower than 8 Gy, although a significant increase in flow signal was observed in animals treated with 8 or 16 Gy. (From El Kaffas, A. et al., *Angiogenesis*, 16, 443, 2013.)

Studies published to date have provided evidence that it may involve leakage of a circulating factor, a bystander effect secondary to endothelial damage, or transient local ischemia/reperfusion produced by the acute microvascular dysfunction and its rapid reversal (Gaugler et al., 2007).

Early critiques of the Garcia-Barros et al. study, which first demonstrated tumor response to ionizing radiation being linked to endothelial cell apoptosis, came through the form of editorial letters (Brown et al., 2003; Suit and Willers, 2003). Garcia-Barros et al. demonstrated that in wild-type *asmase* +/+ mice, tumors responded. The tumor cells in the two groups (*asmase* –/– vs *asmase* +/+) were identical with the only significant intratumoral difference in these mouse groups being the host vasculature in the tumors. One argued that their results were in conflict with previously published data demonstrating that the radiosensitivity of tumors is primarily determined by the inherit sensitivity of the tumor cells, and not host-derived stroma (using immunodeficient mice that are two- to threefold more radiosensitive than regular mice) (Gerweck et al., 2006; Ogawa et al., 2007). However, a 2011 follow-up study utilizing *asmase* knockout non-immunodeficient mice demonstrated that indeed averting endothelial apoptosis would increase tumor radiation resistance. A second editorial letter argued that findings by Garcia-Barros et al. may be caused by an immune response; however, this was quickly refuted in 2004 where it was shown that the cell lines in question (melanomas) do not elicit a host immune response in wild-type mice and that the *asmase –/–* phenotype is not deficient in antitumor immunity. Specifically, these experiments were repeated in *Rag –/–* mice, which lack T and B cells, and *MEF –/–* cells, which lack natural killer (NK) and NK T cells, with no demonstrable immune effect, which could account for results with *asmase –/–* mice (Garcia-Barros et al., 2004).

A number of additional studies have been conducted by Kolesnick and Fuks demonstrating the importance of the ceramide pathway in radiation response and establishing the role of ceramide integral to such responses (Garcia-Barros et al., 2003, reviewed in Pena et al. 2000; Folkman and Camphausen, 2001; Kolesnick, 2002, 2003; Sathishkumar et al., 2005). These include the first work by Garcia-Barros et al. (2003), which used *asmase* knockout mice with melanoma and fibrosarcoma tumors demonstrating a functional deficit in ceramide signaling that made tumors resistant to high doses (15 Gy) of single-dose radiation. Research conducted thereafter in *Caenorhabditis elegans* involved loss-of-function mutants of conserved genes of sphingolipid metabolism. Inactivation of ceramide synthase led to the loss of ionizing radiation–induced apoptosis of germ cells. Microinjection of long-chain natural ceramide led to the restoration of germ cell apoptosis. Radiation-induced increases in ceramide were localized to the mitochondrial membrane indicating that is where signaling pathways integrate to regulate stress-induced apoptosis (Deng et al., 2008). Further molecular-based research demonstrated that mitochondrial ceramide-rich macrodomains functionalize Bax upon irradiation. More specifically, the study indicates that ceramide, generated in the mitochondrial outer membrane of mammalian cells in response to ionizing radiation, forms a platform into which Bax inserts, oligomerizes, and functionalizes as a pore (Lee et al., 2011). This permeabilizes mitochondria in endothelial cells leading to cell death as part of cell stress responses to large radiation doses as a working molecular mechanism.

In summary, the question remains, but perhaps better rephrased given the new scientific understanding developed over the past decade: to what degree are tumor responses to radiation regulated by the inherit radiosensitivity of the tumor cells or by the host-derived stroma? Numerous research groups have published evidence supporting both sides of the argument (Langley et al., 1997; Brown, 2009; El Kaffas et al., 2012, 2013, 2014; Martin, 2013; Tran et al., 2013). Recognizing that the role of blood vessels in regulating tumor response to radiation response is likely complicated. However, a paradigm shift is nonetheless underway that is giving a more prominent role to the tumor endothelium component in radiation response with implications for radiation planning and delivery. This concept is summarized in Figure 2.2. In addition, various newly developed vascular targeting agents have been demonstrated to act synergistically with radiation therapy to target tumor blood vessels. A number of reviews have been published on proposed combinations (O'Reilly, 2006; Senan and Smit, 2007; El Kaffas et al., 2014). In the next section, we briefly review recently developed agents that can biomechanically activate the ceramide pathway described earlier to significantly increase tumor responses to radiation doses far well in excess of that demonstrated by chemical radiation sensitizing agents.

2.3 ULTRASOUND-STIMULATED MICROBUBBLES FOR INDUCING LOW-DOSE VASCULAR EFFECTS

2.3.1 VASCULAR EFFECTS OF RADIATION

As summarized earlier, recent data indicate that radiation-induced apoptosis of endothelial cells can lead to vascular destruction and subsequent secondary tumor cell death. In this endothelial cell–driven response paradigm, tumor cells die secondarily and, as a result, of the damage caused by radiation to the microvasculature. Recent investigations have demonstrated the activation of a ceramide-dependent endothelial cell death pathway in terms of vascular responses to large doses of radiation.

In the research reviewed here, we hypothesized that this vascular-driven tumor cell death could be enhanced using ultrasound-based biophysical methods. These are founded in the delivery of mechanical energy to endothelial cells and complement ionizing-based energy effects. The approach summarized here is

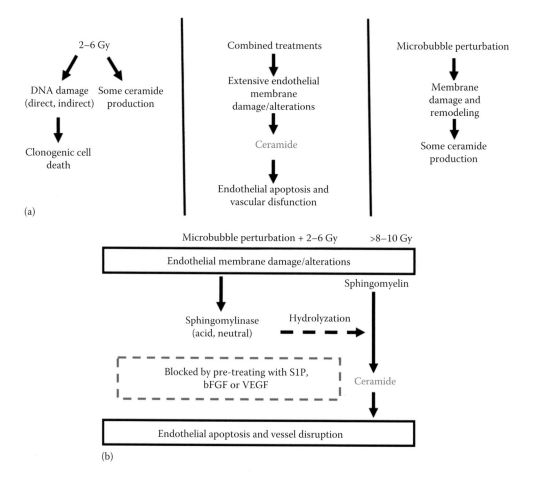

Figure 2.2 Posited mechanism of ceramide production following radiation and microbubbles. (a) Radiation doses lesser than 2–6 Gy will predominantly cause DNA damage. Microbubbles alone will result in some ceramide. Combined, these will produce ceramide levels similar to high doses of radiation to activate apoptosis. (b) Mechanism of ceramide-based cell death resulting from radiation doses greater than 8 Gy or ultrasound microbubbles combined with 2–6 Gy radiation doses. (From El Kaffas, A. and Czarnota, G.J., *Future Oncol.*, 11, 1093, 2015.)

one to perturb the endothelial lining of the vasculature, not with antiangiogenic or pharmacological agents that have mostly angiostatic effects with limited clinical success and impact, but with highly effective biophysical ultrasound-stimulated microbubble-mediated perturbations. These appear to activate the same ceramide-dependent cell death in endothelial cells that high doses of radiation also cause. The research is consistent with the hypothesis that more effective perturbation of endothelial cells (causing them primarily to die) leads to increased microvascular destruction that can enhance radiation effects (Kolesnick, 2002).

2.3.2 MICROBUBBLES AND ULTRASOUND

Ultrasound microbubble contrast agents are microspheres of gas, usually air or a perfluorocarbon, stabilized by a thin shell of biocompatible material such as protein or lipid. A number of agents are approved for clinical use worldwide. In the research described in the succeeding text, one example is Definity (Lantheus Medical, North Billerica, MA), which is perfluoropropane gas within a lipid shell. For these microbubbles, the typical median bubble diameter is 3–4 μm permitting their passage to the systemic circulatory system following peripheral venous injection. These microbubbles permit the visualization of vasculature (Foster et al., 2000; Simpson et al., 2001; Goertz et al., 2005a,b) with no notable clinically detected damage to tissues or significant adverse effects. In addition to serving as vascular imaging contrast agents, acoustic exposure of bubbles at or near their resonant frequency can perturb the function of nearby cells with effects including a reversible increase of cell membrane permeability. Biological effects related to acoustic bubble disruption, such as those due to the formation of local microjets and shockwaves, are capable of permeabilizing, as well as destroying a cell (Karshafian et al., 2010). Other investigations (Cosgrove, 2006; Haag et al., 2006; Karshafian et al., 2009, reviewed in Caissie et al., 2011) have also demonstrated microbubble-enhanced reversible drug uptake, with stable bubble oscillation and acoustic microstreaming causally involved. In the research summarized here, the use of such microbubble agents with ultrasound has been demonstrated to enhance the effects of radiation. Exposure parameters are similar to that used with diagnostic ultrasound scans, with the exception that greater quantities of microbubbles are used to elicit an effect. On their own, the ultrasound effects described here result in minimal damage from which tissues recover but, when combined with radiation, result in striking amounts of cell death in tumors (Al-Mahrouki et al., 2012; Czarnota et al., 2012a,b; Tran et al., 2012; Nofiele et al., 2013).

2.3.3 MICROBUBBLE-BASED ENHANCEMENT OF RADIATION CELL DEATH EFFECTS

It has been recently demonstrated that ultrasound and microbubble-mediated endothelial cell perturbation can significantly enhance radiation effectiveness. This proof of principle research has been demonstrated *in vitro* (Al-Mahrouki et al., 2012; Nofiele et al., 2013) and *in vivo* in a prostate cancer model (Czarnota et al., 2012a,b), in a bladder cancer model *in vivo* (Tran et al., 2012), and in a breast cancer model (unpublished results).

This effect represents a paradigm shift in the way tumors can be driven to respond to radiation. The proposed approach, the biology of which is described further in the succeeding text, enhances radiation effects far and beyond any chemical agent. An extensive ischemic event is triggered by the approach that goes beyond transient hypoxia with subtle radiobiological effects. In tissues the treatment actually causes significant hypoxia leading to anoxia, leading to ischemic cell death, which is a mix of apoptosis and necrosis. Classical radiobiological concerns that vessel destruction may inhibit further radiotherapy are obviated in this method by the complete cell death detected histologically and the superior cure rates seen experimentally (Czarnota et al., 2012a,b). The rationale behind the approach is that areas with induced anoxic cell death do not require any further therapeutic doses of radiation due to the massive destruction already caused. The minimal tumor areas without such massive anoxic effects that continue to have partially functioning vasculature surrounding tumor cells that are oxygenated are driven to tumor cell destruction by repeated treatments. These areas are characterized by endothelial cells, likely with increased radiosensitivity due to microbubble-mediated membrane damage. The method represents a paradigm shift away from traditional radiation oncology as radiation effects here are enhanced due to massive vascular destruction and a physiological synergy. Data indicated 40%–60% cell death within 24 hours after single, combined treatments of ultrasound-stimulated microbubbles and 2–8 Gy doses of radiation. Treatments with multiple combined ultrasound and radiation fractions demonstrated a whole tumor effect, causing tumor regression (Czarnota et al., 2012a; Figures 2.3 and 2.4).

2.3.4 MICROBUBBLE-BASED ENHANCEMENT OF RADIATION VASCULAR EFFECTS

Endothelial cells as the primary target for these microbubble-based radiation-enhancing treatments have been identified using immunohistochemistry. Results indicated grossly apparently vascular collapse and treatment-related vascular leakage such as factor VIII (von Willebrand factor). Endothelial cell apoptosis was evident as indicated by immunohistochemistry using CD31 and TUNEL staining. Vascular effects related to these treatments have been investigated noninvasively using high-frequency power Doppler imaging. The same parameters of radiation dose and ultrasound-stimulated microbubble concentration affected the vascular index in tumor xenografts coincident with increases in cell death (Figure 2.5). Specifically, power Doppler data obtained at 20 MHz indicated blood flow disruption with treatments consisting of a 20% ± 37% decrease with microbubble ultrasound treatment, an 18% ± 22% decrease with 8 Gy radiation alone, and a 65% ± 8% decrease with the combined treatments (mean ± standard error [$P < 0.05$]) (Czarnota et al., 2012a). Immunohistochemistry staining *in vitro* under similar conditions indicated ceramide production (Figure 2.6).

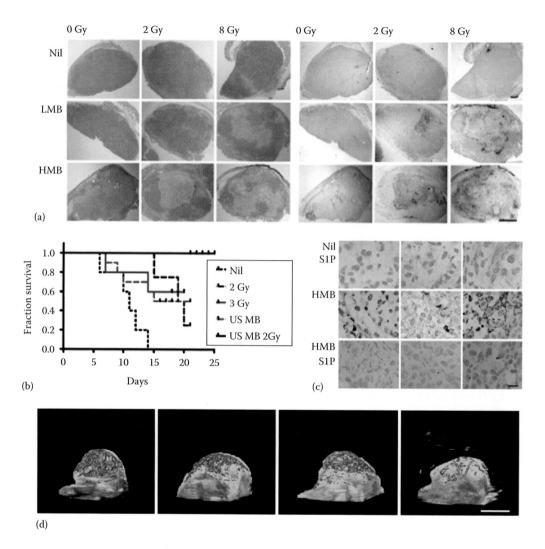

Figure 2.3 Panels indicate tumors treated with ultrasound-stimulated microbubbles in PC3 tumor xenografts. (a) Top row: Nil, no microbubbles; middle row: LMB, low microbubble concentration; bottom row: HMB, high microbubble concentration. Left panels present hematoxylin and eosin staining. Right panels present TUNEL staining. The scale bar indicates 2 mm. (b) Response assessments for multiple fraction experiments—survival. Kaplan–Meier survival curves are presented for cohorts of mice treated with 2 Gy fractions (24 Gy in 12 fractions over 3 weeks) (BED10 = 28.8 Gy), 2 Gy fractions with two ultrasound-stimulated microbubble treatments weekly, 3 Gy fractions (45 Gy in 15 fractions over 3 weeks) (BED10 = 58.5 Gy), and ultrasound-stimulated microbubble treatments weekly (twice weekly for 3 weeks). Endpoints were tumor at least doubling in size and reaching modified humane care endpoints (lack of ambulation, tumor greater than 2 cm in diameter). (c) Representative high-magnification views of ISEL staining of sections of PC3 prostate tumors treated with radiation and/or ultrasound-activated microbubbles in the presence of sphingosine-1-phosphate. Note the presence of ISEL staining with bubble exposure (middle) with increasing doses of radiation. This cell death was inhibited with diminished staining in the presence of S1P (bottom). Scale bar, 50 μm. (d) Representative power Doppler flow data acquired 24 hours after treatment. From left to right, representative specimens are presented for no treatment, microbubbles alone, 8 Gy radiation alone, and ultrasound-stimulated microbubbles and 8 Gy radiation. There is marked blood flow disruption by the combined treatment. The scale bar indicates 5 mm. (Adapted from Czarnota, G.J. et al., *Proc. Natl. Acad. Sci. U.S.A.*, 109, E2033, 2012a; Czarnota, G.J. et al., *Proc. Natl. Acad. Sci. U.S.A.*, 109, 11904, 2012b.)

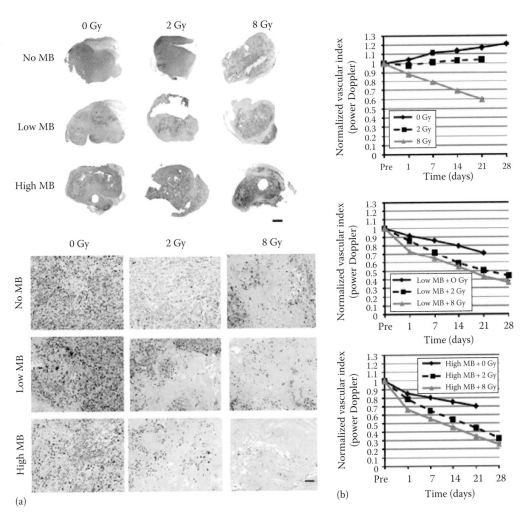

Figure 2.4 (a) Bladder xenografts tumor results. Top panels: TUNEL-staining 24 hours after treatment indicates tumor cell death resulting from local treatment. Radiation treatments alone show condensed necrotic area in multiple foci around the tumor area, which is most apparent after 8 Gy. Low- and high-concentration microbubble treatments with varying doses of radiation are also presented. Severe necrotic areas observed at tumor periphery and development of cystic spaces with treatment were observed. Scale bar indicates 2 mm. Bottom panels: Hematoxylin and eosin (H&E) staining 3–4 weeks after treatment. Top row: H&E staining through the tumor specimens demonstrated expected results with single-fraction radiation treatment of 2 Gy showed no obvious regions of cell death. Tumor cells were homogenous and evenly dispersed. With 8 Gy, cells were sparse and showed regions of decreased density, suggesting cell death. Middle row and bottom row: Post irradiation with low and high microbubbles resulting in tumor cell necrosis at the center. Tumors treated with combined microbubble ultrasound and radiation demonstrated the largest areas of cell death. Bar indicates 50 mm. (b) Long-term response monitoring of tumor vasculature. Top: Tumors treated with varying doses of radiation alone ($P < 0.001$ for 0 and 8 Gy showing significant difference in vascularity; $P > 0.05$ for 2 Gy showing insignificant changes). Middle and bottom: Low- and high-concentration ultrasound-stimulated microbubbles in combination with radiation treatment. ($P < 0.001$). Results indicate persistent decreases in power Doppler–detected blood flow, which were greatest in the combination treatments. (Adapted from Nofiele, J.T. et al., *Technol. Cancer Res. Treat.*, 12, 53, 2013.)

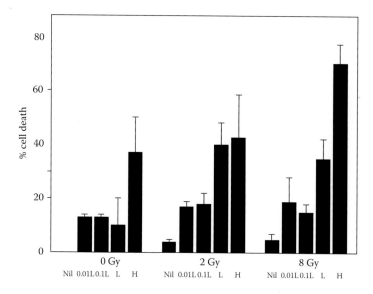

Figure 2.5 Quantitative analysis of cell death in response to microbubble exposure with different radiation doses. Percentage ISEL staining from microbubble concentrations administered to mice. For microbubble concentrations: Nil indicated no treatment; 0.01 and 0.1 L indicate dilutions of the low microbubble concentration (L) and (H) indicates the high microbubble concentration used. Different radiation doses include 0, 2, and 8 Gy, as labeled. (Adapted from Kim, H.C. et al., *PLoS One*, 9, e102343, 2014.)

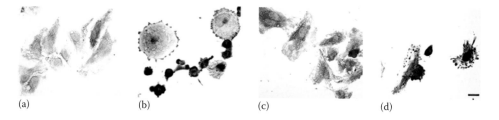

Figure 2.6 Ceramide staining of (human umbilical vein endothelial cell) endothelial cells exposed to (a) Nil, (b) 8 Gy radiation, (c) ultrasound-stimulated microbubbles, and (d) ultrasound-stimulated microbubbles and 8 Gy radiation. The scale bar represents 20 μm. (From Al-Mahrouki, A.A. et al., *Dis. Model Mech.*, 7, 363, 2014.)

Vascular effects in addition to cell death effects were time dependent when the two treatment components (ultrasound-stimulated microbubble treatment and radiation) were separated by time. Experiments were conducted with ultrasound and radiation modalities separately, and also together with a time delay introduced with ultrasound-stimulated microbubble treatment first and then subsequent radiation treatment (at a 0, 3, 6, 12, and 24 hour delay). Treatment effects that were maximally statistically significant were obtained when the two treatments were separated by 6 hours, as assessed by cell death immunohistochemistry. This coincided with a maximal decrease in ultrasound detected blood flow. Radiation given at that time resulted in a maximal effect 24 hours later, in terms of ISEL-detected cell death and disruption of blood flow-linked power Doppler–detected signal. The data implied a 9 hour window for radiation therapy after microbubble exposure with no statistically significant difference between results from 3 to 12 hours (Czarnota et al., 2012a; Figure 2.7).

More comprehensive analyses were conducted in subsequent studies in prostate tumor-bearing models and in mouse models with bladder tumor xenografts (Tran et al., 2012; Kim et al., 2013). Using HT-1376 bladder cancer tumors exposed to 2 and 8 Gy of ionizing radiation as experimental controls, findings indicated that tumor vasculature was diminished in flow (vascular index [VI]) within the first 24 hours of treatment, primarily at the 8 Gy dose (VI 0.88 ± 0.00), and negligibly at the 2 Gy dose

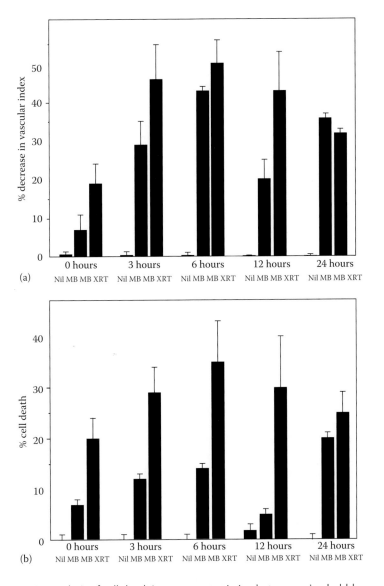

Figure 2.7 (a) Quantitative analysis of cell death in response to timing between microbubble exposure and radiation treatment. Decrease in micropower Doppler data measured vascular index with microbubbles and combined treatment. Nil, no treatment imaged before and 24 hours later; MB, treatment with microbubbles only (low concentration) and killing of mice at the indicated times after microbubble exposure (0, 3, 6, 12, and 24 hours); MBXRT, treatment with microbubbles and interval time as indicated between subsequent radiation treatment (8 Gy). (b) Resulting ISEL+ cell death corresponding to treatments as described in (a). For treatments with microbubbles alone, note the maximal effect on blood flow at 6 hours and cell death after 6 hours. Combined treatments follow a similar trend. (From Czarnota, G.J. et al., *Proc. Natl. Acad. Sci. U.S.A.*, 109, E2033, 2012a; Czarnota, G.J. et al., *Proc. Natl. Acad. Sci. U.S.A.*, 109, 11904, 2012b.)

(VI 0.98 ± 0.00). Long-term 21-day results indicated a negligible detected increase in flow at 2 Gy (VI of 1.04 ± 0.00) and with 8 Gy (VI of 0.60 ± 0.01). Results indicated that combination treatments with microbubbles and radiation similarly led to synergistic antivascular effects as observed by power Doppler and immunohistochemistry. At a low microbubble concentration, there was a decrease in detectable flow with 2 Gy administration (VI 0.86 ± 0.00) and with 8 Gy (VI 0.73 ± 0.00). At 8 Gy, these numbers were 0.79 ± 0.01 and 80.66 ± 0.00, respectively. At 21 days with 2 Gy combined with the low microbubble concentration treatment, a VI of 0.50 ± 0.01 was observed and, at 8 Gy with the same concentration of microbubbles, a VI of 0.43 ± 0.00 was observed. With a higher bubble concentration, VI values of 0.44 ± 0.01 and 0.34 ± 0.01 were observed for the same radiation doses at 21 days (Tran et al., 2012). Work conducted by Kwok et al. (2013) in PC3 tumors indicated similar effects with coincident decreases in blood flow and increases in cell death. In that study, tumor specimens were subjected to VEGF and CD31 staining after treatments in order to assess the effects on endothelial cells (Kim et al., 2013). These immunohistochemistry methods revealed decreases in vascular content with different treatments. For radiation alone 0, 2, and 8 Gy treatments, the quantified proportions of CD31-stained vessels were $100\% \pm 3\%$, $77\% \pm 4\%$, and $50\% \pm 20\%$. In the presence of ultrasound-stimulated microbubbles, these proportions were $60\% \pm 10\%$, $50\% \pm 10\%$, and $20\% \pm 10\%$.

2.3.5 MICROBUBBLE-BASED ENHANCEMENT OF CELL DEATH EFFECTS

Initial experiments were conducted with PC3 prostate tumor xenografts in SCID mice and indicated with single doses of radiation ranging from 2 to 8 Gy and different concentrations of ultrasound-stimulated microbubbles. The research utilized over 345 tumor-bearing animals. Results indicated supra-additive statistically significant amounts of tumor cell death with the combined treatments. Treatments with radiation alone resulted in minimal apoptotic or necrotic cell death ($4\% \pm 2\%$ death for 2 Gy, mean $\pm$ SE) as did ultrasound-stimulated microbubble treatments alone ($10\% \pm 4\%$ death for the low bubble concentration). In contrast, the combination of these resulted in obvious macroscopic regions of cell death in the area of ultrasound treatment occupying approximately $40\% \pm 8\%$ or more of tumor cross-sectional area for the 2 Gy dose combined with the ultrasound treatments with a low microbubble concentration. The combined 2 Gy and high microbubble concentration ultrasound treatment resulted in more cell death ($44\% \pm 13\%$). The combination of 8 Gy and the high microbubble concentration resulted in even more cell death ($70\% \pm 8\%$). The combined 2 Gy and microbubble treatments were significantly different compared to 2 Gy alone or ultrasound-activated microbubble treatments alone, for the ultrasound-stimulated low and high microbubble concentration treatments, respectively (all P values <0.0001). This was also observed for the combined 8 Gy and microbubble treatments compared to 8 Gy or ultrasound microbubble treatments alone (all P values <0.0001) (19). Experiments with xenografted bladder tumors in mice exhibited similar effects on cell death as did work with breast tumor xenografts in immunocompromised mice (Tran et al., 2012). Representative results are presented in Figures 2.3 and 2.4.

Experiments have been conducted investigating physical parameters in order to understand which parameters can affect the enhancement of radiation. Different peak negative pressures, microbubble concentrations, and radiation doses have been investigated. In Kim et al. (2013), prostate xenograft tumors (PC3) in severe combined immunodeficiency mice were subjected to ultrasound treatment at various peak negative pressures (250, 570, and 750 kPa) at a center frequency of 500 kHz, different microbubble concentrations (8, 80, and 1000 µL/kg), and different radiation doses (0, 2, and 8 Gy). Twenty-four hours after treatment, tumors were excised and assessed for cell death. This study revealed that increases in radiation dose, microbubble concentration, and ultrasound pressure promoted apoptotic cell death and cellular disruption within tumors by up to 21%, 30%, and 43%, respectively. Eventually, increases in pressure or bubble concentration, with other parameters fixed, led to no further increases in cell death (Figure 2.5). Vascular parameters related to such treatments have also been further explored (Kwok et al., 2013).

2.3.6 *ASMASE* PATHWAY AND CERAMIDE EFFECTS ON TREATMENTS

We demonstrated that the synergy between ultrasound-stimulated microbubbles and radiation arises from mechanical damage imparted to endothelial cells caused by cavitating microbubbles in an ultrasound field. Results *in vitro* indicate that this effect is prominent in endothelial cells but universal to many cell types. This activates endothelial cell death when then combined with low (2 Gy) doses of radiation *in vivo*.

Research *in vitro* was first conducted in human umbilical vein endothelial cells (HUVEC), acute myeloid leukemia cells (AML), murine fibrosarcoma cells (KHT-C), prostate cancer cells (PC3), breast cancer cells (MDA-MB-231), and astrocytes. Exposure of these cell types to ultrasound-stimulated bubbles alone resulted in increases in ceramide for all cell types and survivals of 12% ± 2%, 65% ±5%, 83% ± 2%, 58% ± 4%, 58% ±3%, and 18% ± 7% for HUVEC, AML, PC3, MDA, KHT-C, and astrocyte cells, respectively. Effects *in vitro* indicated additive treatment enhancements and increases in intracellular ceramide content as soon as 1 hour after exposure to ultrasound-activated microbubbles and radiation (Figure 2.6). The first results pointing to the important of the *asmase* pathway were that in *asmase +/+* astrocytes, survival decreased from 56% ± 2% after 2 Gy radiation alone and from 17% ± 7% after ultrasound and microbubbles alone to 5% ± 2% when combined. In contrast, results from experiments using ASMase-deficient astrocytes (*asmase –/–*) or S1P indicated less of an effect. *Asmase –/–* cells or +/+ cells were protected from cell death induction in treatment with S1P more than *asmase +/+* cells in response to treatments with ultrasound and radiation. All cell types in this study demonstrated ceramide immunohistochemical labeling in response to ultrasound-stimulated microbubble treatment, radiation, and more with the combination of the two treatment modalities. *Asmase –/–* cells showed little or no ceramide production with the exception of cells treated with ultrasound-stimulated microbubbles and radiation (Nofiele et al., 2013).

Gene expression profiling and quantitative analyses, in addition to further work with immunohistochemistry *in vitro*, revealed key roles again for the *asmase* pathway and genes involved in apoptosis, as well as cell membrane repair. Gene expression analyses revealed an upregulation of genes known to be involved in apoptosis and ceramide-induced apoptotic pathways, including *SMPD2*, *UGT8*, *COX6B1*, *Caspase 9*, and *MAP2K1* with ultrasound-stimulated microbubble exposure, but not *SMPD1* (Al-Mahrouki et al., 2012).

Extensive immunohistochemical analysis was also used to probe further several biomarkers in order to evaluate cell proliferation (Ki67), blood leakage (factor VIII), angiogenesis (cluster of differentiation molecule, CD31), ceramide formation, angiogenesis signaling (VEGF), oxygen limitation (prolyl hydroxylases, PHD2), and DNA damage/repair (gamma H2AX). Data indicated a treatment-related reduced vascularity due to vascular disruption by ultrasound-stimulated microbubbles, resulting in increased ceramide production and increased DNA damage of tumor cells despite decreased tumor oxygenation, with significantly less proliferating cells in the combined treatments. In an investigation of treatment factors related to ultrasound stimulation of microbubbles, a link was observed between cell death detected and ceramide production, which varied with ultrasound peak negative pressure and microbubble concentration, in addition to radiation dose (Al-Mahrouki et al., 2014).

Manipulating the *asmase* pathway either genetically or chemically inhibits the microbubble effect *in vitro* and *in vivo* in endothelial cell and tumor models, respectively. Analyses of experiments using PC3 prostate tumor xenografts but inhibited by S1P exposure indicated that treatment with S1P, given 30 minutes before and 5 minutes after treatments, resulted in a diminishment of detected apoptotic cell death. No statistically significant differences between 0, 2, and 8 Gy treatments in the presence of S1P and ultrasound-stimulated microbubble exposure (Czarnota et al., 2012a), were observed (Figure 2.3c).

The working model for these treatments is that data indicate destruction of blood flow. In this mechanism, the synergy between ultrasound-stimulated microbubbles and radiation arises from mechanical damage imparted to endothelial cells caused by cavitating microbubbles in an ultrasound field. The same pathways activated by high doses of radiation, causing a membrane damage–related *asmase*-dependent increase in ceramide, are stimulated by ultrasound and microbubbles. This activates endothelial cell death when combined with low (2 Gy) doses of radiation (Czarnota et al., 2012a). In this mechanism, radiation effect is converted from tumor cell DNA damage alone, in addition to causing vessel destruction and tumor apoptosis and necrosis. Any resulting hypoxia is actually anoxia, which is associated with cell death. The concern that such vessel destruction may inhibit further radiotherapy is obviated by the complete cell death detected histologically and the superior cure rates seen experimentally. Working models are presented in Figures 2.8 and 2.9.

In this type of ultrasound treatment, microbubbles are administered intravenously and act on blood vessel endothelial cells. Tumor selectivity and a therapeutic ratio are obtained by focusing the ultrasound beam only on tumor. Modern ultrasound technology now available clinically (e.g., Phillips Sonalleve) permits 3D ultrasound fields to be sculpted to 1 mm specificity, thus completely avoiding normal tissue.

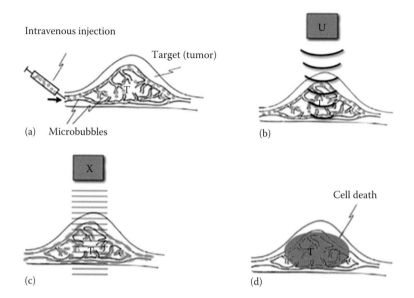

Figure 2.8 Treatment schematic for ultrasound microbubble stimulation of radiation enhancement. (a) Microbubbles are injected into a target (T) tumor through the vasculature. (b) Ultrasound (U) is applied to acoustically stimulated microbubbles, causing their cavitation. (c) This perturbs the endothelial cells lining the blood vessels in the tumors (green). Perturbed blood vessels are then sensitive to low-dose x-ray (X) radiation effects that normally do not cause extensive cell death. (d) One treatment causes extensive cell death within 24 hours in 50%–60% of the tumor volume. (From Czarnota, G.J. et al., *Proc. Natl. Acad. Sci. U.S.A.*, 109, E2033, 2012a; Czarnota, G.J. et al., *Proc. Natl. Acad. Sci. U.S.A.*, 109, 11904, 2012b.)

Whereas tumor treatments with antiangiogenic drugs alone typically leave the rim of a tumor viable (Palmowski et al., 2008), in the treatments being developed here, tumor periphery treatment is possible with ultrasound and encompassed by radiation.

2.3.7 NONINVASIVE MONITORING OF MICROBUBBLE-BASED ENHANCEMENT OF RADIATION EFFECTS

A number of noninvasive quantitative imaging methods have been used to monitor the effects of treatments to study the associated vascular effects and their consequences on tissue microstructure and tumor microenvironment. These consist of power Doppler analyses (with studies described earlier), as well as work with photoacoustic and quantitative ultrasound methods. Photoacoustic systems can be used as functional imaging to measure concentrations of chromophores (Razansky et al., 2009) and blood oxygen saturation levels (Oladipupo et al., 2011) by using multiple optical illumination wavelengths. The main tissue constituent that absorbs light in the near-infrared spectral range (other than melanocytes) is the red blood cell with high hemoglobin absorption (Yao and Wang, 2011). Therefore, typical photoacoustic images based on endogenous contrast depict either resolvable blood vessels, the aggregate signal from many nonresolvable blood vessels, or combinations of both, in a manner similar to ultrasound. Since oxy- and deoxyhemoglobin have different optical absorption spectra, illuminating tissue with a series of different laser wavelengths can help successfully measure photoacoustic signals of each. Spatial maps of the total hemoglobin concentration and the oxygen saturation of hemoglobin can be generated (Wang et al., 2006; Stein et al., 2009; Saha and Kolios, 2011), and the data can be combined with volumetric blood flow to calculate the metabolic rate of oxygen (Yao et al., 2011). Importantly, these functional parameters are thought to be critical in treatment response monitoring (Cerussi et al., 2010; Roblyer et al., 2011; Falou et al., 2012; Sadeghi-Naini et al., 2012a).

For treatments with prostate tumors, photoacoustic data were assessed quantitatively and indicated significant decreases in oxygen saturation across all treatment conditions where radiation alone or ultrasound-stimulated microbubbles or the combination of these was administered ($P < 0.05$)

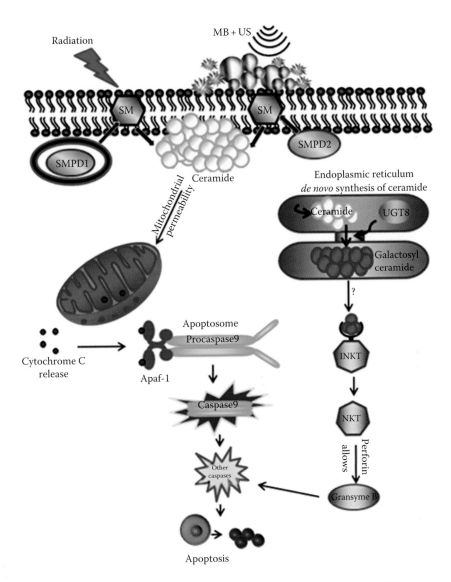

Figure 2.9 A proposed schematic model of apoptotic mechanistic pathways in human umbilical vein endothelial cell based on gene expression changes in the combined treatments using microbubbles, ultrasound, and x-ray radiation. The putative hydrolysis of sphingomyelin (SM) by the sphingomyelinases (SMPD1 and SMPD2) results in the production of ceramide, which can act on mitochondria to facilitate the release of cytochrome c. This could then complex with procaspase 9 and Apaf1. The formed complex (apoptosome) in the presence of ATP leads to the activation of caspase 9, which can then activate other caspases and can ultimately initiate apoptosis. Alternatively, ceramide produced from de novo synthesis in the endoplasmic reticulum can be modified by UGT8 to form galactosyl ceramide, which is a molecule known to activate the natural killer (NK) T cells. Active NK T cells will then secrete perforin, which assembles at the target cell plasma membrane and allows the passage of granzyme B, leading to apoptotic pathways. (From Al-Mahrouki, A. et al., *Ultrasound Med. Biol.*, 38, 1958, 2012.)

(Briggs et al., 2014). Specifically, treatments using 8 Gy and microbubbles resulted in oxygen saturation decreases of 28% ± 10% at 570 kPa and, with 750 kPa, a decrease of 25% ± 29%, which corresponded to 44% ± 9% and 40% ± 14% respective decreases in blood flow as measured with power Doppler ultrasound. Corresponding histology indicated 31% ± 5% at 570 kPa and 37% ± 5% at 750 kPa in terms of tumor cell death in sectional histology.

In general, decreases in oxygen saturation occurred, which became greater with increasing radiation dose and with increasing ultrasound pressure. These were linked to significant changes in vascularization confirmed by one-way CD31 staining results ($P < 0.05$). The strongest changes across all three treatment assessment modalities in Briggs et al. were found when combining 8 Gy of radiation with ultrasound-stimulated microbubbles at both pressures.

That study documented $28\% \pm 10\%$ (570 kPa) and $25\% \pm 29\%$ (750 kPa) decreases in oxygen saturation, with $44\% \pm 9\%$ and $40\% \pm 14\%$ respective decreases in power Doppler measured blood flow. Corresponding immunohistochemistry indicated $15\% \pm 2\%$ at 570 kPa ($P \pm 0.033$) and $20\% \pm 2\%$ at 750 kPa ($P \pm 0.015$) decreases in intact tumor vasculature, and $31\% \pm 5\%$ at 570 kPa ($P = 0.005$) and $37\% \pm 5\%$ at 750 kPa (Briggs et al., 2014; Figure 2.10).

Quantitative ultrasound methods have also been used to noninvasively monitor treatment effects. These are methods that probe tissue microstructure and have been established as methods to track tumor cell death (Banihashemi et al., 2008; Sadeghi-Naini et al., 2012b, 2013a, Sannachi et al., 2014, reviewed in Czarnota and Kolios, 2010, Sadeghi-Naini et al., 2012b). In these studies, prostate tumors treated with ultrasound-stimulated microbubbles and radiation were investigated using both high-frequency high-resolution quantitative ultrasound in addition to conventional frequency-based quantitative ultrasound. Quantitative ultrasound indicated changes in scatterer size and "effective" scatterer concentration which were consistent with cell death (Lee et al., 2012; Kim et al., 2013; Sadeghi-Naini et al., 2013b; Al-Mahrouki et al., 2014).

2.3.8 MULTIPLE REPEATED TREATMENT EFFECTS

Since the treatments here promote hypoxia but have an increased therapeutic efficacy when used as single treatment, multiple treatment experiments were undertaken to compare tumor cure rates. Experiments were conducted using fractionated radiotherapy and multiple weekly ultrasound-stimulated microbubble treatments in order to investigate animal survivals. Treatments with multiple combined ultrasound and radiation fractions demonstrated a whole tumor effect causing tumor regression (Figure 2.1b). In survival experiments (Figure 2.1b), the combination of ultrasound-stimulated microbubbles with radiation turned a noncurative dose of radiation into one with more effect than a curative dose of radiation. Animals treated with 24 Gy of radiation and bubbles (BED10 = 28.8 Gy) had a superior survival to animals with a more curative dose (BED10 = 58.5 Gy) (Caissie et al., 2011). Also, in these combined ultrasound and radiation treatments, there was no evidence of a viable rim seen with chemical antiangiogenic agents, but rather a destruction of tumor vasculature due to endothelial cell death, vascular collapse, and subsequent tumor cell death (Czarnota et al., 2012a).

2.3.9 MICROBUBBLE-BASED ENHANCEMENT OF RADIATION EFFECTS AND ANTIANGIOGENIC AGENTS

It is recognized that antiangiogenics can interact with radiation potentiating its effects, and factors like bFGF, which also modulate endothelial cell survival, can protect tissues from radiation-based antivascular effects. Similarly, it may be possible to further enhance the effects of these ultrasound-stimulated microbubble treatments through the use of antiangiogenics (El Kaffas et al., 2012, reviewed in El Kaffas et al., 2014).

El Kaffas et al. investigated the combination of two novel complementary vascular targeting agents with radiation therapy in a strategy aiming to sustain vascular disruption. Delta-like ligand 4 (Dll4) blockade (angiogenesis deregulator) treatment has been administered in conjunction with ultrasound-stimulated microbubble and radiation treatments. Results indicate a significant tumor response in animals treated with ultrasound-stimulated microbubble treatments combined with radiation, and Dll4 mAb, leading to a synergistic tumor growth delay. Triple combination treatments caused the greatest tumor growth delay, with a decrease in tumor size evident at 5 days after the start of treatment, followed by an interruption in growth lasting for nearly 15 days. An average tumor growth delay of 24 days was observed ($P < 0.05$) for the triple combination treatment. It is posited that in this context, the initial treatments lead to a destruction of vasculature through the mechanism discussed earlier and the addition of antiangiogenic agents prevents the regrowth of normal vasculature that would otherwise stimulate tumor regrowth (El Kaffas et al., 2013).

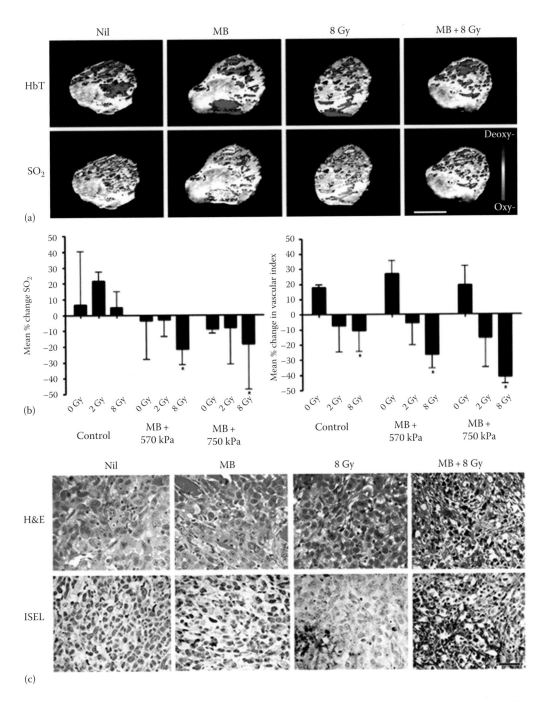

Figure 2.10 (a) Photoacoustic data for microbubble radiation enhancement. Nil, no treatment; MB, microbubble alone treatment; 8 Gy, x-ray treatment alone; MB+8 Gy, treatment with ultrasound-stimulated microbubbles and radiation. Top row indicates total hemoglobin (HbT) signal, whereas the bottom row indicates oxygen saturation (SO_2) with color coding indicating relative signals from deoxyhemoglobin and oxyhemoglobin. (b) Quantitative photoacoustic results indicating changes in oxygen saturation and the power Doppler vascular index. Data indicate combined ultrasound-stimulated microbubble and ultrasound treatments cause cell death by disrupting vasculature. (c) Histology for ultrasound-stimulated microbubble therapy. The top row demonstrates hematoxylin and eosin staining and the bottom *in situ* nick end labeling for cell death. Nil, no treatment; MB, ultrasound-stimulated microbubbles; 8 Gy, x-ray radiation treatment; MB+ 8 Gy, combined therapy. Note the increased cell death in the combined therapy with vascular stranding. Scale bar indicates 50 μm. (Adapted from Briggs, K. et al., *Technol. Cancer Res. Treat.*, 13, 435, 2014.)

2.3.10 IMPLICATIONS FOR MODERN DELIVERY OF RADIATION THERAPY

The biological mechanisms describing the effects of radiation on cells are well understood at conventional radiation doses (1.8–2 Gy). High doses seem, however, to interact with biology through different molecular pathways (i.e., endothelial membrane, lipid metabolism signaling, and the mitochondria) and thus require thorough investigations to describe their cellular and molecular responses in both tumor and host-derived stromal cells. If endothelial cells are conclusively found to be a secondary tumor target, particularly with high-dose radiotherapy, then these cannot be ignored in the advent of radiation oncology as a field. This would of course have implications for the classic radiobiology and the use of its linear–quadratic mathematical model developed over the past decades, which describes dose-dependent radiation response (assuming DNA damage only) of specific cancer cell lines as a function of measurable parameters (the alpha–beta ratio). The classic radiobiological model has had its share of success in describing low-dose radiation response but has not held out as well at higher radiation doses. It is possible that at high doses, deviations from the linear–quadratic response could be accounted for in a model that considers radiation-based stromal effects and anoxia-based acute cell death. Various investigators are reexamining the role of this model in the context of modern radiotherapy practices as described earlier. Whatever the findings over the next decade, new technologies that open the door to new therapeutic strategies require a reinvestigation into conventional knowledge. Findings may highlight the poorly guided therapeutic approaches of conventional therapy delivery to date and open the door to novel therapeutic strategies including the vascular targeting strategies discussed here.

In summary, the studies described earlier delineating the role of ceramide in both membrane-based mechanical and radiation damage imparted on cells, particularly with endothelial cells, set the scene for further investigations in tumor models, as a function of radiation dose and ultimately for translation to patients.

REFERENCES

Al-Mahrouki A, Karshafian R, Giles A, Czarnota GJ (2012) Bioeffects of ultrasound-stimulated microbubbles on endothelial cells: Gene expression changes associated with radiation enhancement in vitro. *Ultrasound Med Biol* 38:1958–1969.

Al-Mahrouki AA, Iradji S, Tran WT, Czarnota GJ (2014) Cellular characterization of ultrasound-stimulated microbubble radiation enhancement in a prostate cancer xenograft model. *Dis Model Mech* 7:363–372.

Banihashemi B, Vlad R, Debeljevi, B, Giles A, Kolios MC, Czarnota GJ (2008) Ultrasound imaging of apoptosis in tumour response: Novel preclinical monitoring of photodynamic therapy effects. *Cancer Res* 68:8590–8596.

Briggs K, Al-Mahrouki A, Nofiele J, El-Falou A., Stanisz M., Kim HC, Kolios MC, Czarnota GJ (2014) Non-invasive monitoring of ultrasound-stimulated microbubble radiation enhancement using photoacoustic imaging. *Technol Cancer Res Treat* 13:435–444.

Brown JM, Carlson DJ, Brenner DJ (2014) The tumor radiobiology of SRS and SBRT: Are more than the 5R's involved? *Int J Radiat Oncol Biol Phys* 88:254–262.

Brown M (2009) Controversy section: What causes the radiation GI syndrome? *Int J Radiat Oncol Biol Phys* 70:799–800.

Brown M, Bristow R, Glazer P, Hill R, McBride W, McKenna G, Muschel R (2003) Tumor response to radiotherapy regulated by endothelial cell apoptosis (II). *Science* 302:1894.

Caissie A, Karshafian R, Hynynen K, Czarnota GJ (2011) Ultrasound contrast microbubbles: In vivo imaging and potential therapeutic applications. In: Goins B and Phillips W (eds.), *Nanoimaging*, pp. 1–25. Singapore: Pan Stanford Publishing.

Carpinteiro A, Dumitru C, Schenck M, Gulbins E (2008) Ceramide-induced cell death in malignant cells. *Cancer Lett* 264:1–10.

Cerussi AE, Tanamai VW, Mehta RS, Hsiang D, Butler J, Tromberg BJ (2010) Frequent optical imaging during breast cancer neoadjuvant chemotherapy reveals dynamic tumour physiology in an individual patient. *Acad Radiol* 17:1031–1039.

Chen J, Lin Y, Chiang C, Hong J, Yeh C (2010) Characterization of tumor vasculature derived from angiogenesis and vasculogenesis by high-frequency three-dimensional Doppler ultrasound. *Symp A Q J Mod Foreign Lit* 1:2319–2322.

Chometon G, Jendrossek V (2009) Targeting the tumour stroma to increase efficacy of chemo- and radiotherapy. *Clin Transl Oncol* 11:75–81.

Cosgrove D (2006) Ultrasound contrast agents: An overview. *Eur J Radiol* 60:324–330.

Czarnota GJ, Karshafian R, Burns PN, Wong CS, Al-Mahrouki A, Lee J, Caissie A et al. (2012a) Tumour radiation response enhancement by acoustical stimulation of the vasculature. *Proc Natl Acad Sci USA* 109:E2033–E2041.

Czarnota GJ, Karshafian R, Burns PN, Wong CS, Al-Mahrouki A, Lee J, Caissie A et al. (2012b) Tumour radiation response enhancement by acoustical stimulation of the vasculature. *Proc Natl Acad Sci USA Plus* 109:11904–11905.

Czarnota GJ, Kolios MC (2010) Ultrasound detection of cell death. *Imaging Med* 2:17–28.

Deng X, Yin X, Allan R, Lu DD, Maurer CW, Haimovitz-Friedman A, Fuks Z, Shaham S, Kolesnick R (2008) Ceramide biogenesis is required for radiation-induced apoptosis in the germ line of *C. elegans*. *Science* 322:110–115.

Dewhirst MW, Cao Y, Moeller B (2008) Cycling hypoxia and free radicals regulate angiogenesis and radiotherapy response. *Nat Rev Cancer* 8:425–437.

Dvorak HF (1986) Tumors: Wounds that do not heal: Similarities between tumor stroma generation and would healing. *N Engl J Med* 315:1650–1659.

El Kaffas A, Al-Mahrouki A, Tran WT, Giles A, Czarnota GJ (2014) Sunitinib effects on the radiation response of endothelial and breast tumor cells. *Microvasc Res* 92:1–9.

El Kaffas A, Czarnota GJ (2015) Biomechanical effects of microbubbles: From radiosensitization to cell death. *Future Oncol* 11:1093–1108.

El Kaffas A, Giles A, Czarnota GJ (2013) Dose-dependent response of tumor vasculature to radiation therapy in combination with Sunitinib depicted by three-dimensional high-frequency power Doppler ultrasound. *Angiogenesis* 16:443–454.

El Kaffas A, Nofiele J, Giles A, Cho S, Liu SK, Czarnota GJ (2014) Dll4-notch signalling blockade synergizes combined ultrasound-stimulated microbubble and radiation therapy in human colon cancer xenografts. *PLoS One* 9:e93888.

El Kaffas A, Tran W, Czarnota GJ (2012) Vascular strategies for enhancing tumour response to radiation therapy. *Technol Cancer Res Treat* 11:421–432.

Falou O, Soliman H, Sadeghi-Naini A, Iradji S, Lemon-Wong S, Zubovits J, Spayne J et al. (2012) Diffuse optical spectroscopy evaluation of treatment response in women with locally advanced breast cancer receiving neoadjuvant chemotherapy. *Trans Oncol* 5:238–346.

Fan F, Schimming A, Jaeger D, Podar K (2012) Targeting the tumor microenvironment: Focus on angiogenesis. *J Oncol* 2012:281261.

Folberg R, Hendrix MJ, Maniotis AJ (2000) Vasculogenic mimicry and tumor angiogenesis. *Am J Pathol* 156:361–381.

Folkman J (1996) Tumor angiogenesis and tissue factor. *Nat Med* 2:167–168.

Folkman J, Camphausen K (2001) What does radiotherapy do to endothelial cells? *Science* 293:227–228.

Foster FS, Burns PN, Simpson DH, Wilson SR, Christopher DA, Goertz DE (2000) Ultrasound for the visualization and quantification of tumour microcirculation. *Cancer Metastasis Rev* 19:131–138.

Fuks Z, Kolesnick R (2005) Engaging the vascular component of the tumor response. *Cancer Cell* 8:89–91.

Fukumura D, Jain RK (2007) Tumor microvasculature and microenvironment: Targets for anti-angiogenesis and normalization. *Microvasc Res* 74:72–84.

Fukumura D, Jain RK (2008) Imaging angiogenesis and the microenvironment. *APMIS* 116:695–715.

Garcia-Barros M, Lacorazza D, Petrie H, Haimovitz-Friedman A, Cardon-Cardo C, Nimer S, Fuks Z, Kolesnick R (2004) Host acid sphingomyelinase regulates microvascular function not tumor immunity. *Cancer Res* 64:8285–8291.

Garcia-Barros M, Paris F, Cordon-Cardo C, Lyden D, Rafii S, Haimovitz-Friedman A, Fuks Z, Kolesnick R (2003) Tumor response to radiotherapy regulated by endothelial cell apoptosis. *Science* 300:1155–1159.

García-Barros M, Thin TH, Maj J, Cordon-Cardo C, Haimovitz-Friedman A, Fuks Z, Kolesnick R (2010) Impact of stromal sensitivity on radiation response of tumors implanted in SCID hosts revisited. *Cancer Res* 70:8179–8186.

Gaugler M-H, Neunlist M, Bonnaud S, Aubert P, Benderitter M, Paris F (2007) Intestinal epithelial cell dysfunction is mediated by an endothelial-specific radiation-induced bystander effect. *Radiat Res* 167:185–193.

Gerweck LE, Vijayappa S, Kurimasa A, Ogawa K, Chen DJ (2006) Tumor cell radiosensitivity is a major determinant of tumor response to radiation. *Cancer Res* 66:8352–8355.

Goel S, Wong AH, Jain RK (2012) Vascular normalization as a therapeutic strategy. *Cold Spring Harb Perspect Med* 2:1–24.

Goertz DE, Cherin E, Needle A, Karshafian R, Brown AS, Burns PN, Foster FS (2005a) High Frequency nonlinear B-scan imaging of microbubble contrast agents. *IEEE Trans Ultrason Ferroelectr Freq Control* 52:65–79.

Goertz DE, Needles A, Burns PN, and Foster FS (2005b) High-frequency nonlinear flow imaging of microbubble contrast agents. *IEEE Trans Ultrason Ferroelectr Freq Control* 52:495–502.

Gordon MS, Mendelson DS, Kato G (2010) Tumor angiogenesis and novel antiangiogenic strategies. *Int J Cancer* 126: 1777–1787.

Gorski DH, Beckett MA, Jaskowiak NT, Calvin DP, Mauceri HJ, Salloum RM, Seetharam S et al. (1999) Advances in brief blockade of the vascular endothelial growth factor stress response increases the antitumor effects of ionizing radiation. *Cancer Res* 59:3374–3378.

Haag P, Frauscher F, Gradl J, Seitz A, Schafer G, Lindner JR, Klibanov AL, Bartsch G, Klocker H, Eder IE (2006) Microbubble-enhanced ultrasound to deliver an antisense oligodeoxynucleotide targeting the human androgen receptor into prostate tumours. *J Steroid Biochem Mol Biol* 102:103–113.

Hanahan D, Weinberg RA (2000) The hallmarks of cancer. *Cell* 100:57–70.

Karshafian R, Bevan PD, Williams R, Samac S, Burns PN (2009) Sonoporation by ultrasound-activated microbubble contrast agents: Effect of acoustic exposure parameters on cell membrane permeability and cell viability. *Ultrasound Med Biol* 35:847–860.

Karshafian R, Samac S, Bevan PD, Burns PN (2010) Microbubble mediated sonoporation of cells in suspension: Clonogenic viability and influence of molecular size on uptake. *Ultrasonics* 50:691–697.

Kerbel RS (2000) Tumor angiogenesis: Past, present and the near future. *Carcinogenesis* 21:505–515.

Kim HC, Al-Mahrouki A, Gorjizadeh A, Karshafian R, Czarnota GJ (2013) Effects of biophysical parameters in enhancing radiation responses of prostate tumours with ultrasound-stimulated microbubbles. *Ultrasound Med Biol* 39:1376–1387.

Kim HC, Al-Mahrouki A, Gorjizadeh A, Sadeghi-Naini A, Karshafian R, Czarnota GJ (2014) Quantitative ultrasound characterization of tumor cell death: Ultrasound-stimulated microbubbles for radiation enhancement. *PLoS One* 9:e102343.

Kirkpatrick JP, Meyer JJ, Marks LB (2008) The linear-quadratic model is inappropriate to model high dose per fraction effects in radiosurgery. *Semin Radiat Oncol* 18:240–243.

Kirsch DG, Santiago PM, di Tomaso E, Sullivan JM, Hou W-S, Dayton T, Jeffords LB et al. (2010) P53 Controls radiation-induced gastrointestinal syndrome in mice independent of apoptosis. *Science* 327:593–596.

Kolesnick R (2002) The therapeutic potential of modulating the ceramide/sphingomyelin pathway. *J Clin Invest* 110:3–8.

Kolesnick R (2003) Response to comments on "Tumor response to radiotherapy regulated by endothelial cell apoptosis." *Science* 302:1894.

Kwok SJJ, El Kaffas A, Lai P, Al Mahrouki A, Lee J, Iradji S, Tran WT et al. (2013) Ultrasound-mediated microbubble enhancement of radiation therapy studied using three-dimensional high-frequency power Doppler ultrasound. *Ultrasound Med Biol* 39:1983–1990.

Langley RE, Bump EA, Quartuccio SG, Medeiros D, Braunhut SJ (1997) Radiation-induced apoptosis in microvascular endothelial cells. *Br J Cancer* 75:666–672.

Lee H, Rotolo JA, Mesicek J, Penate-Medina T, Rimner A, Liao WC, Yin X et al. (2011) Mitochondrial ceramide-rich macrodomains functionalize bax upon irradiation. *PLoS One* 6:e19783.

Lee J, Karshafian R, Papanicolau N, Giles A, Kolios MC, Czarnota, GJ (2012) Quantitative ultrasound for the monitoring of novel microbubble and ultrasound radiosensitization. *Ultrasound Med Biol* 38:1212–1221.

Lehnert S (2007) *Biomolecular Action of Ionizing Radiation*. New York: Taylor & Francis.

Mahadevan V, Hart IR (1990) Metastasis and angiogenesis. *Acta Oncol* 29(1):97–103.

Maj JG, Paris F, Haimovitz-Friedman A (2003) Microvascular function regulates intestinal crypt response to radiation microvascular function regulates intestinal crypt response to radiation. *Cancer Res* 63:4338–4341.

Martin BJ (2013) Inhibiting vasculogenesis after radiation: A new paradigm to improve local control by radiotherapy. *Semin Radiat Oncol* 23:281–287.

Moeller BJ, Dewhirst MW (2006) HIF-1 and tumour radiosensitivity. *Br J Cancer* 95:1–5.

Moeller BJ, Dreher MR, Rabbani ZN, Schroeder T, Cao Y, Li CY, Dewhirst MW (2005) Pleiotropic effects of HIF-1 blockade on tumor radiosensitivity. *Cancer Cell* 8:99–110.

Nofiele JT, Karshafian R, Furukawa M, Al Mahrouki A, Giles A, Wong S, Czarnota GJ (2013) Ultrasound-activated microbubble cancer therapy: Ceramide production leading to enhanced radiation effect in vitro. *Technol Cancer Res Treat* 12:53–60.

O'Reilly MS (2006) Radiation combined with antiangiogenic and antivascular agents. *Semin Radiat Oncol* 16:45–50.

Ogawa K, Boucher Y, Kashiwagi S, Fukumura D, Chen D, Gerweck LE (2007) Influence of tumor cell and stroma sensitivity on tumor response to radiation. *Cancer Res* 67:4016–4021.

Oladipupo S, Hu S, Kovalski J, Yao J, Santeford A, Sohn RE, Shohet R, Maslov K, Wang LV, Arbeit JM (2011) VEGF is essential for hypoxia-inducible factor-mediated neovascularization but dispensable for endothelial sprouting. *Proc Natl Acad Sci USA* 108:13264–13269.

Palmowski M, Huppert J, Hauff P, Reinhardt M, Schreiner K, Socher MA, Hallscheidt P, Kauffmann GW, Semmler W, Kiessling F (2008) Vessel fractions in tumour xenografts depicted by flow- or contrast-sensitive three-dimensional high-frequency Doppler ultrasound respond differently to antiangiogenic treatment. *Cancer Res* 68:7042–7049.

Paris F, Fuks Z, Kang A, Capodieci P, Juan G, Ehleiter D, Haimovitz-Friedman A, Cordon-Cardo C, Kolesnick R (2001) Endothelial apoptosis as the primary lesion initiating intestinal radiation damage in mice. *Science* 293:293–297.

Pena L, Fuks Z, Kolesnick R (2000) Radiation-induced apoptosis of endothelial cells in the murine central nervous system: Protection by fibroblast growth factor and sphingomyelinase deficiency. *Cancer Res* 60:321–327.

Prise KM, Schettino G, Folkard M, Held KD (2005) New insights on cell death from radiation exposure. *Lancet Oncol* 6:520–528.

Razansky D, Distel M, Vinegoni C, Ma R, Perrimon N, Koster RW, Ntziachristos V (2009) Multispectral opto-acoustic tomography of deep-seated fluorescent proteins in vivo. *Nat Photon* 3:412–417.

Roblyer D, Ueda S, Cerussi A, Tanamai W, Durkin A, Mehta R, Hsiang D et al. (2011) Optical imaging of breast cancer oxyhemoglobin flare correlates with neoadjuvant chemotherapy response one day after starting treatment. *Proc Natl Acad Sci USA* 108:14626–14631.

Rotolo JA, Kolesnick R, Fuks Z (2009) Timing of lethality from gastrointestinal syndrome in mice revisited. *Int J Radiat Oncol Biol Phys* 73:6–8.

Rotolo JA, Maj JG, Feldman R, Ren D, Haimovitz-Friedman A, Cordon-Cardo C, Cheng EH-Y, Kolesnick R, Fuks Z (2008) Bax and Bak do not exhibit functional redundancy in mediating radiation-induced endothelial apoptosis in the intestinal mucosa. *Int J Radiat Oncol Biol Phys* 70:804–815.

Rotolo JA, Mesicek J, Maj J, Truman J-P, Haimovitz-Friedman A, Kolesnick R, Fuks Z (2010) Regulation of ceramide synthase-mediated crypt epithelium apoptosis by DNA damage repair enzymes. *Cancer Res* 70:957–967.

Sadeghi-Naini A, Falou O, Hudson JM, Bailey C, Burns PN, Stanisz G, Kolios MC, Czarnota GJ (2012a). Imaging innovations for cancer therapy response monitoring. *Future Med* 4:311–327.

Sadeghi-Naini A, Falou O, Hudson JM, Bailey C, Burns PN, Yaffe MJ, Stanisz GJ, Kolios MC, Czarnota GJ (2012b). Imaging innovations for cancer therapy response monitoring. *Imaging Med* 4:311–327.

Sadeghi-Naini A, Falou O, Tadayyon H, Al-Mahrouki A, Tran W, Papanicolau N, Kolios MC, Czarnota GJ (2013a). Conventional frequency ultrasonic biomarkers of cancer treatment response in vivo. *Transl Oncol* 6:234–243.

Sadeghi-Naini A, Papanicolau N, Falou O, Tadayyon H, Lee J, Zubovits J, Sadeghian A et al. (2013b). Low-frequency quantitative ultrasound imaging of cell death in vivo. *Med Phys* 40:082901.

Saha RK, Kolios MC (2011) Effects of erythrocyte oxygenation on optoacoustic signals. *J Biomed Opt* 16:115003.

Sannachi L, Tadayyon H, Sadeghi-Naini A, Kolios MC, Czarnota GJ (2014) Personalization of breast cancer chemotherapy using noninvasive imaging methods to detect tumour cell death responses. *Breast Cancer Manage* 3:31–35.

Sathishkumar S, Boyanovsky B, Karakashian AA, Rozenova K, Giltiay NV, Kudrimoti M, Mohiuddin M, Ahmed MM, Nikolova-Karakashian M (2005) Elevated sphingomyelinase activity and ceramide concentration in serum of patients undergoing high dose spatially fractionated radiation treatment. *Cancer Biol Ther* 4:979–986.

Senan S, Smit EF (2007) Design of clinical trials of radiation combined with antiangiogenic therapy. *Oncologist* 12:465–477.

Siemann DW (2006) *Vascular-Targeted Therapies in Oncology*. Chichester, UK: John Wiley.

Simpson DH, Burns PN, Averkiou MA (2001) Techniques for perfusion imaging with microbubble contrast agents. *IEEE Trans Ultrason Ferroelectr Freq Control* 48:1483–1494.

Song CW, Kim M-S, Cho LC, Dusenbery K, Sperduto PW (2014) Radiobiological basis of SBRT and SRS. *Int J Clin Oncol* 19:570–578.

Stein EW, Maslov K, Wang LV (2009) Noninvasive, in vivo imaging of the mouse brain using photoacoustic microscopy. *J Appl Phys* 105:102027.

Suit HD, Willers H (2003) Comment on "Tumor response to radiotherapy regulated by endothelial cell apoptosis" (I). *Science* 302:1894.

Tahergorabi Z, Khazaei M (2012) A review on angiogenesis and its assays. *Int J Basic Med Sci* 15:1110–1126.

Tran WT, El Kaffas A, Al-Mahrouki A, Gillies C, Czarnota GJ (2013) A review of vascular disrupting agents as a concomitant anti-tumour modality with radiation. *J Radiother Pract* 12:255–262.

Tran WT, Iradji S, Sofroni E, Giles A, Eddy D, Czarnota GJ (2012) Microbubble and ultrasound radioenhancement of bladde cancer. *Br J Cancer* 107:469–476.

Wang X, Xie X, Ku G, Wang LV, Stoica G (2006) Noninvasive imaging of hemoglobin concentration and oxygenation in the rat brain using high-resolution photoacoustic tomography. *J Biomed Opt* 11:024015.

Yao J, Maslov KI, Zhang Y, Xia Y, Wang LV (2011) Label-free oxygen-metabolic photoacoustic microscopy in vivo. *J Biomed Opt* 16:076003.

Yao J, Wang LV (2011) Photoacoustic tomography: Fundamentals, advances and prospects. *Contrast Media Mol Imaging* 6:332–345.

3

Gamma Knife: From single-fraction SRS to IG-HSRT

Daniel M. Trifiletti, Jason P. Sheehan, and David Schlesinger

Contents

3.1   INTRODUCTION

Radiosurgery has traditionally been a high-dose, single-fraction treatment technique that has been found to be extremely effective for a large spectrum of malignant and benign neurosurgical conditions (Leksell, 1951). Delivery of high-dose radiotherapy in a single session leaves very little room for error, and as a result, radiosurgery maintains a requirement for rigorous accuracy and precision management in treatment delivery. Gamma Knife® (GK) radiosurgery (Elekta Instruments AB, Stockholm, Sweden) traditionally achieves this through the use of isocentric convergence of many small beamlets (201 or 192, depending on the model of the device) to create large dose gradients. A rigid head frame immobilizes the patient's head and also defines a stereotactic coordinate system with a direct mechanical linkage between the patient's head and the isocenter of the GK, that is, suitable for localization and targeting (Lunsford et al., 1988; Lindquist and Paddick, 2007). Image guidance is based on up-front imaging of the patient using fiducial systems mounted to the patient's head frame to localize anatomy relative to the stereotactic frame of reference.

The development of radiosurgery did not end with the invention of the GK, however. As experience accrued using linear accelerators for radiosurgery as an alternative to the GK, it became apparent that for certain clinical situations (for instance, tumors larger than is typically indicated for radiosurgery, or tumors directly adjacent to sensitive organs at risk [OARs]), delivery of the total dose over several fractions (hypofractionated stereotactic radiotherapy [HSRT]) creates some potential advantage, with similar tumoricidal effectiveness paired with further reduced normal tissue toxicity. The experience with linear accelerators also demonstrated the potential advantages to be gained by the use of in-room imaging techniques, making possible accurate and precise patient localization without the use of a rigid head frame and thereby making practical hypofractionated regimes.

Several techniques for both patient immobilization and image guidance have been developed that make possible hypofractionation on the GK without compromising the historic precision characteristic of gamma knife radiosurgery (GKRS). This chapter explores the logistics, limitations, and future potential of frameless image-guided HSRT using the Gamma Knife Perfexion with the eXtend™ system (Elekta Instruments AB, Stockholm, Sweden). Additional potential HSRT platforms are explored in subsequent chapters.

3.2   TRADITIONAL GAMMA KNIFE RADIOSURGERY: SINGLE-FRACTION A PRIORI IMAGE-GUIDED RADIOSURGERY

3.2.1   IMMOBILIZATION

Traditional GK radiosurgery is performed using a rigid stereotactic frame (the Leksell® Coordinate Frame), which is placed around the patient's head and fixed using four pins inserted to the outer table of the patient's skull. Frame placement is often performed in a small procedure room near the radiosurgery center, using a local anesthetic to numb the pin sites as needed. Other centers prefer to administer light sedation in addition to the local anesthetic.

By design and definition, the stereotactic frame defines a coordinate system called the Leksell Coordinate System®, which has an origin superior, posterior, and right of the patient's head and increments toward the patient's left (+X), anterior (+Y), and inferior (+Z). The coordinates of the frame system near the

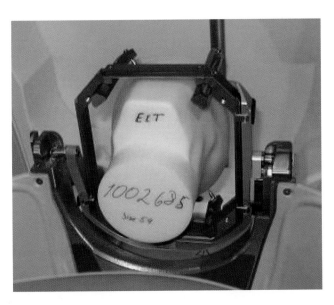

Figure 3.1 The Leksell stereotactic G-frame attached to an anthropomorphic head phantom and docked to the Gamma Knife treatment bed. In addition to providing rigid immobilization, the stereotactic frame defines a coordinate system throughout the patient's head that is mechanically linked to the machine coordinate system.

center of the head are (100 mm, 100 mm, 100 mm) in the Leksell coordinate system and the X-, Y-, and Z-coordinates for any intracranial target are all positive in their values. The Leksell frame mounts mechanically to the treatment table on the GK using an adapter in the case of the Perfexion (Figure 3.1), and therefore there is a mechanical correspondence between the stereotactic space defined by the frame and the coordinates of the GK itself.

3.2.2 IMAGE GUIDANCE

Image guidance for traditional radiosurgery occurs a priori of the procedure itself. Immediately following frame placement, patients are generally transported for treatment planning imaging. The modalities involved may include MR, CT, and/or biplane angiography depending on the indication. Images are linked to the stereotactic coordinate system using modality-specific indicator boxes which are attached to the stereotactic frame during the imaging procedure. The indicator boxes result in fiducial markers in the resulting images, which, once they are registered with the treatment planning system, allow any point in the patient's brain anatomy to be referenced in stereotactic coordinate space (Figure 3.2a,b). Nonstereotactic images (often including MR and/or PET) acquired prior to the frame placement are possible but must be coregistered to one of the stereotactic image studies to be useful.

3.3 LIMITATIONS OF TRADITIONAL GAMMA KNIFE SRS TECHNIQUES WHEN APPLIED TO IG-HSRT

There are several limitations to the traditional GK stereotactic radiosurgery (SRS) frame that limit its utility in HSRT. Most apparent is that the process of frame placement is invasive: pins are inserted into the outer table of the skull to create a rigid mechanical interface between the patient's head and the treatment machine. Moreover, this rigid association between the frame and the patient's skull is critical for the creation of the coordinate system used for localization and targeting, and any change in it invalidates the existing treatment plan.

A second limitation is that image guidance is a priori. This means that any change in the rigid association between frame and patient skull requires a new imaging study to reestablish the location of the patient anatomy relative to the coordinate system.

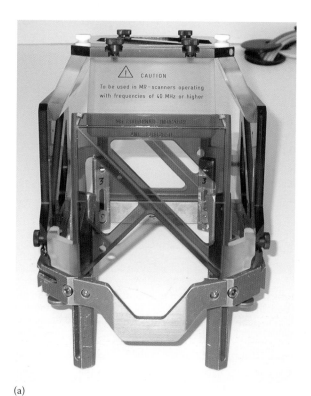

(a)

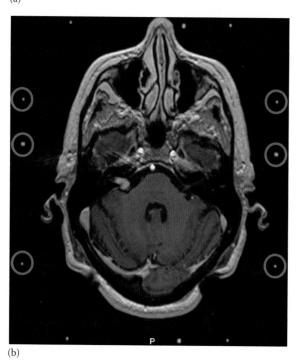

(b)

Figure 3.2 (a) An MR indicator box mounted to a Leksell stereotactic frame. The indicator box channels are filled with copper sulfate solution that appear bright in MR images. (b) The fiducial marks (marked in red) on an MR image acquired with a stereotactic frame and MR indicator box.

Beyond the inherent disadvantages of patient satisfaction, there are technical limitations to leaving a rigid head frame on for days as well. The presence alone of a rigid head frame does not ensure rigid fixation beyond 1 day. Subtle shifts in the position of the head frame over the treatment course are possible, and daily measurements and retorquing of the pins by half to one turn would be prudent to rule out these subtle systematic errors (by either digital probe or imaging-based measurements).

3.4 REQUIREMENTS FOR GAMMA KNIFE IG-HSRT

3.4.1 ACCURACY AND PRECISION REQUIREMENTS

The most critical component of a hypofractionated immobilization system is that it can reliably and repeatedly localize the isocenter in three-dimensional space for each treatment fraction and that this localization must remain valid over the course of each treatment fraction. This tenant is the basis by which SRT is safe and feasible. A commonly acceptable tolerance for isocenter displacement is a nonsystematic error of less than 1 mm. Although this tolerance is somewhat arbitrarily defined, there is evidence that adherence results in superior local control (Treuer et al., 2006). For the single-fraction case, the "gold standard" assumption has been that the rigid stereotactic frame provides superior immobilization performance over the relatively short time frame required to deliver a radiosurgical treatment. Given the small geometric distances between tumor and sensitive OARs in the brain, any IG-HSRT system cannot deviate far from the single-fraction standard.

3.4.2 REQUIREMENT FOR PATIENT ACCEPTANCE

A possibly overlooked component of rigid immobilization is that it must be reasonably well tolerated by patients. Patient satisfaction has become a critical component of medical care, and other fractionated radiotherapy approaches (i.e., gynecologic brachytherapy) have led to psychosocial disorders in some patients, thought to be related to the discomfort of the applicator left in place between fractions (Kirchheiner et al., 2014). Any IG-HSRT system that limits patient discomfort would also lead to less inter- and intrafraction motion, require less intrafraction treatment breaks, and have faster daily patient setup.

3.4.3 REQUIREMENT FOR SOME POTENTIAL TO EXPAND INDICATIONS

Traditional single-fraction GK radiosurgery has been a remarkably successful technique, with over 880,000 patients treated worldwide between 1991 and 2014, with 74,000 treated in 2014 alone (Leksell Gamma Knife Society, 2014). A successful system for hypofractionated GK treatments requires a rationale that creates an expansion of indications beyond those already effectively managed with the current system.

There is great potential in a method of reliable GK immobilization and hypofractionation. As described in other chapters, HSRT would increase the scope of SRT by permitting radiosurgery in anatomic locations that have previously been treated with conventionally fractionated radiotherapy because of concern of adjacent normal tissue tolerance. Additionally, it is possible that some intracranial tumors have a biology that would demonstrate improved local control with multifraction radiosurgery (Jee et al., 2014; Minniti et al., 2014; Toma-Dasu et al., 2014; Casentini et al., 2015).

3.5 HISTORICAL ATTEMPTS TO HYPOFRACTIONATE RADIOSURGERY TREATMENTS

There have been a variety of historical attempts both within and outside the GK subspecialty to create methods that could allow for hypofractionation. This section summarizes some historical attempts which form the basis for modern GK-HSRT.

3.5.1 PROTRACTED FRAME APPLICATION

The feasibility of this method was first reported by Simonová in the early 1990s as a method to achieve HST while only utilizing available devices (Simonová et al., 1995). They reported on 48 patients that underwent head frame placement and then returned for once daily treatments for 2–6 days. The method was considered feasible, well tolerated, and relatively safe. However, patients were admitted for the duration of therapy, making this an expensive treatment option. Additionally, patient reported outcomes were not included in this report. A similar "split-dose" approach was reported where the total SRS dose was divided into two equal fractions. Patients underwent frame placement, imaging for treatment planning, and then first treatment fraction in the evening of the first day, followed by a second fraction delivered approximately 14–15 hours later. The authors of the study reported that the treatment was well tolerated and showed a small survival benefit for patients receiving 2 fractions as compared to an earlier cohort receiving single-fraction SRS. However, the authors cautioned against the possibility of a frame becoming dislodged over the total time of the procedure (Davey et al., 2007).

3.5.2 REMOVABLE FRAME SYSTEMS

The TALON cranial fixation system (NOMOS Corp., Sewickley, PA) is a removable frame system that permits rigid fixation of the skull to a head frame through attachment to base screws inserted into the patient's skull. These screws are attached to the TALON system and permit minute adjustments of the cranium after fixation. The screws are left in place between fractions (usually 2–5 days). Salter et al. reported on the TALON system's positional accuracy and estimated that 95% of true isocenter position between fractions would fall within 1.55 mm of the planned isocenter position. The TALON system was well tolerated by patients; however, three of nine patients included developed infections at the screw sites, and two patients had loosening of the screws between fractions requiring retightening (Salter et al., 2001). The TALON system was not attempted in a GK SRT context.

3.5.3 RELOCATABLE FRAME SYSTEMS

Multiple relocatable head frame systems have been developed over the past 15 years. This includes rigid frames used for radiosurgery registration that are not invasively attached to the patient (Reisberg et al., 1998; Alheit et al., 1999; Ryken et al., 2001; Baumert et al., 2005; Minniti et al., 2010; Ruschin et al., 2010). Examples include systems that have utilized bite blocks, head straps, thermoplastic masks, optical tracking, or some combination. In all cases, the important characteristics include a relatively simple, noninvasive method for placing the patient in a repeatable treatment position corresponding to the position at the time of treatment planning.

3.6 HISTORICAL DEVELOPMENT OF ON-BOARD IMAGE GUIDANCE FOR RADIOSURGERY

The development of in-room image-guidance systems for radiotherapy was a significant development that enhanced the accuracy and precision by which a patient could be set in the correct treatment position. These systems, designed primarily for linear accelerator–based radiotherapy, were quickly adapted for use in radiosurgery contexts. Systems evolved from simple 2D MV portal imaging systems that used film (and later flat-panel detectors) that were exposed by the treatment beam to allow clinicians to verify the target was within the collimated field (Dong et al., 1997). The invention of amorphous-silicon flat-panel detectors motivated attempts to use the treatment machine itself as a megavoltage energy cone-beam CT (MV-CBCT) system (Pouliot et al., 2005). Kilovoltage cone-beam CT (kV-CBCT) systems were developed using x-ray tubes and detectors mounted orthogonally from the linac treatment beam (Jaffray, 2007). Dual ceiling-/floor-mounted stereoscopic kV x-ray systems were developed specifically for radiosurgery applications.

The above developments for linear accelerators were motivated as much for extracranial stereotactic and nonstereotactic indications as they were for intracranial indications, for which the stereotactic frame was a well-established and well-validated technique for intracranial radiosurgery. However, as noted above, in certain clinical situations an enhanced ability to hypofractionate is considered advantageous. To that end, David Jaffray's group at Princess Margaret Hospital developed a kV-CBCT system that they successfully integrated to a GK Perfexion. The system uses a conventional 90 kVp rotating anode x-ray tube and an opposing detector. The system is supported by a set of vertical supports which allow the system to translate from a parked position above the shield doors of the Perfexion to an imaging position between the patient and the shield doors. A rotational axis allows the system to rotate 210° for imaging; 1 mmor 0.5 mm isotropic voxel resolutions are achievable with a reconstruction field of view of 25.6 × 25.6 × 19.3 cm (Ruschin et al., 2013).

3.7 eXtend™ SYSTEM FOR THE GAMMA KNIFE® PERFEXION™

While the previous section summarizes work that has been performed to explore options for GK IG-HSRT, in practice the unitary commercial solution to allow for hypofractionated GK radiosurgery treatments is the eXtend™ system. The GK eXtend system makes possible reproducible, frameless stereotactic fixation of the head through a suctioned dental mold of the hard palate and maxillary teeth. The system removes the requirement for surgical intervention needed for frame placement, and no devices are left *in situ* between fractions that could cause pain or serve as a nidus for infection (Ruschin et al., 2010).

3.7.1 MAIN COMPONENTS

The eXtend frame system consists of a carbon fiber front plate to which a dental impression/mouthpiece can be attached, a base plate to which the frontpiece can be attached, and a vacuum cushion on which the patient's head sits (Figure 3.3). The eXtend frame rigidly docks with the GK patient positioning system (PPS). The mouthpiece of the frame is attached via plastic tubing to the patient control unit (PCU). The PCU consists of a vacuum pump and tubing that connects to the mouthpiece and interfaces with the patient and the treatment unit (Figure 3.4). The reposition check tool (RCT) consists of an acrylic measurement template and an associated set of digital measurements probes. The RCT fits into slots on the eXtend frame. Measurement holes in the RCT template are used for measurement of head position to confirm three dimensional positioning between fractions (Figure 3.5).

3.7.2 DENTAL MOLD CREATION

The first step in the use of the eXtend system is the selection of the mouthpiece and the creation of the dental mold. A dental impression is created using standard impression material (vinyl polysiloxane) using a mixing gun. A plastic spacer placed between the mold and the hard palate before inserting the mouthpiece into the patient's mouth to create the impression. The spacer allows for an airspace in which the vacuum can suction the mold to the palate, aligned by dental anatomy (Figure 3.6). Once the mouthpiece is placed in the patient's mouth, even pressure must be maintained along the palate to allow the impression material

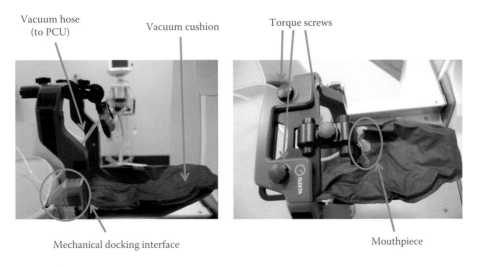

Vacuum hose (to PCU)   Vacuum cushion   Torque screws

Mechanical docking interface   Mouthpiece

Figure 3.3 The eXtend frame system and its components.

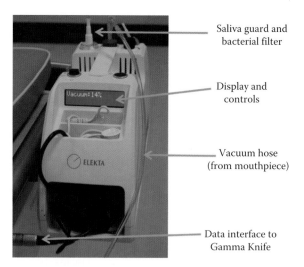

Saliva guard and bacterial filter

Display and controls

Vacuum hose (from mouthpiece)

Data interface to Gamma Knife

Figure 3.4 The patient control unit (PCU) for the Gamma Knife eXtend system. The PCU creates a vacuum that is used to monitor patient movement and sends data during treatment to the Gamma Knife control system to interrupt treatment if the vacuum level falls below a set threshold.

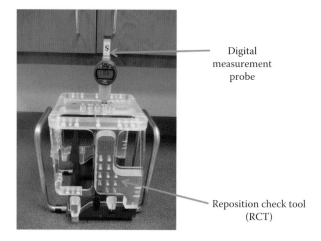

Digital measurement probe

Reposition check tool (RCT)

Figure 3.5 The reposition check tool (RCT) template and associated digital measurement probe. The red carrier doubles as a quality-assurance (QA) tool for the RCT.

to cure (Figure 3.7). If there is insufficient material between the teeth and the mouthpiece or between the hard and soft palate, then reliable suction may be difficult.

Creation of the dental impression may be completed by the treatment team, but often benefits from the experience of a dentist as the process is identical to that used to create dental impressions for crowns and other dental interventions. An important caveat is that the dental impression is critical to ensure proper daily alignment and minimize intrafraction motion. Edentulous patients or patients without adequate dentition are not eligible for eXtend immobilization.

3.7.3 SETUP AT GAMMA KNIFE

Creation of the dental impression is followed by setup at the GK and the construction of the eXtend frame system using the completed mouthpiece.

Figure 3.6 Creation of an eXtend system dental impression. The impression material fills a mouthpiece, and the plastic spacer (in purple) creates a vacuum space within the dental material.

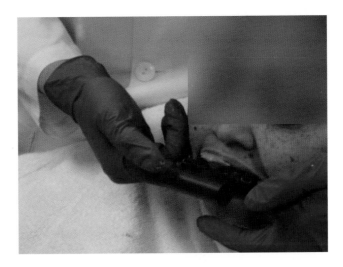

Figure 3.7 Placement of an eXtend system mouthpiece in a patient's mouth. Even pressure must be applied for several minutes while the impression material cures. (Patient's face blurred for confidentiality).

3.7.3.1 Initial patient positioning

The patient positioning process begins with the patient on the GK couch, supine with the head on the vacuum cushion and the head frame removed. Ensuring that the patient is comfortable at this point is important because after dental mold insertion, they will be unable to speak clearly. This includes head elevation and angle, leg positioning/flexion, ambient temperature, and introduction of the team members in the room.

3.7.3.2 Dental mold insertion/vacuum test

After the patient is in a comfortable position, the dental mold connection with the spacer and vacuum tubing is confirmed and is guided into the patient's mouth and abutted to the hard palate and maxillary dentition. The PCU vacuum is then tested with the mouthpiece in place to a vacuum level of

30%–40% (as a percentage of atmospheric pressure). To test for suction quality, gently tugging on the mouthpiece can ensure a good seal and ensure patient comfort with suction level. The PCU contains a saliva and bacteria filter that can help maintain sterility and steady suction over time. Additionally, the PCU has a safety alarm that will detect a loss of suction (defined as a 10% change in suction from the set point).

3.7.3.3 Frame creation

With the dental mold in place and the vacuum activated, the head frame can be secured. This is first done by attaching the frontpiece to the mold and then by locking the frontpiece to the docking area (which is locked to the GK couch). When patient comfort is again confirmed, the mouthpiece and head frame are hand tightened and then secured with a torque wrench (Figure 3.8).

3.7.3.4 Vacuum cushion creation

In the supine position with the head frame attached, the vacuum cushion is molded to the scalp and the PCU is used to evacuate air from the cushion. As the vacuum level in the cushion increases, the cushion becomes increasingly rigid and molded to the shape of the patient's head. When complete, the result is a rigid cushion containing a firm impression of the dorsal aspect of the scalp that will be maintained for each fraction.

This time point defines the stereotactic alignment of the patient's head with the couch. Any change in the vacuum pressure of the mouth piece, of the vacuum cushion, or of the tension in the screws of the head frame would result in compromise of the rigidity and reproducibility of the eXtend system. If these changes occur, the system should be reset from the beginning.

3.7.3.5 Test measurements/measurement hole selection

To confirm proper alignment of the head within the frame, daily reference measurements are made to using the RCT. These are compared to the measurements taken at the time of simulation imaging. The RCT consists of four plastic panels that surround the patients head in the eXtend frame (Figure 3.9). Measurements are taken with a pair of electronic linear measurement probes that are included with the eXtend system (C150XB Digimatic Indicator, Mitutoyo Corp.). The probes measure the distance between preset holes in the RCT and the scalp. At least one aperture (and ideally more than one) must be chosen for each panel of the RCT. Apertures should be chosen to ideally allow normal incidence of the probe tips to the patient's head. Choosing apertures far apart from each other and avoiding areas of loose skin or fat can improve the precision and reproducibility of measurements. During this initial setup step, each aperture chosen and the distance to the head are recorded on a worksheet.

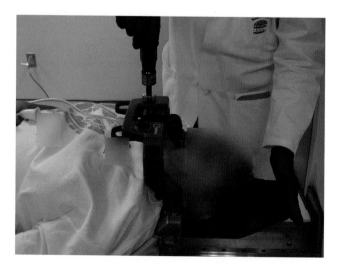

Figure 3.8 Creation of a patient-specific eXtend frame by tightening the locking screws on the frame front plate with a torque wrench (patient's face blurred for confidentiality).

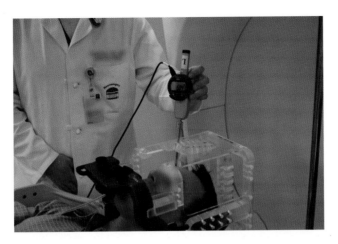

Figure 3.9 Physician acquiring reposition measurements using the reposition check tool and the digital probe system.

3.7.3.6 Ancillary setup information and patient instructions

Some centers may wish to augment the RCT position measurements with ancillary measurements that can guide subsequent patient setup. These include measuring the distance from the earlobe laterally to the side of the frame and from the frame inferiorly to the shoulder. Measuring the height of the cushion on the couch can be particularly useful when considering that the treatment couch used during subsequent CT or MRI may be a different shape and thickness as compared to the GK couch. These measurements can serve to approximate patient positioning before the more detailed measurements are made with the RCT. Patient instruction between fractions is relatively simple but critical. Patients are advised to avoid any dental manipulation between fractions and to avoid cutting or braiding hair.

3.7.4 SIMULATION (CT) IMAGING

Following initial setup at the GK, patients proceed to simulation imaging which will serve as the reference stereotactic images for treatment planning.

3.7.4.1 Simulation imaging setup and reference RCT measurements

The basic principle of the eXtend system is that the patient position at the time of treatment must match (to within a small uncertainty threshold) the patient position at the time of simulation imaging. Therefore, at the time of stereotactic CT imaging, reference measurements are collected that will serve as the standard to compare future measurements to (prior to each treatment delivery). The process begins with the stereotactic immobilization of the patient as outlined earlier, but it is done on the CT couch as opposed to the GK couch. During any period that the head frame is assumed to be rigidly fixed, the PCU should be set to alarm for changes in vacuum, and the patient should be visually monitored to ensure patient comfort, as hand signals are preferred while the mouthpiece is in place. Measurements proceed as described earlier, using the measurement apertures chosen at the time of initial setup at the GK. These measurements are read off of the display on the PCU and recorded on a worksheet for later use.

3.7.4.2 Simulation stereotactic CT imaging

After proper immobilization is achieved and RCT measurements confirmed and recorded, the eXtend CT indicator box is mounted to the frame. The CT indicator is a transparent box with implanted fiducial markers that can serve as rigid points in the GK treatment planning software (GammaPlan, Elekta AB, Stockholm, Sweden). CT images of the head are then obtained from vertex to midframe and with a field of view wide enough to include the entire CT indicator and corresponding fiducial markers (Figure 3.10). Intravenous contrast can be utilized as clinically indicated.

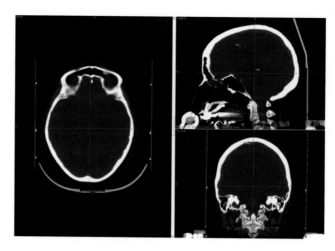

Figure 3.10 Stereotactic CT of an eXtend patient. The CT field of view must cover the entire head and include the lateral fiducial markers.

3.7.4.3 Post-CT measurements

Immediately after the CT sequences are obtained and before releasing the vacuum suction, post-CT measurements with the RCT are important to verify that the patient did not shift during CT imaging. This is done by the same method as described earlier, though the same apertures as were used in the pre-CT measurements. Any difference of more than 0.5 mm from the pre-CT value should prompt the team to remove and reposition the eXtend head frame, remeasure, reobtain CT images, and then confirm measurements. After the repeat measurements are verified, the suction can be released and the head frame removed. Images are then transferred to the GK treatment planning system.

3.7.5 INTEGRATION OF NONSTEREOTACTIC SCANS

The eXtend system requires that a stereotactic CT be used as a stereotactic reference. CT images are less likely to suffer from localized geometric distortion and the eXtend frame does not fit within all MR head coils, two considerations which may be the source for this requirement. However, multimodality images (especially MR) are critical for the visualization of most intracranial indications, so nonstereotactic images may be incorporated into treatment planning via image registration. The GK treatment planning system includes cross-modality rigid coregistration algorithms for this purpose (Viola and Wells III, 1997).

3.7.6 TREATMENT PLANNING

After all stereotactic and nonstereotactic imaging have been imported into the GK treatment planning system, the stereotactic CT images are registered to stereotactic space using the fiducial markers visible in the CT images. MR and other modality images are then coregistered to the reference CT images as described earlier. The details of target visualization and delineation vary by institution; however, the planning includes functionality to delineate target volume(s) and adjacent OARs similar to traditional SRT planning. Isocenter-based "shots" are placed and customized based on target size, shape, and adjacent OARs (Figure 3.11). A total dose and number of fractions are entered and the plan is reviewed by the neurosurgeon, medical physicist, and radiation oncologist.

3.7.7 TREATMENT PROCEDURE

3.7.7.1 Entering reference measurements (first fraction)

When a patient treatment is opened at the GK console prior to delivery of the first treatment fraction, the system requires that the reference position measurements acquired at the time of CT imaging be entered into the system. Accuracy at this step is critical because it will create a reference point to which each

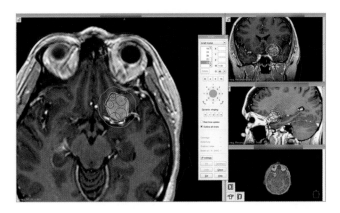

Figure 3.11 A treatment plan for a hypofractionated Gamma Knife radiosurgery case using the eXtend system.

fraction will be directly compared. As such, the reference measurements should be carefully double checked (preferably by a separate member of the team) prior to moving forward.

3.7.7.2 Repositioning measurements

When the patient is prepared to proceed with therapy, they should enter the GK vault and have the eXtend treatment head attached in a similar manner to when it was attached prior to CT-simulation (supine, vacuum cushion in place, mouthpiece, vacuum, frontpiece, secure to couch). Then new measurements should be collected using the electronic probe though the same apertures that were used for the reference measurements. This process is guided by the GK eXtend console, which automatically captures each probe measurement and compares to the previously entered reference measurements. Once all selected RCT apertures have been measured, the console software will calculate a three dimensional translational vector of the difference in patient position as compared to the reference measurement. The system will warn the operator if the radial positional difference is greater than 1.0 mm and suggest the clinician consider repositioning to achieve a more favorable patient position. However, ultimately clinical judgment is involved in determining what level of positional uncertainty is acceptable.

3.7.7.3 Treatment

Following patient positioning, each individual SRT treatment is administered in the same way as single-fraction GK treatment. The GK treatment couch translates the patient's head into the center of the radiation body of the unit, placing the patient's head at a location corresponding to the coordinates of each shot of the treatment plan in turn. Each position is maintained for a dwell duration as calculated by the treatment planning system in order to achieve the desired overall dose distribution. Treatments are monitored by the operator of the machine via video and audio surveillance. The patient is provided with a call button to alert the operator if they require assistance as the patient cannot speak with the eXtend mouthpiece in place. It can be useful for the treatment team (usually a radiation therapist, medical physicist, neurosurgeon, and radiation oncologist) to develop a set of hand gestures that provide general communication.

3.7.7.4 Intrafraction position monitoring

Patient immobilization is monitored using the PCU vacuum surveillance system. Patient motion beyond a very small threshold will trigger a loss of suction in the eXtend mouthpiece. Any loss of vacuum greater than 10% of vacuum level set at time of patient positioning will trigger an interrupt which will shield the GK ^{60}Co sources and pause the treatment, automatically withdrawing the patient from the treatment position in the machine. This occurrence requires that the patient position be remeasured using the RCT and probe system and repositioned if required before treatment can resume. If adequate repositioning is impossible, then a new stereotactic CT may be acquired and the treatment plan shifted to accommodate the new patient position.

3.7.8 ACCURACY COMPARED TO SIMILAR SYSTEMS

The mean setup uncertainty of the GK eXtend system has been shown to be reproducibility on the order of 0.4–1.3 mm (Ruschin et al., 2010; Sayer et al., 2011; Schlesinger et al., 2012; Ma et al., 2014). Table 3.1 reports the mean displacement of the patient after setup by a representative variety of proposed immobilization systems applicable to fractionated radiosurgery. As demonstrated, the setup uncertainty of the eXtend system is comparable to other available relocatable frame systems.

3.7.9 LIMITATIONS OF eXtend SYSTEM

The eXtend system provides a reliable, noninvasive method for reproducible immobilization of patients for HSRT. However, as compared to HSRT systems for other treatment devices, the eXtend system may have some potential limitations.

3.7.9.1 Complicated workflow

One drawback to the eXtend system is the complex logistic aspect of the mouthpiece creation, application, and RCT measurement system. The mouthpiece can be bulky and must fit snuggly and firmly for the frame to be reliably fixed. Generally, patients consider the first day of treatment arduous because dental mold creation, head frame fitting, CT, planning, first fraction, and numerous of precise measurements can take hours to complete with the head frame in place. However, subsequent treatments are considered relatively convenient (Sayer et al., 2011).

3.7.9.2 Vacuum as a proxy for motion

The basis of the real-time monitoring of the eXtend system is in the vacuum alert. The system relies on the assumption that any change in the vacuum pressure of greater than 10% equates to a displaced target. It does not detect possible patient motion in which the vacuum level changes by less than 10%. Conversely, the vacuum alert assumes that any change in vacuum of greater than 10% means that the patient moved systematically (as opposed to no movement or temporary movement). Additionally, after the alert is activated, it is impossible to tell if the change in pressure was indeed related to patient movement or potentially equipment failure in the vacuum or tubing, and confirmation of functional equipment is

Table 3.1 **Reports of the residual setup uncertainty of a variety of relocatable immobilization systems for fractionated radiation treatments**

AUTHOR	DEVICE	SETUP DISPLACEMENT (SD) (mm)
Sweeney et al. (1998)	Bite block + vacuum assist	<1.02[a]
Rosenberg et al. (1999)	GTC frame	1.1(0.6)[b]
Ryken et al. (2001)	Mask + optically tracked bite block	0.16 (0.04)[c]
Baumert et al. (2005)	Mask + bite block	2.2 (1.1)[d]
Kunieda et al. (2009)	Bite block + vacuum assist	0.93–1.09 (0.52–0.88)[d]
Minniti et al. (2010)	Relocatable frame + upper jaw support	0.5 (0.4)[d]
Ruschin et al. (2010)	eXtend prototype	1.0[e]/1.3[c]
Schlesinger et al. (2012)	eXtend clinical system	0.64 (0.25)[e]

Source: Adapted from Schlesinger, D. et al., *J. Neurosurg.*, 117(Suppl.), 217, 2012.
Superscripted letters indicate basis for setup displacement measurement.
[a] Fiducials vs surface landmarks
[b] Orthogonal radiograph landmarks
[c] Fiducials vs CBCT
[d] Simulation CT vs QA CT
[e] Probe/depth measurements

warranted for unexpected vacuum alarm. In any case, activation of the vacuum alarm results in a treatment pause, and the entire head frame should be removed, replaced, and remeasured to ensure proper placement and treatment accuracy.

3.7.9.3 Patient contraindications (dentition/performance status/gag)

For patients to be eligible for HST using the GK eXtend system, they should be otherwise fit for radiosurgery (limited number of intracranial targets, adequate performance score, etc.). Although it should be noted that there are some aspects of the eXtend system that may require additional patient cooperation compared to a standard head frame, patients with a sensitive gag reflex may not be willing or able to tolerate a bulky mouthpiece for the duration of multiple treatments.

Additionally, adequate dentition is critical to the reproducibility of the placement of the dental mold. There is no strict definition of "adequate dentition," but factors to consider are the number, size, health, and distribution of maxillary teeth. Any palatal surgery or reconstruction (i.e., cleft palate) is a relative contraindication to the eXtend system and should be utilized cautiously in such patients. We have found that our dental colleagues can not only aid in determining which patients would provide "adequate dentition," but they can also provide expertise in the creation of high-quality, well-fitting dental molds.

3.8 THE FUTURE OF GAMMA KNIFE IG-HSRT: THE GAMMA KNIFE® ICON™

The eXtend system has proven to be a practical, functional method for achieving hypofractionation using the GK. However, the system is limited in scope and functionality relative to comparable systems routinely used in linac-based radiosurgery. Recognizing that the eXtend system would not be an optimal solution on its own, Elekta Instruments, the manufacturer of the GK, began a development cycle intending to address some of the shortcomings of the eXtend system for IG-HSRT. In particular, they built on the work of Ruschin et al. (2013) and created an integrated CBCT system sufficient in quality to provide on-board verification of patient position and correction. They also integrated an optical tracking system intended to monitor intrafraction patient position and gate the treatment delivery during positional excursions from the planned treatment position (Figure 3.12).

At the time of this publication, the prototype of the CBCT system features a double-hinged arm design that is mounted in front of the shield doors of the unit. The arm unfolds from a parked position to an imaging position and can then complete a partial rotation around the patient's head. Final imaging

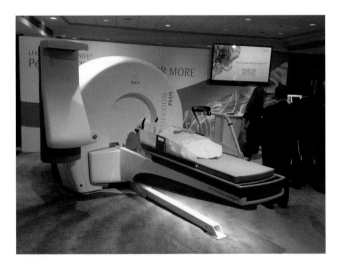

Figure 3.12 A display prototype of the Gamma Knife Perfexion Plus (since rebranded the Gamma Knife Icon) at the *17th International Leksell Gamma Knife Society Meeting*. The prototype includes on-board CBCT imaging and an optical motion-tracking system.

Figure 3.13 A closeup of the reflective markers used for the optical motion-tracking system. Marker locations are circled.

specifications have yet to be released by the manufacturer; however, preliminary data presented at the *17th International Leksell Gamma Knife Society Meeting* (2014) in New York suggest that the system will include two preset imaging modes, a high-dose and low-dose mode. Each will have a 200° rotation, an imaging volume of 448 cm³, 0.5 mm voxel size, and a resolution of 8 line pairs/cm using 332 projections at 90 kVp. The high-dose mode will have a CT dose index (CTDI) of approximately 7.0 mGy and will have a higher signal to noise ratio as compared to the low-dose mode with a CTDI of 2.8 mGy. Preliminary tests on localization uncertainty using a phantom suggest a mean positional uncertainty of <0.2 mm (Eriksson and Nordström, 2014; Eriksson et al., 2014).

The optical tracking system consists of a folding infrared camera system mounted near the foot of the GK PPS that is aimed toward the head of the unit. Optical markers are placed on two posts built into the superior end of the PPS. A third adhesive optical marker is placed on the patient's nose (Figure 3.13).

The workflow of the new device is ultimately planned to support traditional G-frame treatments, eXtend frame treatments, as well as thermoplastic masks. Treatment plans can be created for patients based on prior volumetric MR or CT imaging. At the time of treatment, patients are set up on the GK PPS. A CBCT of the patient is obtained, determining the position of the patient relative to the GK machine coordinates. Coregistration of the treatment planning images to the CBCT creates a stereotactic coordinate system transformation, which allows the planned dose distribution to be mapped to the GK coordinate space. Because the CBCT has a known spatial relationship with the rest of the treatment machine, a fixed stereotactic frame is no longer required, making possible treatments with devices including the eXtend frame and even a thermoplastic mask. The optical tracking system will ideally ensure that the patient is treated only when in a position within an acceptable level of uncertainty for radiosurgery. The optical tracker acquires a baseline relative position between the post markers and the patient marker and then tracks the relative marker position. If the relative positions deviate more than a threshold amount, the treatment can be temporarily suspended by moving the GK source sectors to a blocked position. In this manner, the treatment can be efficiently gated to match the patient position. At the time of this writing, final specifications of the proposed system are not available, so it is not known to what tolerance the final system will be able to achieve.

Unofficially announced as the GK Perfexion Plus, at the *17th International Leksell Gamma Knife Society Meeting*, Elekta rebranded the new treatment device as the "Gamma Knife® Icon™" in 2015 (at the *3rd European Society for Radiotherapy and Oncology Forum*). The new device is not yet cleared by the US Food and Drug Administration (FDA) in the United States, and it is pending CE mark in Europe.

3.9 CONCLUSIONS

Radiosurgery has evolved remarkably compared to previous decades, led by advances in existing surgical delivery platforms including the GK platform. Radiosurgery has become more flexible; expanding from a strictly single-fraction modality, now to the cusp of establishing itself as a multifraction delivery system. Future advancements to GK will give it true image-guided HSRT capabilities without sacrificing the precision and accuracy that have established GK as the "gold standard" for radiosurgery.

ACKNOWLEDGMENTS

David Schlesinger receives research support from Elekta Instruments AB, Stockholm, Sweden.

REFERENCES

Alheit H, Saran FH, Warrington AP, Rosenberg I, Perks J, Jalali R, Shepherd S, Beardmore C, Baumert B, Brada M (1999) Stereotactically guided conformal radiotherapy for meningiomas. *Radiother Oncol* 50:145–150.

Baumert BG, Egli P, Studer S, Dehing C, Davis JB (2005) Repositioning accuracy of fractionated stereotactic irradiation: Assessment of isocentre alignment for different dental fixations by using sequential CT scanning. *Radiother Oncol* 74:61–66.

Casentini L, Fornezza U, Perini Z, Perissinotto E, Colombo F (2015) Multisession stereotactic radiosurgery for large vestibular schwannomas. *J Neurosurg* 122:818–824.

Davey P, Schwartz M, Scora D, Gardner S, O'Brien PF (2007) Fractionated (split dose) radiosurgery in patients with recurrent brain metastases: Implications for survival. *Br J Neurosurg* 21:491–495.

Dong L, Shiu A, Tung S, Boyer A (1997) Verification of radiosurgery target point alignment with an electronic portal imaging device (EPID). *Med Phys* 24:263–267.

Eriksson M, Nordström H (2014) Design and performance characteristics of a cone beam CT System for the LGK Perfexion. In: *17th International Leksell Gamma Knife Society Meeting*, New York.

Eriksson M, Nutti B, Hennix M, Malmberg A, Nordström H (2014) Position accuracy analysis of the stereotactic reference defined by CBCT on LGK Perfexion. In: *17th International Leksell Gamma Knife Society Meeting*, New York.

Jaffray DA (2007) Kilovoltage volumetric imaging in the treatment room. *Front Radiat Ther Oncol* 40:116–131.

Jee TK, Seol HJ, Im YS, Kong DS, Nam DH, Park K, Shin HJ, Lee JI (2014) Fractionated gamma knife radiosurgery for benign perioptic tumors: Outcomes of 38 patients in a single institute. *Brain Tumor Res Treat* 2:56–61.

Kirchheiner K, Czajka-Pepl A, Ponocny-Seliger E, Scharbert G, Wetzel L, Nout RA, Sturdza A, Dimopoulos JC, Dörr W, Pötter R (2014) Posttraumatic stress disorder after high-dose-rate brachytherapy for cervical cancer with 2 fractions in 1 application under spinal/epidural anesthesia: Incidence and risk factors. *Int J Radiat Oncol Biol Phys* 89:260–267.

Kunieda E, Oku Y, Fukada J, Kawaguchi O, Shiba H, Takeda A, Kubo A (2009) The reproducibility of a HeadFix relocatable fixation system: Analysis using the stereotactic coordinates of bilateral incus and the top of the crista galli obtained from a serial CT scan. *Phys Med Biol* 54:N197–N204.

Leksell L (1951) The stereotaxic method and radiosurgery of the brain. *Acta Chir Scand* 102:316–319.

Leksell Gamma Knife Society (ed.) (2014) Leksell Gamma Knife: Indications treated 1968 to 2014.

Lindquist C, Paddick I (2007) The Leksell Gamma Knife Perfexion and comparisons with its predecessors. *Neurosurgery* 61:130–140; discussion 140–131.

Lunsford LD, Flickinger JC, Steiner L (1988) The gamma knife. *JAMA* 259:2544.

Ma L, Pinnaduwage D, McDermott M, Sneed PK (2014) Whole-procedural radiological accuracy for delivering multi-session gamma knife radiosurgery with a relocatable frame system. *Technol Cancer Res Treat* 13:403–408.

Minniti G, Esposito V, Clarke E, Scaringi C, Bozzao A, Falco T, De Sanctis V et al. (2014) Fractionated stereotactic radiosurgery for patients with skull base metastases from systemic cancer involving the anterior visual pathway. *Radiat Oncol (Lond, U.K.)* 9:110.

Minniti G, Valeriani M, Clarke E, Montagnoli R, Saporetti F, Enrici RM, D'Arienzo M, Ciotti M (2010) Fractionated stereotactic radiotherapy for skull base tumors: Analysis of treatment accuracy using a stereotactic mask fixation system. *Radiat Oncol (Lond, U.K.)* 5:1.

Pouliot J, Bani-Hashemi A, Chen AJ, Svatos M, Ghelmansarai F, Mitschke M, Aubin M et al. (2005) Low-dose megavoltage cone-beam CT for radiation therapy. *Int J Radiat Oncol Biol Phys* 61:552–560.

Reisberg DJ, Shaker KT, Hamilton RJ, Sweeney P (1998) An intraoral positioning appliance for stereotactic radiotherapy. *J Prosthet Dent* 79:226–228.

Rosenberg I, Alheit H, Beardmore C, Lee KS, Warrington AP, Brada M (1999) Patient position reproducibility in fractionated stereotactic radiotherapy: An update after changing dental impression material. *Radiother Oncol* 50:239–240.

Ruschin M, Komljenovic PT, Ansell S, Ménard C, Bootsma G, Cho YB, Chung C, Jaffray D (2013) Cone beam computed tomography image guidance system for a dedicated intracranial radiosurgery treatment unit. *Int J Radiat Oncol Biol Phys* 85:243–250.

Ruschin M, Nayebi N, Carlsson P, Brown K, Tamerou M, Li W, Laperriere N et al. (2010) Performance of a novel repositioning head frame for gamma knife perfexion and image-guided linac-based intracranial stereotactic radiotherapy. *Int J Radiat Oncol Biol Phys* 78:306–313.

Ryken TC, Meeks SL, Pennington EC, Hitchon P, Traynelis V, Mayr NA, Bova FJ, Friedman WA, Buatti JM (2001) Initial clinical experience with frameless stereotactic radiosurgery: Analysis of accuracy and feasibility. *Int J Radiat Oncol Biol Phys* 51:1152–1158.

Salter BJ, Fuss M, Vollmer DG, Sadeghi A, Bogaev CA, Cheek DA, Herman TS, Hevezi JM (2001) The TALON removable head frame system for stereotactic radiosurgery/radiotherapy: Measurement of the repositioning accuracy. *Int J Radiat Oncol Biol Phys* 51:555–562.

Sayer FT, Sherman JH, Yen CP, Schlesinger DJ, Kersh R, Sheehan JP (2011) Initial experience with the eXtend System: A relocatable frame system for multiple-session gamma knife radiosurgery. *World Neurosurg* 75:665–672.

Schlesinger D, Xu Z, Taylor F, Yen CP, Sheehan J (2012) Interfraction and intrafraction performance of the Gamma Knife eXtend system for patient positioning and immobilization. *J Neurosurg* 117(Suppl):217–224.

Simonova G, Novotny J, Novotny J, Jr., Vladyka V, Liscak R (1995) Fractionated stereotactic radiotherapy with the Leksell Gamma Knife: Feasibility study. *Radiother Oncol* 37:108–116.

Sweeney R, Bale R, Vogele M, Nevinny-Stickel M, Bluhm A, Auer T, Hessenberger G, Lukas P (1998) Repositioning accuracy: Comparison of a noninvasive head holder with thermoplastic mask for fractionated radiotherapy and a case report. *Int J Radiat Oncol Biol Phys* 41:475–483.

Toma-Dasu I, Sandstrom H, Barsoum P, Dasu A (2014) To fractionate or not to fractionate? That is the question for the radiosurgery of hypoxic tumors. *J Neurosurg* 121(Suppl):110–115.

Treuer H, Kocher M, Hoevels M, Hunsche S, Luyken K, Maarouf M, Voges J, Muller RP, Sturm V (2006) Impact of target point deviations on control and complication probabilities in stereotactic radiosurgery of AVMs and metastases. *Radiother Oncol* 81:25–32.

Viola P, Wells III WM (1997) Alignment by maximization of mutual information. *Int J Comput Vis* 24:137–154.

4 CyberKnife image-guided hypofractionated stereotactic radiotherapy

Christopher McGuinness, Martina Descovich, and Igor Barani

Contents

4.1 OVERVIEW OF THE CYBERKNIFE SYSTEM

The CyberKnife system is a linear accelerator (linac) mounted on a robotic arm specifically designed for stereotactic radiosurgery (SRS) and stereotactic body radiation therapy (SBRT). The robotic arm enables the delivery of radiation from hundreds of noncoplanar, nonisocentric beams around the patient achieving highly conformal dose distributions. Stereotactic targeting accuracy is achieved by combining real-time orthogonal x-ray images with advanced image recognition software for automatic tracking of bony landmarks, implanted fiducials, or clearly distinguishable tumors within the lung. This allows delivery of highly conformal hypofractionated treatments in the entire body without the need for rigid fixation devices. Figure 4.1 shows a CyberKnife treatment vault with the important components labeled.

The CyberKnife system came to market in the late 1990s. It was designed as a frameless alternative to existing SRS systems for the treatment of brain and upper spine lesions. Over the years, the development of new tracking methods expanded the clinical applications of CyberKnife to several extracranial sites (including the spine, lung, liver, pancreas, and prostate) and reduced the targeting accuracy to below 1 mm for both static and dynamic tracking methods (Kilby et al., 2010). In 2014 the CyberKnife M6 Series was released. The major change introduced by the M6 system was the addition of a micro-multileaf collimator (MLC) to the collimator system, which is anticipated to significantly reduce treatment time while maintaining or improving treatment quality. Technical specifications for the M6 provided here are based upon the vendor's documentation; early publications from users will be included when available.

4.1.1 SYSTEM SPECIFICATIONS

The CyberKnife system consists of an X-band cavity magnetron and a side-coupled standing wave linac mounted on a robotic manipulator (Kuka Roboter GmbH, Augsburg, Germany). The linac produces an unflattened 6 MV photon beam with a dose rate up to 1000 cGy/min. The beam is collimated using either

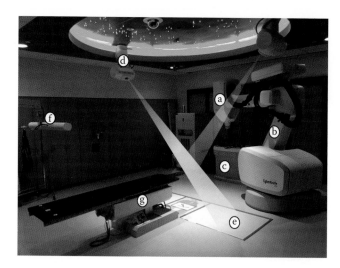

Figure 4.1 Image of a CyberKnife treatment suite. (a) Linear accelerator. (b) Robotic manipulator arm. (c) Exchange table with all 12 fixed collimators and the Iris™ variable aperture collimator. (d) X-ray imaging source. (e) Flat-panel detector. (f) Synchrony® camera array. (g) Patient positioning couch.

one of 12 fixed circular tungsten cones with diameters ranging from 5 to 60 mm or the Iris™ (Accuray, Inc., Sunnyvale, CA) variable aperture collimator with the same set of field sizes (Echner et al., 2009). Field size is defined at a source-to-axis distance (SAD) of 800 mm. The Iris collimation system consists of two hexagonal banks of tungsten, producing a twelve-sided aperture. The mechanical uncertainty of the Iris field sizes is 0.2 mm, which affects the output factor for the smallest field size (5, 7.5, and 10 mm). The uncertainty in output factor for the 5 mm aperture can be up to 10% and is approximately 1.4% for the 10 mm aperture. While the manufacturer restricts the use of the 5 mm aperture, we do not recommend using either 5 or 7.5 mm Iris aperture for clinical cases.

An automated exchange table system enables switching between the Iris and the fixed collimator housing and changes the tungsten cones automatically during treatment. Plans generated with multiple apertures typically result in better quality (dose conformity, homogeneity) and require a lower number of monitor units (MUs) (Pöll et al., 2008). However, using multiple fixed cones is not practical because it requires multiple path traversals and results in excessively long treatment times. The Iris collimator allows using multiple apertures without these limitations.

The M6 system has the option to use either the fixed or Iris collimators or the micro-MLC. The exchange table system enables one to automatically pick up the three housing (fixed, Iris, and MLC assembly). However, due to space limitation on the exchange table, the automatic exchange of the 12 fixed cones is no longer available. The micro-MLC consists of 41 pairs of fully interdigitated tungsten leaves 2.5 mm thick and 100 mm wide at 800 mm SAD allowing a maximum field size of 120 mm (leaf motion direction) by 100 mm. The vendor specification for interleaf leakage is less than 0.5% maximum (Accuray Inc., 2013). The addition of the micro-MLC is expected to expand the clinical applications of CyberKnife, allowing the treatment of large lesions (>6 cm) and the use of conventionally fractionated dose regimens. Early investigations have shown that plans generated with the micro-MLC result in more homogeneous dose distribution, a lower number of MU, and significantly shorter treatment time (Van De Water et al., 2011; McGuinness et al., 2015).

Treatments are delivered from hundreds of beams arranged around the target. Each beam is defined by a source point, called a node, a direction, and a field size. Plans with the micro-MLCs may have several segments with different MLC leaf patterns for each beam. The complete set of nodes is called the path set and can range from 23 to 133 depending on the site and patient position. The image-guided system consists of two diagnostic x-ray sources mounted in the ceiling and two amorphous silicon flat-panel detectors embedded in the floor, imaging the patient from two orthogonal views at ±45° oblique angles. Target localization during patient setup and treatment delivery is achieved by comparing the live camera images with a library of digitally reconstructed radiographs (DRRs) generated from the planning CT at 45° angles through the imaging center. Based on this comparison, the tracking software calculates the differences in the three translational and three rotational directions between simulation and treatment positions (couch correction parameters). Patients are positioned on a motorized treatment table with either 5 or 6 degrees of freedom, depending on couch model. If the couch correction parameters are below the threshold sets for treatment, the robot retargets the radiation beams, without the need to stop the treatment and move the patient couch.

The CyberKnife system includes an integrated treatment planning system (TPS) called MultiPlan™ (Accuray, Inc.). It has capabilities for contouring, treatment plan optimization, and image guidance structure definition that will be discussed further in this chapter.

4.2 PATIENT SETUP AND TREATMENT SIMULATION

Proper patient setup and simulation is important to allow the full capabilities of the system during treatment planning and delivery. It is particularly important to ensure the patient is comfortable at the time of simulation since treatment times for CyberKnife can be up to an hour, though this is anticipated to be shorter for the M6. For brain lesions, a thermoplastic head mask with a headrest should be used. For cervical spine lesions, a head and shoulder mask should be used to minimize motion of the head and neck. For thoracic or lumbar spine lesions, a vacuum bag or foam cradle can be used to immobilize the thoracic, abdominal, or pelvic regions. Alternatively, patients can be positioned just on a foam pad to improve comfort, as patients positioned comfortably are less likely to move during treatment. For thoracic cases, the patient can be placed on a thick pad so their arms fall below the level of the body, thereby increasing the potential number of lateral beams that can be used without concern for beams passing through the arms. This is preferred instead of raising the arms overhead to prevent extending outside of the patient safety zone. For lumbar and pelvic cases, the arms can rest on the patient's chest.

CT simulation is usually performed with the patient in the supine position. A CT scan with slice thickness between 1 and 1.5 mm is recommended. The slice thickness is important as finer slices result in higher-resolution DRRs and ultimately in better tracking accuracy (Adler et al., 1999). The scan should be centered on the target extending 10–15 cm above and below the superior and inferior border of the target and/or encompassing all the organs at risk (OAR) such as the bowels or stomach. Lateral and AP topograms can be used prior to the full CT scan to determine the desired extent of the CT image. All patient anatomy where desired beams enter the body must be included in the scan. The primary CT used for treatment planning must be a noncontrast CT as the contrast-enhancing agents might distort the quality of the DRR and impact tracking accuracy.

4.3 VOLUME DEFINITION AND TREATMENT PLANNING

A CT image is required for dose calculation during treatment planning and to generate DRRs used for setup and tracking during treatment delivery. Other imaging modalities such as MRI, PET, or additional CT scans can be incorporated directly in the MultiPlan software and registered to the primary CT image. Image registration and fusion can be performed manually or semiautomatically using fiducial marker positions or maximization of mutual information (Maurer and West, 2006).

MultiPlan provides an autosegmentation feature for cranial structures using an atlas-based approach. Due to variability of cranial anatomy, the system matches the patient image with multiple atlas images and chooses one optimal CT and three optimal MR atlas images. It uses a nonrigid registration algorithm to map the atlas image onto the patient's CT and T1 weighted MR image (Studholme et al., 1996). A set of warped contours is generated for the patient image from the set of atlas contours following the registration process.

Brain lesions are contoured using gadolinium-enhanced T1- and T2-weighted fluid-attenuated inversion recovery (FLAIR) MRI sequences. The MR images are fused onto CT images for treatment planning,

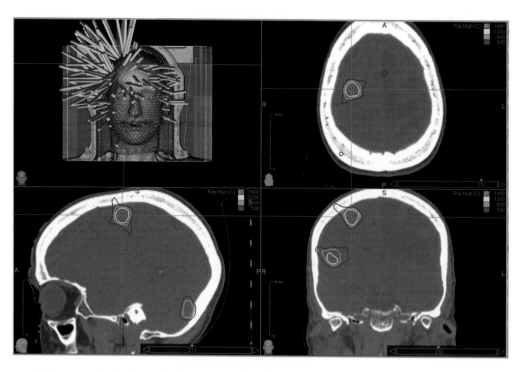

Figure 4.2 An example of a plan with three separate brain metastases. The prescription dose for this case was 19 Gy shown in red. The 5 Gy isodose is shown in blue.

dose calculation, and DRR generation. Primary brain lesions, postoperative resection cavities, single and multiple brain metastasis, and benign disease (such as trigeminal neuralgia) can be treated on the CyberKnife. Ma et al. evaluated the relationship of number of targets and radiosurgery platform to the dose to normal brain (Ma et al., 2011a) and developed an optimization technique to improve the planning quality of multiple metastasis treatments (Ma et al., 2011b). Small lesions are typically treated in a single fraction as in Gamma-Knife radiosurgery. Larger lesions, or lesion located in critical areas (near the optic structures or the brain stem), are treated in 3–5 fractions. The high conformality and steep dose falloff for a case with multiple brain metastases can be seen in Figure 4.2. Brain metastases are usually treated with small planning target volume (PTV) margin (0–1 mm). For postoperative brain cases, the surgical resection cavity is usually expanded by 2 mm to create the clinical target volume (CTV)/PTV (Murphy, 2009).

Target volume definition for spinal SBRT is described in Radiation Therapy Oncology Group (RTOG) 0631 (Ryu, 2011) and more recently in a consensus report (Cox et al., 2012). To summarize, MR and CT images are fused to help define the target volume and spinal cord. The CTV should encompass any abnormal marrow signal and adjacent normal bone. Single and multilevel spinal lesions can be treated on the CyberKnife. Figure 4.3 shows an example of dose distribution for a single thoracic spine lesion. A highly conformal dose distribution can be achieved with sharp dose falloff near the spinal cord. Notably, even the low dose isodose line (i.e., 5 Gy) bends away from the spinal cord. Sahgal et al. developed a treatment planning approach to improve the dose distribution in multiple consecutive vertebral body metastases (Sahgal et al., 2008). The spinal cord must be taken into special consideration for these treatments. Often a 2 mm expansion is included on the contoured spinal cord volume, and the expanded volume is subtracted from the PTV adding a safety margin to compensate for contouring and registration uncertainties and possible misalignment during treatment. Chuang et al. investigated the effects of residual target motion in CyberKnife radiosurgery and calculated patient-specific residual target motions on the order of 2 mm (Chuang et al., 2007). In another study, Fürweger et al. (2010) evaluated the targeting accuracy and residual motion in 260 patients treated with single-fraction CyberKnife radiosurgery and concluded that submillimeter targeting accuracy could be achieved despite patient motion.

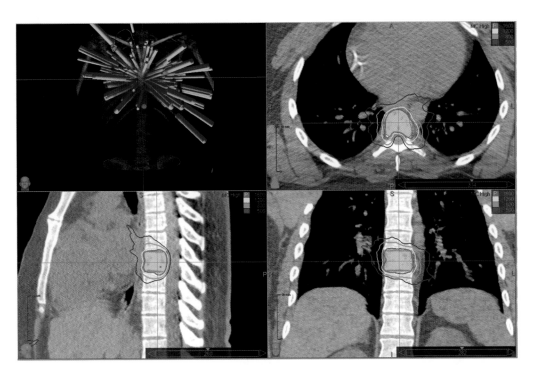

Figure 4.3 The dose distribution for this thoracic spine lesion demonstrates the conformality and sharp dose falloff near the spinal cord that can be achieved with a large number of noncoplanar beams available on CyberKnife. The prescription isodose for this case was 16 Gy shown in red. The 8 and 5 Gy isodose levels are shown in green and blue, respectively.

4.4 PLAN OPTIMIZATION AND DOSE CALCULATION

Treatment plans are generated using one of three optimization methods: isocentric, conformal, or sequential. The optimization problem is framed given thousands of possible beams defined by node position, beam angle, and collimator size (for fixed or Iris plans) or segment shape (for plans with the micro-MLC). Segment shapes are defined as eroded, perimeter, or random. Eroded shapes include a fraction of the entire PTV as seen from a bird's-eye view. Perimeter shapes are narrow fields around the perimeter of the target which help to achieve highly conformal dose distributions. Random shapes are random, as the name implies, within the PTV chosen automatically by the TPS. Once the beam parameters are chosen, the user can define maximum dose objectives, minimum dose objectives, mean dose objectives, dose volume lower limits, dose volume upper limits, optimal coverage, or homogeneous dose to specific volumes, and the optimization algorithm finds the best subset of beams and beam weights to meet the given constraints and objectives with the minimal MUs possible.

Isocentric plans use only isocentric beams and are adequate only for small spherical brain lesions. Conformal plans include both isocentric and nonisocentric beams to create plans that prioritize conformality. Isocentric and conformal plans are optimized using either a simplex or iterative optimization algorithm. For both of these approaches, the user defines dose–volume constraints and objectives and weights for the objectives. The optimizer finds the best arrangement of beams and beam weights that meet the given set of constraints and objectives.

The sequential optimization algorithm proposed by Schlaefer and Schweikard (2008) was developed to more closely mimic the decision-making process of a clinician and is by far the most commonly used optimization algorithm for CyberKnife planning. Dose constraints and objectives are defined by the user, similar to isocentric or conformal plans. But rather than setting weights to prioritize importance, the objectives are defined in order of decreasing importance. The optimizer manipulates beams and beam weights until the first objective is met and then proceeds to the next objectives sequentially. The solution for each prior step becomes a constraint with a user-defined relaxation factor as the optimizer moves to subsequent objectives. In this way target coverage can be guaranteed before minimizing dose to OAR.

For plans using fixed collimators, or the Iris collimator, one of two dose calculation algorithms is available: ray tracing and Monte Carlo. For plans using the MLC, a finite-size pencil beam (FSPB) algorithm is available (Jeleń et al., 2005). The ray-tracing algorithm accounts for heterogeneity corrections along the primary path only. It computes an effective path length based on the electron density in the CT image but does not include effects of tissue inhomogeneity on scattered radiation. A contour correction is applied to the ray-tracing algorithm to estimate the effective depth of off-axis points. The beam for a given collimator size is divided into 12 equally spaced rays at 30° intervals around the perimeter of the cone, which are calculated using a trilinear interpolation with the nearest four rays. Contour correction improves the accuracy of dose calculation for oblique beam incidence and should be selected in the case of superficial targets.

The Monte Carlo algorithm includes the effect of tissue inhomogeneity on the scattered dose, which can be quite significant at air–tissue interfaces and somewhat significant at bone–tissue interfaces. Differences in dose calculations can be quite significant in lung cases when planning with ray tracing versus Monte Carlo (Wilcox et al., 2010). It is recommended to use Monte Carlo for final dose calculation in all thoracic cases and for targets near the sinuses or other air cavities.

The FSPB dose calculation algorithm is used to calculate the dose from irregularly shaped fields by deconvolving the field into a subset of pencil beams. The dose is calculated for each individual pencil beam and then combined to calculate the final dose. There is a lateral scattering correction option which accounts for the effect of tissue inhomogeneity on the scattered dose.

4.5 TREATMENT DELIVERY AND IMAGE GUIDANCE

The CyberKnife system is capable of delivering highly conformal dose distributions with stereotactic imaging accuracy making it well suited for hypofractionated treatments. To ensure the conformal dose distribution is being delivered to the desired target volume while sparing adjacent OAR, highly accurate target localization and real-time tracking capabilities are implemented using sophisticated image guidance. A pair of orthogonal kilovoltage x-ray sources and detectors provides high contrast images of bony landmarks or fiducial markers which can be used for patient setup and accurate motion tracking in real time throughout the treatment. Images can be taken every 15–150 seconds (typical imaging frequency is 30–60 seconds, depending on treatment site).

4.5.1 FIDUCIAL TRACKING

Fiducial tracking uses radio-opaque markers for positioning. Ideally, three or more separate markers should be used to provide 6D corrections (three translation and three rotations). This is most commonly used for prostate and liver lesions where fiducial markers are implanted directly into the organ. It can also be used for lung cases though risk of pneumothorax due to fiducial implantation must be considered for this approach. There is a fiducial-free option for tracking lung lesions that can be clearly distinguished on orthogonal x-ray images. Screws or pins fixed to the vertebral body can also be used for fiducial tracking though this is rare as other fiducial-less tracking methods have been developed for spine lesions.

4.5.2 6D SKULL TRACKING

Skull tracking is used for intracranial cases, or any site that is considered fixed with respect to the skull. The patient's skull is imaged with 2D orthogonal images and a transformation algorithm determines the best linear transformation between the image and the DRR. The transformations are combined and back projected to determine the 6D transformation that best aligns the current skull position to the original planning CT skull position. The algorithm is described by Fu and Kuduvalli (2008).

4.5.3 Xsight SPINE TRACKING

Xsight spine tracking is used for spine lesions—cervical, thoracic, lumbar, and sacral—or any sites that are considered fixed with respect to the spine. Image registration is based on the differential contrast between bony features in the vertebral bodies. During planning, the user defines an imaging center that is just anterior to the spinal cord and midline relative to the vertebral body. A grid of 81 (9 × 9) nodes, shown in Figure 4.4, is displayed over each of the two orthogonal DRRs usually encompassing several vertebral bodies. The user can adjust the overall size of the grid to maximize the number of nodes containing bony features. A box matching algorithm computes local displacement vectors for each node point between the image taken of the patient during treatment and the original DRR and computes a final translation and rotation vector used to register the patient (Fu et al., 2006). The algorithm has been demonstrated to be very robust with a total system accuracy of 0.61 mm (Ho et al., 2007).

4.5.4 Xsight SPINE TRACKING IN THE PRONE POSITION

Spinal treatments delivered in the prone position can benefit from decreased dose to anterior organs such as the heart and bowels (Descovich et al., 2012). This is due to the increased number of beams available from posterior directions that are unavailable when the patient is positioned supine due to physical limitation of the robot and couch. However, breathing motion becomes a significant problem for spine treatments when the patient is prone (Füerweger et al., 2011). Even if breathing motion is compensated, a 2 mm margin should be added to the CTV to account for the reduced accuracy of respiration-compensated tracking. This additional margin nullifies the potential dosimetric gain of prone treatments for spine lesions, and careful consideration criteria for patient selection should be applied (Füerweger et al., 2014).

Synchrony™ (Accuray, Inc.) is a motion management system that accounts for breathing motion. The robot position is continuously readjusted to follow a moving target. A correlation model between the position of a set of infrared light-emitting diodes (LEDs) on the patient's body that are imaged many times a second and the vertebral body that are imaged using the orthogonal x-ray imaging system. Prior to treatment, a series of x-ray images are used to develop a correlation model between the LEDs and the target. The model is updated during treatment as x-ray images are taken approximately every 60 seconds. As the patient breathes, the beams are adjusted to follow the motion. Overall tracking accuracy of <0.95 mm are possible using this tracking method (http://www.cyberknifelatin.com/pdf/brochure-tecnico.pdf).

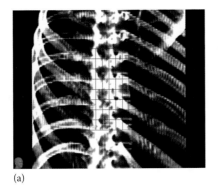 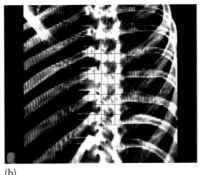

(a)   (b)

Figure 4.4 Digitally reconstructed radiographs (DRR) for the two orthogonal views are shown in panel (a) and (b). Xsight spine tracking compares features in orthogonal x-ray images taken during patient setup with DRR generated in the planning computed tomography to determine 6D corrections. The algorithm compares bony features within the blue grid shown in the figure. The user determines the grid size and location during the planning process.

CHECKLIST: KEY POINTS FOR CLINICAL PRACTICE

- CyberKnife enables accurate delivery of IG-HSRT treatment to patients with intracranial and spinal lesions
- The system consists of a compact linac attached to a robotic manipulator
- CyberKnife IG-HSRT is frameless as image guidance is performed constantly throughout the treatment
- Plans consist of hundreds of highly focused, noncoplanar radiation beams, which enable one to achieve highly conformal dose distribution with steep-dose gradient
- An understanding of the system operating principles is essential to plan simulation and delivery procedures appropriately

REFERENCES

Accuray Inc. Equipment specification (2009) http://www.cyberknifelatin.com/pdf/brochure-tecnico.pdf.

Accuray Inc. (2013) CyberKnife M6 series technical specifications. Technical specification web page: http://www.printable.p1nnacle.com/accuray2/501047.A 20M6 20Spec 20Brochure.pdf.

Adler JR, Jr., Murphy MJ, Chang SD, Hancock SL (1999) Image-guided robotic radiosurgery. *Neurosurgery* 44(6):1299–1306.

Chuang C, Sahgal A, Lee L, Larson D, Huang K, Petti P, Verhey L, Ma L (2007) Effects of residual target motion for image-tracked spine radiosurgery. *Med Phys* 34(11):4484.

Cox BW, Spratt DE, Lovelock M, Bilsky MH, Lis E, Ryu S, Sheehan J et al. (2012) International spine radiosurgery consortium consensus guidelines for target volume definition in spinal stereotactic radiosurgery. *Int J Radiat Oncol Biol Phys* 83(5):e597–e605.

Descovich M, Ma L, Chuang CF, Larson DA, Barani IJ (2012) Comparison between prone and supine patient setup for spine stereotactic body radiosurgery. *Technol Cancer Res Treat* 11(3):229–236. http://www.ncbi.nlm.nih.gov/pubmed/22468994.

Echner GG, Kilby W, Lee M, Earnst E, Sayeh S, Schlaefer A, Rhein B et al. (2009) The design, physical properties and clinical utility of an iris collimator for robotic radiosurgery. *Phys Med Biol* 54(18):5359–5380.

Fu D, Kuduvalli G (2008) A fast, accurate, and automatic 2D–3D image registration for image-guided cranial radiosurgery. *Med Phys* 35(5):2180.

Fu D, Kuduvalli G, Maurer CR, Adler JR (2006) 3D target localization using 2D local displacements of skeletal structures in orthogonal x-ray images for image-guided spinal radiosurgery. *Int J Comput Assist Radiol Surg* 1:198–200.

Fürweger C, Drexler C, Kufeld M, Muacevic A, Wowra B, Schlaefer A (2010) Patient motion and targeting accuracy in robotic spinal radiosurgery: 260 single-fraction fiducial-free cases. *Int J Radiat Oncol Biol Phys* 78(3):937–945.

Füerweger C, Drexler C, Kufeld M, Wowra B (2011) Feasibility of fiducial-free prone-position treatments with cyberknife for lower lumbosacral spine lesions. *Cureus* 3(1):e21.

Fürweger C, Drexler C, Muacevic A, Wowra B, de Klerck EC, Hoogeman MS (2014) CyberKnife robotic spinal radiosurgery in prone position: Dosimetric advantage due to posterior radiation access? *J Appl Clin Med Phys/Am College Med Phys* 15(4): 4427.

Ho AK, Fu D, Cotrutz C, Hancock SL, Chang ST, Gibbs IC, Maurer CR, Adler JR (2007) A study of the accuracy of cyberknife spinal radiosurgery using skeletal structure tracking. *Neurosurgery* 60(2 Suppl 1):ONS147–ONS156.

Jeleń U, Söhn M, Alber M (2005) A finite size pencil beam for IMRT dose optimization. *Phys Med Biol* 50(8):1747–1766.

Kilby W, Dooley JR, Kuduvalli G, Sayeh S, Maurer CR (2010) The cyberknife robotic radiosurgery system in 2010. *Technol Cancer Res Treat* 9(5):433–452.

Ma L, Petti P, Wang B, Descovich M, Chuang C, Barani IJ, Kunwar S, Shrieve DC, Sahgal A, Larson DA (2011a) Apparatus dependence of normal brain tissue dose in stereotactic radiosurgery for multiple brain metastases. *J Neurosurg* 114(6):1580–1584.

Ma L, Sahfal A, Hwang A, Hu W, Descovich M, Chuang C, Barani I, Sneed PK, McDermott M, Larson DA (2011b) A two-step optimization method for improving multiple brain lesion treatments with robotic radiosurgery. *Technol Cancer Res Treat* 10(4):331–338.

Maurer CR, Jr., West JB (2006) Medical image registration using mutual information. In: Heilbrun MP (ed.), *CyberKnife Radiosurgery: Practical Guide*, 2nd ed. Sunnyvale, CA: The CyberKnife Society.

McGuinness C, Gottschalk AR, Lessard E, Nakamura J, Pinnaduwage D, Pouliot J, Sims C, Descovich M (2015) Investigating the clinical advantages of a robotic linac equipped with a multi-leaf collimator in the treatment of brain and prostate cancer patients. *J Appl Clin Med Phys* 16(5). doi:10.1120/jacmp.v16i5.5502.

Murphy MJ (2009) Intrafraction geometric uncertainties in frameless image-guided radiosurgery. *Int J Radiat Oncol Biol Phys* 73(5):1364–1368.

Pöll JJ, Hoogeman MS, Prévost JB, Nuyttens JJ, Levendag PC, Heijmen BJ (2008) Reducing monitor units for robotic radiosurgery by optimized use of multiple collimators. *Med Phys* 35(6):2294–2299.

Ryu S (2011) Radiation Oncology Group. RTOG 0631 protocol information. www.rtog.org/clinicaltrials/protocoltable/ studydetails.aspx?study=0631, Open to accrual date 2011.

Sahgal A, Chuang C, Larson D, Huang K, Petti P, Weinstein P, Ma L (2008) Split-volume treatment planning of multiple consecutive vertebral body metastases for cyberknife image-guided robotic radiosurgery. *Med Dosim* 33(3):175–179.

Schlaefer A, Schweikard A (2008) Stepwise multi-criteria optimization for robotic radiosurgery. *Med Phys* 35(5):2094.

Studholme C, Hill DL, Hawkes DJ (1996) Automated 3-D registration of MR and CT images of the head. *Med Image Anal* 1(2):163–175.

Van De Water S, Hoogeman MS, Breedveld S, Nuyttens JJME, Schaart DR, Heijmen BJM (2011) Variable circular collimator in robotic radiosurgery: A time-efficient alternative to a mini-multileaf collimator? *Int J Radiat Oncol Biol Phys* 81(3):863–870.

Wilcox EE, Daskalov GM, Lincoln H, Shumway RC, Kaplan BM, Colasanto JM (2010) Comparison of planned dose distributions calculated by monte carlo and ray-trace algorithms for the treatment of lung tumors with cyberknife: A preliminary study in 33 patients. *Int J Radiat Oncol Biol Phys* 77(1):277–284.

5 Linac-based IG-HSRT technology

Richard Popple

Contents

5.1 INTRODUCTION

Traditionally, high-precision, high-dose radiotherapy has required a stereotactic headframe to transfer the target coordinates from the imaging frame of reference to the treatment delivery system frame of reference and to maintain the patient position during treatment. Because affixing a headframe is an invasive procedure, stereotactic radiotherapy (SRT) was typically limited to single-fraction treatment of intracranial targets. Furthermore, imaging, planning, quality assurance, and treatment delivery had to be accomplished in a single day, making stereotactic procedures resource intensive. Finally, the planning and delivery techniques were limited to spherical dose distributions (shots), making planning and delivery challenging for large and complex target shapes. In 1994, Brada and Laing described the Royal Marsden Hospital experience using SRT for brain tumors and, despite these limitations, concluded "The technology of stereotactic radiotherapy is evolving, and it is likely that SRT will be integrated into conventional radiotherapy practice to become simply a high-precision technique of radiotherapy delivery in everyday use" (Brada and Laing, 1994). Their prediction is rapidly becoming a reality.

Brada and Laing identified four requirements for SRT: precise patient fixation, accurate target delineation, target localization, and means of delivery. Accurate target delineation has long been available in the form of diagnostic MR and CT imaging procedures. Precision patient fixation and target localization without a stereotactic frame have been enabled by the advent of image guidance. In-room megavoltage (MV) and kilovoltage (kV) imaging has made localization of a target volume possible without the need for a frame. Furthermore, image guidance in combination with appropriate patient immobilization has made stereotactic treatment possible for extracranial sites as well (Ryu et al., 2001).

Six-degree-of-freedom (DOF) patient positioning systems accurately align the treatment planning and linear accelerator (linac) coordinate systems, further improving target localization. Intrafraction motion monitoring systems have been developed to monitor the patient position during treatment, eliminating the need for a frame to maintain targeting accuracy. Finally, the linac as a means of delivery has significantly improved since the report of the Royal Marsden Hospital experience. Multileaf collimators (MLCs) coupled with computer-optimized planning have made intensity-modulated radiation therapy (IMRT) and modulated arc therapy (MAT) possible. IMRT and MAT can generate dose distributions that are highly conformal and have rapid dose falloff, even for complex target shapes. IMRT and MAT are more efficient to deliver than shot-based techniques, reducing treatment times. Flattening filter–free (FFF) beams, now entering widespread clinical use, have dose rates up to 2.4 times higher than conventional flattened beams. FFF beams further reduce the time required to deliver high-dose per fraction treatments.

5.2 LINEAR ACCELERATOR TECHNOLOGY

5.2.1 MULTILEAF COLLIMATOR

The MLC provides beam shaping for modern C-arm linacs and is used to shape the aperture for 3D conformal radiation therapy (3D-CRT) and dynamic conformal arc (DCA) therapy and to produce beam modulation for IMRT and MAT. MLC technology has been extensively described elsewhere (see, e.g., Van Dyk, 1999). Briefly, an MLC is comprised of high-density leaves with various positions arranged in opposing pairs such that the leaf edges are aligned parallel to the beam divergence. In the direction of leaf travel, there are two designs: double focused, for which the leaf end remains parallel to the beam divergence throughout the range of motion, and single focused, for which the leaf moves perpendicular to the central axis. Single-focused MLCs have rounded leaf ends to maintain coincidence between the light and radiation fields and a nearly constant penumbra over the entire range of leaf motion. An MLC either replaces one set of secondary collimators or is a tertiary collimator located below the secondary collimators.

MLCs are the core technology enabling the C-arm linac delivery techniques used for hypofractionated radiation therapy. 3D-CRT is a technique in which field apertures are designed to conform to the target shape for each of a number of beam directions, typically 3–9. 3D-CRT has been in use for several decades and has been described extensively elsewhere (see, e.g., Khan and Gerbi, 2012). DCA is an arc technique analogous to 3D-CRT in that the aperture shape varies to conform to the target shape as the gantry rotates. IMRT uses multiple computer-optimized apertures for each beam direction to generate nonuniform incident radiation fluence. The hallmark of IMRT is the ability to create concave dose distributions to better spare normal tissues. There are two types of MLC-based IMRT, static and dynamic. For static MLC IMRT, the MLC leaves are stationary while radiation is being delivered. For dynamic MLC (DMLC) IMRT, the MLC leaves move while radiation is being delivered. IMRT is in widespread use and has also been extensively described elsewhere (see, e.g., Khan and Gerbi, 2012, and volume 2 of Van Dyk, 1999). MAT is a rotational technique in which the MLC aperture shape varies with gantry angle. In most modern applications, the dose rate and gantry speed also vary. MAT differs from conformal arc therapy in that the aperture shapes and weights are inverse planned to meet dosimetric objectives, rather than conforming to the target shape at all gantry angles.

The dosimetric properties of MLCs are similar between designs (Huq et al., 2002). The average of inter- and intraleaf transmission is typically in the range 1%–2.5% and, for modulated techniques, has the effect of limiting the range of modulation achievable within the field defined by the secondary collimator. The effect of leaf leakage can be reduced by moving the secondary collimators to conform to the extrema of the leaf positions, a technique commonly referred to as jaw tracking or jaw following. For spine radiosurgery, jaw tracking has been shown to decrease the dose to the spinal cord for both DMLC IMRT and MAT delivery techniques, although the effect for MAT was small and not of clinical significance (Snyder et al., 2014). The magnitude of the jaw tracking effect on DMLC relative to MAT was likely due to the different optimization techniques. The DMLC planning technique was fluence map optimization followed by leaf sequencing and resulted in leaf settings having small apertures, whereas the direct aperture optimization used for MAT resulted in larger apertures. Consequently, the modulation factor (a measure of the monitor units required to deliver the prescribed dose) was higher for DMLC relative to MAT, resulting in more dose due to leakage. Jaw tracking reduces the leakage dose and so had more effect on the DMLC plans than the MAT plans. The impact of jaw tracking on critical structure sparing is proportional to the degree of modulation.

MLC leaf widths, as projected to isocenter, typically range from 2.5 to 10 mm. The effect of leaf width on dosimetry is dependent on delivery technique, target volume, and target shape, and is greatest for nonmodulated techniques. For DCA, Dhabaan et al. (2010) found improvement in both conformity and normal tissue sparing for a 2.5 mm leaf width MLC compared to a 5 mm leaf width. Jin et al. (2005) compared MLC leaf widths of 3, 5, and 10 mm for DCA and IMRT radiosurgery. For DCA, the conformity improved as the leaf width decreased with the largest improvement observed for the smallest targets. For IMRT, the conformity index was essentially the same for the 3 and 5 mm leaf widths; however, the narrower leaf width provided modestly better sparing of small critical structures for intracranial cases. Monk et al. (2003) compared MLCs having 3 and 5 mm leaf width for 3D conformal radiosurgery and found that although the conformity of the 5 mm MLC was not as good as that for the 3 mm MLC, the plans for the 5 mm leaf width met the Radiation Therapy Oncology Group (RTOG) clinical criteria. They concluded that the differences were small enough that the improvement in conformity index was not large enough to dictate equipment choice. Serna et al. (2015) found no improvement in conformity for a 2.5 mm width relative to a 5 mm leaf width but did find an improvement in dose falloff, particularly for targets smaller than 10 cm^3. Wu et al. (2009a) examined the effect of leaf width on the sparing of critical structures adjacent to the target volume, comparing a 2.5 and 5 mm leaf widths for stereotactic radiosurgery (SRS) of targets abutting the brain stem or the spinal cord. They found that the 2.5 mm leaf width had significantly improved sparing of adjacent critical structures, particularly for small targets and complex geometries. Chae et al. (2014) came to the same conclusion for spine targets, finding that target coverage was improved for both IMRT and MAT for a 2.5 mm leaf width, particularly for complex shape. The general conclusion of studies on the effect of leaf width is that a narrower leaf improves plan quality but that the degree of improvement is dependent on the technique, the target size, and the target shape. The effect of leaf width on plan quality is greatest for DCA and least for MAT. A smaller leaf width has the largest effect on plan quality for small targets and for targets with complex geometries.

5.2.2 MODULATED ARC THERAPY

MAT is a rotational technique in which the MLC aperture shape varies with gantry angle. In most modern applications, the dose rate and gantry speed also vary. MAT differs from conformal arc therapy in that the aperture shapes and weights are inverse planned to meet dosimetric objectives, rather than conforming to the target shape at all gantry angles.

MAT was first described by Yu in 1995. Limited clinical implementation followed (Ma et al., 2001; Yu et al., 2002; Duthoy et al., 2004; Wong et al., 2005); however, due to lack of robust planning tools, it primarily remained a technique of academic interest. In 2008, the introduction of variable dose rate during gantry rotation by linac manufacturers in conjunction with clinically usable planning tools resulted in the widespread adoption of MAT. There are two major types of MAT, although the terminology for the two is often used interchangeably. Yu described intensity-modulated arc therapy (IMAT), in which multiple arcs were delivered at a constant dose rate and gantry speed. The apertures of the arcs overlap to generate a nonuniform fluence from any given gantry angle. As originally conceived, the arcs were planned using a two-step process. The first step was fluence map optimization with no consideration of machine constraints, followed by a second step in which apertures were generated to approximate the optimal fluence map while considering the speed constraints of the MLC leaves and the gantry. For IMAT, multiple arcs were necessary to deliver the optimized dose distribution. Otto (2008) described volumetric modulated arc therapy (VMAT) in which a single arc was combined with variable dose rate and gantry speed, and the aperture shapes and weights were directly optimized. In current use, the terms IMAT and VMAT are often used interchangeably to describe any type of inverse-planned therapy comprised of one or more arcs during which the aperture shape changes during gantry rotation.

Conformal arc therapy has been widely used for linac intracranial radiosurgery (Shiu et al., 1997; Cardinale et al., 1998; Leavitt, 1998; Solberg et al., 2001), so it was a natural extension to use MAT for stereotactic applications. The primary advantage of MAT relative to nonmodulated techniques is the ability to improve conformity and/or sparing of critical structures. Multiple studies have compared MAT to DCA (Wu et al., 2009a; Audet et al., 2011; Huang et al., 2014; Salkeld et al., 2014; Serna et al., 2015; Zhao et al., 2015). In general, these studies found that MAT produced more conformal dose distributions while maintaining similar low dose spill, as measured by the volume of healthy brain receiving more than 50% of the prescription dose. Plan quality was found to be better for multiple noncoplanar arcs relative to a single arc. In addition to malignant disease, treatment of large intracranial arteriovenous malformations has been reported using MAT and hypofractionated radiation therapy (RT) with promising results (Subramanian et al., 2012). Other nonmalignant lesions that are typically treated using nonmodulated techniques are also amenable to MAT and can be treated with similar plan quality and higher efficiency (Abacioglu et al., 2014).

MAT has been applied to stereotactic spine radiotherapy, for which conformal techniques are not adequate (Wu et al., 2009a). Compared to conventional IMRT, MAT results in equivalent target dosimetry and reduced treatment time. The conformity of MAT is slightly improved relative to IMRT because the

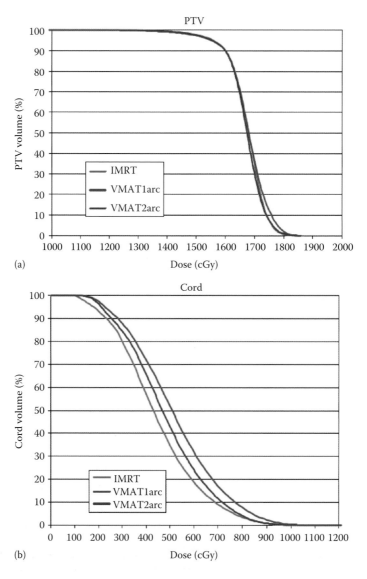

Figure 5.1 Average dose volume histogram curves for the (a) planning target volume and (b) spinal cord. (From Wu, Q.J. et al., *Int. J. Radiat. Oncol. Biol. Phys.*, 75, 1596, 2009b.)

limited number of entrance directions for IMRT results in more spillage of the prescription dose outside of the target volume; however, this difference is probably not of clinical significance. For the spinal cord, achieving equivalence of MAT to IMRT requires two or more arcs. Wu et al. found that a single arc was inferior to both two arcs and to IMRT. For example, for a prescription dose of 16 Gy, 1% of the spinal cord received more than 9 Gy for a single arc, whereas it received 8.5 Gy for two arcs and for conventional IMRT. The average dose volume histograms for the planning target volume and spinal cord are shown in Figure 5.1. Interestingly, Wu et al. found that two arcs both improved plan quality and reduced treatment delivery time (8.56 min for one arc vs 7.88 min for two). This seemingly counterintuitive result is because the increased freedom of two aperture shapes at each gantry position allows more efficient coverage of the target volume, resulting in fewer total monitor units from the two-arc plan.

Plan quality for MAT and IMRT is similar, and so the primary advantage of MAT is reduced treatment time, although the degree of reduction depends on the specific techniques being compared. For example, for spine SBRT, Wu et al. found a treatment time reduction, whereas Kuijper found similar treatment times because Kuijper used more arcs for MAT and fewer beams for IMRT than Wu (Wu et al., 2009b; Kuijper et al., 2010). The largest component of the difference in treatment time between conventional IMRT and MAT is the time spent in between beams waiting for data transfer and for gantry positioning. Increased automation of IMRT delivery, becoming available on modern linacs, will likely reduce the time difference between IMRT and MAT. Coupled with planning techniques that combine fixed and arc modulated fields (Matuszak et al., 2013), automated delivery systems will blur the distinction between fixed field IMRT and MAT (Popple et al., 2014).

5.2.3 FLATTENING FILTER–FREE BEAMS

Linacs generate MV photon beams by bombarding a high-Z target with MV electrons. The photon production is forward peaked and results in a nonuniform dose distribution. To ameliorate this issue, a nonuniform filter is introduced to differentially absorb photons such that the resulting dose distribution is uniform at a reference depth such as 10 cm. Introduction of the flattening filter comes at the penalty of reducing the dose rate throughout the field to that of the periphery. However, for small field sizes such as those encountered in stereotactic irradiation, the field is reasonably uniform without the flattening filter. The use of fluence modulation techniques (IMRT and MAT) further reduces the utility of a flattening filter. Removing the flattening filter increases the dose rate near central axis by a factor of two to four, depending on the beam energy, which can lead to significantly shorter delivery times for hypofractionated treatment regimens. In 1991, O'Brien reported removing the flattening filter from a 6 MV linac, achieving a dose rate increase of 2.75 and acceptable flatness (O'Brien et al., 1991). In 2007, Bayouth reported on image-guided radiosurgery, both intra- and extracranial, using an FFF linac (Bayouth et al., 2007). Planning studies demonstrated feasibility for body radiosurgery as well (Vassiliev et al., 2009). C-arm linacs without a flattening filter became commercially available in 2010, and early clinical experience demonstrated treatment times for CNS radiosurgery approaching those of conventional fractionation, as shown in Figure 5.2.

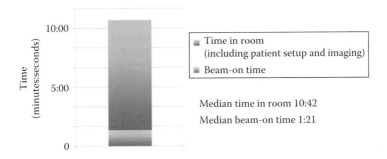

Figure 5.2 Graphical representation of treatment time for central nervous system SRS using flattening filter–free linear accelerator. The median radiation beam-on time was 1:21, while the median time the patient spent in the treatment room, including treatment setup and imaging, was 10:42. (From Prendergast, B.M. et al., *J. Radiosurg. BRT*, 1, 117, 2011.)

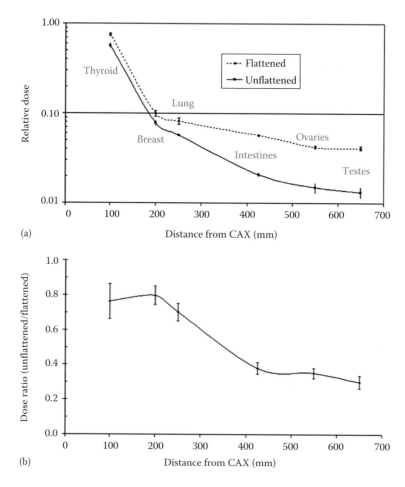

(a)

(b)

Figure 5.3 (a) Sample data illustrating the drop in peripheral dose for the unflattened beam delivery in the clinical intensity-modulated radiotherapy plans. (b) Plot of the average dose ratio (unflattened/flattened) for all of the plans delivered. CAX, central axis. (From Cashmore, J. et al., *Int. J. Radiat. Oncol. Biol. Phys.*, 80, 1220, 2011.)

The flattening filter is a source of scattered photons that contribute significantly to leakage and peripheral dose. FFF beams reduce the peripheral dose by eliminating this source of scatter (Kry et al., 2010; Cashmore et al., 2011; Kragl et al., 2011). Furthermore, FFF beams require less current at the target to deliver a given dose, thus reducing other sources of head leakage as well (Vassiliev et al., 2006). For 6 MV, removing the flattening filter reduces leakage radiation by 60% and the dose outside of the field edge by 11% (Cashmore, 2008). The impact of reduced leakage dose is illustrated in Figure 5.3, which compares the out-of-field dose for flattened and unflattened 6 MV beams due to intracranial irradiation in children (Cashmore et al., 2011). For SBRT, the use of FFF beams has been shown to reduce the dose 20 cm from the field edge by 23% for 6 MV and 31% for 10 MV (Kragl et al., 2011).

An FFF beam delivers a higher surface dose than a flattened beam of the same energy due to the lower energy components in the beam spectrum. However, the surface dose for a flattened beam has stronger field size dependence because the scatter from a flattening filter, which increases with increasing field size, is considerably lower in energy than the primary beam and increases the dose at shallow depths. The surface dose for large field sizes (>~30 cm) is similar between flattened and unflattened beams (Vassiliev et al., 2006; Cashmore, 2008; Kragl et al., 2009); however, at the smaller fields typical of hypofractionated radiation therapy, the surface dose is larger for filter-free beams. For a 4×4 cm^2 field at 3 mm depth, the dose increase relative to a flattened beam of the same energy is ~20% for 6 MV (Vassiliev et al., 2006; Kragl et al., 2009) and 25% for 10 MV (Kragl et al., 2009). Surface dose decreases with increasing energy, and for a 10 MV, unflattened beam is about 25% less than a 6 MV unflattened beam and similar to that of a 6 MV flattened beam (Kragl et al., 2009). The surface dose is the same for fields collimated by MLCs or by the secondary collimators (Wang et al., 2012). Although the surface dose for FFF beams is modestly higher than flattening filter, the increased dose is not likely of clinical significance, and utilizing a sufficient number of treatment fields or arcs, as is good SRS/SBRT practice, should be sufficient to mitigate the modestly higher relative surface dose of FFF beams.

With respect to calibration, FFF beams have two characteristics different from conventional flattened beams. First, the beam spectrum has more low energy components compared to a flattened beam. However, the relationship between depth dose and stopping power ratio given in the American Association of Physicists in Medicine (AAPM) Task Group 51 protocol can be used (McEwen et al., 2014). The higher dose rate leads to concern about ion recombination; however, it has been demonstrated that the two-voltage technique remains valid for the dose rates produced by FFF beams (Kry et al., 2012). If the reference chamber is sufficiently long, a small correction factor is needed to account for the change in dose rate over the length of the chamber; otherwise, the calibration of FFF beams is straightforward and uses the same protocol as a flattened beam (McEwen et al., 2014).

The dose rate delivered by FFF beams is approximately two to four times the dose rate of flattened beams, leading to concerns about differing radiobiological effect relative to conventional flattened beams. A number of in vitro studies have reported on cell survival in flattening filter beams. Most studies have found that there is no difference in cell survival over the dose rate range encompassed by flattened and unflattened beams (Sorensen et al., 2011; King et al., 2013; Verbakel et al., 2013); however, at least one study has reported that FFF beams reduced clonogenic survival of cancer cells (Lohse et al., 2011). A review of the dose rate effects in external beam radiation therapy concluded that the dose rate effect is governed by the total fraction delivery time, not by the average linac dose rate or by the instantaneous dose per pulse (Ling et al., 2010). For the dose range typical of hypofractionated radiation therapy, the radiobiological effect is expected to increase when treatment time is reduced. As illustrated in Figure 5.4, late-responding

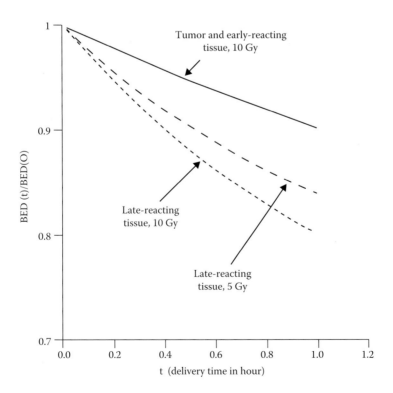

Figure 5.4 Normalized biological equivalent dose as a function of overall beam-on time for delivering 10 Gy to the tumor and early-responding tissue and 5 or 10 Gy to the late-reacting tissue. (From Ling, C.C. et al., *Radiother. Oncol.*, 95, 261, 2010.)

normal tissues have a larger increase than tumor and early-responding normal tissues (Ling et al., 2010), suggesting caution when implementing any technology that significantly shortens overall treatment time. Early reports of experience with FFF beams, typically in combination with MAT, have shown low rates of acute toxicity (Scorsetti et al., 2011, 2014, 2015; Alongi et al., 2012, 2013; Prendergast et al., 2013; Wang et al., 2014). There is limited long-term experience with FFF beams and so the data on late effects are still maturing; however, preliminary results have not demonstrated unexpected toxicity.

5.2.4 IGRT

Image guidance has become an essential component of modern radiation therapy (De Los Santos et al., 2013). The goal of image-guided radiation therapy (IGRT) is improvement of targeting accuracy to permit the use of smaller setup margins and consequently reduce irradiation of normal tissue. In the context of stereotactic treatment techniques, image guidance replaces the stereotactic frame for aligning the target with the linac coordinate system. To achieve stereotactic accuracy using relocatable fixation devices instead of a frame, image guidance is necessary before every treatment (Masi et al., 2008; Murphy, 2009). In-room image guidance began with MV electronic portal imagers (EPIDs) that were developed as a direct replacement for film. In recent years, MV EPIDs have been largely supplanted by kV imaging. There are a variety of kV technologies available for image guidance (De Los Santos et al., 2013), but the most common for stereotactic applications using C-arm linacs are gantry-mounted (on board) kV imagers and room-fixed stereoscopic kV imagers.

An onboard kV imaging system is composed of an x-ray source mounted on the gantry opposite an amorphous silicon flat-panel detector. Both the source and detector are retractable. Presently available systems are mounted to the gantry orthogonal to the treatment beam axis (Varian Medical Systems, Palo Alto, CA; Elekta Oncology Systems, Crawley, UK), although one system that is no longer available was mounted parallel to the beam axis (Siemens, Concord, CA). The imager can be used to obtain radiographs from stationary gantry positions or can be rotated to acquire a large number of projection images to construct a cone-beam CT (CBCT). Planar radiographs can be used for stereoscopic image guidance, typically using orthogonal views. Onboard imagers and CBCT have been extensively described elsewhere (De Los Santos et al., 2013).

The source and detector of an onboard imaging system undergo small motions as the gantry rotates, resulting in small misalignments with the MV radiation isocenter. Systematic motions are compensated for by measuring them as a function of gantry angle using one or more high-density balls having a known geometry with respect to the MV isocenter. The resulting relationship between image offset and gantry angle, referred to as a flexmap, is stored and used to realign each image to the MV isocenter. Realignment is accomplished either by physically shifting the detector position to compensate for the offset or by correcting the image coordinate system in software. Both methods have been shown to have submillimeter residual positioning error (Bissonnette et al., 2008), resulting in radiosurgery positioning accuracy comparable to that reported for stereotactic frames (Chang et al., 2007).

Stereoscopic x-ray imagers use a pair of x-ray sources and corresponding flat-panel detectors mounted in the treatment room. The central rays of the sources intersect at machine isocenter and are separated by an angle sufficient for stereoscopic visualization of patient anatomy (Verellen et al., 2003). Stereoscopic systems are capable of submillimeter accuracy (Verellen et al., 2003), comparable to a stereotactic frame (Gevaert et al., 2012b). Stereoscopic x-ray imaging serves as the primary image guidance system for robotic radiosurgery systems (Adler et al., 1997), whereas it often complements the onboard imaging system for modern C-arm linacs. Presently, the BrainLab ExacTrac system (BrainLAB, Feldkirchen, Germany) is the only stereoscopic x-ray system commercially available for use with C-arm linacs.

One advantage of room-fixed stereo-imaging systems relative to gantry-mounted systems is that images can be obtained over a wider range of gantry and couch positions. Gantry-mounted systems have significant limitations when the couch is not near zero (International Electrotechnical Commission coordinate system) because the imager or the source will collide with the couch. Although this is not a problem for pretreatment imaging, it limits the utility of an onboard imaging system to verify correct patient positioning after a couch rotation.

Clinical use of both onboard and room-fixed image guidance systems should include routine testing to evaluate the coincidence of the MV radiation isocenter with the isocenter of the imaging system, particularly for stereotactic applications. Daily end-to-end testing using a phantom containing markers in known positions can be used to test localization accuracy. The AAPM Task Group 142 report recommends that IGRT systems used for stereotactic techniques demonstrate an accuracy of ±1 mm (Klein et al., 2009).

5.2.5 INTRAFRACTION POSITION MONITORING

For frameless image-guided stereotactic treatment techniques, pretreatment image guidance should be accompanied by intrafraction motion monitoring. Hoogeman et al. (2008) studied intrafraction motion for intracranial targets and, in both prone and supine positions, extracranial targets in patients. The frequency distribution of the displacement vector magnitude at several intervals from the initial IGRT position correction is shown in Figure 5.5. For the intracranial targets, 95% of displacements were less than 1.6 mm over a 15 minute interval. For extracranial targets treated in a supine position, 95% of displacements were within 2.8 mm over the same interval. However, for prone treatments, the displacement was significantly larger, with 5% of displacements being larger than 3.1 mm after only 1 minute. The larger intrafraction

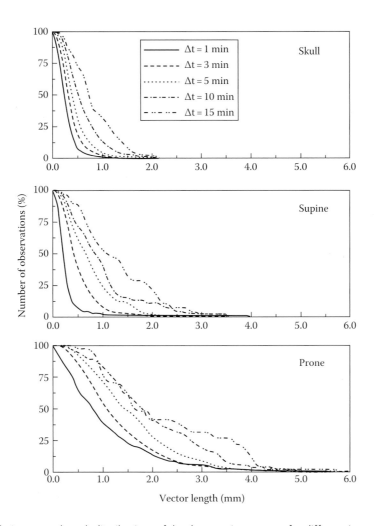

Figure 5.5 Cumulative vector length distributions of the three patient groups for different intervals. For all groups, the width of the cumulative distributions increased with time. For the skull and supine group, intervals of 1–3 minutes yielded distributions with a width in the 1 mm or even submillimeter range. The width of the distributions of the prone group was larger than of the other groups, which was partly caused by respiration-induced motion of the fiducial markers. The horizontal scale of the graphs was set to 6.0 mm to keep them readable. Note that the maximum value observed for the prone group was 12.3 mm, exceeding the scale of the graph. (From Hoogeman, M.S. et al., *Int. J. Radiat. Oncol. Biol. Phys.*, 70, 609, 2008.)

motion in the prone group was attributed in part to respiration. Hoogeman et al. (2008) recommended repeat imaging and patient setup correction at 5 minute intervals and suggested a 0.6 mm margin to account for residual motion for intracranial targets and 1.0 and 2.0 mm for extracranial targets treated in the supine and prone position, respectively. Similarly, Murphy et al. (2003) found that intrafraction motion in the skull and spine was typically within 0.8 mm over 1–5 minute intervals and concluded that a tracking interval of 1–2 minutes is necessary for most intracranial and spinal radiosurgery applications. If the target position is monitored at less frequent intervals, the margin must be increased to assure that the target receives the prescribed dose (Murphy et al., 2003; Hoogeman et al., 2008; Murphy, 2009; Kang et al., 2013).

X-ray image guidance systems can be used for intrafraction motion monitoring, but onboard systems have limited applicability for noncoplanar geometries because of potential collision of the imaging system with the treatment table or the patient. Room-fixed (stereoscopic) systems do not have collision risks; however, there may be configurations in which the view of one of the detectors is obstructed by the gantry or the couch. Furthermore, x-ray systems are not real time, and the time required for processing and decision making limits the frequency for which such systems are practical. In addition, the small, but nonnegligible, x-ray dose may be a consideration.

One solution to these concerns for cranial RT is optical systems that monitor the patient position in real time. There are two approaches to optical monitoring. The first is monitoring of infrared (IR) reflective or emissive targets affixed to the patient or a surrogate such as a bite block (Meeks et al., 2000; Wang et al., 2010). The second approach is an optical surface imaging (OSI) system that uses a set of cameras to capture a 3D rendering of the patient surface.

Optically guided systems that use fiducial markers are comprised of IR LEDs (Meeks et al., 2000), IR markers attached to a bite block (Wang et al., 2010), or IR markers attached to the patient's skin (Wang et al., 2001). Using a stereoscopic IR camera, the targets are detected in images, and the 3D position is calculated and compared with the expected position. IR fiducial monitoring systems can achieve submillimeter accuracy (Gevaert et al., 2012b; Tagaste et al., 2012). While sufficient for position monitoring, the intrafraction positioning accuracy is inadequate for pretreatment positioning and x-ray-based image guidance is necessary to achieve stereotactic accuracy (Wang et al., 2010).

OSI systems use surface imaging cameras oriented to view the isocenter (Li et al., 2006). The cameras obtain a 3D point cloud that is compared with a reference surface to determine patient position and orientation. Multiple cameras are typically used both to improve accuracy and to ensure that sufficient cameras have an unobstructed line of sight over the entire range of gantry motion. When used for patient setup, the reference surface is derived from the treatment planning CT, whereas for intrafraction motion monitoring, the reference surface is obtained directly from the OSI following patient setup. For intracranial SRS, OSI of the face can achieve submillimeter motion tracking accuracy at frame rates of approximately 5/s (Peng et al., 2010; Li et al., 2011; Wiersma et al., 2013). The face must be exposed, so open-faced masks or other immobilization devices are needed when using OSI. OSI can maintain positioning accuracy similar to a stereotactic frame during treatment; however, to achieve that accuracy, CBCT or stereoscopic x-ray imaging is necessary for initial positioning (Peng et al., 2010; Li et al., 2011; Wiersma et al., 2013).

5.2.6 SIX DEGREE-OF-FREEDOM POSITIONING

Conventional treatment tables have 4DOF: three translations and one rotation about the vertical axis. Any rotational misalignment of the patient around the longitudinal axis (roll) or the lateral axis (pitch) cannot be corrected with a 4DOF table. Correcting rotational misalignment is not important for small, spherical targets that are easily localized; however, rotational misalignment can have significant dosimetric consequences for targets that are complex in shape, such as paraspinal lesions; do not have nearby localization anatomy, such as targets in the center of the skull; or are nonisocentric, such as multiple intracranial targets treated using a single isocenter. Robotic and manual systems have been developed that correct for pitch and roll. Robotic systems comprise a pitch and roll stage that is sandwiched between the couch translation stage and the couch top. These systems correct for all 6DOF under computer control. Manual positioners generally consist of an extension to the treatment couch to support the head that can be tilted around the longitudinal and lateral axes and have inclinometers to display pitch and roll (Dhabaan et al., 2012).

Phantom studies have demonstrated that robotic 6DOF systems have accuracies better than 0.5 mm and 0.5° (Meyer et al., 2007; Wilbert et al., 2010; Gevaert et al., 2012b). In vivo studies suggest similar accuracy as assessed by CBCT following 6DOF correction (Dhabaan et al., 2012; Lightstone et al., 2012), with one study reporting that 97% of patients were within 1 mm and 0.5° (Gevaert et al., 2012a). A number of investigations have reported on the improvement of target coverage for 6DOF relative to 4DOF or 3DOF setup. Gevaert et al. found that 4DOF correction resulted in 5% less target volume receiving the prescription dose relative to 6DOF correction, with an extreme case having 35% less target volume coverage relative to 6DOF (Gevaert et al., 2012a). Other groups (Schreibmann et al., 2011; Dhabaan et al., 2012) have found similar results. For spinal radiosurgery, ignoring rotational corrections (3DOF) can result in significant underdosage of the tumor. The degree of underdose is dependent on the magnitude of rotational error and the size and shape of the tumor. Small tumors and those that are irregularly shaped are more likely to suffer an underdose (Schreibmann et al., 2011).

5.3 TREATMENT PLANNING

5.3.1 TREATMENT PLAN EVALUATION

Treatment planning objectives can be classified along a spectrum ranging from conformity to organ avoidance. For a target volume embedded in isotropic normal tissue, conformity of the prescription isodose volume to the target and rapid dose falloff in all directions are the treatment planning goals. A lesion located in brain or lung parenchyma and having no nearby critical structures, such as the brain stem or proximal bronchial tree, is a typical example. Conversely, if a target volume is located in close proximity to critical structures that have dose tolerances significantly lower than the prescription dose and the tolerance of the surrounding soft tissue, the avoidance of the critical structures is more important than dose spill into the surrounding soft tissue. A typical example is a paraspinal lesion, for which avoidance of the spinal cord is most important, and dose is allowed to spread into the adjacent soft tissue to preferentially spare the cord. Most treatment planning situations fall somewhere between the two extremes, although the majority of intracranial hypofractionated RT and SRS cases can be classified as having conformity as the primary objective and spine SBRT cases as having organ avoidance as the primary objective.

For intracranial targets, the primary metrics used to evaluate linac stereotactic plans are conformity index, gradient index, and healthy brain dose volumes, including the volume of the brain receiving more than 12 Gy (V12Gy) (Levegrun et al., 2004; Blonigen et al., 2010; Minniti et al., 2011). The simplest conformity index, introduced by the RTOG, is the planning isodose volume to target volume (PITV) (Shaw et al., 1993):

$$PITV = \frac{PIV}{TV}$$

where
  PIV is the volume enclosed by the prescription isodose surface
  TV is the target volume

An ideal plan has PITV unity, whereas PITV less than one indicates undertreatment and PITV greater than one overtreatment. This index is simple to calculate and is in widespread use. The flaw in the PITV index is that it does not take into account the location of the prescription isodose. A dose distribution having a PIV equal to the TV but located outside the target volume (a geometric miss) will nevertheless have an idea PITV = 1. To remedy this problem, Paddick and Lippitz proposed an index, often referred to as the Paddick-CI, that captures both the volume and position of the PIV relative to the TV. The Paddick-CI uses the target volume encompassed by the prescription isodose (TVPIV) and is defined as

$$Paddick\ CI = \frac{TVPIV^2}{TV \times PIV}$$

The conformity indices are designed to describe how well the prescription isodose volume conforms to the target. The gradient index is designed to quantify the dose falloff away from the target volume. Paddick and Lippitz defined the gradient index GI as the ratio of the volume of half the prescription isodose (V50) to the volume of the prescription isodose (the PIV) (Paddick and Lippitz, 2006):

$$GI = \frac{V50}{PIV}$$

Caution must be exercised when using the gradient index to compare plans having different conformity indices and thus different PIV values in the denominator of the GI definition. Two plans having the same V50 but different PIVs will also have different gradient indices. When conformity is improved without change in V50, the GI increases and thus appears to be worse. This situation is illustrated in Figure 5.6. Note that for two plans having the same V50, the ratio of the gradient indices is the inverse of the ratio of the PIVs.

Stereotactic treatment of the spine is an organ avoidance problem, and as such gradient and conformity indices have a less important role in evaluating plan quality. Although gradient index is rarely reported, conformity index is often reported as a means to evaluate target coverage relative to spill of the prescription dose outside of the target. However, conformity index is generally secondary to direct dose volume measures of target coverage DV, the dose for which volume V of the target receives at least the dose D. The volume V can be expressed either as a percentage or an absolute value. If no units are given, the value is typically percentage volume. Evaluation of the spinal cord dose generally focuses on the high-dose portion of the dose-volume histogram, and common metrics include the maximum point dose, D0.35cc, and D10% (Ryu et al., 2014). When evaluating spinal cord dose using percentage volume, rather than absolute volume, it is important to be aware of the contouring guidelines used for the spinal cord, because differing lengths of cord will yield different results for the same dose distribution. The RTOG-0631 protocol, for example, specified D10% in relation to the segment of spinal cord extending 5–6 mm beyond the target. Other organs at risk depend on the location of the target along the spine and include the esophagus, lungs, and kidneys (Schipani et al., 2012). These structures are usually not in close proximity to the target, and limiting the dose received by these structures is generally not difficult.

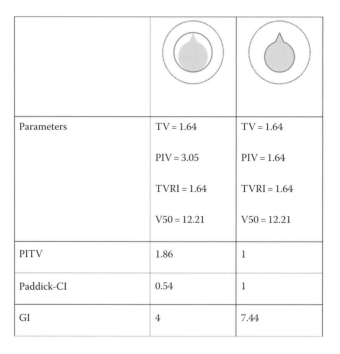

Parameters	TV = 1.64	TV = 1.64
	PIV = 3.05	PIV = 1.64
	TVRI = 1.64	TVRI = 1.64
	V50 = 12.21	V50 = 12.21
PITV	1.86	1
Paddick-CI	0.54	1
GI	4	7.44

Figure 5.6 Gradient index for dose distributions having the same V50 but different conformity indices. Solid red line represents the prescription dose and the dashed blue line half of the prescription dose.

5.3.2 TREATMENT PLANNING TECHNIQUES

5.3.2.1 Intracranial

Treatment planning for the brain has conformity and rapid dose falloff (gradient index) as the primary objectives. For 3D-CRT, and DCA, the aperture size, number of fixed fields or arcs, and the field placement dictate conformity and gradient. However, for inverse-planned IMRT or MAT, conformity and gradient must be included in the objective function. Some planning systems have tools that incorporate dose falloff into the objective function, such as the normal tissue objective included in the Eclipse treatment planning system (Varian Medical Systems). Alternatively, conformity and dose falloff can be included as objectives by assigning dose limits to a series of nested shells constructed around the target, as illustrated in Figure 5.7 (Clark et al., 2010, 2012; Audet et al., 2011). Each shell is created by expanding the target volume and then using Boolean operations to remove the target and the inner shells. Three shells are

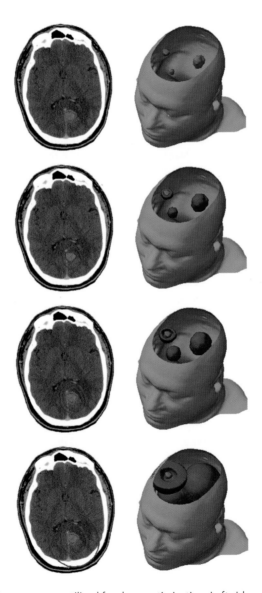

Figure 5.7 Dose control tuning structures utilized for dose optimization. Left side = 2D visualization for a single target patient. Right side = 3D visualization for a multitarget patient. From top to bottom (both sides): target(s) (red), inner control tuning structure (blue), middle control tuning structure (blue), and outer control tuning structure (blue). (From Clark, G.M. et al., *Pract. Radiat. Oncol.*, 2, 306, 2012.)

typically sufficient to provide good control over the dose distribution (Clark et al., 2010, 2012). The inner and middle shells are used to optimize the conformity index and the middle and outer shells to control the gradient index. The maximum dose limits for the inner and middle shells are the prescription dose and half the prescription dose, respectively. These limits force the dose to decrease from the prescription dose at the surface of the target volume to half of the prescription dose at the external surface of the inner shell, forcing the prescription dose to conform to the target. The outer shell is used to confine the 50% isodose volume to the interior of the middle shell, thus minimizing V50 and the gradient index. The upper limit for the outer shell must be less than half of the prescription dose but is otherwise somewhat arbitrary. One group has reported using 40% of the prescription dose (Clark et al., 2012). The dimensions of the shells are based on the desired dose falloff. The dimensions should be small to maximize falloff; however, if the dimensions are too small, the optimizer will not be able to achieve the desired falloff, resulting in suboptimal target coverage. Fortunately, there is a body of literature describing the achievable dose falloff for linac radiosurgery that can provide guidance. The expansion of the target volume to obtain the inner shell should be R50, the distance from the prescription isodose line to the 50% isodose line. The expected value of R50 is dependent on technique and target volume, with R50 increasing as target volume increases. For a 10 mm diameter circular collimator, the R50 distance has been reported for a variety of arc techniques as ranging from approximately 3 to 5 mm (Pike et al., 1990). For MLC 3D-CRT and DCA, an average R50 of 4.1, 5.5, and 6.5 mm has been reported for tumors having diameters less than 20 mm, between 20 and 30 mm, and greater than 30 mm, respectively (Hong et al., 2011). The middle and outer shell thicknesses are somewhat arbitrary. The middle shell is designed to contain the 50% isodose surface, and the thickness determines the minimum dose gradient in the neighborhood of the 50% isodose surface. The purpose of the outer shell is to constrain the 50% isodose line to the interior of the inner shell. One group has reported using 10 and 30 mm expansions of the target, resulting in 5 and 20 mm thickness, for the middle and outer shells, respectively (Clark et al., 2012).

For 3D-CRT or DCA, the volume of normal tissue receiving less than V50 is generally not an explicit planning objective because it is determined by the field geometry. However, for computer-optimized techniques, low dose spill will not be minimized if it is not included as an objective (Thomas et al., 2014). Thomas et al. showed that for treatment of multiple targets using MAT, the volume of normal brain receiving more than 25% of the prescription dose can be limited to between 25 and 250 cm^3 and that a mean brain dose in the range 3%–11% can be achieved (Thomas et al., 2014). Ultimately, the number, size, and locations of the targets will determine what is achievable.

Some groups report including target dose uniformity as a dosimetric objective for radiosurgical planning (Mayo et al., 2010; Audet et al., 2011; Subramanian et al., 2012). For single-fraction radiosurgery using the Gamma Knife, maximum doses as high as twice the prescription dose are common (Paddick and Lippitz, 2006). For Gamma Knife, plans having more dose uniformity had a worse gradient index (Paddick and Lippitz, 2006). This effect is also observed for cones (Meeks et al., 1998), 3DCRT (Hong et al., 2011), and DCA (Hong et al., 2011; Tanyi et al., 2012). Although this effect has not been reported for IMRT or MAT radiosurgery, it is reasonable to expect that forcing uniformity within the target will produce results, because forcing uniformity within the target necessarily results in a reduced dose gradient at the surface of the target. There is a trade-off between target dose uniformity and dose gradient in the normal tissue, so the clinical benefits of each must be evaluated for each patient.

When conformity and gradient are the primary treatment planning objectives, field arrangement is not critical. Achieving a conformal dose distribution with rapid, uniform falloff simply requires a large number of entrance angles, either using fixed beams or a sufficient number of noncoplanar arcs. A small number of fixed beam angles or arcs results in a nonuniform gradient having a faster falloff in some directions at the expense of slower falloff in other directions. For example, a single arc has a rapid dose falloff perpendicular to the arc plane, but a less rapid falloff in plane, whereas a four-arc geometry has less variation in dose gradient (Pike et al., 1990). Multiple, noncoplanar arcs are generally necessary for stereotactic applications (Podgorsak et al., 1989). For 3DCRT, 7–10 noncoplanar fields produces dose distributions similar to multiple noncoplanar arcs, but with slightly higher peripheral dose. Increasing the number of fields reduces the peripheral dose (Bourland and McCollough, 1994). Hong et al. (2011) recommended one beam per 2 Gy of prescribed dose to limit the dose spill at 3–4 Gy. When organ avoidance is an objective, field arrangement is more important. Bouquets of nine beams with optimized arrangements have been reported to achieve good results (Wagner et al., 2001). Similar field geometries are appropriate for IMRT and MAT (Benedict et al., 2001; Clark et al., 2010; Nath et al., 2010; Audet et al., 2011) (9–11 Nath et al. 2010; 4 arcs Audet). If body dose is a concern, geometries that are nearly parallel to the longitudinal body axis should be avoided.

5.3.2.2 Spine

In the spine, avoidance of the spinal cord is the primary goal. The relationship between the target volume and the spinal cord is complex because the target is typically wrapped around the spinal cord. Consequently, DCA and 3DCRT techniques are not able to meet the treatment planning goals for spine SBRT and so intensity-modulated techniques are required (Yenice et al., 2003; Wu et al., 2009a). For fixed-beam IMRT, Yenice et al. (2003) demonstrated that seven beams directed posteriorly and obliquely having a spacing of 20°–30° result in acceptable dosimetry. Kuijper et al. (2010) described similar beam arrangements, shown in Figure 5.8, which were designed based on the complexity of the target volume (vertebral body or entire vertebra) and the spinal level of the involved vertebral body. For MAT, two coplanar arcs having a full rotation are generally sufficient (Wu et al., 2009b; Kuijper et al., 2010); however, Kuijper et al. (2010) have reported using a third arc for cases in which the entire vertebral body is targeted. Using a single arc is not recommended because the achievable dosimetry in the spinal cord is inferior to static IMRT fields (Wu et al., 2009b).

The distance from the target volume to the spinal cord and the degree to which the spinal cord is enclosed by the target governs the achievable target coverage. The target often abuts the spinal cord, necessitating underdosage at the periphery in order to limit the spinal cord dose to an acceptable level. Kuijper et al. (2010) found that when vertebral body is the target, 95% of the target received at least the prescription dose 16 Gy, whereas for the entire vertebral body, only 85% received 16 Gy. Yenice et al. (2003) achieved typical target coverage of 95% for a 20 Gy prescribed dose. A dose gradient of about 10%/mm should be achievable in the region between the target and the spinal cord (Yenice et al., 2003; Kuijper et al., 2010) and can be used to estimate the minimum distance between the cord and the prescription isodose surface prior to treatment planning. It is important to note that, similar to intracranial SRS, forcing dose uniformity within the target will likely reduce the gradient at the boundary of the target, reducing the achievable dose coverage. Maximum doses in the target between 115% and 140% have been reported (Yenice et al., 2003; Wu et al., 2009b; Kuijper et al., 2010).

Although conformity is not a primary objective for spine radiosurgery, the dose to the unspecified normal tissue outside of the target volume should be controlled. The stringent limitation of the spinal cord dose can cause unacceptable dose spillage outside the target, particularly for fixed IMRT techniques, which are particularly susceptible to streaks of high dose extending outside of the target. One approach to controlling dose spill is to create an annulus around the target volume similar to the inner control shell previously described for intracranial targets. Limiting the maximum dose inside the control structure to no more than the prescription dose will generally limit dose spill to acceptable levels. In this manner, a conformity index in the range of 0.9–1.1 can be achieved (Kuijper et al., 2010).

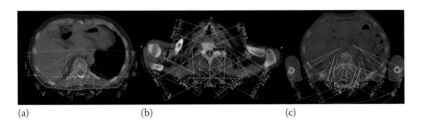

(a) (b) (c)

Figure 5.8 Three beam configurations used for the conventional intensity-modulated radiation therapy plans. Standard beam configuration of (a) the vertebral body, (b) the entire cervical vertebra case, and (c) the entire lumbar vertebra case. (From Kuijper, I.T. et al., *Radiother. Oncol.*, 94, 224, 2010.)

Depending on the location of the lesion within the spine, consideration of other critical structures, such as the lungs, esophagus, kidneys, bowel, heart, and liver, may be necessary. Explicitly including dose limits during inverse planning for critical structures other than the cord is usually unnecessary; however, it is important to review all critical structure doses. Because of the complex modulation required to avoid the spinal cord, unexpected high dose can be delivered to structures that are distant from the target volume. In cases for which the tolerance dose is exceeded, incorporation of explicit dose constraints into the optimization system typically reduces the dose to an acceptable level without compromising target coverage.

5.4 CONCLUSION

In the opinion of Brada and Laing, fractionated SRT was the way forward for the treatment of brain tumors, and C-arm linacs are the preferred platform to accomplish that goal. Their predictions regarding the evolution of linac technology have been realized. With the combination of MLCs, FFF beams, IGRT, 6-DOF tables, and intrafraction motion monitoring, frameless single-fraction or fractionated stereotactic techniques are possible in both the skull and spine. By combining these technologies, highly conformal dose distributions can be delivered with submillimeter accuracy.

The future promises further developments. The increasing sophistication of linac control systems and planning techniques will enable hybrid MAT and IMRT (Matuszak et al., 2013) and the efficient use of a large number of table orientations (Nguyen et al., 2014), further improving plan quality and decreasing treatment time. Knowledge-based planning systems (Shiraishi et al., 2015) have the potential to reduce the expertise required to deliver high-quality radiosurgery. By reducing the learning barriers, knowledge-based systems will facilitate the dissemination of hypofractionated SRT into more centers, fulfilling the prediction of Brada and Ling that SRT will become a routine tool in the radiotherapy armamentarium.

REFERENCES

Abacioglu U, Ozen Z, Yilmaz M, Arifoglu A, Gunhan B, Kayalilar N, Peker S, Sengoz M, Gurdalli S, Cozzi L (2014) Critical appraisal of RapidArc radiosurgery with flattening filter free photon beams for benign brain lesions in comparison to GammaKnife: A treatment planning study. *Radiat Oncol* 9:119.

Adler JR, Jr., Chang SD, Murphy MJ, Doty J, Geis P, Hancock SL (1997) The cyberknife: A frameless robotic system for radiosurgery. *Stereotact Funct Neurosurg* 69:124–128.

Alongi F, Cozzi L, Arcangeli S, Iftode C, Comito T, Villa E, Lobefalo F et al. (2013) Linac based SBRT for prostate cancer in 5 fractions with VMAT and flattening filter free beams: Preliminary report of a phase II study. *Radiat Oncol* 8:171.

Alongi F, Fogliata A, Clerici E, Navarria P, Tozzi A, Comito T, Ascolese AM et al. (2012) Volumetric modulated arc therapy with flattening filter free beams for isolated abdominal/pelvic lymph nodes: Report of dosimetric and early clinical results in oligometastatic patients. *Radiat Oncol* 7:204.

Audet C, Poffenbarger BA, Chang P, Jackson PS, Lundahl RE, Ryu SI, Ray GR (2011) Evaluation of volumetric modulated arc therapy for cranial radiosurgery using multiple noncoplanar arcs. *Med Phys* 38:5863–5872.

Bayouth JE, Kaiser HS, Smith MC, Pennington EC, Anderson KM, Ryken TC, Buatti JM (2007) Image-guided stereotactic radiosurgery using a specially designed high-dose-rate linac. *Med Dosim* 32:134–141.

Benedict SH, Cardinale RM, Wu Q, Zwicker RD, Broaddus WC, Mohan R (2001) Intensity-modulated stereotactic radiosurgery using dynamic micro-multileaf collimation. *Int J Radiat Oncol Biol Phys* 50:751–758.

Bissonnette JP, Moseley D, White E, Sharpe M, Purdie T, Jaffray DA (2008) Quality assurance for the geometric accuracy of cone-beam CT guidance in radiation therapy. *Int J Radiat Oncol Biol Phys* 71:S57–S61.

Blonigen BJ, Steinmetz RD, Levin L, Lamba MA, Warnick RE, Breneman JC (2010) Irradiated volume as a predictor of brain radionecrosis after linear accelerator stereotactic radiosurgery. *Int J Radiat Oncol Biol Phys* 77:996–1001.

Bourland JD, McCollough KP (1994) Static field conformal stereotactic radiosurgery: Physical techniques. *Int J Radiat Oncol Biol Phys* 28:471–479.

Brada M, Laing R (1994) Radiosurgery/stereotactic external beam radiotherapy for malignant brain tumours: The Royal Marsden Hospital experience. *Recent Res Cancer Res* 135:91–104.

Cardinale RM, Benedict SH, Wu Q, Zwicker RD, Gaballa HE, Mohan R (1998) A comparison of three stereotactic radiotherapy techniques; ARCS vs. noncoplanar fixed fields vs. intensity modulation. *Int J Radiat Oncol Biol Phys* 42:431–436.

Cashmore J (2008) The characterization of unflattened photon beams from a 6 MV linear accelerator. *Phys Med Biol* 53:1933–1946.

Cashmore J, Ramtohul M, Ford D (2011) Lowering whole-body radiation doses in pediatric intensity-modulated radiotherapy through the use of unflattened photon beams. *Int J Radiat Oncol Biol Phys* 80:1220–1227.

Chae SM, Lee GW, Son SH (2014) The effect of multileaf collimator leaf width on the radiosurgery planning for spine lesion treatment in terms of the modulated techniques and target complexity. *Radiat Oncol* 9:72.

Chang J, Yenice KM, Narayana A, Gutin PH (2007) Accuracy and feasibility of cone-beam computed tomography for stereotactic radiosurgery setup. *Med Phys* 34:2077–2084.

Clark GM, Popple RA, Prendergast BM, Spencer SA, Thomas EM, Stewart JG, Guthrie BL, Markert JM, Fiveash JB (2012) Plan quality and treatment planning technique for single isocenter cranial radiosurgery with volumetric modulated arc therapy. *Pract Radiat Oncol* 2:306–313.

Clark GM, Popple RA, Young PE, Fiveash JB (2010) Feasibility of single-isocenter volumetric modulated arc radiosurgery for treatment of multiple brain metastases. *Int J Radiat Oncol Biol Phys* 76:296–302.

De Los Santos J, Popple R, Agazaryan N, Bayouth JE, Bissonnette JP, Bucci MK, Dieterich S et al. (2013) Image guided radiation therapy (IGRT) technologies for radiation therapy localization and delivery. *Int J Radiat Oncol Biol Phys* 87:33–45.

Dhabaan A, Elder E, Schreibmann E, Crocker I, Curran WJ, Oyesiku NM, Shu HK, Fox T (2010) Dosimetric performance of the new high-definition multileaf collimator for intracranial stereotactic radiosurgery. *J Appl Clin Med Phys* 11:3040.

Dhabaan A, Schreibmann E, Siddiqi A, Elder E, Fox T, Ogunleye T, Esiashvili N, Curran W, Crocker I, Shu HK (2012) Six degrees of freedom CBCT-based positioning for intracranial targets treated with frameless stereotactic radiosurgery. *J Appl Clin Med Phys* 13:3916.

Duthoy W, De Gersem W, Vergote K, Boterberg T, Derie C, Smeets P, De Wagter C, De Neve W (2004) Clinical implementation of intensity-modulated arc therapy (IMAT) for rectal cancer. *Int J Radiat Oncol Biol Phys* 60:794–806.

Gevaert T, Verellen D, Engels B, Depuydt T, Heuninckx K, Tournel K, Duchateau M, Reynders T, De Ridder M (2012a) Clinical evaluation of a robotic 6-degree of freedom treatment couch for frameless radiosurgery. *Int J Radiat Oncol Biol Phys* 83:467–474.

Gevaert T, Verellen D, Tournel K, Linthout N, Bral S, Engels B, Collen C et al. (2012b) Setup accuracy of the Novalis ExacTrac 6DOF system for frameless radiosurgery. *Int J Radiat Oncol Biol Phys* 82:1627–1635.

Hong LX, Garg M, Lasala P, Kim M, Mah D, Chen CC, Yaparpalvi R et al. (2011) Experience of micromultileaf collimator linear accelerator based single fraction stereotactic radiosurgery: Tumor dose inhomogeneity, conformity, and dose fall off. *Med Phys* 38:1239–1247.

Hoogeman MS, Nuyttens JJ, Levendag PC, Heijmen BJ (2008) Time dependence of intrafraction patient motion assessed by repeat stereoscopic imaging. *Int J Radiat Oncol Biol Phys* 70:609–618.

Huang Y, Chin K, Robbins JR, Kim J, Li H, Amro H, Chetty IJ, Gordon J, Ryu S (2014) Radiosurgery of multiple brain metastases with single-isocenter dynamic conformal arcs (SIDCA). *Radiother Oncol* 112:128–132.

Huq MS, Das IJ, Steinberg T, Galvin JM (2002) A dosimetric comparison of various multileaf collimators. *Phys Med Biol* 47:N159–N170.

Jin JY, Yin FF, Ryu S, Ajlouni M, Kim JH (2005) Dosimetric study using different leaf-width MLCs for treatment planning of dynamic conformal arcs and intensity-modulated radiosurgery. *Med Phys* 32:405–411.

Kang KM, Chai GY, Jeong BK, Ha IB, Lee S, Park KB, Jung JM, Lim YK, Yoo SH, Jeong H (2013) Estimation of optimal margin for intrafraction movements during frameless brain radiosurgery. *Med Phys* 40:051716.

Khan FM, Gerbi BJ (2012) *Treatment Planning in Radiation Oncology*, 3rd ed., p. xiii, 773pp. Philadelphia, PA: Wolters Kluwer/Lippincott Williams & Wilkins Health.

King RB, Hyland WB, Cole AJ, Butterworth KT, McMahon SJ, Redmond KM, Trainer C, Prise KM, McGarry CK, Hounsell AR (2013) An in vitro study of the radiobiological effects of flattening filter free radiotherapy treatments. *Phys Med Biol* 58:N83–N94.

Klein EE, Hanley J, Bayouth J, Yin FF, Simon W, Dresser S, Serago C et al. (2009) Task Group 142 report: Quality assurance of medical accelerators. *Med Phys* 36:4197–4212.

Kragl G, af Wetterstedt S, Knausl B, Lind M, McCavana P, Knoos T, McClean B, Georg D (2009) Dosimetric characteristics of 6 and 10MV unflattened photon beams. *Radiother Oncol* 93:141–146.

Kragl G, Baier F, Lutz S, Albrich D, Dalaryd M, Kroupa B, Wiezorek T, Knoos T, Georg D (2011) Flattening filter free beams in SBRT and IMRT: Dosimetric assessment of peripheral doses. *Z Med Phys* 21:91–101.

Kry SF, Popple R, Molineu A, Followill DS (2012) Ion recombination correction factors (P(ion)) for Varian TrueBeam high-dose-rate therapy beams. *J Appl Clin Med Phys* 13:3803.

Kry SF, Vassiliev ON, Mohan R (2010) Out-of-field photon dose following removal of the flattening filter from a medical accelerator. *Phys Med Biol* 55:2155–2166.

Kuijper IT, Dahele M, Senan S, Verbakel WF (2010) Volumetric modulated arc therapy versus conventional intensity modulated radiation therapy for stereotactic spine radiotherapy: A planning study and early clinical data. *Radiother Oncol* 94:224–228.

Leavitt DD (1998) Beam shaping for SRT/SRS. *Med Dosim* 23:229–236.

Levegrun S, Hof H, Essig M, Schlegel W, Debus J (2004) Radiation-induced changes of brain tissue after radiosurgery in patients with arteriovenous malformations: Correlation with dose distribution parameters. *Int J Radiat Oncol Biol Phys* 59:796–808.

Li G, Ballangrud A, Kuo LC, Kang H, Kirov A, Lovelock M, Yamada Y, Mechalakos J, Amols H (2011) Motion monitoring for cranial frameless stereotactic radiosurgery using video-based three-dimensional optical surface imaging. *Med Phys* 38:3981–3994.

Li S, Liu D, Yin G, Zhuang P, Geng J (2006) Real-time 3D-surface-guided head refixation useful for fractionated stereotactic radiotherapy. *Med Phys* 33:492–503.

Lightstone AW, Tsao M, Baran PS, Chan G, Pang G, Ma L, Lochray F, Sahgal A (2012) Cone beam CT (CBCT) evaluation of inter- and intra-fraction motion for patients undergoing brain radiotherapy immobilized using a commercial thermoplastic mask on a robotic couch. *Technol Cancer Res Treat* 11:203–209.

Ling CC, Gerweck LE, Zaider M, Yorke E (2010) Dose-rate effects in external beam radiotherapy redux. *Radiother Oncol* 95:261–268.

Lohse I, Lang S, Hrbacek J, Scheidegger S, Bodis S, Macedo NS, Feng J, Lutolf UM, Zaugg K (2011) Effect of high dose per pulse flattening filter-free beams on cancer cell survival. *Radiother Oncol* 101:226–232.

Ma L, Yu CX, Earl M, Holmes T, Sarfaraz M, Li XA, Shepard D, Amin P, DiBiase S, Suntharalingam M, Mansfield C (2001) Optimized intensity-modulated arc therapy for prostate cancer treatment. *Int J Cancer* 96:379–384.

Masi L, Casamassima F, Polli C, Menichelli C, Bonucci I, Cavedon C (2008) Cone beam CT image guidance for intracranial stereotactic treatments: Comparison with a frame guided set-up. *Int J Radiat Oncol Biol Phys* 71:926–933.

Matuszak MM, Steers JM, Long T, McShan DL, Fraass BA, Romeijn HE, Ten Haken RK (2013) FusionArc optimization: A hybrid volumetric modulated arc therapy (VMAT) and intensity modulated radiation therapy (IMRT) planning strategy. *Med Phys* 40:071713.

Mayo CS, Ding L, Addesa A, Kadish S, Fitzgerald TJ, Moser R (2010) Initial experience with volumetric IMRT (RapidArc) for intracranial stereotactic radiosurgery. *Int J Radiat Oncol Biol Phys* 78:1457–1466.

McEwen M, DeWerd L, Ibbott G, Followill D, Rogers DW, Seltzer S, Seuntjens J (2014) Addendum to the AAPM's TG-51 protocol for clinical reference dosimetry of high-energy photon beams. *Med Phys* 41:041501.

Meeks SL, Bova FJ, Wagner TH, Buatti JM, Friedman WA, Foote KD (2000) Image localization for frameless stereotactic radiotherapy. *Int J Radiat Oncol Biol Phys* 46:1291–1299.

Meeks SL, Buatti JM, Bova FJ, Friedman WA, Mendenhall WM (1998) Treatment planning optimization for linear accelerator radiosurgery. *Int J Radiat Oncol Biol Phys* 41:183–197.

Meyer J, Wilbert J, Baier K, Guckenberger M, Richter A, Sauer O, Flentje M (2007) Positioning accuracy of cone-beam computed tomography in combination with a HexaPOD robot treatment table. *Int J Radiat Oncol Biol Phys* 67:1220–1228.

Minniti G, Clarke E, Lanzetta G, Osti MF, Trasimeni G, Bozzao A, Romano A, Enrici RM (2011) Stereotactic radiosurgery for brain metastases: Analysis of outcome and risk of brain radionecrosis. *Radiat Oncol* 6:48.

Monk JE, Perks JR, Doughty D, Plowman PN (2003) Comparison of a micro-multileaf collimator with a 5-mm-leaf-width collimator for intracranial stereotactic radiotherapy. *Int J Radiat Oncol Biol Phys* 57:1443–1449.

Murphy MJ (2009) Intrafraction geometric uncertainties in frameless image-guided radiosurgery. *Int J Radiat Oncol Biol Phys* 73:1364–1368.

Murphy MJ, Chang SD, Gibbs IC, Le QT, Hai J, Kim D, Martin DP, Adler JR, Jr. (2003) Patterns of patient movement during frameless image-guided radiosurgery. *Int J Radiat Oncol Biol Phys* 55:1400–1408.

Nath SK, Lawson JD, Simpson DR, Vanderspek L, Wang JZ, Alksne JF, Ciacci J, Mundt AJ, Murphy KT (2010) Single-isocenter frameless intensity-modulated stereotactic radiosurgery for simultaneous treatment of multiple brain metastases: Clinical experience. *Int J Radiat Oncol Biol Phys* 78:91–97.

Nguyen D, Rwigema JC, Yu VY, Kaprealian T, Kupelian P, Selch M, Lee P, Low DA, Sheng K (2014) Feasibility of extreme dose escalation for glioblastoma multiforme using 4pi radiotherapy. *Radiat Oncol* 9:239.

O'Brien PF, Gillies BA, Schwartz M, Young C, Davey P (1991) Radiosurgery with unflattened 6-MV photon beams. *Med Phys* 18:519–521.

Otto K (2008) Volumetric modulated arc therapy: IMRT in a single gantry arc. *Med Phys* 35:310–317.

Paddick I, Lippitz B (2006) A simple dose gradient measurement tool to complement the conformity index. *J Neurosurg* 105(Suppl):194–201.

Peng JL, Kahler D, Li JG, Samant S, Yan G, Amdur R, Liu C (2010) Characterization of a real-time surface image-guided stereotactic positioning system. *Med Phys* 37:5421–5433.

Pike GB, Podgorsak EB, Peters TM, Pla C, Olivier A, Souhami L (1990) Dose distributions in radiosurgery. *Med Phys* 17:296–304.

Podgorsak EB, Pike GB, Olivier A, Pla M, Souhami L (1989) Radiosurgery with high energy photon beams: A comparison among techniques. *Int J Radiat Oncol Biol Phys* 16:857–865.

Popple RA, Balter PA, Orton CG (2014) Point/counterpoint. Because of the advantages of rotational techniques, conventional IMRT will soon become obsolete. *Med Phys* 41:100601.

Prendergast BM, Dobelbower MC, Bonner JA, Popple RA, Baden CJ, Minnich DJ, Cerfolio RJ, Spencer SA, Fiveash JB (2013) Stereotactic body radiation therapy (SBRT) for lung malignancies: Preliminary toxicity results using a flattening filter-free linear accelerator operating at 2400 monitor units per minute. *Radiat Oncol* 8:273.

Prendergast BM, Popple RA, Clark GM, Guthrie BL, Markert JM, Spencer SA, Fiveash JB (2011) Improved clinical efficiency in CNS stereotactic radiosurgery using a flattening filter free linear accelerator. *J Radiosurg BRT* 1:117–122.

Ryu S, Pugh SL, Gerszten PC, Yin FF, Timmerman RD, Hitchcock YJ, Movsas B et al. (2014) RTOG 0631 phase 2/3 study of image guided stereotactic radiosurgery for localized (1–3) spine metastases: Phase 2 results. *Pract Radiat Oncol* 4:76–81.

Ryu SI, Chang SD, Kim DH, Murphy MJ, Le QT, Martin DP, Adler JR, Jr. (2001) Image-guided hypo-fractionated stereotactic radiosurgery to spinal lesions. *Neurosurgery* 49:838–846.

Salkeld AL, Unicomb K, Hayden AJ, Van Tilburg K, Yau S, Tiver K (2014) Dosimetric comparison of volumetric modulated arc therapy and linear accelerator-based radiosurgery for the treatment of one to four brain metastases. *J Med Imaging Radiat Oncol* 58:722–728.

Schipani S, Wen W, Jin JY, Kim JK, Ryu S (2012) Spine radiosurgery: A dosimetric analysis in 124 patients who received 18 Gy. *Int J Radiat Oncol Biol Phys* 84:e571–e576.

Schreibmann E, Fox T, Crocker I (2011) Dosimetric effects of manual cone-beam CT (CBCT) matching for spinal radiosurgery: Our experience. *J Appl Clin Med Phys* 12:3467.

Scorsetti M, Alongi F, Castiglioni S, Clivio A, Fogliata A, Lobefalo F, Mancosu P et al. (2011) Feasibility and early clinical assessment of flattening filter free (FFF) based stereotactic body radiotherapy (SBRT) treatments. *Radiat Oncol* 6:113.

Scorsetti M, Alongi F, Clerici E, Comito T, Fogliata A, Iftode C, Mancosu P et al. (2014) Stereotactic body radiotherapy with flattening filter-free beams for prostate cancer: Assessment of patient-reported quality of life. *J Cancer Res Clin Oncol* 140:1795–1800.

Scorsetti M, Comito T, Cozzi L, Clerici E, Tozzi A, Franzese C, Navarria P et al. (2015) The challenge of inoperable hepatocellular carcinoma (HCC): Results of a single-institutional experience on stereotactic body radiation therapy (SBRT). *J Cancer Res Clin Oncol* 141(7):1301–1309.

Serna A, Puchades V, Mata F, Ramos D, Alcaraz M (2015) Influence of multi-leaf collimator leaf width in radiosurgery via volumetric modulated arc therapy and 3D dynamic conformal arc therapy. *Phys Med* 31:293–296.

Shaw E, Kline R, Gillin M, Souhami L, Hirschfeld A, Dinapoli R, Martin L (1993) Radiation therapy oncology group: Radiosurgery quality assurance guidelines. *Int J Radiat Oncol Biol Phys* 27:1231–1239.

Shiraishi S, Tan J, Olsen LA, Moore KL (2015) Knowledge-based prediction of plan quality metrics in intracranial stereotactic radiosurgery. *Med Phys* 42:908.

Shiu AS, Kooy HM, Ewton JR, Tung SS, Wong J, Antes K, Maor MH (1997) Comparison of miniature multileaf collimation (MMLC) with circular collimation for stereotactic treatment. *Int J Radiat Oncol Biol Phys* 37:679–688.

Snyder KC, Wen N, Huang Y, Kim J, Zhao B, Siddiqui S, Chetty IJ, Ryu S (2014) Use of jaw tracking in intensity modulated and volumetric modulated arc radiation therapy for spine stereotactic radiosurgery. *Pract Radiat Oncol* 5:e155–e162.

Solberg TD, Boedeker KL, Fogg R, Selch MT, DeSalles AA (2001) Dynamic arc radiosurgery field shaping: A comparison with static field conformal and noncoplanar circular arcs. *Int J Radiat Oncol Biol Phys* 49:1481–1491.

Sorensen BS, Vestergaard A, Overgaard J, Praestegaard LH (2011) Dependence of cell survival on instantaneous dose rate of a linear accelerator. *Radiother Oncol* 101:223–225.

Subramanian S, Srinivas C, Ramalingam K, Babaiah M, Swamy ST, Arun G, Kathirvel M et al. (2012) Volumetric modulated arc-based hypofractionated stereotactic radiotherapy for the treatment of selected intracranial arteriovenous malformations: Dosimetric report and early clinical experience. *Int J Radiat Oncol Biol Phys* 82:1278–1284.

Tagaste B, Riboldi M, Spadea MF, Bellante S, Baroni G, Cambria R, Garibaldi C et al. (2012) Comparison between infrared optical and stereoscopic x-ray technologies for patient setup in image guided stereotactic radiotherapy. *Int J Radiat Oncol Biol Phys* 82:1706–1714.

Tanyi JA, Doss EJ, Kato CM, Monaco DL, Meng LZ, Chen Y, Kubicky CD, Marquez CM, Fuss M (2012) Dynamic conformal arc cranial stereotactic radiosurgery: Implications of multileaf collimator margin on dose-volume metrics. *Br J Radiol* 85:e1058–e1066.

Thomas EM, Popple RA, Wu X, Clark GM, Markert JM, Guthrie BL, Yuan Y, Dobelbower MC, Spencer SA, Fiveash JB (2014) Comparison of plan quality and delivery time between volumetric arc therapy (RapidArc) and Gamma Knife radiosurgery for multiple cranial metastases. *Neurosurgery* 75:409–417; discussion 417–408.

Van Dyk J (1999) *The Modern Technology of Radiation Oncology: A Compendium for Medical Physicists and Radiation Oncologists*. Madison, WI: Medical Physics Publishing.

Vassiliev ON, Kry SF, Chang JY, Balter PA, Titt U, Mohan R (2009) Stereotactic radiotherapy for lung cancer using a flattening filter free Clinac. *J Appl Clin Med Phys* 10:2880.

Vassiliev ON, Titt U, Ponisch F, Kry SF, Mohan R, Gillin MT (2006) Dosimetric properties of photon beams from a flattening filter free clinical accelerator. *Phys Med Biol* 51:1907–1917.

Verbakel WF, van den Berg J, Slotman BJ, Sminia P (2013) Comparable cell survival between high dose rate flattening filter free and conventional dose rate irradiation. *Acta Oncol* 52:652–657.

Verellen D, Soete G, Linthout N, Van Acker S, De Roover P, Vinh-Hung V, Van de Steene J, Storme G (2003) Quality assurance of a system for improved target localization and patient set-up that combines real-time infrared tracking and stereoscopic x-ray imaging. *Radiother Oncol* 67:129–141.

Wagner TH, Meeks SL, Bova FJ, Friedman WA, Buatti JM, Bouchet LG (2001) Isotropic beam bouquets for shaped beam linear accelerator radiosurgery. *Phys Med Biol* 46:2571–2586.

Wang JZ, Rice R, Pawlicki T, Mundt AJ, Sandhu A, Lawson J, Murphy KT (2010) Evaluation of patient setup uncertainty of optical guided frameless system for intracranial stereotactic radiosurgery. *J Appl Clin Med Phys* 11:3181.

Wang LT, Solberg TD, Medin PM, Boone R (2001) Infrared patient positioning for stereotactic radiosurgery of extracranial tumors. *Comput Biol Med* 31:101–111.

Wang PM, Hsu WC, Chung NN, Chang FL, Jang CJ, Fogliata A, Scorsetti M, Cozzi L (2014) Feasibility of stereotactic body radiation therapy with volumetric modulated arc therapy and high intensity photon beams for hepatocellular carcinoma patients. *Radiat Oncol* 9:18.

Wang Y, Khan MK, Ting JY, Easterling SB (2012) Surface dose investigation of the flattening filter-free photon beams. *Int J Radiat Oncol Biol Phys* 83:e281–e285.

Wiersma RD, Tomarken SL, Grelewicz Z, Belcher AH, Kang H (2013) Spatial and temporal performance of 3D optical surface imaging for real-time head position tracking. *Med Phys* 40:111712.

Wilbert J, Guckenberger M, Polat B, Sauer O, Vogele M, Flentje M, Sweeney RA (2010) Semi-robotic 6 degree of freedom positioning for intracranial high precision radiotherapy; first phantom and clinical results. *Radiat Oncol* 5:42.

Wong E, D'Souza DP, Chen JZ, Lock M, Rodrigues G, Coad T, Trenka K, Mulligan M, Bauman GS (2005) Intensity-modulated arc therapy for treatment of high-risk endometrial malignancies. *Int J Radiat Oncol Biol Phys* 61:830–841.

Wu QJ, Wang Z, Kirkpatrick JP, Chang Z, Meyer JJ, Lu M, Huntzinger C, Yin FF (2009a) Impact of collimator leaf width and treatment technique on stereotactic radiosurgery and radiotherapy plans for intra- and extracranial lesions. *Radiat Oncol* 4:3.

Wu QJ, Yoo S, Kirkpatrick JP, Thongphiew D, Yin FF (2009b) Volumetric arc intensity-modulated therapy for spine body radiotherapy: Comparison with static intensity-modulated treatment. *Int J Radiat Oncol Biol Phys* 75:1596–1604.

Yenice KM, Lovelock DM, Hunt MA, Lutz WR, Fournier-Bidoz N, Hua CH, Yamada J et al. (2003) CT image-guided intensity-modulated therapy for paraspinal tumors using stereotactic immobilization. *Int J Radiat Oncol Biol Phys* 55:583–593.

Yu CX (1995) Intensity-modulated arc therapy with dynamic multileaf collimation: An alternative to tomotherapy. *Phys Med Biol* 40:1435–1449.

Yu CX, Li XA, Ma L, Chen D, Naqvi S, Shepard D, Sarfaraz M, Holmes TW, Suntharalingam M, Mansfield CM (2002) Clinical implementation of intensity-modulated arc therapy. *Int J Radiat Oncol Biol Phys* 53:453–463.

Zhao B, Yang Y, Li X, Li T, Heron DE, Saiful Huq M (2015) Is high-dose rate RapidArc-based radiosurgery dosimetrically advantageous for the treatment of intracranial tumors? *Med Dosim* 40:3–8.

6 Advanced MRI for brain metastases

Michael Chan, Paula Alcaide Leon, Sten Myrehaug,
Hany Soliman, and Chris Heyn

Contents

6.1 INTRODUCTION

Brain metastases are a common manifestation of metastatic tumors with significant implications regarding morbidity and mortality. In the past, the role of imaging was limited to diagnosis and detection of disease and, at most, a semiquantitative assessment of the response to therapy. However, in the age of personalized medicine with increasing technologic advancements and the development of targeted cancer therapies, there is not only an increasing need for rigorous evaluation of the anatomic details but also the functional details such as tumor metabolism, oxygenation, and perfusion. These factors are often the target of novel therapeutic agents or strategies and have been shown to have prognostic implications. In addition, there has been a growing need for accurate and precise quantitative imaging tools, especially for predicting patient outcomes and assessing early response to therapy, including differentiating treatment-related changes from tumor recurrence.

The focus of this chapter will be on magnetic resonance imaging (MRI) techniques, which are the current gold standard for imaging brain metastases and have been shown to be superior to other modalities, such as computed tomography (CT), for this purpose. While basic MRI techniques provide unparalleled tissue contrast that is essential in the brain, advanced MR techniques—including MR spectroscopy (MRS), perfusion imaging, diffusion-weighted imaging (DWI), and blood-oxygen-level-dependent (BOLD) MRI—provide the opportunity to explore important functional aspects of tumors. We finish our discussion with some future applications. By manipulating various parameters, MR represents an extremely adaptable technique with essentially limitless opportunities for generating information on tumor biomarkers that can be used in the future to guide and direct management.

6.2 MR SPECTROSCOPY

MRS is a technique used to quantify the chemical composition of tissues. Using this methodology, the presence of molecules containing MR-detectable nuclei—such as protons (^{1}H), phosphorus (^{31}P), sodium (^{23}Na), carbon (^{13}C), or fluorine (^{19}F)—can be measured and mapped. MRS is most commonly performed on ^{1}H owing to the large natural abundance of this isotope, the large biological abundance of protons in organic tissues, and the favorable gyromagnetic ratio of protons, which results in better signal-to-noise ratio (SNR) compared to other nuclear species. To detect different nuclear species, a radiofrequency (RF) coil tuned to the specific frequency of that nucleus is needed. Proton MRS can be performed using the RF coils used for clinical MRI, whereas MRS of other species requires a specially built and tuned RF coil.

6.2.1 BASIC PHYSICS AND TECHNIQUE

The physics of MRS is based on the fact that nuclei within a molecule will resonate at a frequency that is proportional to the gyromagnetic ratio (a constant for a given nuclear species) and the local magnetic field experienced by the nucleus. Changes in the local magnetic field result in changes in the resonance frequency of the nucleus, a phenomenon known as chemical shift. This is dependent on the magnetic shielding of nuclei by electrons around the nucleus, which in turn is related to the chemical bonding of the nucleus within the molecule. Depending on the chemical bonding, nuclei within a molecule will resonate at specific frequencies that are characteristic for a given molecule and result in a unique signature that can be detected using MRI. In MRS, the resonance frequencies for a given molecule are expressed in terms of parts per million (ppm), which is a unit of frequency that is the same regardless of the field strength at which the measurement took place. Preferably, MRS should be performed at as high a magnetic field strength as possible because of the SNR advantage gained at higher field.

In the brain, the most abundant proton species is from water, which resonates at a frequency of 4.7 ppm. To improve the detection of brain metabolites, which have concentrations that are orders of magnitude smaller than water, different methodologies are used to selectively suppress the signal from water. There are many ways to accomplish this, but a commonly employed strategy is through the application of RF pulses centered on the resonance frequency of water protons with the aim of saturating and decreasing the signal from water.

The echo time (TE) that is used in MRS is selected based on the specific application and the metabolites of interest. Generally, this can be done with short TE (e.g., 20–40 ms) or long TE (e.g., >135 ms). At lower TE, higher SNR can be achieved allowing the detection of many more metabolites. Visually, the spectra are more complex as many more peaks are visible. Longer TE spectra have lower SNR but are visually simpler to interpret.

Two major MRS techniques are used clinically: single-voxel spectroscopy (SVS) and chemical spectroscopic imaging. SVS allows the characterization and quantification of metabolites in a defined region of interest (ROI) or volume of interest (VOI). This can be achieved by using magnetic field gradients and sequential slice-selective RF excitation to interrogate protons in the VOI (typically a cube with dimensions on the order of centimeters). The two most commonly employed methodologies are point-resolved spectroscopy (PRESS) and stimulated echo acquisition mode (STEAM) (Drost et al., 2002). Of these, PRESS is used more often. The sequence consists of RF pulses to initially suppress water followed by one 90° and two sequential 180° RF pulses. Each RF pulse is applied with a sequential orthogonal magnetic field gradient to selectively excite and refocus signal within the ROI. The sequence is a spin echo sequence that maximizes signal to noise through spin echo refocusing. Compared to STEAM, which does not employ spin echo refocusing, PRESS can achieve higher signal to noise but suffers from longer TE and less precise localization of signal from the VOI. SVS is faster to acquire than MRSI, which allows separate acquisitions of short and long TE data. Additionally, quantification of tissue metabolites, which can be performed with various modeling such as LC model, is more robust with SVS (Provencher, 1993).

MRSI allows for the spatial localization of metabolites. The basic design of MRSI sequences is similar to SVS with format based on PRESS or STEAM, for instance. The addition of phase encode gradients to these sequences is used to encode spatial frequency information, which are used to fill a spectroscopy grid. The spatial frequency information can then be used to reconstruct the spatial distribution of spectra via a Fourier transform. One of the main differences between MRSI and MRI is that frequency encoding gradients are not used with MRSI, as the frequency dimension holds the information related to chemical shift. The need for a separate phase encode step to encode each point within the spectroscopy grid is time consuming, which is one of the limitations of the technique. The major advantage over SVS is the ability to map the spatial distribution of metabolites, which can vary from region to region within a heterogeneous tumor.

6.2.2 INTERPRETATION

The three most abundant species in the brain are N-acetylaspartate (NAA), creatine (Cr), and choline (Cho) (Soares and Law, 2009). The most abundant brain metabolite is NAA located upfield from the water at approximately 2.0 ppm. It is synthesized in the mitochondria of neurons and is therefore a neuronal cell marker and a marker of neuronal cell viability, although it can also be found in glial cells. Pathologies that lead to neuronal cell death or replacement will result in a decrease in NAA. Cr is the second most abundant brain metabolite and is found at approximately 3.0 ppm. The Cr peak is derived from protons found on Cr and phosphocreatine, which are products of energy metabolism. Generally, the level of Cr in the brain is less affected by pathology and can be used as an internal reference. For example, the ratio of a brain metabolite such as NAA can be crudely normalized measuring the NAA to Cr ratio, thereby allowing the NAA levels in a pathological portion of the brain to be compared to normal brain, for example. Cho is found at approximately 3.2 ppm and is the third most abundant metabolite after Cr. The signal is derived from protons on Cho and metabolites of Cho, which are generally found in cell membranes. Cho is therefore a marker of cell membrane turnover and cellular proliferation. Consequently, Cho is typically elevated in brain neoplasms including metastases and generally decreased in areas of tumor necrosis. Elevated Cho is also demonstrated in brain inflammation and other pathologies and is therefore not specific. Figure 6.1 demonstrates a typical SVS proton spectrum for normal brain demonstrating these three major brain metabolites.

Other important metabolites which can be detected include lipids (Lip, 0.9–1.2 ppm), myoinositol (Myo, 3.56 ppm), and amino acid peaks such as glutamine/glutamate (~2.0–2.5 ppm) and alanine (~1.48 ppm). Lactate (Lac) is a doublet centered at 1.3 ppm, which undergoes inversion (seen below the baseline) at TE = 135 ms, thereby allowing it to be seen separately from Lip peaks, which occur at this location at short TE. While Lac can be found in healthy adult brains, it is generally not detectable at physiologic concentrations using most commonly employed proton MRS technique. Increases in brain Lac and detection with MRS indicate increasing underlying anaerobic metabolism, which can be seen with brain neoplasms, as well as a wide range of pathologies including hypoxia, acute inflammation, infection, and metabolic brain disease. Lip can be seen as a result of contamination by fat from adjacent structures (e.g., subcutaneous fat). It is also present in pathologies resulting in cell membrane degradation/necrosis. Myo is a degradation product of myelin and found in glial cells. It can be increased with glial cell proliferation as is seen with inflammation, gliosis, and gliomas. Ala can be seen in meningiomas, and Glx peaks can be seen in certain metabolic brain diseases such as hepatic encephalopathy.

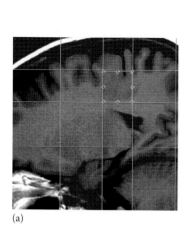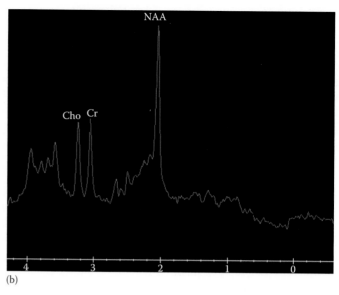

(a)   (b)

Figure 6.1 T1-weighted images of normal white matter with overlay of single-voxel spectroscopy region of interest (box delimited with diamonds) (a) and corresponding MR spectroscopy (point-resolved spectroscopy, echo time 35 ms) (b) showing the typical appearance of the three major metabolic peaks (choline, creatine, and N-acetylaspartate).

6.2.3 CLINICAL APPLICATION

Brain tumors generally show decreased NAA and elevated Cho compared to normal brain (Soares and Law, 2009). Metastases can show an absence of NAA as they are not neuronal derived, although the small size of most metastases and low resolution of SVS or MRSI result in partial volume averaging of adjacent brain containing NAA, so this observation is not commonly made. With tumor necrosis, Lip peaks can be demonstrated. Furthermore, Lac may be elevated in metastases as a result of derangements of tumor cell energy metabolism (e.g., Warburg effect). Generally, there are no specific metabolic differences on MRS for metastases arising from different primary tumors, although one small study did report elevated Lip in untreated colorectal metastases (Chernov et al., 2006).

The limited number of studies examining the utility of proton MRS for predicting response of brain metastases to stereotactic radiosurgery (SRS) has been disappointing. A study involving a small cohort of 26 patients with intracranial metastases (predominantly lung, breast, colorectal) treated with SRS with follow-up at least 3 months posttreatment examined the relationship between baseline proton MRS and treatment response (Chernov et al., 2007). In this cohort, approximately 50% of patients had treatment response (defined as 50% reduction in tumor volume) and 46% had local progression (defined as a 25% increase in tumor volume). The authors found no correlation between the metabolic profile of tumors at baseline and treatment response.

Shortly after SRS, metastases undergo metabolic changes that can be observed by proton MRS even before changes in tumor size are observed. In a small cohort of 81 patients with 85 brain metastases, proton MRS performed within 16–18 hours post-SRS showed significant reductions in Cho/Cr ratios compared to baseline measurements despite a lack of morphologic change in size of metastases at this early time point (Chernov et al., 2004). This decrease was greater for tumors with initially high Cho/Cr ratios. The authors attributed this metabolic change to a reduction in cell proliferation and cell death.

A more longitudinal study of SRS-treated brain metastases demonstrated increase in NAA/Cr ratio and decreases in Cho and Lip content within the first month after treatment in responders (Chernov et al., 2009). No significant alteration was observed in patients with stable tumors, and decreases in NAA/Cr and increases in Lip and Cho were observed in progressors.

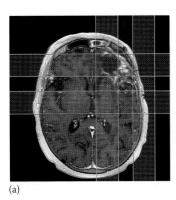
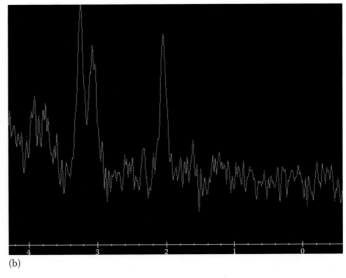

(a)   (b)

Figure 6.2 A 73-year-old woman with history of breast cancer metastasis to the left frontal lobe treated with surgery and stereotactic radiotherapy. This patient subsequently developed an enlarging enhancing component along the posterior treatment margin. T1-weighted postcontrast images with overlay of single-voxel spectroscopy region of interest (box delimited with diamonds) (a) and corresponding MR spectroscopy (point-resolved spectroscopy, echo time 144 ms) (b) demonstrates slightly elevated choline to creatine ratio and reduction in *N*-acetylaspartate, which was suspicious for tumor recurrence.

For differentiating between radiation necrosis and tumor recurrence, most studies show high specificity and suboptimal sensitivity in distinguishing these two entities. After radiation therapy, regions of necrosis within a tumor undergo progressive increases in Lac and Lip peaks, as well as a transient increase in a Cho peak leading to an eventual decrease in Cho with ongoing necrosis. In contrast, tumor recurrence is characterized by a persistently increasing Cho peak (Chernov et al., 2005). Multiple metabolite peak ratios have been proposed in the literature for differentiating between tumor recurrence and radiation-induced necrosis, particularly combinations of the Cho, NAA, and Cr peaks. For example, in a study of 33 intra-axial metastases treated with SRS, or fractionated radiation therapy, the best MRS parameter for predicting tumor progression was the Cho/nCho (Cho in tumor to Cho in normal contralateral brain) ratio, which yielded a sensitivity of 33% and specificity of 100% for Cho/nCho of >1.2 (Huang et al., 2011). The same study also compared MR perfusion which showed superior performance with an area under curve (AUC) of 0.802 versus 0.612 for MRS. Figure 6.2 shows SVS findings suspicious for disease recurrence after SRS, and Figure 6.3 shows SVS findings in a patient with radiation necrosis. Metabolic imaging with MRS has shown some interesting trends, but in everyday practice, the performance of MRS in predicting treatment response and differentiating radiation injury in treated brain metastases has been mediocre at best.

6.3 PERFUSION IMAGING

Perfusion MRI provides information about the tumor microcirculation. The importance of measuring the hemodynamic characteristics in tumors arises from the notion that more aggressive tumors are characterized by endothelial hyperplasia and neovascularization. This technique can be performed using a variety of methods relying on the use of intravascular tracers. In MR, this is most commonly accomplished with intravenous administration of gadolinium-based contrast agents. The two dominant contrast-enhanced MR perfusion techniques are dynamic susceptibility contrast (DSC)-enhanced and dynamic contrast-enhanced (DCE) perfusion imaging. In addition to contrast-enhanced techniques, there are emerging non-contrast-enhanced methodologies for evaluating tissue perfusion. The dominant technique is called arterial spin labeling (ASL).

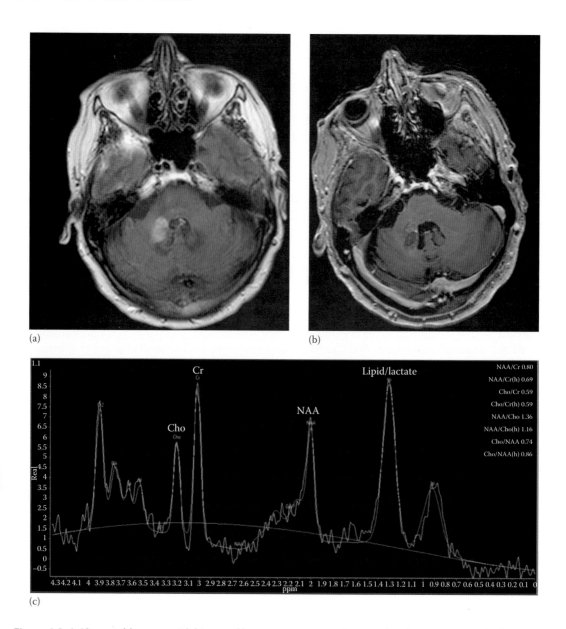

Figure 6.3 A 68-year-old woman with history of lung cancer metastasis treated with stereotactic radiotherapy. Axial (a) FLAIR and (b) T1-weighted postcontrast images demonstrate a minimally enhancing mass in the right cerebellum and brachium pontis. Single-voxel MR spectroscopy analysis (echo time 144 ms) (c) demonstrates a large lipid/lactate peak, normal choline peak and decreased *N*-acetylaspartate peak. While these findings are not specific for any entity in particular, a large lipid/lactate peak can be seen in the setting of radiation-induced necrosis. This mass was found to decrease in size on subsequent follow-up imaging without evidence of recurrence.

MR contrast agents work in a fundamentally different way from CT contrast agents, and understanding this difference is important for appreciating how various contrast-enhanced MR perfusion techniques work and their advantages and limitations. CT contrast agents work by increasing the amount of x-rays absorbed (attenuated) by tissues containing the contrast agent. For CT perfusion, as a bolus of contrast enters the microcirculation of a tissue, the attenuation of x-rays is increased proportionally to the effective concentration of the contrast agent within the tissue of interest. MRI works through the detection of signal generated by magnetization of protons within the tissues being imaged. Gadolinium contrast agents alter the MR signal via two major mechanisms. There is the effect of gadolinium contrast agents on the T1 of water interacting with the contrast agent. For this to occur, the water molecules must directly interact with the gadolinium ion within the chelate (e.g., this occurs over molecular distances). This interaction results in the shortening of T1 and enhancement of MR signal on T1-weighted images, which is not a simple relationship that scales linearly with dose. Competing with the signal enhancement effects caused by T1 shortening are the effects of the gadolinium contrast agent causing shortening of tissue T2 and T2*. These effects occur over larger distances compared to the T1-shortening effects of gadolinium. The T2 and T2* effects are caused by the effects of gadolinium bolus as it passes through the microcirculation. The gadolinium within the vasculature results in a differential magnetic susceptibility of blood and surrounding tissue, leading to micro- and macroscopic magnetic field inhomogeneities causing signal loss via T2 and T2* effects. The effect on T2 or T2* is complex and will depend on a number of parameters including the concentration of gadolinium in the intravascular space, extravascular extravasation of contrast, the configuration and size of the vasculature, diffusion of water through the bulk magnetic susceptibility field of the vasculature, and specific imaging parameters. The different flavors of contrast-enhanced MR perfusion exploit either the T1-shortening effects (e.g., DCE) or T2/T2*-shortening effects (DSC) of gadolinium contrast agents.

6.3.1 DYNAMIC CONTRAST-ENHANCED PERFUSION

In the healthy brain, the blood–brain barrier precludes contrast leakage into the interstitial space. In brain metastases, vessels are abnormal and leaky, allowing for contrast extravasation and enhancement. DCE perfusion uses the T1-shortening signal enhancement effect of extravasated gadolinium on the interstitial tumor water pool. The technique is therefore based on the acquisition of sequential T1-weighted images with temporal resolution of a few seconds. The analysis of DCE perfusion data is performed through pharmacokinetic modeling of tumor gadolinium contrast concentration estimated by measuring dynamic signal intensity changes on T1-weighted images. Other authors use a simpler approach assessing changes in signal intensity directly and obtaining semiquantitative parameters (Kuhl et al., 1999; Padhani et al., 2000; Engelbrecht et al., 2003; Arasu et al., 2011). Commonly used semiquantitative parameters include initial area under the gadolinium concentration curve and wash-in and wash-out rates (Leach et al., 2005; Lankester et al., 2007). These parameters are easier to obtain than model-derived parameters but have lower reproducibility and no direct physiologic meaning. For these reasons, the model-derived approach is preferred (Leach et al., 2005). It is currently unclear which model is most suitable for the various tumor types, tumor sites, and treatment methods. The most widespread model is the modified Tofts model (Tofts et al., 1999). Parameters obtained with this model are described in Table 6.1.

Table 6.1 **Standard kinetic parameters derived from the Tofts model**

PARAMETER	PHYSIOLOGICAL MEANING
V_p (plasma volume fraction)	The blood plasma volume fraction of the whole tissue.
K^{trans} (transfer constant of contrast from the plasma to the tissue extracellular extravascular space)	Related to the balance between capillary permeability and blood flow. When capillary permeability is high, the amount of contrast that leaks out of the vessels depends on the amount of contrast that gets to the capillaries per unit of time. In this case, K^{trans} reflects the blood plasma flow. In cases of low permeability, K^{trans} equals permeability surface area product.
K_{ep} (rate constant)	Rate constant between the interstitial space and blood plasma.
V_e (fraction of extravascular extracellular space)	Fraction on tissue volume corresponding to extracellular extravascular space.

Source: Tofts, P.S. et al., *J. Magn. Reson. Imaging (US)*, 10, 223, 1999.

DCE-MR has been shown to be useful in radiated brain metastases for monitoring treatment response and radiation side effects. The use of pretreatment DCE-MR as a tool for predicting response to radiotherapy has also been widely investigated in other regions outside the brain.

6.3.1.1 Predicting tumor response

Many clinical studies have correlated DCE-MRI-derived parameters with important histopathological features that are related to tumor radiosensitivity (Zahra et al., 2007). An important determinant of tumor response to radiotherapy is tumor oxygenation, with hypoxic tumors being more resistant. A preclinical study in xenografted tumors of eight human melanoma lines showed an inverse relationship between perfusion parameters (K^{trans} and V_e) and tumor hypoxia (Egeland et al., 2012). A similar study in brain gliomas also showed promising results (Jensen et al., 2014).

6.3.1.2 Monitoring response to treatment

DCE-MR has been shown to be useful in monitoring response after radiation in brain metastases. A study in 20 patients with brain metastases treated with whole-brain radiotherapy found that an early change in the specific subvolume of a brain metastatic tumor showing pretreatment high relative cerebral blood volume (CBV) and high K^{trans} is a better predictor for postradiation therapy response than a tumor volumetric change (Farjam et al., 2013). A later study on 26 patients with cerebral metastases treated with SRS showed an overall reduction in K^{trans} values of the cerebral metastases in the early posttreatment period. Furthermore, an increase in K^{trans} values was predictive of tumor progression (Almeida-Freitas et al., 2014). Non-model-derived parameters have also shown value in predicting tumor response after whole-brain radiotherapy in brain metastases (Farjam et al., 2014b).

6.3.1.3 Distinguishing radionecrosis from recurrence

Differentiation between radionecrosis and recurrent metastasis can be very challenging. Radionecrosis consists of fibrinoid necrosis of the blood vessel walls followed by necrosis of the surrounding parenchyma. Capillaries of brain metastases are different from those of the brain as they usually resemble the organ from which the cancer arose and do not have a blood–brain barrier (Long, 1979). On conventional T1-weighted images, both radiation necrosis and recurrent metastases demonstrate enhancement and nonenhancing areas of necrosis. However, several studies have shown a difference in the dynamic nature of the enhancement reflecting different permeability, vessel morphology, and vessel density. Hazle et al. (1997) were able to distinguish between recurrence, radiation necrosis, and a combination of both as they found that radiation necrosis and tumors enhance at different rates.

A study in patients with gliomas and brain metastases describes the use of delayed enhancement maps after 75 minutes of contrast administration to distinguish between active tumor and nontumoral tissues. The former characterized by faster clearance than accumulation, and the later characterized by slower clearance than accumulation (Zach et al., 2012).

6.3.2 DYNAMIC SUSCEPTIBILITY CONTRAST-ENHANCED PERFUSION

DSC relies on the T2/T2*-shortening effect of a bolus of intravascular gadolinium as it passes through the microcirculation of a tumor. Usually a gradient echo T2-weighted echo planar sequence is used, which allows for whole-brain coverage with temporal resolution less than 2 seconds (Petrella and Provenzale, 2000). The drop in MR signal, which occurs as the bolus passes through the tumor microcirculation, is used to calculate the CBV, which reflects microvascular density, and has proven to be very useful for the differentiation of lesions on the basis of their microvasculature. Figure 6.4 illustrates CBV map for a typical clinical case. Another parameter derived from the T2 signal–time curves is the percentage of signal recovery (Barajas et al., 2009). However, this parameter has been shown to be highly dependent on acquisition parameters with very low reproducibility between different centers (Boxerman et al., 2013). There are some limitations of DSC-MR, the main one being the presence of susceptibility artifact in the ROI. This is most commonly due to hemorrhage, but also caused by air and bone in the inferior temporal and frontal regions. This is a very common source of false negatives when evaluating CBV maps from DSC-MR perfusion. The presence of fast contrast leakage during the first pass of the gadolinium bolus is another source of artifact as extravascular gadolinium tends to increase the signal of the tissue, competing

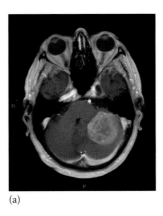
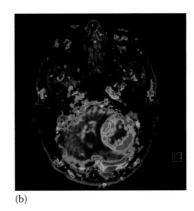

(a)   (b)

Figure 6.4 A 72-year-old man with esophageal cancer and a 2-week history of headaches, dizziness, nausea, and vomiting. Axial (a) T1-weighted postgadolinium images demonstrate a large enhancing mass in the left cerebellar hemisphere. Axial (b) dynamic susceptibility contrast perfusion-derived CBV map overlaid on T1-weighted postgadolinium image shows a ring of high CBV within the lesion. Resection of the lesion revealed metastatic adenocarcinoma.

with the T2 shortening caused by intravascular gadolinium. This problem also causes artificially low CBV values. The effect of contrast extravasation can be solved by contrast preloading or by postprocessing correcting algorithms.

With regard to radiated brain metastases, DSC-MR perfusion has proven to be useful in the early posttreatment period predicting patient outcomes as well as in the long-term follow-up distinguishing between recurrent metastases and radiation necrosis.

6.3.2.1 Early posttreatment phase

Essig studied 18 patients with cerebral metastases by using DSC-MR to assess if preradiation and early time point CBV measurements of metastases could predict outcome in patients treated with SRS (Essig et al., 2003). Measurements of regional CBV changes in metastases and normal brain after treatment were also performed. The authors found that pretreatment CBV was not able to predict outcome. Measurements of CBV at 6 weeks posttreatment were a more sensitive and specific biomarker of treatment outcome. In particular, a reduction in regional CBV was predictive of treatment outcome with sensitivity in excess of 90% compared to sensitivity of 64% for a change in tumor volume alone. CBV values of normal brain were in line with expected physiologic range and were unchanged with therapy (Essig et al., 2003). Similar results were obtained by Weber with a slightly larger sample (25 patients) (Weber et al., 2004).

6.3.2.2 Late posttreatment phase

Hoefnagels investigated the ability of perfusion MRI to differentiate between tumor recurrence and radiation necrosis in patients showing radiological progression of disease. They concluded that when lesions display a CBV greater than 1.85 times the normal grey matter, necrosis can be excluded (specificity of 100% and sensitivity of 70%) (Hoefnagels et al., 2009). A similar study in patients treated with fractionated radiation therapy revealed 56% sensitivity and 100% specificity for CBV as a predictor of tumor progression (Huang et al., 2011). A sensitivity of 91% and a specificity of 72% for a relative CBV to white matter threshold of 1.54 for identifying recurrent tumor versus radionecrosis has been described (Barajas et al., 2009). In another study, a relative CBV to white matter greater than 2.1 provided 100% sensitivity and 95.2% specificity (Mitsuya et al., 2010).

In contrast to the above studies, which demonstrated that an increased CBV after SRS serves as a marker for tumor recurrence, a recent study found an association between early 1-month posttreatment CBV reductions and tumor progression (Jakubovic et al., 2014). The authors postulated that these apparently contradicting findings may be explained by the time-dependent evolution of vascular changes after radiotherapy. In fact, a study exploring postradiation vascular changes in a tumor model demonstrated a transient switch between two different types of angiogenesis—sprouting angiogenesis to a nonsprouting (intussusceptive) angiogenesis (Hlushchuk et al., 2008, 2011). This "angiogenic switch" was hypothesized to correspond with early reductions in CBV. As time elapses, tumor recurrence is accompanied by a switch back to sprouting angiogenesis with increasing tumor vasculature and increasing CBV, whereas a favorable treatment response is associated with further CBV reduction.

6.3.3 ARTERIAL SPIN LABELING

ASL is an emerging clinical tool that allows the characterization of blood flow without the administration of exogenous contrast. In ASL, intravascular water is magnetically labeled using RF pulses. By comparing the MR signal of tissue imaged with and without magnetic intravascular water labeling, the blood flow to the tissue can be quantified. There are a number of specific methodologies for accomplishing this including continuous ASL (CASL), pulsed ASL (PASL), and a hybrid of the two methods termed pseudocontinuous ASL (pCASL). In CASL of the brain, a slice positioned at the level of the extracranial internal carotid arteries is excited using an inversion RF pulse, and a separate acquisition of the brain is acquired. The inverted magnetization from in-flowing blood from the neck will reduce the signal in the brain image. By comparing the signal intensity of the brain image to an acquisition without intravascular water labeling, the amount of signal resulting from blood flow can be quantified. Generally, CASL has had limited clinical utility because of the significant hardware demands required to achieve the proper conditions for blood water labeling and other effects (e.g., magnetization transfer), which complicate the quantification of blood flow. One type of PASL methodology called flow-sensitive alternating inversion recovery is based on an acquisition scheme, which employs two acquisitions of the target tissue: one acquisition with slice-selective inversion RF pulse and one with a nonselective inversion pulse. By comparing the signal from the tissue using these two acquisitions, the amount of signal resulting from flowing blood in the slice-selective acquisition can be calculated. The advantage of PASL is that the labeling of blood water is more easily implemented; however, the signal to noise of the methodology is inferior to CASL. pCASL overcomes many of the shortcomings of CASL and PASL and is now a commonly used ASL methodology.

ASL has been used in the study of brain metastases treated with SRS. In a small cohort of 25 patients (28 total brain metastases), a decrease in relative CBF for metastases at 6 weeks posttherapy measured using a pCASL methodology was predictive of treatment response (Weber et al., 2004).

As ASL becomes more widely available on clinical systems, the application of this methodology in neuro-oncology will increase. A number of challenges remain, however, including the accuracy of CBF measurements made with ASL in tissues with longer blood transit times, for example, where the effects of T1 relaxation on the blood water pool become significant. Furthermore, the reproducibility of the methodology across vendors as well as repeatability within subjects requires further study. Nonetheless, the ability to evaluate tumor physiology without the administration of IV contrast will certainly be a tremendous advantage, for example, in evaluating patients with brain metastases particularly in studies that require multiple examinations over a short time interval where the administration of repeated gadolinium contrast is undesirable and impractical.

6.4 DIFFUSION-WEIGHTED IMAGING

DWI is a promising technique for evaluating treatment response. Using this methodology, the Brownian motion of free water molecules within a tissue can be estimated, and a map of the apparent diffusion coefficient (ADC) of the tissue can be produced. By measuring the microscopic motion of water within a tissue, the structure of the tissue at the cellular and subcellular level can be inferred. For example, water that is primarily compartmentalized within the cells of a tissue will have movement through space that is more restricted compared to water in the extracellular compartment. Thus, areas of high cellularity will demonstrate more restricted water motion and lower ADC values than areas of low cellularity or necrosis.

However, there are limitations to this simplistic interpretation of ADC values. Immediately after surgery or treatment, a low ADC within a tumor could indicate cytotoxic injury, which is characterized by cell swelling and water shifting from the extracellular compartment that allows for free movement to the motion-restricted intracellular compartment. Furthermore, a low ADC does not always imply cellularity or cellular water, as it can also be seen in the context of interactions between water and proteins or other substances that hinder its motion. For example, blood clot, highly proteinaceous material, or pus can restrict water motion and result in low ADC.

6.4.1 CLINICAL APPLICATION

A number of studies have explored how ADC changes after radiation therapy and how these changes correlate with response. In comparing results from the literature, there are important differences between studies that have to be considered. First, the timing of when diffusion measurements are made is an important consideration. Generally, most studies using DWI to predict treatment response have DWI measurements made days to a few weeks after treatment. Furthermore, there are a number of other variables that may limit comparison between studies and a universal interpretation of results, including what imaging parameters were used, how quantification was performed, whether it was done using an ROI tool or quantified on a pixel-by-pixel (voxel-by-voxel) basis, as well as how nonsolid, necrotic, or cystic components of the tumor were addressed.

One of the first studies examining the utility of DWI was performed on a small cohort of eight patients (six with brain metastases) (Mardor et al., 2003). These tumors were mainly treated with single-fraction SRS (16 Gy), with MR performed at 0.5 T. Measurements were made at baseline, 1, 7, and 14 days posttreatment. ROI analysis was performed on the solid portion of the metastasis (no necrosis was seen in any of the metastases studied). Outcome was tumor volume measured as volume of enhancing tumor approximately 48 days after treatment. The authors found a statistically significant moderate correlation between tumor volume change and change in ADC calculated from baseline and 7 days after therapy. Specifically, tumors demonstrating greater volume reduction and positive response to treatment showed greater increases in ADC at 1 week posttreatment compared with tumors that did not respond.

A trend of increasing ADC in treated metastases has been confirmed in other larger studies. For example, in a prospective observational study on 86 patients (including 38 brain metastases) treated with SRS and MRI performed at 1.5 T, ROI analysis performed on solid portion demonstrated increases in ADC measured at 1-month posttreatment (Huang et al., 2010). An analysis of the relationship between ADC changes and treatment response was not carried out. A larger study performed on 107 patients with brain metastases treated with SRS demonstrated a statistically significant decreases in ADC for progressors compared to patients with stable or responsive disease (Lee et al., 2014).

A major limitation of ROI-based analysis is the question of how best to deal with metastases that have a heterogeneous imaging appearance pretreatment and/or develop imaging heterogeneity after treatment. Imaging heterogeneity is presumably related to underlying histological heterogeneity in a tumor. Decisions on where to place the ROI on the pretreatment scan or posttreatment scans can lead to significant measurement differences. Furthermore, a ROI-based analysis is statistically weakened when multiple voxels within the ROI are grouped together during the analysis. A number of voxel-based analyses have emerged, which can potentially overcome these problems and provide a very powerful picture of imaging changes that occur after radiation therapy. One such technique is functional diffusion mapping in which a voxel-by-voxel calculation of ADC changes between pre- and posttreatment scans is performed by spatially coregistering the two imaging datasets (Moffat et al., 2005). The power in this type of analysis is that it does not make an assumption that all regions of the tumor are initially the same or respond the same after treatment. It also effectively deals with tumor heterogeneity issues by analyzing and treating each voxel within the analysis volume independently. By examining the distribution of ADC changes across the entire tumor, functional diffusion mapping has been shown to be a promising biomarker for assessing treatment response in primary brain tumors. One of the limitations of this technique is related to problems associated with coregistering imaging data after significant changes in tumor volume or tissue distortion. Other voxel-based quantitative diffusion techniques without a reliance on image coregistration—such as the diffusion abnormality index—have recently emerged and have shown utility in predicting treatment response in brain metastases (Farjam et al., 2014a).

6.5 INTRAVOXEL INCOHERENT MOTION

Intravoxel incoherent motion (IVIM) is a diffusion MR technique that allows approximation of tumor blood flow and blood volume without the administration of IV contrast. Essentially, the motion of water within the microcirculation of a tumor can be modeled as a random (incoherent) process, which is similar to molecular water diffusion but occurs at a macroscopic scale (Le Bihan et al., 1986, 1988). By performing a diffusion MR experiment using measurements with several weak magnetic gradients (low b values) and stronger gradients (high b values), the contribution of water motion from blood flow and molecular water diffusion can be separated. Instead of reporting an "apparent" diffusion coefficient which has contributions from water motion at both scales, the ADC can be unpacked into a value which approaches the true molecular water diffusion (D), a parameter called the perfusion fraction (f), which is the fraction of water within the intravascular compartment (similar to blood volume), and a parameter called the pseudodiffusion coefficient (D^*), which has been compared to blood flow (Le Bihan and Turner, 1992). Figure 6.5 illustrates a treated metastasis analyzed using IVIM methodology.

While the initial description of the IVIM methodology is quite old, only recently has the methodology found increasing use in the CNS and in studying brain tumors (Federau et al., 2014; Kim et al., 2014b). A recent study examining IVIM in treated brain metastases showed improved accuracy of IVIM combined

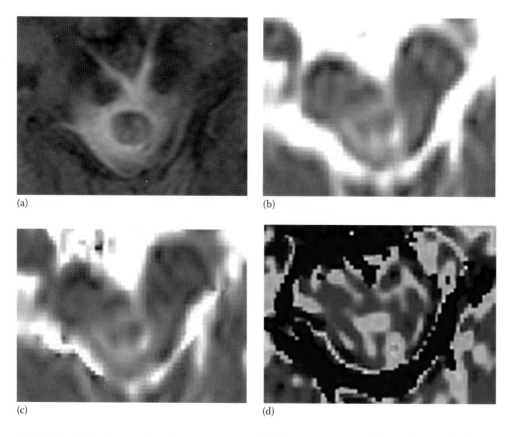

(a)

(b)

(c)

(d)

Figure 6.5 Patient with stereotactic radiosurgery–treated brain stem metastasis. Cropped magnified images of the brain stem showing round T2 FLAIR hypointense midbrain lesion with surrounding vasogenic edema (a). Diffusion MRI conventional apparent diffusion coefficient (ADC) map (b) shows tumor to have diffusion characteristics similar to the brain but less water diffusion than surrounding edema. The diffusion coefficient map (c) and perfusion fraction/f map (d) calculated from intravoxel incoherent motion model. The diffusion coefficient map is similar to ADC as expected. The f map shows elevated perfusion fraction in the treated metastasis (cyan) compared to surrounding edema. The cerebral aqueduct is displaced posteriorly and laterally to the left (white arrow) and shows high CSF flow (red), which is captured by this technique.

with DSC perfusion over DWI+DSC, IVIM, and DSC alone in distinguishing between treatment response and tumor recurrence for brain metastases (Kim et al., 2014a).

6.6 BLOOD OXYGEN LEVEL-DEPENDENT MRI

Hypoxia is a key driver of an array of malignant processes including angiogenesis, alterations in energy metabolism, and cancer cell metastasis and invasion. Tumor hypoxia is also correlated with a higher risk of metastatic disease, radioresistance, and mortality (Harada, 2011). Thus, an ability to noninvasively quantify tumor hypoxia could provide valuable insight into the significance of this factor and may also be important in tailoring therapy, particularly to tumors demonstrating higher levels of hypoxia. While there are PET imaging techniques for evaluating tissue hypoxia such as with direct oxygen extraction measurements with $15O_2$ or with hypoxia markers such as 18F-FAZA, the availability of these radiotracers and inherent technical limitations of PET (e.g., low spatial resolution) limit their widespread use for evaluating metastatic disease. MRI techniques based on the principle of blood oxygen level-dependent (BOLD) imaging may offer a methodology for evaluating tumor hypoxia using technology that is widely accessible.

The BOLD effect is most commonly used in functional MRI (fMRI) to map areas of brain activation. It is predicated on the concept of neurovascular coupling in which blood flow to regions of activated brain are transiently increased resulting in a local decrease in deoxyhemoglobin within the microcirculation. Whereas oxyhemoglobin is diamagnetic with a negligible effect on the local magnetic field, deoxyhemoglobin is paramagnetic, and its presence within blood vessels results in magnetic susceptibility effects on the intravascular compartment as well as local tissues in the extravascular compartment. These effects on the MR signal are best detected using T2*-weighted sequences. In the case of brain activation, the lower deoxyhemoglobin results in a very small increase in MR signal that can be characterized using statistical methods. By measuring these small signal changes and correlating them with a specific task (e.g., finger tapping), areas of the brain activation can be mapped.

Magnetic susceptibility effects in any given tissue are complex and are affected by a number of factors that can confound BOLD measurements. Recently, there has been progress in modeling and estimating tissue oxygenation based on measurements of tissue T2. The work by He et al. has shown a relationship between tissue oxygenation, tissue transverse relaxation, hematocrit, and blood volume fraction. Using this methodology, tissue oxygenation in normal brain tissue has been measured and validated (He et al., 2008; Christen et al., 2011). The application of this theory in neoplasia has been accomplished in a rat glioma model but has yet to be reproduced by others and also applied in brain metastases (Christen et al., 2012).

Overall, quantitative BOLD techniques are promising, but much work has yet to be done to validate the methodology and the assumptions underlying the theory, which may be valid in normal tissue but violated in disease states such as tumors. Additionally, the methodology in its present state is complex and time consuming, which will limit its implementation as a tool in clinical practice.

6.7 FUTURE APPLICATIONS

6.7.1 CHEMICAL EXCHANGE SATURATION TRANSFER

Chemical exchange saturation transfer (CEST) is a technique that allows the detection of molecular targets using MR (Ward et al., 2000). The technique is based on selective excitation of a mobile proton pool (typically hydroxyl, amine, or amide protons) on the molecular target, which chemically exchanges with a larger free water pool. The end result is saturation or reduction in signal intensity of the free water pool MR signal. Because of the exchange of protons with the larger free water pool, an amplification of the signal response occurs, and the sensitivity of CEST is therefore far greater than what would be possible via the direct detection of proton species on the molecular target. By comparing preexcitation and postexcitation images, the distribution of the molecular target can be mapped. Figure 6.6 shows a representative CEST map of a brain metastasis, which was produced by computing the magnetic transfer ratio at the frequency of the amine CEST peak at 2 ppm.

The CEST contrast mechanism can be used to detect intrinsic molecular species and chemical environments such as amino acids, peptides, proteins, and tissue pH (Ward and Balaban, 2000;

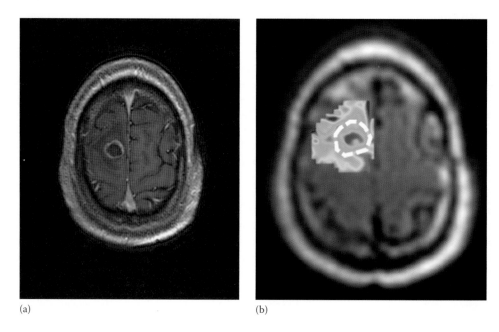

(a)   (b)

Figure 6.6 Patient with brain metastases. Axial (a) T1-weighted postgadolinium images demonstrate a rim-enhancing mass in the right frontal lobe. Axial (b) chemical exchange saturation transfer (CEST)-weighted images superimposed on FLAIR images shows a colored region (with red representing the highest and blue representing the lowest values) indicating the extent of the lesion (including the tumor and edema), and the circled region indicates the boundary of enhancement. The CEST-weighted map was produced by computing the magnetization transfer ratio at the frequency of the amine CEST peak at 2 ppm.

Jin et al., 2012). The technique has, for example, been used to evaluate the distribution of peptides in primary brain tumors, which correlate with tumor grade (Wen et al., 2010). CEST can also be used to detect exogenously administered molecular species, which act as contrast agents or molecular tracers. This can be done with naturally occurring molecules or molecular constructs which incorporate paramagnetic (Aime et al., 2002), diamagnetic (McMahon et al., 2008), or hyperpolarized species (Schroder et al., 2006) (PARACEST, DIACEST, HYPERCEST, respectively). Recently, exogenously administered glucose and glucose analogs (e.g., 19F-fluorodeoxyglucose [19FDG] and 2-deoxyglucose) have been mapped using CEST (Rivlin et al., 2013; Walker-Samuel et al., 2013). These results indicate the potential of MR for metabolic imaging with capability similar to that of 18FDG-PET.

6.7.2 HYPERPOLARIZED ^{13}C

Hyperpolarized ^{13}C MRS is a functional imaging technique that uses exogenous ^{13}C-labeled molecules, such as pyruvate or fumarate, and MRS methodology to investigate metabolic processes in normal and pathologic tissues. In MRI, SNR depends on a number of factors, which include the polarization of the nuclei generating the signal. Polarization refers to the difference in the number of nuclei with spins that are parallel with the magnetic field and the number of nuclei with spins that are antiparallel. Whereas with nonhyperpolarized ^{13}C nuclei in normal thermal equilibrium, there is only a slightly higher proportion of nuclei with spins aligned in parallel compared to antiparallel, the process of hyperpolarizing ^{13}C nuclei—a process that leads to the number of nuclear spins aligned in parallel being several orders of magnitude higher than antiparallel nuclear spins—results in a significantly higher signal. In fact, the recent development of a novel process using dynamic nuclear polarization for polarizing nuclear spins in solution has allowed for a 10,000-fold signal increase over other conventional MRS methods (Ardenkjaer-Larsen et al., 2003; Golman et al., 2003).

To date, hyperpolarized [1–^{13}C] pyruvate is the most widely studied and used hyperpolarized substrate. Following intravenous injection and uptake of [1–^{13}C] pyruvate by cells, the hyperpolarized [1–^{13}C] pyruvate molecules may undergo metabolism in one of three pathways: conversion to Lac catalyzed by

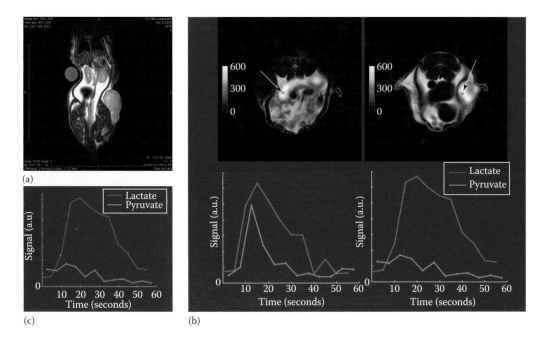

Figure 6.7 A rat model with a MDA-MB-231 breast cancer xenograft tumor treated with 16 Gy radiation. (a) A coronal T2-weighted image demonstrates the tumor in the left flank. (b) An axial image with lactate signal overlayed on a T2-weighted anatomic image following intravenous injection with hyperpolarized [1–^{13}C] pyruvate shows increased signal within the tumor. (c) Time-resolved metabolic data demonstrate changes in the lactate and pyruvate signal within the voxel indicated in (b) following the injection.

lactate dehydrogenase (LDH), transamination to form alanine catalyzed by alanine aminotransferase, or decarboxylation to form carbon dioxide catalyzed by pyruvate dehydrogenase (Kurhanewicz et al., 2011). The kinetics of these metabolic reactions and relative concentrations of the metabolites—which differ depending on the state of the tissue—can be measured and monitored in real time, thus providing important metabolic information about the tissue of interest. In tumors, the metabolism of hyperpolarized [1–^{13}C] pyruvate molecules within cancer cells depends on a number of factors, including how much of the agent is delivered to the tumor, how much is taken up by the cancer cells, as well as the LDH concentration and activity (Brindle, 2012). Figure 6.7 shows Lac signal within a breast cancer xenograft in a rat model following intravenous injection with hyperpolarized [1–^{13}C] pyruvate. Time-resolved metabolic data demonstrate changes in the Lac and pyruvate signal following the injection.

The use of hyperpolarized [1–^{13}C] pyruvate MRS has been shown in multiple preclinical studies to have a potential for predicting treatment outcomes and assessing early response to therapy (Brindle et al., 2011; Kurhanewicz et al., 2011; Yen et al., 2011; Brindle, 2012). High reproducibility in differentiating malignant glioma tumors from normal brain tissues in a mouse and nonhuman primate tumor model by measuring pyruvate and Lac signals using hyperpolarized ^{13}C MRS imaging has been demonstrated (Park et al., 2010, 2014). Changes in pyruvate metabolism following radiotherapy in rat glioma models using a similar technique have also been shown (Day et al., 2011). Hyperpolarized [1–^{13}C] pyruvate has also been shown to have a role in detecting response to the PI3K inhibitor in both glioblastoma and breast cancer mouse models. In these models, reductions in the hyperpolarized Lac signal correlated with reductions in LDH activity due to PI3K pathway inhibition (Ward et al., 2010; Chaumeil et al., 2012).

6.8 MULTIPARAMETRIC APPROACHES

An emerging technique in oncology imaging is the application of multiparametric approaches for predicting treatment response or distinguishing between tumor recurrence and pseudoprogression. A summary of the MR techniques discussed in this chapter with respect to differentiating tumor recurrence

Table 6.2 Summary of magnetic resonance techniques for differentiating tumor recurrence from radiation necrosis

MR TECHNIQUE	TUMOR RECURRENCE	RADIATION NECROSIS
Conventional MRI	Persistent increase in the size of the enhancing lesion and degree of vasogenic edema	Transient increase in lesion size after SRS with eventual decrease in the size of the enhancing lesion and degree of vasogenic edema
	May demonstrate nonenhancing areas of necrosis	Peripheral rim enhancement with central decrease enhancement
MR spectroscopy	Persistently increasing Cho peak on serial imaging	Transient increased Cho peak with eventual decreasing Cho peak on serial imaging
	Nonspecific decreased NAA peak with or without increased lipid/lactate peaks	Decreased NAA peak and increased lipid/lactate peaks
DCE-MRI	Increased K^{trans} compared to necrosis	Reduced K^{trans} compared to recurrence
DSC-MRI	Time-dependent changes, but overall higher CBV compared necrosis	Time-dependent changes, but overall lower CBV compared recurrence
ASL	Increase in relative CBF	Decrease in relative CBF
DWI	Decreased ADC values compared to radiation necrosis	Increased ADC values following SRS compared to recurrence
IVIM	Increased f value and decreased D value	Decreased f value and increased D value
BOLD MRI, CEST, and hyperpolarized ^{13}C MRS	More research is required	More research is required

from radiation necrosis is provided in Table 6.2. While the preceding sections illustrate the progress in advanced MR techniques to answer these questions, it is unlikely that any one single imaging technique will provide a definitive answer for these complex clinical questions. Multiparametric approaches, which use multiple imaging measures to predict specific outcomes, may be one possible solution. In this methodology, a specific clinical outcome is correlated with a set of imaging measurements in a training patient dataset. The combination of imaging parameters determined from the training set are then used to predict outcomes prospectively in a test patient dataset. For example, a multiparametric analysis based on measurements of ADC, normalized CBV, and initial AUC from tumor perfusion was used to predict early tumor progression from pseudoprogression in glioblastoma (Park et al., 2015). The study confirmed a superior diagnostic accuracy of the multiparametric approach compared to the performance of the individual imaging parameters alone. Such multiparametric approaches have also found utility in predicting glioma grade for instance. The application of multiparametric techniques in treated brain metastases is an emerging area of research.

6.9 CONCLUSION

As cancer treatment strategies improve and patients live longer with systemic disease, the burden of brain metastases will continue to grow, as will the need for accurate and precise imaging tools for evaluating this disease. MRI has emerged as a powerful tool not only for diagnosing and detecting brain metastases but also for predicting patient outcomes and evaluating early response to therapy. The techniques described earlier provide only a brief introduction into the vast possibilities offered by MRI for generating information regarding tumor biomarkers that can be used to guide and direct therapy.

ACKNOWLEDGMENTS

The authors would like to thank Dr. Greg Stanisz (Professor at the University of Toronto and Senior Scientist at the Sunnybrook Research Institute) for contributing the CEST images from his work. The authors would also like to express gratitude to Dr. Charles Cunningham (Associated Professor at the University of Toronto and Senior Scientist at the Sunnybrook Research Institute) for contributing the hyperpolarized ^{13}C MRS images from his work.

REFERENCES

Aime S, Delli Castelli D, Fedeli F, Terreno E (2002) A paramagnetic MRI-CEST agent responsive to lactate concentration. *J Am Chem Soc* 124:9364–9365.

Almeida-Freitas DB, Pinho MC, Otaduy MC, Braga HF, Meira-Freitas D, da Costa Leite C (2014) Assessment of irradiated brain metastases using dynamic contrast-enhanced magnetic resonance imaging. *Neuroradiology* 56:437–443.

Arasu VA, Chen RC, Newitt DN, Chang CB, Tso H, Hylton NM, Joe BN (2011) Can signal enhancement ratio (SER) reduce the number of recommended biopsies without affecting cancer yield in occult MRI-detected lesions? *Acad Radiol* 18:716–721.

Ardenkjaer-Larsen JH, Fridlund B, Gram A, Hansson G, Hansson L, Lerche MH, Servin R, Thaning M, Golman K (2003) Increase in signal-to-noise ratio of >10,000 times in liquid-state NMR. *Proc Natl Acad Sci USA* 100:10158–10163.

Barajas RF, Chang JS, Sneed PK, Segal MR, McDermott MW, Cha S (2009) Distinguishing recurrent intra-axial metastatic tumor from radiation necrosis following gamma knife radiosurgery using dynamic susceptibility-weighted contrast-enhanced perfusion MR imaging. *Am J Neuroradiol* 30:367–372.

Boxerman JL, Paulson ES, Prah MA, Schmainda KM (2013) The effect of pulse sequence parameters and contrast agent dose on percentage signal recovery in DSC-MRI: Implications for clinical applications. *Am J Neuroradiol* 34:1364–1369.

Brindle K (2012) Watching tumours gasp and die with MRI: The promise of hyperpolarised 13C MR spectroscopic imaging. *Br J Radiol* 85:697–708.

Brindle KM, Bohndiek SE, Gallagher FA, Kettunen MI (2011) Tumor imaging using hyperpolarized 13C magnetic resonance spectroscopy. *Magn Reson Med* 66:505–519.

Chaumeil MM, Ozawa T, Park I, Scott K, James CD, Nelson SJ, Ronen SM (2012) Hyperpolarized 13C MR spectroscopic imaging can be used to monitor everolimus treatment in vivo in an orthotopic rodent model of glioblastoma. *Neuroimage* 59:193–201.

Chernov M, Hayashi M, Izawa M, Nakaya K, Ono Y, Usukura M, Yoshida S et al. (2007) Metabolic characteristics of intracranial metastases, detected by single-voxel proton magnetic resonance spectroscopy, are seemingly not predictive for tumor response to gamma knife radiosurgery. *Minim Invasive Neurosurg* 50:233–238.

Chernov M, Hayashi M, Izawa M, Ochiai T, Usukura M, Abe K, Ono Y et al. (2005) Differentiation of the radiation-induced necrosis and tumor recurrence after gamma knife radiosurgery for brain metastases: Importance of multi-voxel proton MRS. *Minim Invasive Neurosurg* 48:228–234.

Chernov MF, Hayashi M, Izawa M, Abe K, Usukura M, Ono Y, Kubo O, Hori T (2004) Early metabolic changes in metastatic brain tumors after gamma knife radiosurgery: 1H-MRS study. *Brain Tumor Pathol* 21:63–67.

Chernov MF, Hayashi M, Izawa M, Nakaya K, Tamura N, Ono Y, Abe K et al. (2009) Dynamics of metabolic changes in intracranial metastases and distant normal-appearing brain tissue after stereotactic radiosurgery: A serial proton magnetic resonance spectroscopy study. *Neuroradiol J* 22:58–71.

Chernov MF, Ono Y, Kubo O, Hori T (2006) Comparison of 1H-MRS-detected metabolic characteristics in single metastatic brain tumors of different origin. *Brain Tumor Pathol* 23:35–40.

Christen T, Lemasson B, Pannetier N, Farion R, Remy C, Zaharchuk G, Barbier EL (2012) Is T2* enough to assess oxygenation? Quantitative blood oxygen level-dependent analysis in brain tumor. *Radiology* 262:495–502.

Christen T, Lemasson B, Pannetier N, Farion R, Segebarth C, Remy C, Barbier EL (2011) Evaluation of a quantitative blood oxygenation level-dependent (qBOLD) approach to map local blood oxygen saturation. *NMR Biomed* 24:393–403.

Day SE, Kettunen MI, Cherukuri MK, Mitchell JB, Lizak MJ, Morris HD, Matsumoto S, Koretsky AP, Brindle KM (2011) Detecting response of rat C6 glioma tumors to radiotherapy using hyperpolarized [1–^{13}C] pyruvate and 13C magnetic resonance spectroscopic imaging. *Magn Reson Med* 65:557–563.

Drost DJ, Riddle WR, Clarke GD, AAPM MR Task Group #9 (2002) Proton magnetic resonance spectroscopy in the brain: Report of AAPM MR task group #9. *Med Phys* 29:2177–2197.

Egeland TA, Gulliksrud K, Gaustad JV, Mathiesen B, Rofstad EK (2012) Dynamic contrast-enhanced-MRI of tumor hypoxia. *Magn Reson Med* 67:519–530.

Engelbrecht MR, Huisman HJ, Laheij RJ, Jager GJ, van Leenders GJ, Hulsbergen-Van De Kaa CA, de la Rosette JJ, Blickman JG, Barentsz JO (2003) Discrimination of prostate cancer from normal peripheral zone and central gland tissue by using dynamic contrast-enhanced MR imaging. *Radiology* 229:248–254.

Essig M, Waschkies M, Wenz F, Debus J, Hentrich HR, Knopp MV (2003) Assessment of brain metastases with dynamic susceptibility-weighted contrast-enhanced MR imaging: Initial results. *Radiology* 228:193–199.

Farjam R, Tsien CI, Feng FY, Gomez-Hassan D, Hayman JA, Lawrence TS, Cao Y (2013) Physiological imaging-defined, response-driven subvolumes of a tumor. *Int J Radiat Oncol Biol Phys* 85:1383–1390.

Farjam R, Tsien CI, Feng FY, Gomez-Hassan D, Hayman JA, Lawrence TS, Cao Y (2014a) Investigation of the diffusion abnormality index as a new imaging biomarker for early assessment of brain tumor response to radiation therapy. *Neuro Oncol* 16:131–139.

Farjam R, Tsien CI, Lawrence TS, Cao Y (2014b) DCE-MRI defined subvolumes of a brain metastatic lesion by principle component analysis and fuzzy-c-means clustering for response assessment of radiation therapy. *Med Phys* 41:011708.

Federau C, Meuli R, O'Brien K, Maeder P, Hagmann P (2014) Perfusion measurement in brain gliomas with intravoxel incoherent motion MRI. *Am J Neuroradiol* 35:256–262.

Golman K, Olsson LE, Axelsson O, Mansson S, Karlsson M, Petersson JS (2003) Molecular imaging using hyperpolarized 13C. *Br J Radiol* 76 Spec No 2:S118–S127.

Harada H (2011) How can we overcome tumor hypoxia in radiation therapy? *J Radiat Res* 52:545–556.

Hazle JD, Jackson EF, Schomer DF, Leeds NE (1997) Dynamic imaging of intracranial lesions using fast spin-echo imaging: Differentiation of brain tumors and treatment effects. *J Magn Reson Imaging* 7:1084–1093.

He X, Zhu M, Yablonskiy DA (2008) Validation of oxygen extraction fraction measurement by qBOLD technique. *Magn Reson Med* 60:882–888.

Hlushchuk R, Makanya AN, Djonov V (2011) Escape mechanisms after antiangiogenic treatment, or why are the tumors growing again? *Int J Dev Biol* 55:563–567.

Hlushchuk R, Riesterer O, Baum O, Wood J, Gruber G, Pruschy M, Djonov V (2008) Tumor recovery by angiogenic switch from sprouting to intussusceptive angiogenesis after treatment with PTK787/ZK222584 or ionizing radiation. *Am J Pathol* 173:1173–1185.

Hoefnagels FW, Lagerwaard FJ, Sanchez E, Haasbeek CJ, Knol DL, Slotman BJ, Vandertop WP (2009) Radiological progression of cerebral metastases after radiosurgery: Assessment of perfusion MRI for differentiating between necrosis and recurrence. *J Neurol* 256:878–887.

Huang CF, Chiou SY, Wu MF, Tu HT, Liu WS, Chuang JC (2010) Apparent diffusion coefficients for evaluation of the response of brain tumors treated by gamma knife surgery. *J Neurosurg* 113(Suppl):97–104.

Huang J, Wang AM, Shetty A, Maitz AH, Yan D, Doyle D, Richey K et al. (2011) Differentiation between intra-axial metastatic tumor progression and radiation injury following fractionated radiation therapy or stereotactic radiosurgery using MR spectroscopy, perfusion MR imaging or volume progression modeling. *Magn Reson Imaging* 29:993–1001.

Jakubovic R, Sahgal A, Soliman H, Milwid R, Zhang L, Eilaghi A, Aviv RI (2014) Magnetic resonance imaging-based tumour perfusion parameters are biomarkers predicting response after radiation to brain metastases. *Clin Oncol (R Coll Radiol)* 26:704–712.

Jensen RL, Mumert ML, Gillespie DL, Kinney AY, Schabel MC, Salzman KL (2014) Preoperative dynamic contrast-enhanced MRI correlates with molecular markers of hypoxia and vascularity in specific areas of intratumoral microenvironment and is predictive of patient outcome. *Neuro Oncol* 16:280–291.

Jin T, Wang P, Zong X, Kim SG (2012) Magnetic resonance imaging of the amine-proton EXchange (APEX) dependent contrast. *Neuroimage* 59:1218–1227.

Kim DY, Kim HS, Goh MJ, Choi CG, Kim SJ (2014a) Utility of intravoxel incoherent motion MR imaging for distinguishing recurrent metastatic tumor from treatment effect following gamma knife radiosurgery: Initial experience. *Am J Neuroradiol* 35:2082–2090.

Kim HS, Suh CH, Kim N, Choi CG, Kim SJ (2014b) Histogram analysis of intravoxel incoherent motion for differentiating recurrent tumor from treatment effect in patients with glioblastoma: Initial clinical experience. *Am J Neuroradiol* 35:490–497.

Kuhl CK, Mielcareck P, Klaschik S, Leutner C, Wardelmann E, Gieseke J, Schild HH (1999) Dynamic breast MR imaging: Are signal intensity time course data useful for differential diagnosis of enhancing lesions? *Radiology* 211:101–110.

Kurhanewicz J, Vigneron DB, Brindle K, Chekmenev EY, Comment A, Cunningham CH, Deberardinis RJ et al. (2011) Analysis of cancer metabolism by imaging hyperpolarized nuclei: Prospects for translation to clinical research. *Neoplasia* 13:81–97.

Lankester KJ, Taylor JN, Stirling JJ, Boxall J, d'Arcy JA, Collins DJ, Walker-Samuel S, Leach MO, Rustin GJ, Padhani AR (2007) Dynamic MRI for imaging tumor microvasculature: Comparison of susceptibility and relaxivity techniques in pelvic tumors. *J Magn Reson Imaging* 25:796–805.

Le Bihan D, Breton E, Lallemand D, Aubin ML, Vignaud J, Laval-Jeantet M (1988) Separation of diffusion and perfusion in intravoxel incoherent motion MR imaging. *Radiology* 168:497–505.

Le Bihan D, Breton E, Lallemand D, Grenier P, Cabanis E, Laval-Jeantet M (1986) MR imaging of intravoxel incoherent motions: Application to diffusion and perfusion in neurologic disorders. *Radiology* 161:401–407.

Le Bihan D, Turner R (1992) The capillary network: A link between IVIM and classical perfusion. *Magn Reson Med* 27:171–178.

Leach MO, Brindle KM, Evelhoch JL, Griffiths JR, Horsman MR, Jackson A, Jayson GC et al. (2005) The assessment of antiangiogenic and antivascular therapies in early-stage clinical trials using magnetic resonance imaging: Issues and recommendations. *Br J Cancer* 92:1599–1610.

Lee CC, Wintermark M, Xu Z, Yen CP, Schlesinger D, Sheehan JP (2014) Application of diffusion-weighted magnetic resonance imaging to predict the intracranial metastatic tumor response to gamma knife radiosurgery. *J Neurooncol* 118:351–361.

Long DM (1979) Capillary ultrastructure in human metastatic brain tumors. *J Neurosurg* 51:53–58.

Mardor Y, Pfeffer R, Spiegelmann R, Roth Y, Maier SE, Nissim O, Berger R et al. (2003) Early detection of response to radiation therapy in patients with brain malignancies using conventional and high b-value diffusion-weighted magnetic resonance imaging. *J Clin Oncol* 21:1094–1100.

McMahon MT, Gilad AA, DeLiso MA, Berman SM, Bulte JW, van Zijl PC (2008) New "multicolor" polypeptide diamagnetic chemical exchange saturation transfer (DIACEST) contrast agents for MRI. *Magn Reson Med* 60:803–812.

Mitsuya K, Nakasu Y, Horiguchi S, Harada H, Nishimura T, Bando E, Okawa H, Furukawa Y, Hirai T, Endo M (2010) Perfusion weighted magnetic resonance imaging to distinguish the recurrence of metastatic brain tumors from radiation necrosis after stereotactic radiosurgery. *J Neurooncol* 99:81–88.

Moffat BA, Chenevert TL, Lawrence TS, Meyer CR, Johnson TD, Dong Q, Tsien C et al. (2005) Functional diffusion map: A noninvasive MRI biomarker for early stratification of clinical brain tumor response. *Proc Natl Acad Sci USA* 102:5524–5529.

Padhani AR, Gapinski CJ, Macvicar DA, Parker GJ, Suckling J, Revell PB, Leach MO, Dearnaley DP, Husband JE (2000) Dynamic contrast enhanced MRI of prostate cancer: Correlation with morphology and tumour stage, histological grade and PSA. *Clin Radiol* 55:99–109.

Park I, Larson PE, Tropp JL, Carvajal L, Reed G, Bok R, Robb F et al. (2014) Dynamic hyperpolarized carbon-13 MR metabolic imaging of nonhuman primate brain. *Magn Reson Med* 71:19–25.

Park I, Larson PE, Zierhut ML, Hu S, Bok R, Ozawa T, Kurhanewicz J et al. (2010) Hyperpolarized 13C magnetic resonance metabolic imaging: Application to brain tumors. *Neuro Oncol* 12:133–144.

Park JE, Kim HS, Goh MJ, Kim SJ, Kim JH (2015) Pseudoprogression in patients with glioblastoma: Assessment by using volume-weighted voxel-based multiparametric clustering of MR imaging data in an independent test set. *Radiology* 275:792–802.

Petrella JR, Provenzale JM (2000) MR perfusion imaging of the brain: Techniques and applications. *Am J Roentgenol* 175:207–219.

Provencher SW (1993) Estimation of metabolite concentrations from localized in vivo proton NMR spectra. *Magn Reson Med* 30:672–679.

Rivlin M, Horev J, Tsarfaty I, Navon G (2013) Molecular imaging of tumors and metastases using chemical exchange saturation transfer (CEST) MRI. *Sci Rep* 3:3045.

Schroder L, Lowery TJ, Hilty C, Wemmer DE, Pines A (2006) Molecular imaging using a targeted magnetic resonance hyperpolarized biosensor. *Science* 314:446–449.

Soares DP, Law M (2009) Magnetic resonance spectroscopy of the brain: Review of metabolites and clinical applications. *Clin Radiol* 64:12–21.

Tofts PS, Brix G, Buckley DL, Evelhoch JL, Henderson E, Knopp MV, Larsson HB et al. (1999) Estimating kinetic parameters from dynamic contrast-enhanced T(1)-weighted MRI of a diffusable tracer: Standardized quantities and symbols. *J Magn Reson Imaging* 10:223–232.

Walker-Samuel S, Ramasawmy R, Torrealdea F, Rega M, Rajkumar V, Johnson SP, Richardson S et al. (2013) In vivo imaging of glucose uptake and metabolism in tumors. *Nat Med* 19:1067–1072.

Ward CS, Venkatesh HS, Chaumeil MM, Brandes AH, Vancriekinge M, Dafni H, Sukumar S et al. (2010) Noninvasive detection of target modulation following phosphatidylinositol 3-kinase inhibition using hyperpolarized 13C magnetic resonance spectroscopy. *Cancer Res* 70:1296–1305.

Ward KM, Aletras AH, Balaban RS (2000) A new class of contrast agents for MRI based on proton chemical exchange dependent saturation transfer (CEST). *J Magn Reson* 143:79–87.

Ward KM, Balaban RS (2000) Determination of pH using water protons and chemical exchange dependent saturation transfer (CEST). *Magn Reson Med* 44:799–802.

Weber MA, Thilmann C, Lichy MP, Gunther M, Delorme S, Zuna I, Bongers A et al. (2004) Assessment of irradiated brain metastases by means of arterial spin-labeling and dynamic susceptibility-weighted contrast-enhanced perfusion MRI: Initial results. *Invest Radiol* 39:277–287.

Wen Z, Hu S, Huang F, Wang X, Guo L, Quan X, Wang S, Zhou J (2010) MR imaging of high-grade brain tumors using endogenous protein and peptide-based contrast. *Neuroimage* 51:616–622.

Yen YF, Nagasawa K, Nakada T (2011) Promising application of dynamic nuclear polarization for in vivo (13)C MR imaging. *Magn Reson Med Sci* 10:211–217.

Zach L, Guez D, Last D, Daniels D, Grober Y, Nissim O, Hoffmann C et al. (2012) Delayed contrast extravasation MRI for depicting tumor and non-tumoral tissues in primary and metastatic brain tumors. *PLoS One* 7:e52008.

Zahra MA, Hollingsworth KG, Sala E, Lomas DJ, Tan LT (2007) Dynamic contrast-enhanced MRI as a predictor of tumour response to radiotherapy. *Lancet Oncol* 8:63–74.

7

From frame to frameless: Brain radiosurgery

Young Lee and Steven Babic

Contents

7.1 INTRODUCTION

The term *stereotactic* ("stereo" from the Greek word στερεός [*stereos*], which means "solid," and "-taxis", which means "arrangement" or "order") defines the 3D localization of a point in space by a unique set of coordinates that correspond to a fixed, external reference frame. Historically, such a frame has acted as a support for hollow probes that guide electrodes or biopsy needles to precise locations within either an animal or human brain.

Since the 1950s, the stereotactic principle has been adopted by neurosurgeons in the technique of stereotactic radiosurgery (SRS), in which narrow beams of cobalt-60 gamma rays (Gamma Knife) are focused to a small target within the brain. Since the mid-1980s, adaptations of medical linear accelerators (linacs) to produce similarly precise megavoltage x-ray beams have made this technique accessible to many hospitals. With the development of relocatable head frames used with CT-based fiducial marker systems, precisely planned, accurately deliverable single-fraction SRS and fractionated stereotactic radiation therapy (SRT) of the brain are readily available to many oncology centers.

Lesions suitable for SRS, generally less than 3 cm in diameter, range from small malignant brain metastases, to semibenign lesions such as acoustic neuromas, and to benign arteriovenous malformations (AVMs). To reduce toxicity to sensitive structures such as the brain stem and optic nerves, some lesions may be treated with SRT. Whole brain treatments can be an effective option but at the expense of neurocognitive deficits, and when other metastases appear post whole brain radiation therapy, treatment options are limited (Sahgal et al., 2015). Because of the highly focused nature of SRS and SRT, these techniques are suitable for retreating small recurrent disease, and as a result, treating multiple metastases separately is becoming more commonly practiced. In order to deliver high radiation doses to small, multiple targets, accurate treatment positioning is crucial. Figure 7.1 shows different aspects of the SRS system and how each element affects the accuracy.

SRS has conventionally utilized an invasive nonmoveable head frame to immobilize and position the patient with the highest degree of accuracy and precision. With frame-based systems, the tumor's stereotactic location within the rigid frame is defined, and it can be reproduced at the time of treatment using a stereotactic coordinate localization device and the lasers within the treatment room. It is imperative

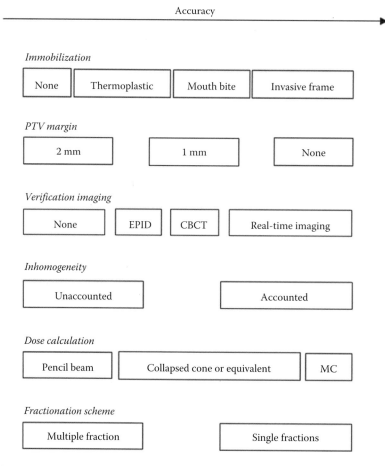

Figure 7.1 The dependence of different aspects of stereotactic radiosurgery delivery on accuracy. The accuracy increases from left to right, and the effect on planning target volume (PTV) margin is illustrated. Other factors can influence accuracy such as imaging and fusion. However, imaging and fusion are not mutually exclusive in the influence on accuracy. EPID, electronic portal imaging device; CBCT, cone-beam computed tomography; MC, Monte Carlo.

that the spatial relationship of the frame relative to the skull is upheld since any slippage of the frame prior to treatment will result in positioning error and if not corrected for, may lead to geographic miss of the target and or normal critical structures being overdosed. Although stereotactic frame slippage is not common, it can be one significant drawback of utilizing rigid nonmoveable head frames. In addition to this, a number of other disadvantages include the need for a neurosurgeon for the purpose of frame placement; once the frame is attached, simulation, imaging, planning, quality assurance (QA) checks, and treatment need to be completed all within the same day; fractionated treatment is not practical; there is pain and uneasiness for the patient; and there exists a risk of bleeding and infection at the site of frame attachment. As a result of these limitations and with the advancement of image-guided radiation therapy (IGRT), where setup verification images can be utilized to precisely check the isocenter location using bony anatomy, there has been a move toward noninvasive moveable frames and frameless systems for SRS and SRT.

The transition of "frame" to "frameless" brain radiosurgery is discussed in this chapter with respect to the evolving technology and physical differences between the immobilization systems.

7.2 FRAME-BASED SRS: USING LOCALIZER BOXES

Frame-based SRS can be used with Gamma Knife and linacs equipped with external cones and/or micro-multileaf collimators (micro-MLCs). Compared to the CyberKnife robotic radiosurgery system, real-time imaging is not utilized to correct for intrafractional motion with Gamma Knife–based SRS nor is it commonly used during linac delivery (Wowra et al., 2012). Furthermore, it can be argued that real-time imaging is not necessary with a frame-based system as the intrafractional motion is negligible, especially for invasive, neurosurgical frames. In this subsection, we discuss the frame-based SRS in detail, including the benefits and limitations of the system.

Central to the stereotactic method is the requirement for imageable fiducial reference markers that are attached to the stereotactic system immobilizing the patient. These markers are vital in providing accurate geometrical information on the coordinates of the planned isocenter. They are commonly in the form of "crowns" containing various rod configurations, etchings on the sides of plastic fiducial boxes, and/or wires stretched between rigid spacers. The fundamental and ideal requirements of a fiducial system are (1) no significant scan artifacts are generated that obscure the images used for target and organs-at-risk (OARs) delineation; (2) precise and rapid docking onto the patient's immobilization system; (3) an unambiguous and preferably, simple marker arrangement which enables manual checking of computed target coordinates; and (4) capability of correcting for the effect of imaging slices being nonperpendicular to the scanner couch, which has a tendency to sag from the weight of the patient as it passes through the scan plane. Although calculation software that gives general solutions to the equations of fiducial markers is readily available, it is often preferable to adjust the frame tilt such that the patient CT scans are orthogonal to the couch and parallel to the scan plane. This enables ease of checking and keeps an intuitive feel for the precise isocenter setup geometry.

Some of the commercial systems and the requirements and advantages of each system in the context of SRS are discussed in the following sections.

7.2.1 NEUROSURGICAL INVASIVE FRAMES

Some examples of neurosurgical frames that remain the most reliable and stable platforms include the Cosman–Roberts–Wells (CRW) frame presented in Figure 7.2 and Leksell frame shown in Figure 7.3. Accurate fixation of the stereotactic frame to the patient's head is achieved by means of three to four steel pins that are inserted into tiny holes drilled into the patient's skull. Frame fitting with this procedure consequently requires a local anesthetic and a neurosurgeon to both place and remove the frame. Another limitation is that all of the processes required for the treatment procedure need to be carried out in a single day. Although such frames can be removed and refitted for limited fractionation regimes, this is not ideal as placing the frame in the same position is difficult and can cause significant patient discomfort. The following sections briefly describe the two most widely used commercially available invasive frames.

(a)   (b)

Figure 7.2 (a) Radionics box with plastic templates placed on the box with printed templates and (b) the Cosman–Roberts–Wells system placed on the patient on a CT scanner couch.

7.2.1.1 Brown–Roberts–Wells/Cosman–Roberts–Wells

The original Brown–Roberts–Wells (BRW) system, which consisted of a skull base ring with carbon epoxy head posts that offers minimal CT interference, was created at the University of Utah in 1977. The frame ring is attached to the patient with screws that are tightened into the skull. The localizer unit is secured to the ring with three ball-and-socket interlocks and consists of six vertical posts and three diagonal posts, creating an N-shaped appearance (Figure 7.2a). This "N" construct establishes the axial CT plane relative to the skull base by calculating the relative distance of the oblique to the vertical rods. Target coordinates are established by identifying the axial slice that best features the lesion. The x and y coordinates for each of the nine fiducial rods are identified on the CT or MRI, as are the x and y coordinates for the target. All coordinates are then converted to coordinates in stereotactic space. In the 1980s, Wells and Cosman simplified and improved the BRW by designing an arc guidance frame similar to the Leksell frame. The arc system directs a stereotactic probe isocentrically around the designated target, thus avoiding a fixed entry point.

The CRW system included some of the same design elements as the BRW system, including a phantom frame, the same CT localizer, and the same probe depth fixed at 16 cm. New innovations included the introduction of MRI-compatible frames and localizers and versatility in arc-to-frame applications that enabled inferior trajectories into the posterior fossa or lateral routes into the temporal lobe. One of the compatible SRS treatment planning systems with the CRW frame is the Radionics radiosurgery treatment planning software (Integra NeuroSciences, Burlington, MA). Printouts of stereotactic coordinate templates (Figure 7.2b) and how they fit onto the treatment localizer box is crucial in setting up the patient in the correct position.

When not corrected, caution should be taken using geometrically distorted magnetic resonance images (MRI) for SRS with metal frames. Using images of phantoms with CRW and Leksell frames scanned on a GE Signa 1.5 T machine, Burchiel et al. (1996) showed that the CRW frame caused larger MRI distortions compared to the Leksell frame. The study concluded that properties of frame systems used for stereotactic neurosurgery may greatly influence the accuracy of frame-based stereotactic neurosurgery, and that the accuracy of these frame systems is testable.

Li et al. (2011) had compared the BRW to the frameless PinPoint® system using the VisionRT surface-matching system. On average, for 11 patients (19 lesions), the translational and rotational magnitudes (1 standard deviation) were observed to be 0.3 (0.2) mm and 0.2 (0.2)°, respectively.

7.2.1.2 Leksell

The Leksell Coordinate Frame G, made of titanium, is fixed to the patient's head using four self-tapping screws, which keep the frame firmly and accurately in place. It is lighter than the CRW frame, and it is fully MR compatible. Its small frame size fits in most MR head coils and minimizes distortion. MR, CT, and angiography localizer boxes (Figure 7.3) ensure parallel and equidistant images, and there are table

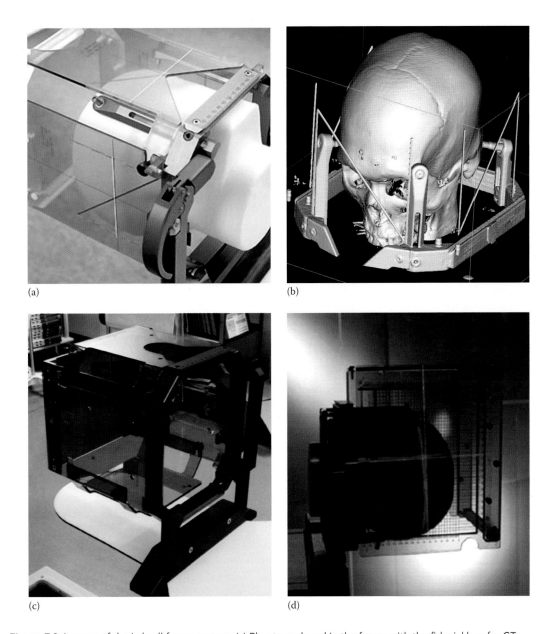

Figure 7.3 Images of the Leksell frame system. (a) Phantom placed in the frame with the fiducial box for CT imaging, (b) treatment planning system rendered bony anatomical image with the CT fiducial box, (c) fiducial box for angiography imaging, and (d) phantom in the frame with the fiducial box used in treatment.

adapters for CT, angiography, and treatment units, including linac and Gamma Knife. These integrated tabletop adapters ensure patient fixation is consistent at all stages of the process, assuring accuracy of target localization and patient setup to submillimeter mechanical accuracy.

The original neurosurgical frame consisted of a semicircular arc with a moveable probe carrier. The arc is fixed to the patient's head in such a manner that its center corresponds with a selected cerebral target. The electrodes are always directed toward the center and, hence, to the target. Rotation of the arc around the axis rods in association with lateral adjustment of the electrode carrier enables any convenient point of entrance of the electrodes to be chosen, independent of the site of the target (Leksell, 1971; Lundsford et al., 1988).

The model G base frame is rectangular and has dimensions of 190 mm by 210 mm. A straight or curved front piece can be used, as it allows airway access in emergencies. The x, y, and z axes on the frame recapitulate the CT and MRI axes. The frame center coordinates are 100, 100, and 100, whereas a hypothetical frame origin (x, y, and $z = 0$) resides in the superior posterior right side of the frame. In the neurosurgical frame, the semicircular arc attached to the base frame has a radius of 190 mm (Louw, 2003).

Leksell frame is supported by many treatment planning systems, but it is the only frame that is currently in clinical use with the Leksell Gamma Knife system.

7.2.2 NONINVASIVE MOVEABLE FRAMES

Unlike rigid stereotactic head frames, these immobilization devices can be easily removed since they are not permanently attached to the patient, thereby significantly reducing patient anxiety and discomfort. There are several different types of noninvasive moveable frames which offer accurate and reproducible cranial immobilization comparable to the previously described nonmoveable frame-based systems. As with the invasive frames, for treatment planning, a stereotactic fiducial-based localizer system can be attached to the moveable head frame at the time of CT simulation, and at the time of treatment delivery, a stereotactic localizer can be utilized for setting the patient up to the treatment isocenter. A few varieties of these noninvasive relocatable frames are described below, and they include the Gill–Thomas–Cosman (GTC) frame, the BrainLAB mask system, and the Laitinen Stereoadapter 5000 frame.

7.2.2.1 Gill–Thomas–Cosman frame

The GTC frame consists of an aluminum alloy base ring that is attached to the treatment couch for rigid immobilization, a dental plate/oral appliance, an occipital headrest pad, and Velcro straps. The dental plate and headrest pad are custom made by, respectively, taking a dental impression of the patient's teeth and posterior skull encompassing the occipital protuberance. Both of these patient-specific devices are securely mounted to the head ring. The Velcro strap is posteriorly fixed to the headrest and is also connected anteriorly to the base ring. Together, the three different components uphold the frame's position on the patient. To enable daily setup reproducibility, the Velcro strap lengths can be marked at each side and a clear plastic hemispherical dome with fixed holes or portals, called a depth confirmation helmet, can be used. The depth confirmation helmet is placed over the head ring, and a rod with a millimeter scale is inserted into each hole to measure the distance to the cranial surface. The distance readings from each hole can be compared to the readings obtained at the time of CT simulation to ensure that the frame has been accurately placed. The GTC frame is a commercially available (Integra, Plainsboro, NJ) relocatable head frame that was adapted to be compatible with the BRW stereotactic coordinate system. Its design is originally based on the Gill–Thomas frame (Gill et al., 1991; Graham et al., 1991).

Using the depth helmet before each treatment fraction, Das et al. (2011) measured the daily relocation error of the GTC frame. Based on 10 patients, they found a mean vector displacement or radial error of 1.03 ± 0.34 mm. In the mediolateral, anteroposterior, and craniocaudal directions, the mean errors were 0.38, 0.15, and 0.17 mm, respectively, and all errors were within ± 2 mm, 97%–99% of all cases. Using the depth confirmation helmet as well, Burton et al. (2002) evaluated the setup reproducibility on 31 patients, and they reported mean errors of 0.1, 0.1, and 0.4 mm in the mediolateral, anteroposterior, and craniocaudal directions, respectively. They also determined a mean displacement vector of 1.2 mm, with 92% of the displacement vectors less than 2 mm and 97% less than 2.5 mm. Utilizing 126 anterior and 123 lateral daily pretreatment portal images coregistered to the digitally reconstructed radiographs (DRRs) from planning CTs (the reference images) of 15 patients, Kumar et al. (2005) determined a total 3D mean displacement vector of 1.8 ± 0.8 mm with a range of 0.3–3.9 mm. In an earlier study from measurements based upon 20 patients, the GTC frame was found to have superior relocalization accuracy on the order of ± 0.4 mm.

7.2.2.2 BrainLAB mask system

Based upon the GTC frame, the BrainLAB mask system (BrainLAB, Munich, Germany) consists of a patient-specific thermoplastic mask, a U-shaped frame, vertical posts, and an optional bite block that attaches to an upper jaw device for additional support. Vertical posts fasten the thermoplastic mask to the head ring, and an adaptor is utilized to attach the head ring to the imaging or treatment couch. At the time of CT simulation, the thermoplastic mask is custom shaped to the patient's head both anteriorly and posteriorly, and a CT localizer box is attached to the head ring for image localization. For the purpose of patient setup at the time of treatment, a stereotactic localizer box or target positioner is fixed to the head ring.

Many authors have investigated the setup errors and intrafraction motion of patients immobilized with the BrainLAB mask (Alheit et al., 2001; Minniti et al., 2011; Theelen et al., 2012). Alheit et al. (2001) utilized simulator films and electronic portal imaging device (EPID) to show that patient position reproducibility using the BrainLAB mask is less than 2 mm. Both Minniti et al. (2011) and Theelen et al. (2012) used serial CT scans for BrainLAB mask-positioning verification and reported a mean 3D displacement of 0.5 ± 0.7 mm (maximum of 2.9 mm) and 1.16 ± 0.68 mm (maximum of 2.25 mm), respectively. Both observed the largest translational deviation to be in the superior–inferior direction.

Using pretreatment kV imaging, Bednarz et al. (2009) investigated the setup accuracy of the BrainLAB mask and found a mean 3D displacement of 3.17 ± 1.95 mm from the isocenter. This result was consistent with those of Willner et al. (1997) who reported a mean 3D vector deviation of 2.4 ± 1.3 mm from the isocenter. Ali et al. (2010) utilized kV onboard imaging and reported mean shifts of 0.1 ± 2.2, 0.7 ± 2.0 and –1.6 ± 2.6 mm in anterior–posterior, medial–lateral, and superior–inferior directions, respectively. Using posttreatment kV imaging, Ramakrishna et al. (2010) investigated the intrafraction motion with the BrainLAB mask and found a mean intrafraction shift of 0.7 ± 0.5 mm.

7.2.2.3 Laitinen Stereoadapter 5000 frame

The Laitinen Stereoadapter 5000 (Sandstrom Trade and Technology Inc., Welland, Ontario, Canada) is a noninvasive relocatable stereotactic frame that is attached to the patient's head using (1) two earplugs that are bilaterally inserted into the external auditory canals and (2) a nasal support assembly that rests against the bridge of the nose (Laitinen et al., 1985; Kalapurakal et al., 2001). The earplugs are respectively attached to two lateral plates. The lateral plates are joined together at the vertex using a connector plate and are also attached to the nasal support assembly through two side arms. The nasal support assembly is equipped with an adjustable thumbscrew, which is used to secure the earplugs against the lateral plates and to push them into the external auditory canals. For further frame stabilization, a strap is attached to the lateral plates and wraps posteriorly around the head. A couch adaptor device secures the frame's connector plate at the vertex to the imaging or treatment couch. Since the frame is made of an aluminum alloy and plastic, it is both MRI and CT compatible. To accommodate patients with varying sizes of external auditory canals, different sizes of earplugs are available. This frame has been found to be well tolerated by both children and adults, although a mild pressure sensation at the ear canals does yield some patient discomfort (Golden et al., 1998; Kalapurakal et al., 2001).

For reproducible setups and repositioning accuracy, a number of components (i.e., connector plate, arms of the nasal support assembly) of the Stereoadaptor 5000 frame have graduated scales in millimeters that together with target plates (these attach to the lateral side plates) can be utilized to establish the frame's reference coordinates and to set up the patient to the treatment isocenter. A number of authors have evaluated the repositioning accuracy of this frame. Using orthogonal portal images, Golden et al. (1998) found the reproducibility to be generally about 2 mm. Utilizing portal images coregistered to CT scout images, Kalapurakal et al. (2001) reported a mean isocenter shift of 1.0 ± 0.7, 0.8 ± 0.8, and 1.7 ± 1.0 mm in the lateral (x), anterior/posterior (y), and superior/inferior (z) directions, respectively. Testing for accuracy and reproducibility of repeated mountings, Delannes et al. (1991) determined a mean distance error of 0.9, 0.6, and 0.9 mm in the x, y, and z coordinates, respectively.

7.2.3 SUMMARY

Frame-based SRS using localizer or fiducial boxes enable accurate setup of patients without the need for patient setup verification images. With invasive frames, patients can be immobilized accurately to less than

1 mm setup errors. Though rare, slippage is possible with invasive frames, and without imaging, these errors may not be easily caught. This highlights the importance of SRS and SRT.

A small decrease in accuracy is inevitable with moveable frame systems. However, the moveable frames can allow similar accuracy in patient immobilization to invasive frames in fractionated treatments without the discomfort to patients. To account for the small decrease in accuracy, a larger margin should be used to create a planning target volume (PTV) (see Figure 7.1).

The loss in accuracy from invasive to moveable frames, however, maybe lessened with the advancement of IGRT, as it is widely available for linac treatments and being incorporated into SRS and SRT. The vital role that IGRT plays in SRS and SRT will be discussed within the subsequent sections.

7.3 NONINVASIVE FRAMELESS SYSTEMS

In order to compensate for the loss of rigid immobilization associated with invasive head frames, a high-precision IGRT must be added to all frameless systems. Imaging at the time of treatment is used to directly determine the position of the target and to correct for any patient movement and or positioning errors. As a result, the accurate correlation between the patient anatomy and immobilization device that is key to the frame-based stereotactic approach is no longer essential.

In general, the daily repositional accuracy of a relocatable frame should vary by less than 1 mm. Relocation checks are best carried out by fusing the bony anatomy of repeat sets of cone-beam CT (CBCT), but if not possible, anterior–posterior and lateral EPID can be taken in order to quantify the displacements of reference markers attached to the frame. Alternatively, physical depth measurements to the surface of the head from a reference "depth helmet," mounted onto the stereotactic frame, may be made. Optical video methods can be used based on fixed geometry wall-mounted cameras in the treatment room and reflective markers on the patient. For frameless positioning, treatment room image-guided systems must include onboard CBCT, onboard MV EPID, or onboard kV images and/or kV x-ray systems mounted on the ceiling and floor. Some commercially available noninvasive frameless systems include the eXtend frame, PinPoint frame, optically guided bite block, and BrainLAB frameless system.

7.3.1 eXtend

The eXtend frame system (Elekta, Stockholm, Sweden) is a noninvasive vacuum bite-block repositioning head frame for cranial immobilization. The main components of this system include a carbon fiber frame body that is attached to the treatment couch, a headrest, and a mouthpiece that is affixed to the frame body using a frontpiece (Ruschin et al., 2010). Prior to treatment, a patient-specific dental impression of the upper mouth is obtained together with a cushion impression of the back of the skull. A vacuum device is used to suction the custom bite block to the patient's upper hard palate. To verify that the patient's head is accurately positioned within the eXtend frame, a spring-loaded digital dial gauge is inserted through slotted holes in a repositioning check tool (RCT) that is attached to the frame. The distance between the frame and the patient's head is measured and compared to reference values measured on the initial day of treatment. The patient can then be repositioned at the time of setup if the difference exceeds a predefined tolerance (e.g., 1 mm). In one study that evaluated four patients immobilized with the eXtend frame and treated on a Gamma Knife machine (Perfexion, Elekta, Stockholm, Sweden), the mean radial positioning error was found to be between 0.33 and 0.84 mm (Sayer et al., 2011). Using the RCT and CBCT image guidance, Ruschin et al. (2010) reported on the setup accuracy and intrafraction motion of the eXtend frame system on 12 patients. Specifically with CBCT, the mean 3D intrafraction motion was found to be 0.4 ± 0.3 mm, and with the RCT, it was 0.7 ± 0.5 mm. For patients treated on a linac and on a Gamma Knife machine (Perfexion, Elekta, Stockholm, Sweden), the mean 3D setup error was 0.8 and 1.3 mm, respectively, thus confirming eXtend frame's excellent immobilization performance.

7.3.2 PinPoint® (AKTINA)

Similar in design to the eXtend frame, the PinPoint (Aktina Medical, Congers, NY) is a commercially available noninvasive frameless system equipped with a vacuum fixation bite-block device for patient localization and fixation (see Figure 7.4). It consists of an internal and external component. The internal

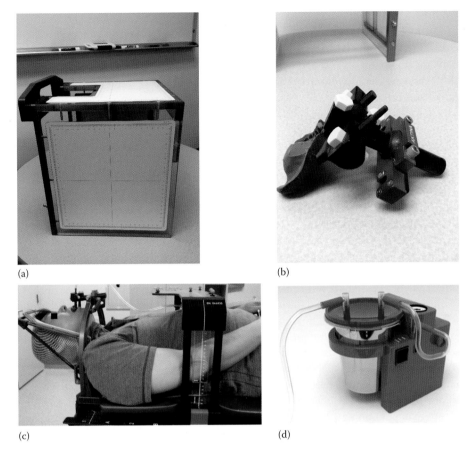

Figure 7.4 PinPoint® frameless system. (a) The cranial localizer box that allows for three-point localization, (b) vacuum fixation mouthpiece with patient-specific dental impression, (c) patient immobilized in the thermoplastic support frame together with mouthpiece attached to an external arch block, and (d) portable suction unit that provides vacuum suction.

component contains a custom-made patient-specific dental mouthpiece with continuous mild vacuum suction to the upper hard palate. A gentle vacuum suction is applied between the dental mouthpiece and the upper hard palate to assure tight contact. The external component consists of the dental mouthpiece secured to a metal arch frame that is in turn locked into a carbon fiber couch board equipped with a thermoplastic head support. The head support is patient specific and is formed by creating an impression of the back of the skull. It limits head motion in the left, right, superior, and posterior directions. An adjustable, rigid connector defines the inferior and anterior limits and confines head rotations particularly nodding (pitch) and shaking (roll). Once the patient is immobilized in the PinPoint frame, the patient's head cannot move without losing suction. To setup the patient to treatment isocenter, a localizer box with three imbedded spherical CT visible markers is attached over the bridge of the PinPoint system.

Li et al. (2011) evaluated the PinPoint frameless system and its ability to immobilize patients and restrict head motion. A video-based 3D optical surface imaging system with three ceiling-mounted camera pods (AlignRT, Vision RT Ltd, London, UK) was utilized to verify treatment setup as well as to monitor and quantify head motion near-real time during treatment. Two hypofractionated (SRT) and two single-fraction (SRS) patients (10 treatment fractions in total) were immobilized with PinPoint. and the magnitude of motion was compared against that attained with 11 SRS patients immobilized with the invasive BRW head frame. In terms of setup verification, the mean 3D translational difference using PinPoint was 0.9 ± 0.3 mm. The mean translations and rotations were 0.3 ± 0.2 mm and 0.2° ± 0.1°, respectively. These values were found to be consistent with the magnitude of motion quantified with the BRW head frame. Although a slow head-drifting motion was observed with the PinPoint frame, for 98% of the time the magnitude of head motion was within 1.1 mm and 1.0°; thus, showing that this frameless system is adequate to tightly limit head motion during stereotactic radiation treatments.

7.3.3 OPTICALLY GUIDED FRAMELESS SYSTEM

The optically guided frameless system consists of several components: a cushion to hold the patient's head posteriorly, a thermoplastic mask that immobilizes the patient anteriorly from the forehead to the upper lip, a custom bite block molded to the patient's upper jaw, an optical array of fiducial markers that attaches to the bite block, and an in-room infrared camera system mounted on the ceiling. This system was developed at the University of Florida (Bova et al., 1997) and is commercially available through Varian Medical Systems (Palo Alto, CA). The unique feature of this system is that immobilization is separated from localization since the bite block is not fixed to the thermoplastic mask, thus making this system also compatible with rigid stereotactic frames (Meeks et al., 2000). The localization is achieved through the optical reference array and the infrared camera system. The infrared camera is equipped with illuminators that emit infrared light. The light is subsequently reflected off of infrared fiducials on the optical array and detected by the camera's charge-coupled device optics to accurately locate the position of the fiducials. The position of the fiducials relative to the treatment isocenter is predefined in a stereotactic coordinate system, and by combining this with a calibration matrix that relates the position of the camera to the treatment machine isocenter, target localization and real-time tracking relative to the isocenter are enabled (Kamath et al., 2005).

Relative to the stereotactic rigid head frame, Meeks et al. (2005) showed that the optic-guided bite-plate system provides a mean patient localization accuracy of 1.1 ± 0.3 mm. Philips et al. (2000) found that with this system, positioning a target point in the radiation field had an accuracy of 1.0 ± 0.2 mm. Using orthogonal kV planar imaging verification, Wang et al. (2010) conducted a retrospective analysis of the setup accuracy of 56 patients using the optically guided frameless system and reported an average 3D isocenter localization error of 0.37 mm with a maximum error of 2 mm. Peng et al. (2010) reported a mean setup error of 1.2 ± 0.7 mm using a combined optical tracking and 3D ultrasound imaging system (SonArray system, Varian, Palo Alto, CA) and CBCT. Based on 15 patients, Ryken et al. (2001) determined the average localization accuracy at isocenter to be submillimeter at 0.82 ± 0.41 mm, and they concluded that in terms of target localization and accuracy, the optically guided frameless system is comparable to frame-based systems.

7.3.4 BrainLAB FRAMELESS MASK

The BrainLAB frameless thermoplastic mask (BrainLAB AG, Feldkirchen, Germany) is a commercially available immobilization system that is strengthened with a custom-made mouthpiece and three reinforcing straps attached under the mask and covering the forehead, chin, and the area below the nose (Gevaert et al., 2012a). For localization and real-time tracking, six infrared markers are placed on top of the thermoplastic mask and are detected by an infrared camera system (ExacTrac) mounted to the ceiling. In conjunction with this optical guidance system, stereoscopic, planar kV x-ray images (Novalis Body) are taken and registered in 6 degrees of freedom (6DOF) with the DRRs from the planning CT images. Once the registration is accepted, any departures from the treatment isocenter are determined. Translational and rotational positioning errors in 6DOF can be corrected for with the treatment couch and a robotic tilt module underneath the tabletop (Verellen et al., 2003; Gevaert et al., 2012b).

Gevaert et al. (2012a) investigated the setup errors and intrafraction motion of 40 patients immobilized with the BrainLAB frameless mask system. Prior to 6DOF correction, the setup errors were found to be significantly larger in the lateral and longitudinal directions, and the mean 3D setup error was 1.91 ± 1.25 mm. Intrafractional errors were found to be significantly larger in the longitudinal direction (mean shift of 0.11 ± 0.55 mm), and the mean 3D intrafractional motion was 0.58 ± 0.42 mm.

The mean intrafractional rotations were comparable for the vertical, longitudinal, and lateral directions and all within ± 0.03°. These results were found to comparable to the mean 3D intrafraction motion of the BRW invasive head frame (considered the gold standard of immobilization) reported by Ramakrishna et al. (2010) to be 0.40 ± 0.30 mm. Verbakel et al. (2010) investigated the positional accuracy of the BrainLAB frameless system using a hidden target test with a head phantom and found the accuracy to be approximately 0.3 mm in each direction (1 standard deviation). Utilizing posttreatment x-ray verification on 43 patients, they also determined the intrafraction motion and reported it to be 0.35 ± 0.21 mm (maximum of 1.15 mm). Based on their findings, Verbakel et al. (2010) concluded that patient setup with the BrainLAB frameless mask together with ExacTrac/Novalis Body is accurate and stable, and intrafraction motion is very small.

7.3.5 IMAGE-GUIDED FRAMELESS SRS

For frameless SRS, image guidance plays a crucial role in (1) localizing/identifying the target that may or may not be within a defined coordinate system and (2) ensuring cranial immobilization comparable to that attained with rigid frames (viewed as the gold standard of immobilization). Image guidance is typically utilized in the initial patient setup, pretreatment verification, intrafraction monitoring, and posttreatment verification. Prior to treatment, the patient is initially aligned to fiducials on the immobilization mask or localization device that is placed during simulation. After this is done, planar and/or volumetric imaging is completed to look for large setup variations. Planar imaging may include kV or EPID, as well as stereoscopic kV imaging (e.g., BrainLAB Novalis Body). Volumetric image acquisition (e.g., kV CBCT) followed by 3D–3D matching with the treatment planning CT offers the greatest amount of information for 6DOF patient setup. The geometric accuracy of planar and volumetric radiation-based imaging systems has been extensively studied and reported to be 1–2 and ≤1 mm, respectively (De Los Santos et al., 2013). To correct for initial patient setup variations that exceed a certain threshold (e.g., 1 mm translation and 1° rotational), robotic 6DOF treatment couches (e.g., Hexapod [Elekta, Stockholm, Sweden], Robotic Tilt Module [BrainLAB AG]) can be utilized to reposition the patient. Pretreatment verification images (again with planar or volumetric imaging) may then be taken to confirm that the patient repositioning was accurate. These may also be used as a reference to evaluate intrafraction motion.

Following treatment, posttreatment images should be taken and compared against the reference pretreatment images to reveal and quantify any unexpected intrafraction motion. Intrafraction motion can also be done in real time so that patient movement is continuously monitored during treatment. Some commercial systems that have been developed for this purpose include a radiation-based system called ExacTrac (BrainLAB AG, Feldkirchen, Germany) that combines infrared and 2D orthogonal kV images for "snapshot" images during treatment and a non-radiation-based system called AlignRT (VisionRT, London, UK), which uses two or more cameras to perform rapid patient surface imaging. Both systems have been shown to have a geometric accuracy of 1–2 mm.

7.3.6 SUMMARY

For frameless immobilization systems, IGRT is crucial for the system to work with accuracies required for SRS and SRT. The system relies on the accuracy and the precision of the IGRT system without the external localization box. It is still crucial that the immobilization is rigid and stable as most IGRT for linac, and Gamma Knife does not account for patient motion intrafractionally, and the treatments can be long especially with increasing multiple metastases treatments. When a linac is used for SRS, extra–QA system must be put into place to maintain the accuracy required.

7.4 STEREOTACTIC TREATMENT PLANNING AND DELIVERY

To successfully treat SRS and SRT, immobilization is only one part of a complicated planning and treatment delivery process outlined in Figure 7.5. The following sections briefly describe the procedures throughout the SRS and SRT process and how the selection of the immobilization system can affect each stage.

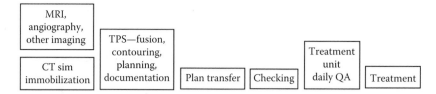

Figure 7.5 A typical stereotactic radiosurgery and stereotactic radiation therapy planning to treatment process.

7.4.1 SIMULATION

CT/MRI or MRI alone are usually required to localize the target volumes as well as the OARs. CT can be used to account for inhomogeneity in the tissue for dose calculation, but in the head, using only MRI with electron density overrides is also acceptable.

In general terms, attempts should be made to minimize distortion in MRI. Any hidden clips buried in long hair should be identified and removed. Impulse generators for deep brain stimulators should have their amplitude reduced to zero to prevent side effects. A protocol should be in place with the radiology department to ensure frequent calibration to minimize field heterogeneities. The Leksell frame is engineered to very high standards, and Burchiel et al. (1996) have indicated that its metallic purity is such that it induces little distortion relative to other commercially available frames. Specific MR pulse sequences are available in many articles and are beyond the scope of this topic. However, it should be noted that T1-weighted sequences with thin cuts reduce spatial distortion, and inversion recovery images sharply demarcate anatomical structures.

For MRI or PET imaging, sealed tubes containing the appropriate liquid and in similar configurations to CT localizers can be used but have been superseded by image fusion techniques. Digital subtraction angiography and MR require well-thought-out QA procedures. For angiography, imaging modality that can aid in localizing AVMs for radiosurgery, small markers or rulers built into the three faces of fiducial plates (one anterior to the patient, and two left and right to the patient) are used. In this application, it is particularly useful to have the flexibility to image the lesion without restrictions to the source-patient-image plane geometry.

For all images, the frame (and metal screws for invasive frames) can cause image distortions so caution should be taken when imaging the patient to try and minimize artifacts from occurring near the target and OAR regions.

7.4.2 TREATMENT PLANNING

Commercially available stereotactic planning systems are often associated with complete stereotactic, surgical, brachytherapy, and radiosurgery packages. The ability of a treatment planning system (TPS) to register or fuse multimodality images is essential as MRI is an important imaging tool for the brain, but CT scans are generally accepted as the most geometrically accurate, distortion-free, tomographic imaging modality.

Manipulation of the virtual patient showing the target volume and sensitive structures has become essential for forward planning, and beam optimization software is important as the complexity of sensitive structure avoidance makes intuitive planning more difficult. With multiple treatment targets becoming more of the norm and also patients returning for more treatments, there is a need for inverse planning modules to help the planning process. Plan analysis tools for the calculation of dose–volume data and assessment of optimized "cost functions" are also required. Automated and optimized inverse planning, which requires specification of dose–volume constraints, is emerging in many radiation therapy clinics.

Stereotactic treatment planning requires the planning of multiple noncoplanar arcs focused to single or several isocenters or multiple, fixed, and noncoplanar conformal beams. The planner must be aware of the limitations of the given treatment machine geometry regarding potential collimator–couch collisions as most of the collision software available on the treatment planning systems are limited. Adjustable beams-eye-views, which show the proximity of sensitive structures such as the brain stem and optic chiasm, are essential features. Overlapping of beam entrance and exits should ideally be avoided, and in general, arc planes should be separated according to the expression 180/n degrees, where n is the number of arcs to be used. The planner needs to be aware and avoid creating beams or arcs that enter through any high-density regions of the immobilization device as this can cause errors in the dose calculation. Avoidance of sensitive, often previously treated structures mean that optimization software and biological tumor control probability and normal tissue complication probability modeling are increasingly important developments. In general the axial, sagittal, coronal, and arc planes are displayed. Dose–volume histogram analyses are used to assist the clinical team in deciding the optimum beam arrangement. The planning system must also have the ability to import and export images and plan information. For frame-based systems, there must be a transfer of the stereotactic coordinates from the planning system to the unit in addition to the normal transfer of information.

Three to five arcs and 7–13 static beams produce sufficient normal tissue sparing for SRT. However, when planning SRT or radiosurgery for benign conditions, particularly in young patients, it is relevant to consider increasing the number of arcs to reduce the exit dose to the brain from any sagittally orientated arcs, although these may be interrupted to avoid possibly sensitive exiting segments. For larger and/or irregular target volumes (35–70 cc), 4–6 fixed, noncoplanar fields are likely to be a more appropriate technique. In this case, either conformal blocking or MLC of each beam's BEVs of target can be used. The added efficiency of the MLC device always has to be balanced against the gold standard of target conformation given by customized lead alloy blocks (corresponding crudely to a micro-MLC having infinitely small leaves).

7.4.3 TREATMENT SETUP AND QUALITY ASSURANCE

Prior to treatment, it is important to be able to confirm the planned setup using the CT scanner or treatment machine itself, depending on the type of lesion to be treated. Automatic or assisted setups are invaluable in limiting potential errors, and imaging can provide the desirable confirmation of accuracy for this type of treatment using exposures displaying the whole-head anatomy.

The linac requires additional QA checks at differing intervals. These checks depend on the precise technique used. The following summary of quality control procedures is typical of those undertaken with typical frequency of SRS checks, with suggested tolerances given in brackets. Note these tests are in addition to the normal machine QA (see American Association of Physicists in Medicine [AAPM] Report No. 54, 142; Solberg et al., 2012).

Daily
1. Frame fixation to patient (<1 mm)
2. Comparison of the left lateral, right lateral, and ceiling lasers with the corresponding light field centers at the cardinal gantry angles (<0.5 mm)
3a. Movement of the light field center cross for 180° collimator rotation (<0.5 mm)
3b. Movement of the radiation field center cross for 180° collimator rotation using onboard EPID (<0.5 mm)
4. Light field symmetry of the selected tertiary collimator about the field center cross (<0.5 mm)

Monthly (frequency may depend on how often treatments are done on unit)
1. A single transverse arc is executed over a water phantom with a calibrated chamber at the isocenter. The chamber dose is compared to a computer plan of the same treatment for representative monitor units used on actual treatments (±2%).

Commissioning and annually
1. EPID (or film) of a radioopaque object set at the laser defined isocenter taken at regular gantry intervals using a tertiary collimator/micro-MLC attached to the treatment head (Lutz et al., 1988; Podgorsak et al., 1989) (<1 mm).
2. *End-to-end test*: With a head phantom mounted on the stereotactic frame (Figure 7.6), CT planning and treatment of a "lesion" in the phantom is given. Thermoluminescent dosimeters or optically stimulated luminescence dosimeters stacked along the three cardinal axes or a film through the

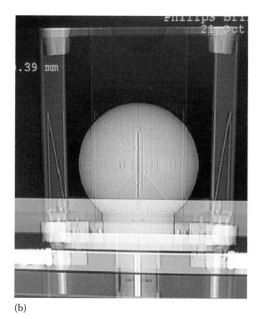

(a)　　　　　　　　　　　　　　　(b)

Figure 7.6 A typical stereotactic radiosurgery phantom localized for end-to-end test. (a) Lucy 3D phantom set up in a Leksell stereotactic frame for CT simulation and (b) CT scout view of a Lucy phantom setup.

isocenter are measured and compared with the corresponding multiple arc plan (less than 5% dose and 1 mm profile displacement). Relative isodose information (or absolute with good calibration) can be obtained from film measurements. Gel dosimetry shows promise for verification of the 3D isodose envelopes treated with SRS; however, it requires expertise in reading out the dose using optical CT or MRI.

Other checks, such as checking that the tertiary collimators are not damaged and any interlocks (such as the barcode reader on the Elekta [Stockholm, Sweden] tertiary cone system) are in working order, should be done. The QA just discussed encompasses general tests recommended for SRS. These listed tests are not exhaustive and may follow a different frequency. Note that within these tests, it is important to establish position and dose at different gantry/couch positions. For SRT, different MLC tests should be added to check position and motion of the MLC in volumetric modulated arc therapy and intensity-modulated radiation therapy.

In general, the daily repositional accuracy of a relocatable frame should vary by less than 1 mm. Relocation checks are best carried out by overlaying the bony anatomy of repeat sets of CBCT images but if not possible, anterior–posterior and lateral EPIDs can be taken in order to quantify the displacement of reference markers attached to the frame. Alternatively, physical depth measurements to the surface of the head from a reference depth helmet mounted onto the stereotactic frame may be made. Optical video methods can be used based upon fixed geometry, wall-mounted cameras in the treatment room, and reflective markers placed on the patient.

7.4.4 OVERALL ACCURACY AND MARGINS

The overall accuracy achievable with a linac-based system should be less than 2 mm under routine conditions, provided sufficient emphasis is placed on the importance of QA. However, localization of the target volume remains the greatest uncertainty in SRS/SRT, and it is important to view the accuracy in its full clinical perspective. Assessing the accuracy of a given system involves rationalizing the cumulative effect of small, typically 0.5–1 mm errors, which generally translate in practice to a 1–3 mm safety margin around the lesion.

With SRS systems, smaller margins may be justified depending on the immobilization system used. The cumulative effect of all the errors in treatment should not be ignored, but each clinic needs to assess and evaluate this error and its effect on the OAR dose. With increasing number of cases of brain metastases presenting to radiation oncology clinics, minimizing dose to the normal brain is essential, especially within regions where radiation dose is linked to a decrease in cognitive function. Keeping cumulative dose to the normal brain low is a major concern as more and more patients are receiving focal irradiation and also returning for further irradiation to other sites.

Often with SRT, a variable PTV margin is required where some compromise is needed close to sensitive organs. This PTV is then covered by the prescribed isodose surface, commonly 90% or 95%. If multiple isocenters are used with arcing circular cross-sectional beam configurations, dose uniformity is sacrificed for a more conformal target coverage. In this case, the prescribed isodose may be 50% of the maximum "hot spots" in the overlapping regions of two spherical dose distributions.

7.5 CONCLUSION

Frame to frameless SRS and SRT requires the assessment and understanding of the whole workflow, from patient demographics, general treatment philosophy and the availability of the imaging techniques, QA, and to PTV margins involved in treatment. SRS and SRT are treatment techniques where improved accuracy is required for treating well-defined targets inside the head. Going frameless allows for more patient comfort, and a minimal reduction in patient setup accuracy can be achieved with good procedures in place, which includes QA and IGRT. See Table 7.1 for summary of all frames discussed in this chapter. In order to account for a reduction in accuracy, margins must be used to create PTV from the target volumes. This process should follow a study by each clinic, which should take into account all the processes that have been highlighted in this section, such as immobilization system, imaging, fusion, planning, delivery system, and IGRT. With the IGRT system in place, there should be a treatment verification process in place such that patient position should be corrected when a threshold of patient position (translational and rotational) is observed (Figure 7.7). This can increase patient treatment time, and therefore, care should be taken with initial patient setup in order to minimize reimaging of the patient.

Table 7.1 Summary of all the frames discussed with achievable accuracy of the immobilization systems

IMMOBILIZATION DESIGN	DEVICE NAME	IMAGE GUIDANCE REQUIRED (PLANAR AND/OR VOLUMETRIC)	ACCURACY (INCLUDING SETUP AND VECTOR DISPLACEMENT) (mm)
Invasive frame	BRW/CRW	No	<1
	Leksell	No	<1
Noninvasive moveable frame	GTC	No	1–3
	BrainLAB mask system	No	1–3
	Laitinen Stereoadapter 5000	No	1–2
Noninvasive frameless	eXtend	Yes	<1.5
	PinPoint®	Yes	<1.5
	Optically guided bite block	Yes	<2
	BrainLAB frameless mask	Yes	<2

Note: Image guidance is not required for "frame" immobilization designs but may add to positioning accuracy if added.

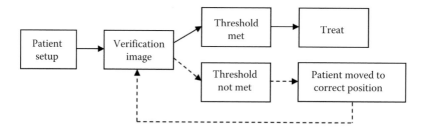

Figure 7.7 An example of image-guided radiation therapy procedure for linear accelerator delivery. Dashed lines indicate a loop which would be repeated until the threshold is met.

With all the correct procedures in place, frameless immobilization promises to deliver similar accuracy in patient setup to both SRS and SRT, and its refined methods of patient fixation, treatment planning, delivery, and verification make it an important component in the evolution of high-precision radiation.

REFERENCES

AAPM Report No. 54 (June 1995). Stereotactic radiosurgery. Report of Task Group 42, Radiation Therapy Committee. American Institute of Physics, Inc. Woodbury, NY.

AAPM Report No. 142 (September 2009) Task Group 142 report: Quality assurance of medical accelerators. Report of Task Group 142, Radiation Therapy Committee.

Alheit H, Dornfeld S, Dawel M, Alheit M, Henzel B, Steckler K, Blank H, Geyer P (2001) Patient position reproducibility in fractionated stereotactically guided conformal radiotherapy using the BrainLab mask system. *Strahlenther Onkol* 177:264–268.

Ali I, Tubbs J, Hibbitts K, Algan O, Thompson S, Herman T, Ahmad S (2010) Evaluation of the setup accuracy of a stereotactic radiotherapy head immobilization mask system using kV on-board imaging. *J Appl Clin Med Phys* 11:3192.

Bednarz G, Machtay M, Werner-Wasik M, Downes B, Bogner J, Hyslop T, Galvin J, Evans J, Curran W, Jr., Andrews D (2009) Report on a randomized trial comparing two forms of immobilization of the head for fractionated stereotactic radiotherapy. *Med Phys* 36:12–17.

Bova FJ, Buatti JM, Friedman WA, Mendenhall WM, Yang CC, Liu C (1997) The University of Florida frameless high-precision stereotactic radiotherapy system. *Int J Radiat Oncol Biol Phys* 38:875–882.

Burchiel KJ1, Nguyen TT, Coombs BD, Szumoski J (1996) MRI distortion and stereotactic neurosurgery using the Cosman-Roberts-Wells and Leksell frames. *Stereotact Funct Neurosurg* 66(1–3):123–136.

Burton KE, Thomas SJ, Whitney D, Routsis DS, Benson RJ, Burnet NG (2002) Accuracy of a relocatable stereotactic radiotherapy head frame evaluated by use of a depth helmet. *Clin Oncol (R Coll Radiol)* 14:31–39.

Das S, Isiah R, Rajesh B, Ravindran BP, Singh RR, Backianathan S, Subhashini J (2011) Accuracy of relocation, evaluation of geometric uncertainties and clinical target volume (CTV) to planning target volume (PTV) margin in fractionated stereotactic radiotherapy for intracranial tumors using relocatable Gill-Thomas-Cosman (GTC) frame. *J Appl Clin Med Phys* 12:3260.

De Los SJ, Popple R, Agazaryan N, Bayouth JE, Bissonnette JP, Bucci MK, Dieterich S et al. (2013) Image guided radiation therapy (IGRT) technologies for radiation therapy localization and delivery. *Int J Radiat Oncol Biol Phys* 87:33–45.

Delannes M, Daly NJ, Bonnet J, Sabatier J, Tremoulet M (1991) Fractionated radiotherapy of small inoperable lesions of the brain using a non-invasive stereotactic frame. *Int J Radiat Oncol Biol Phys* 21:749–755.

Gevaert T, Verellen D, Engels B, Depuydt T, Heuninckx K, Tournel K, Duchateau M, Reynders T, De RM (2012a) Clinical evaluation of a robotic 6-degree of freedom treatment couch for frameless radiosurgery. *Int J Radiat Oncol Biol Phys* 83:467–474.

Gevaert T, Verellen D, Tournel K, Linthout N, Bral S, Engels B, Collen C et al. (2012b) Setup accuracy of the Novalis ExacTrac 6DOF system for frameless radiosurgery. *Int J Radiat Oncol Biol Phys* 82:1627–1635.

Gill SS, Thomas DG, Warrington AP, Brada M (1991) Relocatable frame for stereotactic external beam radiotherapy. *Int J Radiat Oncol Biol Phys* 20:599–603.

Golden NM, Tomita T, Kepka AG, Bista T, Marymont MH (1998) The use of the Laitinen stereoadapter for three-dimensional conformal stereotactic radiotherapy. *J Radiosurg* 1(3):191–200.

Graham JD, Warrington AP, Gill SS, Brada M (1991) A non-invasive, relocatable stereotactic frame for fractionated radiotherapy and multiple imaging. *Radiother Oncol* 21:60–62.

Kalapurakal JA, Ilahi Z, Kepka AG, Bista T, Goldman S, Tomita T, Marymont MH (2001) Repositioning accuracy with the Laitinen frame for fractionated stereotactic radiation therapy in adult and pediatric brain tumors: Preliminary report. *Radiology* 218:157–161.

Kamath R, Ryken TC, Meeks SL, Pennington EC, Ritchie J, Buatti JM (2005) Initial clinical experience with frameless radiosurgery for patients with intracranial metastases. *Int J Radiat Oncol Biol Phys* 61:1467–1472.

Kumar S, Burke K, Nalder C, Jarrett P, Mubata C, A'hern R, Humphreys M, Bidmead M, Brada M (2005) Treatment accuracy of fractionated stereotactic radiotherapy. *Radiother Oncol* 74:53–59.

Laitinen LV, Liliequist B, Fagerlund M, Eriksson AT (1985) An adapter for computed tomography-guided stereotaxis. *Surg Neurol* 23:559–566.

Leksell L (1971) *Stereotaxis and Radiosurgery: An Operative System.* Springfield, IL: Thomas.

Li G, Ballangrud A, Kuo LC, Kang H, Kirov A, Lovelock M, Yamada Y, Mechalakos J, Amols H (2011) Motion monitoring for cranial frameless stereotactic radiosurgery using video-based three-dimensional optical surface imaging. *Med Phys* 38:3981–3994.

Louw (2003) Stereotactic surgery with the Leksell frame. In: Schulder M, Gandhi CD (eds.), *Handbook of Stereotactic and Functional Neurosurgery*, pp. 27–33. New York: Dekker.

Lundsford LD, Leksell D (1988) The Leksell system. In: *Modern Stereotactic Neurosurgery*, pp. 27–46. Boston, MA: Martinus Nijhoff Publishing.

Lutz W, Winston KR, Maleki N (1988) A system for stereotactic radiosurgery with a linear accelerator. *Int J Radiat Oncol Biol Phys* 14:373–381.

Meeks SL, Bova FJ, Wagner TH, Buatti JM, Friedman WA, Foote KD (2000) Image localization for frameless stereotactic radiotherapy. *Int J Radiat Oncol Biol Phys* 46:1291–1299.

Minniti G, Scaringi C, Clarke E, Valeriani M, Osti M, Enrici RM (2011) Frameless linac-based stereotactic radiosurgery (SRS) for brain metastases: Analysis of patient repositioning using a mask fixation system and clinical outcomes. *Radiat Oncol* 6:158.

Peng LC, Kahler D, Samant S, Li J, Amdur R, Palta JR, Liu C (2010) Quality assessment of frameless fractionated stereotactic radiotherapy using cone beam computed tomography. *Int J Radiat Oncol Biol Phys* 78:1586–1593.

Phillips MH, Singer K, Miller E, Stelzer K (2000) Commissioning an image-guided localization system for radiotherapy. *Int J Radiat Oncol Biol Phys* 48:267–276.

Podgorsak EB, Pike GB, Olivier A, Pla M, Souhami L (1989) Radiosurgery with high energy photon beams: A comparison among techniques. *Int J Radiat Oncol Biol Phys* 16:857–865.

Ramakrishna N, Rosca F, Friesen S, Tezcanli E, Zygmanszki P, Hacker F (2010) A clinical comparison of patient setup and intra-fraction motion using frame-based radiosurgery versus a frameless image-guided radiosurgery system for intracranial lesions. *Radiother Oncol* 95:109–115.

Ruschin M, Nayebi N, Carlsson P, Brown K, Tamerou M, Li W, Laperriere N et al. (2010) Performance of a novel repositioning head frame for gamma knife perfexion and image-guided linac-based intracranial stereotactic radiotherapy. *Int J Radiat Oncol Biol Phys* 78:306–313.

Ryken TC, Meeks SL, Pennington EC, Hitchon P, Traynelis V, Mayr NA, Bova FJ, Friedman WA, Buatti JM (2001) Initial clinical experience with frameless stereotactic radiosurgery: Analysis of accuracy and feasibility. *Int J Radiat Oncol Biol Phys* 51:1152–1158.

Sahgal A, Aoyama H, Kocher M, Neupane B, Collette S, Tago M, Shaw P, Beyene J, Chang EL (2015) Phase 3 trials of stereotactic radiosurgery with or without whole-brain radiation therapy for 1 to 4 brain metastases: Individual patient data meta-analysis. *Int J Radiat Oncol Biol Phys* 91:710–717.

Sayer FT, Sherman JH, Yen CP, Schlesinger DJ, Kersh R, Sheehan JP (2011) Initial experience with the eXtend system: A relocatable frame system for multiple-session gamma knife radiosurgery. *World Neurosurg* 75:665–672.

Solberg TD, Balter JM, Benedict SH, Fraass BA, Kavanagh B, Miyamoto C, Pawlicki T, Potters L, Yamada Y (2012) Quality and safety considerations in stereotactic radiosurgery and stereotactic body radiation therapy: Executive summary. *Pract Radiat Oncol* 2(1):2–9.

Theelen A, Martens J, Bosmans G, Houben R, Jager JJ, Rutten I, Lambin P, Minken AW, Baumert BG (2012) Relocatable fixation systems in intracranial stereotactic radiotherapy. Accuracy of serial CT scans and patient acceptance in a randomized design. *Strahlenther Onkol* 188:84–90.

Verbakel WF, Lagerwaard FJ, Verduin AJ, Heukelom S, Slotman BJ, Cuijpers JP (2010) The accuracy of frameless stereotactic intracranial radiosurgery. *Radiother Oncol* 97:390–394.

Verellen D, Soete G, Linthout N, Van AS, De RP, Vinh-Hung V, Van de Steene J, Storme G (2003) Quality assurance of a system for improved target localization and patient set-up that combines real-time infrared tracking and stereoscopic x-ray imaging. *Radiother Oncol* 67:129–141.

Wang JZ, Rice R, Pawlicki T, Mundt AJ, Sandhu A, Lawson J, Murphy KT (2010) Evaluation of patient setup uncertainty of optical guided frameless system for intracranial stereotactic radiosurgery. *J Appl Clin Med Phys* 11:92–100.

Willner J, Flentje M, Bratengeier K (1997) CT simulation in stereotactic brain radiotherapy-analysis of isocenter reproducibility with mask fixation. *Radiother Oncol* 45:83–88.

Wowra B, Muacevic A, Tonn JC (2012) CyberKnife radiosurgery for brain metastases. *Prog Neurol Surg* 25:201–209.

8 Principles of image-guided hypofractionated stereotactic radiosurgery for brain tumors

Kevin D. Kelley, Mihaela Marrero, and Jonathan P.S. Knisely

Contents

8.1 RATIONALE, ADVANTAGES, AND DISADVANTAGES OF HYPOFRACTIONATED STEREOTACTIC RADIOSURGERY

8.1.1 TUMOR RADIOBIOLOGY

A hypofractionated approach has several theoretical advantages in comparison with single-dose radiosurgery. Fractionation serves to exploit fundamental radiobiological principles such as repopulation and reoxygenation. The therapeutic ratio is shifted, leading to improved tumor control and decreased late normal tissue effects through repair of normal tissue sublethal DNA damage, enhanced reoxygenation of tumor cells, and redistribution of tumor cells to more radiosensitive portions of the cell cycle between radiotherapy treatments (Hall and Brenner, 1993). It is important that the physical dosimetric advantages and steep dose falloff that are a hallmark of single-fraction stereotactic radiosurgery (SRS) are maintained with a fractionated approach.

8.1.2 BALANCE BETWEEN TUMOR SIZE AND TOXICITY TO NORMAL BRAIN

When treating large brain tumors, hypofractionation minimizes toxicity to surrounding normal brain compared to single-fraction schedules. The local tumor control rate for tumors larger than ~10 cc (or >2.5 cm in diameter) is usually unsatisfactory when treated in a single fraction. Lower single-fraction radiosurgical doses of <18 Gy are often selected for tumor with diameters >3 cm, based on the Radiation Therapy Oncology Group (RTOG) 90-05 (Shaw et al., 2000) dose-finding trial for single-fraction SRS in recurrent brain tumors, and this can lead to decreased local control rates. This has also been reported with cystic or irregularly enhancing lesions including ring-enhancing metastases treated with single-fraction radiosurgery (Shiau et al., 1997; Mori et al., 1998). Larger intracranial tumors have a well-recognized risk of treatment-related toxicity when SRS treatments are employed (Shaw et al., 2000; Soltys et al., 2015).

Even with a fractionated approach, the risk of developing prolonged reactive edema leading to microischemic and eventual necrotic parenchymal changes is increased when volumes of more than 23 cc of normal brain tissue have received 4 Gy or greater per fraction once the total dose is ≥20 Gy (Ernst-Stecken et al., 2006). Also, early complications after SRS such as seizure or the worsening of neurologic symptoms have been recognized for larger volumes. Most notably, toxicity has been shown to be proportional to the radiation dose delivered to normal brain (Ernst-Stecken et al., 2006). A significant increase in side effects occurs typically if >10 cc of normal brain is treated with more than 10 Gy in a single fraction (Levegrün et al., 2004; Lawrence et al., 2010).

A means of delivering highly conformal ablative doses is necessary when targeting a large volume or an irregularly shaped lesion to circumvent single-fraction dose-limiting toxicity. Hypofractionation has been found to be an effective alternative to single-fraction SRS in these cases. In fact, Ernst-Stecken and colleagues concluded that hypofractionated SRS (HSRS) with 5 fractions of 6–7 Gy is a safe and effective treatment for patients with cerebral metastatic disease not amenable to single-dose radiosurgery due to gross tumor volume (GTV) >3 cc, location involving critical anatomical sites, or if normal brain volume constraints would otherwise be exceeded (Ernst-Stecken et al., 2006).

8.1.3 ELOQUENT AREAS

The clinical manifestations of any radiation-related side effect will strongly depend on the location of the treated volume within the brain. In fact, the frontal and occipital lobes appear to tolerate significantly larger radiosurgical treatment volumes when compared to the temporal lobe, basal ganglia, and brain stem. It has been shown that one or more metastases exceeding a combined volume of more than 3 cc associated with the brain stem, basal ganglia, mesencephalon, or internal capsule will result in significantly more toxicity after single-fraction SRS, thus warranting consideration of a hypofractionated approach when targeting disease within these eloquent areas of the brain (Flickinger et al., 1997).

8.1.4 DISADVANTAGES OF HYPOFRACTIONATION

Disadvantages above and beyond the inconvenience for the patient and staff associated with repetitive delivery of stereotactically directed radiation include problems that have been described in achieving durable local control for radioresistant tumors (melanoma, sarcoma, and renal cell carcinoma) treated with 2–5 fractions compared to more radiosensitive tumors (breast, small-cell lung cancer [SCLC], and non-SCLC) treated with the same fractionation schedule; single-fraction approaches for these diverse histologies did not result in local control differences (Oermann et al., 2013).

8.2 HYPOFRACTIONATED RADIOSURGICAL TREATMENT PLANNING FOR BRAIN TUMORS

8.2.1 PREPLANNING FRAMEWORK

A frameless technique using a thermoplastic immobilization mask allows for an easy and reproducible setup to be achieved with high precision and is ideal for a HSRS approach.

A detailed understanding of the relevant anatomy especially as it relates to patterns of tumor spread and recurrence is absolutely necessary when delineating radiosurgical target volumes (TV). Contrast-enhanced

MR-based diagnostic imaging is essential to evaluate the size and shape of the lesion or cavity for TV determination. The pathology and radiobiology of the tumor type may also be taken into account when selecting hypofractionated dose schedules. It is uncertain that differences in histology should result in differing margin expansions to arrive at optimal clinical and planning target volumes (PTV), but end-to-end testing of the imaging, planning, and delivery approaches used will certainly affect the design of clinical and PTV for a hypofractionated course of therapy (Lightstone et al., 2005). These characteristics will aid in predicting the expected clinical trajectory and appropriateness of treatment technique, dose, and fractionation.

The following questions should be answered when starting the planning process for any radiosurgical case:

1. What is the appropriate dose required to obtain durable tumor control?
2. Are clinical target volume (CTV) margins needed to avoid a marginal recurrence?
3. What are the PTV margins required to account for setup error?
4. Is hypofractionation required to decrease the risk of treatment-related toxicity to an acceptable level?

8.2.2 IMAGING

A contrast-enhanced MRI should be obtained using a standardized SRS MRI protocol. We perform frameless MRI radiosurgery imaging on either a 3 T (Verio 3T, Siemens, Munich, Germany) or 1.5 T (Signa HDxt 1.5T, GE Healthcare, Little Chalfont, UK) MRI system. Our imaging protocol was designed to fit within a half-hour time slot and includes an initial isotropic T1-gradient echo sequence (GRE) noncontrast sequence lasting 4–5 minutes. Then, standard (0.1 mmol/kg) weight-based dosing of a high-relaxivity gadolinium contrast agent is administered via power injection at 2 cc/second followed by a 20cc saline flush. The time of the contrast injection is noted for all patients. Immediately following this, a volumetric T2-weighted acquisition is performed. The timing of this study permits an axial isotropic T1-GRE postcontrast scan to be performed at 10–15 minutes after contrast administration, which we have shown to be beneficial in detection of brain metastases (Kushnirsky et al., 2015). No delay between contrast administration and imaging is needed if not imaging metastatic disease.

The isotropic T1-weighted GRE imaging parameters are brain volume imaging (BRAVO) (GE 1.5 T) T1 weighted (repetition time [TR]/echo time [TE]/inversion time [TI] = 10.7/4.4–14/450 ms, flip angle = 13°) with 1.0 mm slice thickness, bandwidth 25 kHz, matrix 320 × 320, and field of view (FOV) 240 mm. We also employ magnetization prepared rapid acquisition gradient echo (MPRAGE) (Siemens 3 T) (TR/TE/TI = 1900/3.11/900 ms, flip angle = 9°) with 0.9 mm slice thickness (Figure 8.1). Occasionally, additional specialized pulse sequences or acquisitions will be useful to help delineate critical normal structures and TV, and the acquisition of these specialized pulse sequences will be best obtained with close coordination with a neuroradiologist's assistance to ensure studies of quality adequate for planning SRS.

Because MRI may introduce subtle geometric distortion that is inapparent to even a practiced observer, routine, periodic quality assurance of the imaging process with specialized phantoms to identify, reduce, and correct such distortion to acceptably low levels is important to assure the fundamental studies used to plan treatment are as spatially accurate as possible (Sun et al., 2014). Higher–field strength magnets have more issues with geometric distortion than lower–field strength magnets, which should be remembered when choosing a platform for performing imaging for planning SRS.

If the patient is having surgery prior to SRS, the timing of postoperative imaging acquisition for planning radiosurgery is a variable that may require careful consideration. Depending on the size of the cavity and the tumor histology, it may be prudent to allow time for postoperative cavity involution in order to decrease the volume of normal brain treated. Most shrinkage will occur within a few days of surgery, but more gradual shrinkage will occur in many cases over a much more protracted period of time, which may be advantageous for planning stereotactic irradiation, particularly for indolent tumors where there is no need to hasten radiation treatment delivery (Atalar et al., 2013).

8.2.3 SIMULATION AND LOCALIZATION

CT imaging that is requisite for planning HSRS can also help with assessing MRI scans' geometric accuracy, if enough landmarks in the region of interest can be identified in both studies to assure an accurate

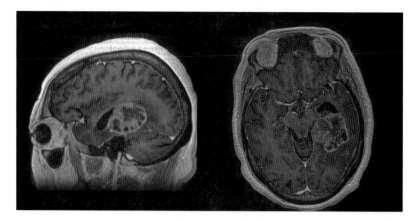

Figure 8.1 A 65-year-old patient with a history of stage III, TxN2M0 adenocarcinoma of the lung initially managed with definitive chemotherapy and radiation was treated with hypofractionated stereotactic radiosurgery (HSRS) after developing metastatic disease involving the CNS. She presented to the emergency room after developing aphasia and was found to have a large medial left temporal lobe metastasis seen clearly on contrast-enhanced T1 sequences as a 4.1 × 4.2 × 3.7 cm cystic peripherally enhancing lesion with central necrosis within the left temporal lobe associated with edema and brain stem compression with apparent mass effect on the left lateral ventricle. The patient was taken to the operating room where a left temporal craniotomy with resection of the metastasis and a partial hippocampectomy was performed. Eight weeks later, she was treated with HSRS to the resection bed and single-fraction stereotactic radiosurgery to a smaller right temporal metastasis that was not visualized at the time of surgery.

rigid registration has been achieved (Poetker et al., 2005). We use CT settings at the time of simulation of 120 kV, 512 mA, and a field of view that includes the entire cranium and any external localizing devices—from the vertex to the bottom of the C3 vertebral body. Helical scanning will permit reformatting to any desired slice thickness—we use reconstructed axial slices of 1.0–1.5 mm thickness. As a supplemental aid to achieving a reproducible setup, BBs (radio-opaque markers) are placed on the isocentric "origin" marks on the thermoplastic mask.

During the simulation, the patient is positioned supine on the simulator couch with arm position dictated by patient comfort. A knee pillow is provided, and the patient is straightened and aligned immediately prior to frameless stereotactic mask placement. A CT localizer base plate and localizer box are applied before simulating the patient with a noncontrast CT scan (Figure 8.2).

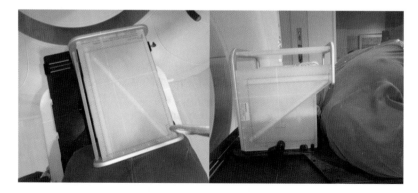

Figure 8.2 An N-type localizer box is used to enable conversion of 2D, planar tomographic coordinates into three dimensions to guide stereotactic targeting. It is essential that the patient is scanned to include the entire localizer box. Helical scanning will permit reformatting to any desired thickness although a 1–1.5 mm slice thickness is acceptable for planning.

8.2.4 3D RIGID IMAGE FUSION

Carrying out accurate rigid image registration is a critical next step. Algorithms in commercial software packages that provide acceptable registrations of mobile extracranial anatomy such as positron emission tomography (PET) and CT fusion for thoracic irradiation often falter at providing the accuracy paramount for intracranial radiosurgery. We have found that not every vendor's product can achieve accurate registrations without repetitive, iterative human interventions to guide the software's algorithmic processes. Algorithms capable of accurate 3D rigid registration can also be found on the Internet. The goal of the process is to carry out appropriate translations of the MRI study so that the voxels of the MRI scan can then be reformatted and displayed with the geometry of the treatment planning CT scan (Bond et al., 2003).

As a quality assurance measure, it is also essential to first check the dates of all scans to be certain the correct CT/MRI combinations are being fused (postop vs preop MRI or correct CT simulation date in the case of multiple past treatments at the same institution). Fusion of the T1-weighted, gadolinium-enhanced volumetric acquisition may be initiated manually with emphasis on alignment at the site of disease (anterior, middle, or posterior cranial fossa). This is not always necessary; tools that delineate a cuboidal region of interest "box" can be placed to encompass all relevant structures including tumor and adjacent bony or parenchymal anatomy and aligned using an automated algorithm such as normalized mutual information to rigidly align the two 3D studies. Fine or coarse adjustments to guide and refine the fusion process can be made accordingly until a 3D alignment has been achieved. For example, for a patient with a vestibular schwannoma, the intracanalicular portion of the tumor should fit perfectly into the bony internal auditory canal (Poetker et al., 2005).

The fusion should be verified in its entirety using one or more tools such as a "spyglass" or contrasting color phases to ensure accurately matching tumor volume and adjacent bony anatomy (Figure 8.3). No matter who performs this process, all aspects of a 3D rigid fusion should be independently verified by the radiation oncologist prior to segmentation of TV and normal structures. Failure to do so may potentially squander valuable time if errors are subsequently identified before treatment, and risks poorer tumor control and an increased risk of injury to the patient if errors inherent in this process are not identified prior to treatment.

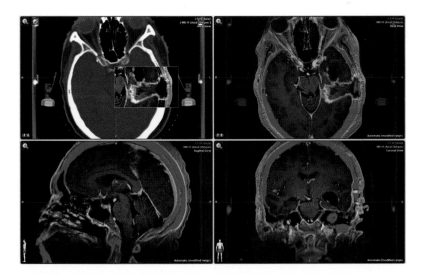

Figure 8.3 Fusion of the T1 post-spoiled gradient recalled (SPGR) diagnostic contrast-enhancing MRI sequence with the CT simulation scan is shown here using iPlan® treatment planning software (BrainLab AG, Munich, Germany). The "image fusion mode" is utilized to verify the image fusion between the stereotactic MRI and the treatment planning CT scan. With tools such as blue/amber phasing and the rectangular spyglass (top left panel), special attention is paid to the degree of bony anatomy alignment between CT (blue) and MRI (amber), which can be faded in and out to verify close agreement especially with respect to the left temporal region where the tumor resection bed is located in this case.

Fusion summary

1. SRS protocol contrast-enhanced MRI
2. T1-weighted postgadolinium volumetric MRI fused to dedicated treatment planning CT obtained using a customized immobilization mask
3. Set localization box to include tumor and/or postoperative cavity to also include adjacent fixed skull base or calvarial anatomy
4. Manually align to roughly approximate scans in all three planes
5. Autoalign
6. Verify fusion at multiple regions surrounding target area with spy glass window and with contrasting color phases before saving under name to include MRI and CT-simulation dates and initials of the operator
7. Physician (time-out) to independently verify dates of all MRIs and CT simulation are again correct and belong to correct patient and that the fusion is geometrically precise in the volume of greatest interest before contouring

8.2.5 TARGET VOLUME DELINEATION

During the target delineation phase, the GTV is outlined by the radiation oncologist and all critical structures (i.e., optic apparatus, brain stem, and spinal cord), whether autosegmented or segmented by residents or physics staff, are also adjusted and verified. It is important, rather than relying on reports alone, to review the proposed GTV with a neurosurgeon and a neuroradiologist before target delineation especially if there is any uncertainty as to the extent of postoperative change and residual disease. In this situation, postoperative pre- and postcontrast scans can be used to define the GTV more accurately. The GTV is defined to include all areas of suspicious enhancement plus the postoperative cavity in the case of prior resection.

A GTV expansion of 1 mm has been shown to be adequate in achieving an effective PTV when using six-degree-of-freedom registration with a frameless radiosurgical technique (Figures 8.4 and 8.5) (Dhabaan et al., 2012; Prabhu et al., 2013). Furthermore, data support the notion that small expansions of 1–2 mm to arrive at a PTV does not result in significantly increased marginal recurrence rates compared to infield recurrences after hypofractionated SRS to treat brain metastases (Eaton et al., 2013). A recently reported phase III trial evaluating 1 and 3 mm PTV expansions for single-fraction SRS has identified higher rates of

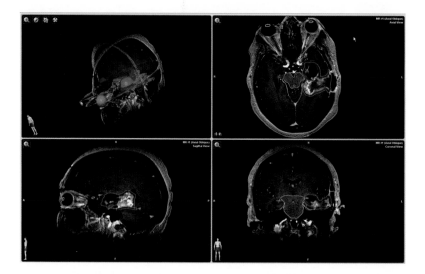

Figure 8.4 Once fusion is independently confirmed, contouring of the gross tumor volume can begin on the T1 postgadolinium MPRAGE (or SPGR) MRI sequence. The contour in red denotes the planning target volume and includes the entire resection bed encompassing all contrast-enhancing postoperative parenchymal tissue and fluid-filled cavity plus an additional 2 mm margin. The brain stem is highlighted in green, the chiasm in orange, and the optic nerves are seen in yellow.

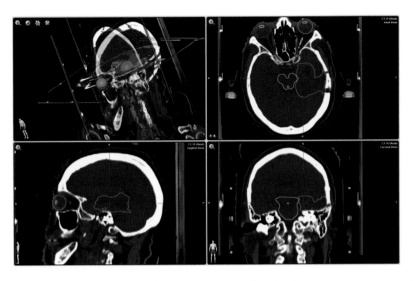

Figure 8.5 Attention is turned to the CT simulation next to verify the planning target volume, especially with respect to the bony anatomy of the skull.

radionecrosis with 3 mm PTV expansions (Kirkpatrick et al., 2015). It should be remembered that volume increases as the cube of the radius ($V = 4/3 \pi r_1 r_2 r_3$), and isometrically increasing the radius of a TV by a few millimeters increases the volume of the final PTV at a startling rate.

Based on studies reported from Stanford University, when treating a postoperative resection bed, we expand the margin by 2 mm; their data appear to show that attempts to be too meticulous in creating a small TV result in a higher rate of local recurrences in this clinical setting (Choi et al., 2012). We always confirm the GTV and PTV with the neurosurgeon involved with the case as a courtesy and to confirm that we have not inadvertently inaccurately defined these volumes.

Target delineation summary
1. GTV = contrast-enhancing gross (or residual disease to include operative cavity if postoperative) based on T1-weighted postcontrast volumetric MRI sequence
2. PTV = GTV + 1 mm (or 2 mm for postoperative resection beds)

8.2.6 PRESCRIPTION AND DOSE–FRACTIONATION SELECTION

Radiosurgical dose selection is primary dictated by tumor (target) volume. Increases in GTV have been shown to negatively correlate with local control (Eaton et al., 2013). As a rule of thumb, for roughly spherical lesions, the diameter from initial diagnostic imaging can be used in selecting an appropriate dose and fractionation schedule. With increasing spatial complexity across the spectrum from spherical toward irregularly shaped lesions, the amount of normal brain treated tends to increase, in part because of the difficulty with generating dose plans that adequately protect normal brain in concave areas of the TV. Some of the adjacent normal parenchymal tissue can be kept out of the treatment volume by relying on highly conformal techniques employing noncoplanar arc-based intensity-modulated treatment delivery. HSRS should be strongly considered once the GTV/cavity surpasses a volume of ~14 cc, which is roughly equivalent to a GTV sphere with a 1.5 cm radius. The corresponding volume for a postoperative PTV (assuming a 2 mm isometric expansion of the GTV) for which HSRS should be used is ~20–21 cc.

8.2.7 BED CALCULATIONS

The currency through which different dose–fractionation schedules are exchanged is the biologically effective dose (BED). It is based on experimentally observed linear–quadratic relationships between dose and tissue effect. BED is a function of the alpha/beta ratio; most malignant brain tumors are assumed to have an α/β of 10, but this is not the case for slower growing, benign neoplasms (Table 8.1).

Table 8.1 Biologically effective dose equivalent table for single-fraction stereotactic radiosurgery to 3 or 5 fraction (fx) hypofractionated stereotactic radiosurgery

Three fractions
15 Gy SRS = 22.5 Gy HSRS (3 fx)
18 Gy SRS = 27.4 Gy HSRS (3 fx)
20 Gy SRS = 30.8 Gy HSRS (3 fx)
22 Gy SRS = 34.1 Gy HSRS (3 fx)
24 Gy SRS = 37.5 Gy HSRS (3 fx)
Five fractions
15 Gy SRS = 26.5 Gy HSRS (5 fx)
18 Gy SRS = 32.6 Gy HSRS (5 fx)
20 Gy SRS = 36.8 Gy HSRS (5 fx)
22 Gy SRS = 41.0 Gy HSRS (5 fx)
24 Gy SRS = 45.3 Gy HSRS (5 fx)

Source: Reprinted from *Int. J. Radiat. Oncol. Biol. Phys.*, Vol 21(3), D.J. Brenner, M.K. Martel, E.J. Hall, Fractionated regimens for stereotactic radiotherapy of recurrent tumors in the brain, pp 819–824, Copyright (1991), with permission from Elsevier.

8.3 DOSIMETRY

8.3.1 BEAM GEOMETRY AND PLANNING

The ultimate goal in radiosurgical treatment planning is simply to cover the PTV with as close to 100% of the prescription dose as possible while, at the same time, maintaining a very sharp dose falloff at the PTV edge to preserve adjacent normal brain and critical structures. Many SRS delivery techniques described here and in other chapters have been devised to further achieving this goal. These include traditional approaches such as step-and-shoot fixed gantry methods and arc-based delivery approaches that have recently been adapted to include intensity modulation.

For smaller, roughly spherical lesions, circular collimators can be used to generate noncoplanar, arc-based plans that have very steep falloff of dose outside the TV. Dynamic arc therapy with high-resolution multileaf collimators can also generate highly conformal plans, but the dose falloff is not quite as steep because of the finite size of collimator leaves in adapting to lesion geometry. Novel techniques such as volumetric intensity-modulated arc-based therapy have been devised to improve conformality while not sacrificing the speed of treatment delivery (Figure 8.6). This approach achieves conformal dose distributions through simultaneously modulating the photon dose rate, gantry rotation speed, and beam shape/aperture through multileaf collimation (Wang et al., 2012; Zhao et al., 2015). This technique has been shown to offer much higher degrees of target conformality and lower dose to normal brain, particularly when noncoplanar arcs are used to deliver treatment (Zhao et al., 2015).

Field/arc placement should be determined with respect to the PTV location and its spatial relationship to critical structures. As an example, it may be prudent when feasible to avoid placing beams that would enter or exit through the eye/optic apparatus, cochlea, or brain stem during treatment, but inverse treatment planning approaches can minimize dose delivery to the critical normal tissues so delineated, or the degrees of arc through which treatment is delivered can be increased so as to render trivial the incidental dose delivered to these structures.

There are several advantages and disadvantages to using coplanar versus noncoplanar arcs. Advantages of noncoplanar arcs include the ability to improve plan conformality and dose homogeneity. Drawbacks include increased treatment delivery time, which in turn could lead to increased setup error. For this reason, we use stereotactic image guidance—either 2D kV x-ray pairs or, more rarely, cone-beam CT

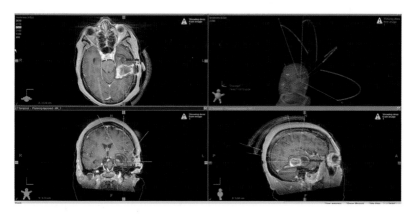

Figure 8.6 A hypofractionated stereotactic radiosurgery treatment plan generated with Aria® treatment planning software (Varian Medical Systems, Palo Alto, CA) using three noncoplanar arcs and 10 MV photons is visualized in axial, sagittal, and coronal planes. This approach was chosen to deliver a conformal dose to the planning target volume (PTV) through modulation of dose rate, gantry speed, as well as the beam aperture. A total dose of 27 Gy in 3 fractions was delivered to the 90% isodose surface using a Novalis TX® stereotactic linear accelerator (Varian Medical Industries, Palo Alto, CA; BrainLab AG, Munich, Germany) to the left temporal lobe PTV.

monitoring (using a tolerance of ≤0.5 mm for digitally reconstructed radiograph to portal registration) before each arc or after each couch position change to improve intrafractional accuracy.

8.3.2 PLAN ASSESSMENT AND QUALITY ASSURANCE

When reviewing a plan, it is the role of the physician to scrutinize the following:
1. Heterogeneity of coverage
2. What isodose surface provides 100% coverage
3. Size and location of "hot" and "cold" spots
4. Dose-volume histogram (DVH) constraints
5. Assess the degree of agreement between PTV and prescription isodose (PI) dose conformity indices

Heterogeneity of coverage (prescribing to a lower isodose surface) may be advantageous in terms of dose falloff outside the PTV and for intensifying dose delivery to tumor cells within the PTV. Similarly, having a small number of target voxels relatively underdosed may be appropriate, depending on the locations of intra-PTV hot spots, the treatment volume that would be irradiated to the prescription dose if these few voxels were included within the PI surface, and the risk of recurrence that these cold spots may introduce. Dose conformity indices can help with selection of PI surfaces. Different dose conformity indices exist, and each has advantages under different scenarios. The RTOG recommends that the PITV, defined as the ratio of the prescription isodose (PI) volume over the target volume (TV), be used; many others have been proposed to address inherent shortcomings of this approach (Knöös et al., 1998; Paddick, 2000; Wagner et al., 2003; Wu et al., 2003; Feuvret et al., 2006; Paddick and Lippitz, 2006).

8.3.3 DOSE VOLUME HISTOGRAM CONSTRAINTS

The "volume receiving 12 Gy" or V_{12} has been widely adopted as the standard method of reporting the dose to the normal brain in single-fraction radiosurgery procedures (Lawrence et al., 2010). This approach has been adapted for reporting outcomes for patients treated with HSRS. As an example, in a case prescribed 5 fractions of 7 Gy, the normal brain volume (normal brain—GTV) receiving >4 Gy/fraction ideally will not exceed 20 cc (Ernst-Stecken et al., 2006). As mentioned earlier, however, different locations will likely have different tolerances, and therefore the location treated should also be defined and reported (e.g., midbrain vs frontal cortex) for subsequent evaluation and refinement of treatment prescription philosophy.

8.3.4 OPTIC APPARATUS CONSTRAINTS

CT-MRI fusion (already performed for TV definition) is recommended as the best approach to define and contour the chiasm (Table 8.2). The optic nerves themselves can be clearly delineated on a CT scan alone given the conspicuity of the optic nerves within the fatty tissues of the orbit before they pass through the optic canals. Using the bony window will help delineate the optic nerve as it traverses the orbital apex. Intracranially, the volumetric MRI pulse sequences can clearly show the chiasm as a structure separate from the cerebrospinal fluid, intracranial arteries, and contiguous normal frontal cortex. It is essential to contour the entire optic apparatus in continuity through the proximal optic tracts.

Consistent agreement has been reached on the low risk of radiation-induced optic neuritis (RION) for a single-fraction D_{max} of <10 Gy, and one major study has indicated an acceptably low-risk RION with a single-fraction D_{max} of <12 Gy (Mayo et al., 2010). In the case of hypofractionation, a maximum dose limit to the chiasm of 19.5 Gy delivered in 3 fractions or 25 Gy delivered in 5 fractions has been quoted in the literature based on retrospective case series (Grimm et al., 2011).

8.3.5 COCHLEAR CONSTRAINTS

When cochlear sparing is at risk based on the proximity of the tumor volume to the petrous bone or in the case of treating vestibular tumors, several factors will determine the risk and thus potential for ipsilateral cochlear sparing (Table 8.3). Prior to treatment, a baseline audiometric evaluation should be obtained. For patients with useful hearing who will receive single-fraction SRS, the prescription dose to the margin of the vestibular schwannoma should be limited to ≤12 Gy to maximize the probability of hearing preservation, but a number of additional items must be attended to for optimizing hearing preservation, including the conformality of the treatment and protection of normal structures, including the brain stem and cochlea (Linskey, 2008).

There are single-fraction SRS data that appear to show that limiting the mean dose to the cochlea to ~4 Gy will improve the probability of hearing preservation. From limited data, hypofractionation (especially in the case of acoustic neuroma) has been shown to provide local durable tumor control and effectively preserve hearing (Bhandare et al., 2010; Hayden Gephart et al., 2013), but there are less data about what cochlear doses may optimally preserve hearing in these small series. After treatment, hearing should be tested with routine audiometry starting 6 months after RT and continuing twice yearly thereafter until any posttreatment changes become manifest and stabilize. This may take several years to occur; early, optimistic reports of hearing preservation with HSRS may become fainter and fainter as surveillance periods lengthen.

Table 8.2 Optic pathway dose constraints for avoidance of ≥ grade 3 optic neuritis

	1 FRACTION	3 FRACTIONS	4 FRACTIONS	5 FRACTIONS
Volume (cc)	<0.2 cc	<0.2 cc	<0.2 cc	<0.2 cc
Volume max (Gy)	8 Gy	15.3 Gy (5.1 Gy/fx)	19.2 Gy (4.8 Gy/fx)	23 Gy (4.6 Gy/fx)
Max point dose (Gy)	10 Gy	17.4 Gy (5.8 Gy/fx)	21.2 Gy (5.3 Gy/fx)	25 Gy (5 Gy/fx)

Source: Reprinted from *Semin Radiat Oncol*, Vol 18(4), R.D. Timmerman, An Overview of Hypofractionation and Introduction to This Issue of Seminars in Radiation Oncology, pp 215–222, Copyright (2008), with permission from Elsevier.

Table 8.3 Cochlear dose constraints to avoid ≥ grade 3 hearing loss

	1 FRACTION	3 FRACTIONS	4 FRACTIONS	5 FRACTIONS
Max point dose (Gy)	9 Gy	17.1 Gy (5.7 Gy/fx)	21.2 Gy (5.3 Gy/fx)	25 Gy (5 Gy/fx)

Source: Reprinted from *Semin Radiat Oncol*, Vol 18(4), R.D. Timmerman, An Overview of Hypofractionation and Introduction to This Issue of Seminars in Radiation Oncology, pp 215–222, Copyright (2008), with permission from Elsevier.

Table 8.4 Brain stem (not medulla) dose constraints to avoid ≥ grade 3 cranial neuropathy

	1 FRACTION	3 FRACTIONS	4 FRACTIONS	5 FRACTIONS
Volume (cc)	<0.5 cc	<0.5 cc	<0.5 cc	<0.5 cc
Volume max (Gy)	10 Gy	18 Gy (6 Gy/fx)	20.8 Gy (5.2 Gy/fx)	23 Gy (4.6 Gy/fx)
Max point dose (Gy)	15 Gy	23.1 Gy (7.7 Gy/fx)	27.2 Gy (6.8 Gy/fx)	31 Gy (6.2 Gy/fx)

Source: Reprinted from *Semin Radiat Oncol*, Vol 18(4), R.D. Timmerman, An Overview of Hypofractionation and Introduction to This Issue of Seminars in Radiation Oncology, pp 215–222, Copyright (2008), with permission from Elsevier.

8.3.6 BRAIN STEM CONSTRAINTS

Constraining dose delivery to the brain stem is essential for some cases during planning (Table 8.4). The dose delivered to the brainstem and the volume of brainstem treated relate importantly to complication probabilities. For single-fraction SRS, a maximum brain stem dose of 12.5 Gy is associated with low (<5%) risk of complications. Higher doses (15.2 Gy) to small volumes have been used with a low reported incidence of complications in patient groups with a poor prognosis for long-term survival (e.g., brain stem metastases) (Kased et al., 2008; Lorenzoni et al., 2009).

REFERENCES

Atalar B, Choi CY, Harsh GR, Chang SD, Gibbs IC, Adler JR, Soltys SG (2013) Cavity volume dynamics after resection of brain metastases and timing of postresection cavity stereotactic radiosurgery. *Neurosurgery* 72:180–185.

Bhandare N, Jackson A, Eisbruch A, Pan CC, Flickinger JC, Antonelli P, Mendenhall WM (2010) Radiation therapy and hearing loss. *Int J Radiat Oncol Biol Phys* 76:S50–S57.

Bond JE, Smith V, Yue NJ, Knisely JPS (2003) Comparison of an image registration technique based on normalized mutual information with a standard method utilizing implanted markers in the staged radiosurgical treatment of large arteriovenous malformations. *Int J Radiat Oncol Biol Phys* 57:1150–1158.

Brenner DJ, Martel MK, Hall EJ (1991) Fractionated regimens for stereotactic radiotherapy of recurrent tumors in the brain. *Int J Radiat Oncol Biol Phys* 21:819–824.

Choi CYH, Chang SD, Gibbs IC, Adler JR, Harsh GR, Lieberson RE, Soltys SG (2012) Stereotactic radiosurgery of the postoperative resection cavity for brain metastases: Prospective evaluation of target margin on tumor control. *Int J Radiat Oncol Biol Phys* 84:336–342.

Dhabaan A, Schreibmann E, Siddiqi A, Elder E, Fox T, Ogunleye T, Esiashvili N, Curran W, Crocker I, Shu HK (2012) Six degrees of freedom CBCT-based positioning for intracranial targets treated with frameless stereotactic radiosurgery. *J Appl Clin Med Phys* 13:3916.

Eaton BR, Gebhardt B, Prabhu R, Shu H-K, Curran WJ, Crocker I (2013) Hypofractionated radiosurgery for intact or resected brain metastases: Defining the optimal dose and fractionation. *Radiat Oncol* 8:135.

Ernst-Stecken A, Ganslandt O, Lambrecht U, Sauer R, Grabenbauer G (2006) Phase II trial of hypofractionated stereotactic radiotherapy for brain metastases: Results and toxicity. *Radiother Oncol* 81:18–24.

Feuvret L, Noël G, Mazeron JJ, Bey P (2006) Conformity index: A review. *Int J Radiat Oncol Biol Phys* 64:333–342.

Flickinger JC, Kondziolka D, Pollock BE, Maitz AH, Lunsford LD (1997) Complications from arteriovenous malformation radiosurgery: Multivariate analysis and risk modeling. *Int J Radiat Oncol Biol Phys* 38:485–490.

Grimm J, LaCouture T, Croce R, Yeo I, Zhu Y, Xue J (2011) Dose tolerance limits and dose volume histogram evaluation for stereotactic body radiotherapy. *J Appl Clin Med Phys* 12:3368.

Hall EJ, Brenner DJ (1993) The radiobiology of radiosurgery: Rationale for different treatment regimes for AVMs and malignancies. *Int J Radiat Oncol Biol Phys* 25:381–385.

Hayden Gephart MG, Hansasuta A, Balise RR, Choi C, Sakamoto GT, Venteicher AS, Soltys SG et al. (2013) Cochlea radiation dose correlates with hearing loss after stereotactic radiosurgery of vestibular schwannoma. *World Neurosurg* 80:359–363.

Kased N, Huang K, Nakamura JL, Sahgal A, Larson DA, McDermott MW, Sneed PK (2008) Gamma Knife radiosurgery for brainstem metastases: The UCSF experience. *J Neurooncol* 86:195–205.

Kirkpatrick JP, Wang Z, Sampson JH, McSherry F, Herndon JE, Allen KJ, DuffyE et al. (2015) Defining the optimal planning target volume in image-guided stereotactic radiosurgery of brain metastases: Results of a randomized trial. *Int J Radiat Oncol* 91:100–108.

Knöös T, Kristensen I, Nilsson P (1998) Volumetric and dosimetric evaluation of radiation treatment plans: Radiation conformity index. *Int J Radiat Oncol Biol Phys* 42:1169–1176.

Kushnirsky M, Nguyen V, Katz JS, Steinklein J, Rosen L, Warshall C, Schulder M, Knisely JP (2015) Time delayed contrast enhanced MRI improves detection of brain metastases and apparent treatment volumes. *J Neurosurg* 11:1–7. [Epub ahead of print].

Lawrence YR, Li XA, El NI, Hahn CA, Marks LB, Merchant TE, Dicker AP (2010) Radiation dose-volume effects in the brain. *Int J Radiat Oncol Biol Phys* 76:S20–S27.

Levegrün S, Hof H, Essig M, Schlegel W, Debus J (2004) Radiation-induced changes of brain tissue after radiosurgery in patients with arteriovenous malformations: Correlation with dose distribution parameters. *Int J Radiat Oncol Biol Phys* 59:796–808.

Lightstone AW, Benedict SH, Bova FJ, Solberg TD, Stern RL (2005) Intracranial stereotactic positioning systems: Report of the American Association of Physicists in Medicine Radiation Therapy Committee Task Group no. 68. *Med Phys* 32:2380–2398.

Linskey ME (2008) Hearing preservation in vestibular schwannoma stereotactic radiosurgery: What really matters? *J Neurosurg* 109(Suppl):129–136.

Lorenzoni JG, Devriendt D, Massager N, Desmedt F, Simon S, Van Houtte P, Brotchi J, Levivier M (2009) Brain stem metastases treated with radiosurgery: Prognostic factors of survival and life expectancy estimation. *Surg Neurol* 71:188–196.

Mayo C, Martel MK, Marks LB, Flickinger J, Nam J, Kirkpatrick J (2010) Radiation dose-volume effects of optic nerves and chiasm. *Int J Radiat Oncol Biol Phys* 76:S28–S35.

Mori Y, Kondziolka D, Flickinger JC, Kirkwood JM, Agarwala S, Lunsford LD (1998) Stereotactic radiosurgery for cerebral metastatic melanoma: Factors affecting local disease control and survival. *Int J Radiat Oncol Biol Phys* 42:581–589.

Oermann EK, Kress MA, Todd JV, Collins BT, Hoffman R, Chaudhry H, Collins SP, Morris D, Ewend MG (2013) The impact of radiosurgery fractionation and tumor radiobiology on the local control of brain metastases. *J Neurosurg* 119:1131–1138.

Paddick I (2000) A simple scoring ratio to index the conformity of radiosurgical treatment plans. Technical note. *J Neurosurg* 93(Suppl 3):219–222.

Paddick I, Lippitz B (2006) A simple dose gradient measurement tool to complement the conformity index. *J Neurosurg* 105(Suppl):194–201.

Poetker DM, Jursinic PA, Runge-Samuelson CL, Wackym PA (2005) Distortion of magnetic resonance images used in gamma knife radiosurgery treatment planning: Implications for acoustic neuroma outcomes. *Otol Neurotol* 26:1220–1228.

Prabhu RS, Dhabaan A, Hall WA, Ogunleye T, Crocker I, Curran WJ, Shu H-JS (2013) Clinical outcomes for a novel 6 degrees of freedom image guided localization method for frameless radiosurgery for intracranial brain metastases. *J Neurooncol* 113:93–99.

Shaw E, Scott C, Souhami L, Dinapoli R, Kline R, Loeffler J, Farnan N (2000) Single dose radiosurgical treatment of recurrent previously irradiated primary brain tumors and brain metastases: Final report of RTOG protocol 90-05. *Int J Radiat Oncol Biol Phys* 47(2):291–298.

Shiau CY, Sneed PK, Shu HK, Lamborn KR, McDermott MW, Chang S, Nowak P et al. (1997) Radiosurgery for brain metastases: Relationship of dose and pattern of enhancement to local control. *Int J Radiat Oncol Biol Phys* 37:375–383.

Soltys SG, Seiger K, Modlin LA, Gibbs IC, Hara W, Kidd EA, Hancock SL, et al. (2015) A phase I/II dose-escalation trial of 3-fraction stereotactic radiosurgery (SRS) for large resection cavities of brain metastases. *Int J Radiat Oncol Biol Phys* 93:S38.

Sun J, Barnes M, Dowling J, Menk F, Stanwell P, Greer PB (2014) An open source automatic quality assurance (OSAQA) tool for the ACR MRI phantom. *Australas Phys Eng Sci Med* 38:39–46. Available at: http://link.springer.com/10.1007/s13246-014-0311-8.

Timmerman RD (2008) An overview of hypofractionation and introduction to this issue of seminars in radiation oncology. *Semin Radiat Oncol* 18:215–222.

Wagner TH, Bova FJ, Friedman WA, Buatti JM, Bouchet LG, Meeks SL (2003) A simple and reliable index for scoring rival stereotactic radiosurgery plans. *Int J Radiat Oncol Biol Phys* 57:1141–1149.

Wang JZ, Pawlicki T, Rice R, Mundt AJ, Sandhu A, Lawson J, Murphy KT (2012) Intensity-modulated radiosurgery with rapidarc for multiple brain metastases and comparison with static approach. *Med Dosim* 37:31–36.

Wu Q-RJ, Wessels BW, Einstein DB, Maciunas RJ, Kim EY, Kinsella TJ (2003) Quality of coverage: Conformity measures for stereotactic radiosurgery. *J Appl Clin Med Phys* 4:374–381.

Zhao B, Yang Y, Li X, Li T, Heron DE, Huq MS (2015) Is high-dose rate RapidArc-based radiosurgery dosimetrically advantageous for the treatment of intracranial tumors? *Med Dosim* 40:3–8.

9 Principles of image-guided hypofractionated radiotherapy of spine metastases

Johannes Roesch, Stefan Glatz, and Matthias Guckenberger

Contents

9.1 INTRODUCTION

With its increasing use in recent years, stereotactic radiotherapy (SRT) has opened up a broad field of indications. In 2011, a survey among radiation oncologists in the United States showed that 39% of the physicians were using stereotactic body radiotherapy (SBRT) for the treatment of spinal tumors. Both SBRT users and nonusers were planning to increase the number of SBRT treatments (Pan et al., 2011). Likewise, a survey in six European countries reported that one-third of the centers were practicing SBRT for vertebral tumors and half of the centers considered the currently available evidence as sufficient for routine treatment outside of clinical trials (Dahele et al., 2015). New techniques and technology have opened the way to improve old and achieve new objectives of spinal irradiation. Pain relief and prevention/reduction of morbidities from bone metastases, like neurological deficits or bony instability, are still the main goals (Lutz et al., 2011). Furthermore, disease control in selected oligometastatic cases and delay of systemic progression are emerging indications. Recent developments and growing experience in this field have led to recommendations particularly pertaining to delineation and image-guided treatment delivery.

9.2 TARGET VOLUME AND ORGANS-AT-RISK DEFINITION

9.2.1 CLINICAL INFORMATION AND IMAGING MODALITIES

Similar to evaluation for other treatment modalities, the patient's general condition, symptomatology, and the physician's assessment provide useful information with regard to the patient's suitability for SBRT. Notably, localization and characteristics of the pain and neurological deficits like paresis, dysesthesia, autonomic dysfunction, or back pain resulting from compression of spinal nerves or the spinal cord itself can add important information on the location and extent of involvement of the vertebral metastases.

Diagnostic imaging is necessary to determine the precise local tumor extension, and the overall disease burden of the spine and the entire body. Furthermore, it can help in patient selection, identification of possible contraindications, and treatment planning. In an international survey of five highly SBRT-experienced centers, spinal instability or neurologic deficit resulting from bony compression of neural structures were regarded as exclusion criteria for SBRT (Guckenberger et al., 2011). Symptomatic spinal cord compression was unanimously regarded to be a contraindication for SBRT by all five centers. Epidural involvement alone impacted the decision for or against SBRT only in two centers. Compression fractures and vertebral instabilities were discussed with surgeons in all five institutions, and there was consensus that surgical intervention would be offered prior to SBRT, especially among the four North American centers. The current Radiation Therapy Oncology Group (RTOG) 0631 study protocol excludes patient with compression fractures causing spinal instability, more than 50% loss of vertebral body height, as well as spinal cord compression or displacement or epidural compression within 3 mm of the spinal cord due to tumor or bony fragments (Ryu et al., n.d.).

The whole spine is frequently assessed by T1- and T2-weighted sagittal magnetic resonance imaging (MRI) with additional pulse sequences and imaging planes, for example, short tau inversion recovery and axial T1/T2 sequences of suspicious vertebrae (Dahele et al., 2011). In cases where spinal cord compression is suspected, MRI is absolutely the preferred imaging technique (Loblaw et al., 2005). Pathological lesions present usually hypointense to fatty bone marrow on native T1-weighted images, and hyperintense when gadolinium-enhanced or T2-weighted imaging is used (Vanel, 2004). Bone scintigraphy, being another instrument to assess the disease burden of the spine, was previously described to be inferior to MRI (Algra et al., 1991).

9.2.2 IMAGING IN TREATMENT PLANNING

The importance of high-quality imaging is critical given the requirement for accurate target and normal tissue delineation and the steep dose gradients between organs at risk (OARs) such as spinal cord and cauda equina and the target volume.

The basic imaging modalities for treatment planning are widely agreed upon. The planning CT scan should be obtained with maximum 2 mm slice thickness in axial direction (Guckenberger et al., 2011; Cox et al., 2012). Contrast enhancement is recommended especially in cases of difficult tumor and OAR discrimination (Dahele et al., 2011). Although delineation of target volume and OAR without registration of additional diagnostic imaging such as MRI has been reported (Ryu et al., 2003; Gibbs et al., 2007), the full extent of the tumor may not be easily visible in the presence of bone destruction or paraspinal and epidural tumor spread. Furthermore, to visualize the spinal cord, CT alone is suboptimal, and spinal cord volumes can be better delineated by using MRI (Geets et al., 2005). Therefore, a dedicated planning MRI is common practice in many centers. T1 sequence with contrast enhancement and a slice thickness of 1–2 mm is reported to be the most commonly used (Guckenberger et al., 2011). Furthermore, T1 sequence without contrast, T2 sequence, and least commonly fluid-attenuated inversion recovery (FLAIR) sequences can help to define the exact tumor extent especially in cases where there is tumor infiltration of the surrounding soft tissue or the epidural space (Loughrey et al., 2000). Moreover, fat suppressed or heavily T2-weighed sequences like fast spin echo (FSE) or constructive interference in steady state are used as myelographic MRI sequences and are used to define the spinal cord.

Standard diagnostic MRI studies are usually suitable for treatment planning, but reimaging is indicated when there are patient position discrepancies impacting the CT-MRI registration accuracy, surgery in the time interval between imaging and radiotherapy, and the development of new symptoms.

CT myelography can be used as an alternative to MRI but is more invasive and less available. Nonetheless, it was shown to provide excellent differentiation of the spinal cord within submillimeter accuracy (Thariat et al., 2009). Therefore, it should be considered in cases where MRI is not feasible due to claustrophobia, or presence of cardiac pacer, or when there is presence of metal implants. Potential side effects include headache, infection, neurological deficit, and allergic reaction. Another problem of CT myelography is that it might be impeded by contrast blockage due to the tumor or postoperative adhesions.

References to 4D CTs to account for breathing-related organ motion are rare. Nelson et al. (2009) reported stable axial position of both the tumor and the spinal cord in respiration-correlated 4D CT and, therefore, did not recommend its use in standard practice. Dieleman et al. (2007) described significant mobility especially of the distal esophagus within 8–9 mm. Whether this has an influence on treatment planning or plan evaluation remains unclear.

In the postoperative setting, reduction of metal-related artifacts in CT and MRI can be achieved through various ways (Stradiotti et al., 2009). Orthopedic metal implants tend to absorb a high percentage of radiation from the CT scanner, causing incomplete or faulty projections. The artifacts are formed during image reconstruction and may be reduced by filter algorithms, adjustment of tube current, and others. These artifacts are related to the materials of x-ray beam attenuation coefficient, favoring titanium before stainless steel or cobalt chrome implants (Stradiotti et al., 2009). Artifacts in MRI occur due to magnetic field inhomogenities, causing geometrical distorsion on the resulting image. Avoidance strategies include optimal patient positioning in the center of the field, the usage of small voxel volumes, and metal artifact reduction sequences, such as FSE.

PET imaging is currently not routinely used for tumor delineation in spinal SBRT. However, its use in retreatment of postoperative spine cases has been reported (Gwak et al., 2006). In three cases, a reduction of the tumor volume by a factor of 2.2 based on the PET-based versus the CT-based planning was described resulting in local control at 6 months in two of three cases. Besides tumor delineation, PET imaging might be a suitable imaging modality to detect recurrence.

9.2.3 IMAGE REGISTRATION

Although not done at every institution, image registration of MRI to the planning CT is common practice and highly recommended if possible, especially in cases of vertebral metastases with a soft tissue component (Dahele et al., 2011; Guckenberger et al., 2012). Image registration errors are reported to be small in most studies. However, the accuracy of rigid image registration of MRI and CT datasets can be affected by different factors. Intrascanning motion, especially when patients suffer from pain, can affect image quality of the CT and the MRI and therefore impede the registration. Motion between different MRI sequences and between acquisitions of different image modalities is another important factor. General technical limitations are reported to be <1 mm in the cranial area (Nakazawa et al., 2014). Ideally, the MRI would be acquired in treatment position by using compatible immobilization devices. Rigid image registration instead of deformable registration is still recommended despite the flexibility of the vertebral column. Nonrigid image registration, an intense focus of research, might play a role in the future (Crum et al., 2004).

9.2.4 FRACTIONATION

Widely accepted doses in SBRT for vertebral metastases are 16–24 Gy delivered in 1 fraction and 12 Gy × 2, 7–10 Gy × 3, or 6–8 Gy × 5 for the multiple-fraction schemes (Pan et al., 2011). Doses are often prescribed to an isodose line of 80% for single-fractionation or 90%–100% for multiple-fractionation schemes, which covers the planning treatment volume.

Although metastatic spinal cord compression is a relative contraindication, SBRT might gain relevance in this situation, and there are data emerging. Ryu et al. reported a series where 85 lesions in 62 patients with epidural compression, but with a motor strength of at least four out of five, were treated with single-fraction SBRT. With reported neurological function improvement of 81%, SBRT appears to have an advantage compared to conventional irradiation in these cases; however, these data are preliminary and retrospective in nature (Ryu et al., 2010). It should be noted that the evaluation of these patients is a challenge as they often are in palliative care and, upon relapse, are less likely to be aggressively evaluated given the severity of the neurologic deficits.

Table 9.1 Spinal cord tolerance doses and corresponding biologically equivalent doses to the epidural tumor

	1 FRACTION (Gy)	3 FRACTIONS (Gy)	5 FRACTIONS (Gy)	10 FRACTIONS (Gy)	20 FRACTIONS (Gy)
Spinal cord tolerance (total physical dose) Equivalent to minimum dose to the epidural tumor component	10	18	23	35	45
Biologically equivalent dose to the epidural tumor component in 2 Gy fractions (EQD2/10)	17	24	29	39	46

Radiobiological modeling may suggest an advantage of moderate hypofractionation compared to single-fraction SBRT in the situation of epidural disease spread. Table 9.1 shows the relationship between number of treatment fractions, accepted spinal cord tolerance doses, and the effective dose to the epidural tumor component, converted in 2 Gy equivalent doses (alpha/beta ratio 10 Gy). Consequently, relative underdosing of the epidural disease may occur with single-fraction SBRT in particular. In fact, a lower dose to the epidural disease may be observed in single-fraction SBRT as compared to higher-dose conventional palliative radiation approaches like 30 Gy in 10 fractions.

9.2.5 ORGANS-AT-RISK DEFINITION

There is wide acceptance that a homogeneous dose of 50 Gy to the spinal cord using conventional fractionation (1.8–2.0 Gy/day) holds a low risk of myelopathy (Emami et al., 1991; Sahgal et al., 2008; Kirkpatrick et al., 2010), whereas the 5% incidence level is reached between 57 and 61 Gy (Schultheiss et al., 1995). In SBRT of the spine, isodose distributions to the spinal cord differ substantially from those of conventional radiotherapy. Smaller volumes are irradiated, and the dose gradient is frequently within the spinal cord itself. In addition to the well-described dose–response effect, a volume effect for spinal cord toxicity is clearly established (Hopewell et al., 1987; Bijl et al., 2002). A nonlocal repair mechanism, acting from nonirradiated to irradiated tissue covering a distance of 2–3 mm was postulated as part of the explanation (Bijl et al., 2002; van Luijk et al., 2005). Based on mouse model, Bijl et al. (2003) reported an increase in paralysis if adjacent spinal cord suffers from a low-dose bath, or treated lesions are too close together.

Despite the common usage of SBRT, the incidence of myelopathy after SBRT is low in the literature. To avoid spinal cord injury, Ryu et al. (2007) proposed that 10 Gy should not exceed 10% of the actual spinal cord volume defined as 6 mm above and below the radiosurgery target. This statement was based on their data of one case of myelopathy out of 86 patients treated with spinal SBRT and a minimum follow-up of at least 1 year. Combined data from Stanford University Medical Center and University of Pittsburgh Medical Center reported a radiation myelopathy incidence of 5 out of 1075 patients, and no specific dosimetric factors contributing to this complication could be identified. The authors concluded that biological equivalent dose estimates were not useful for defining spinal cord tolerance to hypofractionated dose schedules and limiting the volume of spinal cord treated above 8 Gy in a single fraction to 1 cm^3 at maximum was recommended (Gibbs et al., 2009). For stereotactic treatment, incidence of radiation myelopathy is estimated to be less than 1% for a maximum dose in the spinal cord of 13 Gy within 1 fraction or 20 Gy within 3 fractions (Kirkpatrick et al., 2010). Guckenberger et al. (2012) restricted the maximum dose received by 0.1 cm^3 of the planning spinal cord to 23.75 Gy in 5 fractions or 35 Gy in 10 fractions.

Sahgal et al. analyzed a group of nine irradiation naïve patients who underwent spine SBRT in multiple centers and developed radiation-induced myelopathy. In one case, SBRT was given as a boost to conventional hypofractionated radiotherapy of 30 Gy given in 10 fractions. This group was compared to a control cohort of 66 patients, who did not suffer radiation myelopathy after SBRT. A significant difference in small volume analysis and in maximum point volume dose was found between the two groups. Based on these data, a logistic regression model was used to generate a probability profile for human radiation myelopathy (RM) after SBRT treatment of the spine. The authors concluded that a maximum point dose inside the thecal sac of 12.4 Gy for single-fraction 17.0 Gy in 2 fractions, 20.3 Gy in 3 fractions, 23.0 Gy in 4 fractions, and 25.3 Gy in 5 fractions poses a low risk for radiation myelopathy (<5%) (Sahgal et al., 2013). Therefore, this group shares the opinion that dose hot spots in particular affect spinal cord tolerance following SBRT (Sahgal et al., 2012a).

In a similar study, Sahgal et al. compared a group of 5 patients suffering from radiation-induced myelopathy after conventional radiotherapy and SBRT reirradiation of the spine to 14 patients without radiation-induce myelopathy. Again, a significant dosimetric difference was found for the normalized biologically effective point maximum volume dose (P_{max} nBED). Recommendations were given to keep the P_{max} nBED dose within the thecal sac to no more than 20–25 $Gy_{2/2}$, not to exceed a cumulative P_{max} nBED above 70 $Gy_{2/2}$ considering conventional and stereotactic treatment and to have at least a time interval of 5 months between both treatments (Sahgal et al., 2012b). A comprehensive review is provided in Chapter 20.

With regard to delineation, the spinal cord should be defined as seen in the coregistered T2-weighted MRI including at least one vertebral level superior and inferior (SI) to the planning target volume (PTV) (Guckenberger et al., 2011, 2012). The same extension applies for the thecal sac, which serves as the surrogate for the cauda equina. If MRI is not available, the thecal sac can be defined using the treatment planning CT.

Spinal cord motion has been evaluated in a few studies. A cardiac pulse triggered wavelike motion pattern has been described (Figley and Stroman, 2007). Cai et al. (2007) and Figley and Stroman (2007) have described that the mean spinal cord motion is <0.5–0.6 mm with the largest axial amplitude in the anterior–posterior direction. A safety margin or planning-at-risk volume (PRV) expansion of 1–2 mm is often used for treatment planning.

Depending on the dose constraints, the spinal canal as seen on the CT might be used as a surrogate for the spinal cord (Dahele et al., 2011). This accounts for a small safety margin of the organ at risk and at the same time is much better to define on the planning CT than the actual structure (Gerszten et al., 2007). However, this may result in suboptimal coverage of the clinical target volume (CTV), leading to a higher risk of recurrence.

Apart from the spinal cord, other OARs such as esophagus, nerve plexuses, nerve roots, ureters, lung, and liver should also be contoured (Dahele et al., 2011). Compared to SBRT for other body sites, contouring of those structures is not different for spine SBRT and should be done according to existing guidelines (e.g., RTOG contouring atlases).

9.2.6 PATTERNS OF FAILURE

Local and/or clinical tumor recurrence occurs in 6%–23% of cases after spinal SBRT (Chang et al., 2007; Sahgal et al., 2008). Tumor recurrence can be found within, at the edge of, or outside the target volume. Whereas infield recurrence might be more dependent on tumor phenotype (Gerszten et al., 2007), failures occurring contiguous to the target volume may correlate with the target delineation.

Patterns of failure after SBRT for spinal metastases have been reported. Within the immediate superior or inferior vertebrae, neither Gerszten et al. nor Ryu et al. observed any case of tumor progression (Ryu et al., 2004; Gerszten et al., 2007). In a series of 74 cases, Chang et al. (2007) reported tumor progression occurring in the adjacent vertebra in one case. In the same study, the authors identified 17 image-documented cases of progression, and the main sites of recurrence were the epidural space (8/17) (due in part to underdosing as a result of spinal cord constraints) and untreated pedicles, posterior elements (3/17), or posterior parts of the vertebral body (2/17). In this study, the vertebral body alone without the posterior elements was included into the target volume.

Similarly, Nelson et al. (2009) reported on 4/33 recurrences, which were located either in direct relation to the epidural space or the pedicles. In these cases, the PTV included the gross tumor volume (GTV) plus a margin of 3–6 mm. Gerszten et al. (2005) reported six cases of recurrence (6/60) based on a series of renal cell carcinoma spine metastases and SBRT treatments targeting GTV only. Since the recurrences occurred often at the edge of the target volume, the group recommended further inclusion of the adjacent normal-appearing vertebral body. Sahgal et al. (2009) reported of 60 cases treated in a similar way, and the majority of the eight cases of recurrence were situated within 1 mm around the spinal cord. Other sites of recurrence such as within the paraspinal region are rare or directly related to infield recurrence (Chang et al., 2007). Patel et al. (2012) correlated local control rates of 154 extradural spinal lesions to the extent of the vertebral body irradiation. Although not statistically significant, a trend for lower 2-year local tumor progression was found for treatment of the whole vertebral body (21.1%) compared to partial vertebral body irradiation (34.9%).

The question of GTV-to-CTV margins remains open. Neither Sahgal et al. nor Dahele et al. could establish an association between outcome and different delineation concepts in their reviews (Sahgal et al., 2008; Dahele et al., 2011), and differences in opinions about delineation concepts, dose fractionations, and treatment techniques most likely impeded the analysis. However, based on all data available, the main risk for recurrence is at the sites of underdosing or geographic miss of the tumor, either because of proximity to critical OAR or exclusion of imaging-defined uninvolved vertebral segments like the posterior vertebral elements.

9.2.7 TARGET VOLUME DEFINITION

Sahgal et al. (2008) described two fundamentally different concepts of target volume definition. One is to contour only the radiographically visible tumor without application of anatomic margins for potential sites of microscopic disease, as derived from the classical intracranial stereotactic radiosurgery teachings. This GTV may then be directly transformed into the PTV by adding geometrical margins ranging from 0 mm (GTV = PTV) (Gerszten et al., 2007; Sahgal et al., 2007) to 2 mm (Gibbs et al., 2007). The second is to apply a CTV to include anatomic areas at risk within the spinal segment and then apply the PTV (Guckenberger et al., 2012) (Figure 9.1).

The increasing usage of spinal SBRT and the necessity of outcome comparison across different institutions, treatment platforms, and dose fractionation schedules led to formation of consensus definitions to standardize nomenclature and delivery specific to spinal SBRT. In this context, recommendations for target volume definition based on the principal concept of a GTV, CTV, and PTV were established. The International Spine Radiosurgery Consortium published recommendations for target delineation based on a series of cases with the CTV delineated by an expert group of radiation oncologists and neurosurgeons (Cox et al., 2012). Cox et al. suggested the implementation of a modified Weinstein–Boriani–Biagini (WBB) system for evaluation of tumor extent (Figure 9.2). In that study, the target volumes of 10 spinal SBRT cases were independently contoured by 10 physicians and analyzed for consistency by a mathematical model. The GTV represented the complete extent of gross metastatic disease identified by clinical information and all available imaging, including paraspinal and epidural tumor. The contoured CTV was consistently shown to account for tissues suspicious for subclinical microscopic invasion, like regions with abnormal bone marrow signal, by all participating institutions. As a rule, it entirely includes all affected and, if in relation to, adjacent sectors according to the suggested WBB concept. In most cases, the whole involved vertebral body was contoured, and with regard to the GTV extension, the ipsilateral and/or contralateral pedicle was included into the CTV. Direct infiltration of pedicles leads to further enclosure of the lamina. If both pedicles and lamina are involved, the entire posterior element is required to be included. An extraosseous CTV expansion is not necessary if the GTV is strictly confined to the bone.

There is variation in practice as to treating the CTV to a single prescribed dose or to adapt the treatment using a simultaneous integrated boost technique. For example, the target volume concept as published in the dose-intensified image-guided fractionated radiosurgery for spinal metastases (DOSIS) study is a two-dose-level approach with high-risk and low-risk target volumes (Guckenberger et al., 2012). The low-risk PTV includes the entire vertebra of the involved levels with appropriate margin to account for treatment setup errors. The high-risk PTV is defined accordingly to the WBB system described earlier. The survey published by Guckenberger et al. (2011) discussed additional important aspects. No more than three vertebral levels should be treated as one single target volume to avoid uncertainties due to deformation.

Margin expansion of CTV to PTV depends on inter- and intrafraction motion management, treatment platforms, immobilization systems, fractionation, method of prescription, and magnitude of dose. In general, a 3D expansion around the CTV of 3 mm or less wherever applicable is recommended

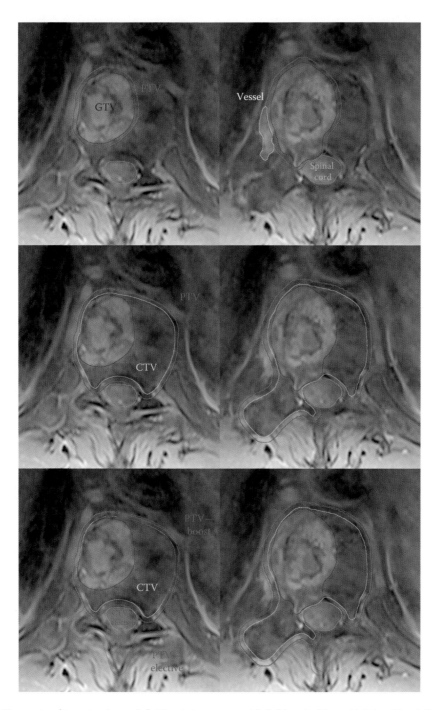

Figure 9.1 Concepts of target volume definition in two cases with (left) and without (right) epidural disease spread; contoured according to an MRI T1 sequence. Concept 1, planning target volume (PTV) = gross tumor volume + 0–1 mm; concept 2, clinical target volume including affected and adjacent parts of vertebral level; concept 3, two-step dose concept according to DOSIS study (Guckenberger et al., 2012) with PTV boost of affected and adjacent parts and PTV elective including whole vertebral level.

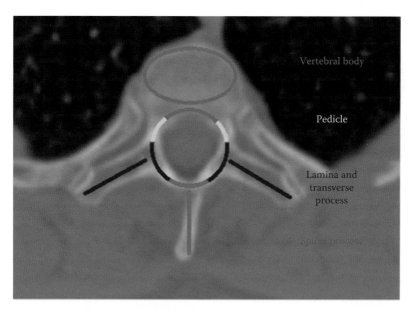

Figure 9.2 Modified Weinstein–Boriani–Biagini model as proposed by the International Spine Radiosurgery Consortium for consensus target volumes for spine radiosurgery.

(Guckenberger et al., 2011; Cox et al., 2012). Some investigators, especially those treating spinal metastases with CyberKnife (Accuray, Sunnyvale, CA), reported that a PTV expansion from GTV was not performed. OAR, particularly the spinal cord and the cauda equina, should always be excluded from PTV. Further PTV cropping may be performed at the discretion of the treating physician to spare the adjacent normal tissues while still ensuring adequate CTV coverage.

9.2.8 TARGET VOLUME DEFINITION IN POSTOPERATIVE STEREOTACTIC BODY RADIOTHERAPY

As stated earlier, the actual target delineation for postoperative radiotherapy remains poorly defined. At the MD Anderson Cancer Center (University of Texas, Houston, TX), particular attention is paid to any residual tumor involving the vertebral artery adjacent to the cervical spine, the postoperative levels of vertebrectomy, the contralateral pedicle, and the bony remnant of thoracic vertebra usually adjacent to the aorta after corpectomy (Sahgal et al., 2008). In comparison, delineation of remaining tumor and the complete resection cavity is typically done at the University of California San Francisco. The close cooperation between the treating spine neurosurgeon and radiation oncologist is critical in the facilitation of accurate target delineation (Sahgal et al., 2011). International consensus guidelines for CTV delineation for postoperative SBRT for spinal metastases are currently being developed (personal communication, Kristin Redmond, Johns Hopkins University).

9.2.9 TARGET VOLUME DEFINITION IN REIRRADIATION

SBRT is a very attractive option in the reirradiation situation, when the radiation tolerance of the spinal cord is at or close to its limit. To the best of our knowledge, no consensus to target volume delineation has been established specifically in the reirradiation situation. While anatomical target volume concepts are well established in the situation without previous radiotherapy, the more conservative approach of omission of elective vertebral volumes may be justified in some reirradiation situations.

9.3 SETUP AND IMAGE GUIDANCE PROCEDURE

The proximity of the spinal tumor to the spinal cord, combined with very steep dose gradients in between, requires highly accurate patient setup and immobilization, most likely the most accurate treatment in extracranial SRT. A prerequisite to a stable patient position, as well as an accurate and suitable treatment, is the patient's comfort in the immobilization device. A comfortable position improves the accuracy of the treatment and prevents involuntary movement, especially with back pain being the most frequent indication for radiotherapy. No specific premedication should be applied, routinely. However, a mild sedative or anxiolytic drug such as lorazepam for agitated patients and medication according to the WHO analgesic ladder for patients in pain may be considered.

9.3.1 IMMOBILIZATION

Traditional stereotactic frame-based immobilization systems fixated the patient invasively to the treatment reference system (Verhey, 1995). Such systems became irrelevant in times of image guidance.

Today, most centers prefer frameless immobilization devices. Treatment delivery without any patient immobilization system is not encouraged, unless thorough active patient monitoring or tracking is applied. Dahele et al. reported excellent spine stability even without rigid patient fixation: posttreatment errors were 1 mm or less and 1° or less in 97% of the treatment fractions. They combined pre- and posttreatment imaging, fast treatment delivery techniques using volumetric modulated arc therapy (VMAT), and support devices like a thin mattress as well as knee supports to improve patient comfort (Dahele et al., 2012).

The immobilization system needs to be suitable for the spinal level to be treated. For cervical spinal tumors, thermoplastic head shoulder masks are frequently used, sometimes combined with vacuum cushions for the caudal part of the body (Lohr et al., 1999). In the thoracic and lumbar region, immobilization systems in spine SBRT are similar to other systems used in high-precision radiotherapy. Most importantly, the patient should be placed in a stable and typically supine position into the immobilization device.

Sahgal and Li et al. analyzed intrafraction patient motion using cone-beam CT (CBCT) imaging and compared three different immobilization devices. They compared the results of evacuated cushions in immobilizing the thoracic and lumbar spine in SBRT to the results of a vacuum device with polyethylene sheet combined with a full-body evacuated cushion in SBRT, and to the results of an S-frame thermoplastic mask immobilizing the cervical spine (Li et al., 2012). They found similar setup errors across the three immobilization systems. Intrafractional patient motion was smallest using the vacuum fixation device, which allowed a small PTV margin of only 2 mm.

9.3.2 PRETREATMENT IMAGE GUIDANCE

Guckenberger et al. (2007) recommended evaluating setup errors in 6 degrees of freedom by CBCT. To keep the dose distribution to the spinal cord within ±5% of the prescribed dose, deviations exceeding 1 mm in the transverse plane, 4 mm in SI direction, or rotations of 3.5° or more should not be tolerated. It is therefore of highest relevance to correct the translational error component online prior to every treatment fraction; the action level should not exceed 1 mm (Kim et al., 2009). Nonrigid deformation might become relevant especially in multilevel SBRT, but such errors cannot be corrected by current image guidance procedures (Langer et al., 2005; Guckenberger et al., 2006; Wang et al., 2008).

Kim et al. (2009) confirmed that the steep dose gradients in spine SBRT, combined with close proximity of the target volume and the spinal cord, require very high treatment accuracy. The implementation of daily image guidance combined with online correction of setup errors in all 6 degrees of freedom provides precise target localization. Wang et al. (2008) recommended a second image-guided radiation therapy (IGRT) verification immediately after correction to ensure that residual positioning errors are ≤1 mm and ≤1° in any direction or axis.

The IGRT procedure can be performed using either CBCT or stereoscopic x-ray systems (e.g., ExacTrac). Chang et al. (2010) performed a comparison of both guidance modalities reporting modest setup discrepancies between CBCT and planar x-ray in spinal SBRT. In their comparison, CBCT achieved more accurate patient positioning information than the stereoscopic x-ray images.

9.3.3 INTRAFRACTION IMAGING

Ma and Sahgal et al. recorded intrafraction target motion every 50–100 seconds using kV x-ray imaging and reported high rates of nonrandom target movements. Cervical spine targets showed the largest motion, whereas thoracic and lumbar locations were less prone to nonrandom motion. Intrafraction motion increased with time and average treatment durations needed to maintain submillimeter and subdegree accuracy were 5.5, 5.9, and 7.1 minutes for cervical, thoracic, and lumbar locations, respectively. Intrafractional imaging combined with periodic interventions to overcome patient-specific target motion was recommended (Ma et al., 2009). Tseng et al. assessed the motion of the spinal cord and cauda equina in patients with spinal metastases using dynamic MRI. They differentiated between physiologic oscillatory motion due to cardiorespiratory activity and random shifts due to patient motion. Physiologic oscillatory motion was found to be small, whereas significant bulk motion was present in all observed 65 patients. Displacements were largest in SI direction (median 0.66 mm/maximum 3.90 mm) compared to the lateral (median 0.59 mm/ maximum 2.87 mm) or anterior–posterior axis (median 0.51 mm/maximum 2.21 mm) (Tseng et al., 2015).

Digital tomosynthesis (DTS) enables spine position tracking and may also increase the treatment accuracy and precision. Verbakel et al. used DTS and triangulation for spine detection (2015). This technique generates volumetric datasets from projection images acquired over a limited gantry angle. CBCT projection images over different gantry rotation angles were used to generate single-slice DTS images. These slices then could be registered to digitally reconstructed DTS images derived from the planning CT. With multiple DTS slices, the 3D position of the spine could be determined through triangulation. The interest in DTS in radiotherapy has grown over the past years. It may offer an interesting addition to the established imaging techniques for intrafractional imaging (Gurney-Champion et al., 2013). These results emphasize the need for sufficient margins, short treatment durations, and/or intrafraction imaging in spine SBRT.

Volumetric modulated arc therapy (VMAT) can potentially shorten treatment times, but its utility depends on the capabilities of the available treatment device. Many studies describe the higher efficiency of VMAT over static intensity-modulated radiotherapy (IMRT) with comparable PTV coverage (Kuijper et al., 2010; Matuszak et al., 2010). Wu et al. evaluated the feasibility of VMAT for spine SBRT with respect to three major goals, namely, to achieve a highly conformal dose distribution, to reduce the dose to the spinal cord, and to shorten the overall treatment time. The plan comparison between IMRT and VMAT using one or two arcs revealed an improved conformity index for VMAT (1.06 vs 1.15). Furthermore, two-arc VMAT and IMRT showed comparable OAR sparing. The mean treatment time of IMRT was 15.86 minutes for IMRT compared to 7.88 minutes for VMAT with two arcs, confirming a substantially improved treatment efficiency (Wu et al., 2009).

Flattening filter-free (FFF) beams with high-dose rates reduce beam-on time and subsequently treatment duration. This technology is increasingly used for SBRT. Stieb et al. analyzed outcome and toxicity after SBRT to different body sites with FFF beams in 84 patients. Median follow-up was 11 months, and no case of severe acute toxicity was observed. Only one patient developed grade 3 late toxicity. Hence, the use of FFF beams in SBRT was concluded to be safe (Stieb et al., 2015). Ong and Verbakel et al. confirmed that SBRT FFF plans provide comparable PTV coverage and OAR sparing. Moreover, the average beam-on time was reduced by a factor up to 2.5 compared to treating with non-FFF plans (Wu et al., 2009).

9.3.4 FIDUCIAL MARKERS

The use of implanted fiducial markers for improved image guidance, which has been well established for lung and prostate radiotherapy, has not been adopted for spine (Gerszten et al., 2007). Pan et al. (2011) reported the proportion of physicians having used fiducial markers for stereotactic treatment of the spine to be 12.4%. Considering the natural immobility of these tumors in relationship to the osseous structures of the spine, and excellent visibility of the bony structures with all available image guidance technologies, the rare usage of fiducial markers is understandable. Nevertheless, there might be an advantage in intrafraction motion compensation or in CT-MRI registration prior to treatment (Mani and Rivazhagan, 2013).

9.4 CONCLUSION

Target delineation in SBRT of the spine varies among different institutions and consensus guidelines are emerging. Complete inclusion of each involved vertebral element into the target volume is recommended. Patterns of recurrence studies showed that special attention should be brought to delineation of the posterior vertebral elements, even more so in cases of epidural disease. The use of high conformal doses and steep dose gradients requires highly accurate set-up and treatment delivery. Therefore, noninvasive immobilization systems and daily CBCT-IG are current treatment standard.

CHECKLIST: KEY POINTS FOR CLINICAL PRACTICE

Planning CT	• Axial slice thickness 1–1.5 mm • IV contrast-enhanced CT may be considered for mass-like lesions
Dedicated planning MRI in treatment position	• Axial slice thickness 1–2 mm • Sequences: T1 native, T1 gadolinium enhanced, T2 FLAIR • Planning CT and MRI acquired ideally on the same day, maximum time interval 2 weeks
Rigid registration of CT and MRI	
Target volume delineation primarily according to contrast-enhanced CT and gadolinium-enhanced T1 MRI	• GTV: morphologic visible tumor • CTV: affected and directly adjacent segments of the vertebral level • PTV: geometrical CTV expansion of 2 mm and exclusion of the spinal cord
Spinal cord definition: various concepts	• Spinal cord as delineated in T2 MRI images with 1–2 mm PRV margin • Thecal sac • Spinal canal
Immobilization	• Patient placed in comfortable position • Sufficient pain medication • Robust patient immobilization and/or continuous active patient monitoring
Image guidance	• Daily image guidance with online correction of setup errors • Action level of maximum 1 mm • Correction of setup errors preferably in 6 degrees of freedom • Verification imaging after the couch shift immediately prior to treatment especially in single fraction radiosurgery • Repeated intrafraction imaging or active patient monitoring
Treatment delivery	• Minimization of treatment delivery time by preferred use of VMAT and FFF beam delivery

REFERENCES

Algra PR, Bloem JL, Tissing H, Falke TH, Arndt JW, Verboom LJ (1991) Detection of vertebral metastases: Comparison between MR imaging and bone scintigraphy. *Radiogr Rev Publ Radiol Soc North Am Inc* 11:219–232.

Bijl HP, van Luijk P, Coppes RP, Schippers JM, Konings AW, van der Kogel AJ (2002) Dose-volume effects in the rat cervical spinal cord after proton irradiation. *Int J Radiat Oncol Biol Phys* 52:205–211.

Bijl HP, van Luijk P, Coppes RP, Schippers JM, Konings AW, van der Kogel AJ (2003) Unexpected changes of rat cervical spinal cord tolerance caused by inhomogeneous dose distributions. *Int J Radiat Oncol Biol Phys* 57:274–281.

Cai J, Sheng K, Sheehan JP, Benedict SH, Larner JM, Read PW (2007) Evaluation of thoracic spinal cord motion using dynamic MRI. *Radiother Oncol* 84:279–282.

Chang EL, Shiu AS, Mendel E, Mathews LA, Mahajan A, Allen PK, Weinberg JS et al. (2007) Phase I/II study of stereotactic body radiotherapy for spinal metastasis and its pattern of failure. *J Neurosurg Spine* 7:151–160.

Chang Z, Wang Z, Ma J, O'Daniel JC, Kirkpatrick J, Yin F-F (2010) 6D image guidance for spinal non-invasive stereotactic body radiation therapy: Comparison between ExacTrac x-ray 6D with kilo-voltage cone-beam CT. *Radiother Oncol J Eur Soc Ther Radiol Oncol* 95:116–121.

Cox BW, Spratt DE, Lovelock M, Bilsky MH, Lis E, Ryu S, Sheehan J et al. (2012) International Spine Radiosurgery Consortium consensus guidelines for target volume definition in spinal stereotactic radiosurgery. *Int J Radiat Oncol Biol Phys* 83:e597–e605.

Crum WR, Hartkens T, Hill DLG (2004) Non-rigid image registration: Theory and practice. *Br J Radiol* 77:S140–S153.

Dahele M, Hatton M, Slotman B, Guckenberger M (2015) Stereotactic body radiotherapy: A survey of contemporary practice in six selected European countries. *Acta Oncol* 54:1237–1241.

Dahele M, Verbakel W, Cuijpers J, Slotman B, Senan S (2012) An analysis of patient positioning during stereotactic lung radiotherapy performed without rigid external immobilization. *Radiother Oncol J Eur Soc Ther Radiol Oncol* 104:28–32.

Dahele M, Zindler JD, Sanchez E, Verbakel WF, Kuijer JPA, Slotman BJ, Senan S (2011) Imaging for stereotactic spine radiotherapy: Clinical considerations. *Int J Radiat Oncol* 81:321–330.

Dieleman EMT, Senan S, Vincent A, Lagerwaard FJ, Slotman BJ, van Sörnsen de Koste JR (2007) Four-dimensional computed tomographic analysis of esophageal mobility during normal respiration. *Int J Radiat Oncol Biol Phys* 67:775–780.

Emami B, Lyman J, Brown A, Coia L, Goitein M, Munzenrider JE, Shank B, Solin LJ, Wesson M (1991) Tolerance of normal tissue to therapeutic irradiation. *Int J Radiat Oncol Biol Phys* 21:109–122.

Figley CR, Stroman PW (2007) Investigation of human cervical and upper thoracic spinal cord motion: Implications for imaging spinal cord structure and function. *Magn Reson Med Off J Soc Magn Reson Med Soc Magn Reson Med* 58:185–189.

Geets X, Daisne J-F, Arcangeli S, Coche E, De Poel M, Duprez T, Nardella G, Grégoire V (2005) Inter-observer variability in the delineation of pharyngo-laryngeal tumor, parotid glands and cervical spinal cord: comparison between CT-scan and MRI. *Radiother Oncol J Eur Soc Ther Radiol Oncol* 77:25–31.

Gerszten PC, Burton SA, Ozhasoglu C, Vogel WJ, Welch WC, Baar J, Friedland DM (2005) Stereotactic radiosurgery for spinal metastases from renal cell carcinoma. *J Neurosurg Spine* 3:288–295.

Gerszten PC, Burton SA, Ozhasoglu C, Welch WC (2007) Radiosurgery for spinal metastases: Clinical experience in 500 cases from a single institution. *Spine* 32:193–199.

Gibbs IC, Kamnerdsupaphon P, Ryu MR, Dodd R, Kiernan M, Chang SD, Adler JR Jr (2007) Image-guided robotic radiosurgery for spinal metastases. *Radiother Oncol* 82:185–190.

Gibbs IC, Patil C, Gerszten PC, Adler JR Jr, Burton SA (2009) Delayed radiation-induced myelopathy after spinal radiosurgery. *Neurosurgery* 64:A67–A72.

Guckenberger M, Hawkins M, Flentje M, Sweeney RA (2012) Fractionated radiosurgery for painful spinal metastases: DOSIS—A phase II trial. *BMC Cancer* 12:530.

Guckenberger M, Meyer J, Vordermark D, Baier K, Wilbert J, Flentje M (2006) Magnitude and clinical relevance of translational and rotational patient setup errors: A cone-beam CT study. *Int J Radiat Oncol Biol Phys* 65:934–942.

Guckenberger M, Meyer J, Wilbert J, Baier K, Bratengeier K, Vordermark D, Flentje M (2007) Precision required for dose-escalated treatment of spinal metastases and implications for image-guided radiation therapy (IGRT). *Radiother Oncol* 84:56–63.

Guckenberger M, Sweeney RA, Flickinger JC, Gerszten PC, Kersh R, Sheehan J, Sahgal A (2011) Clinical practice of image-guided spine radiosurgery—Results from an international research consortium. *Radiat Oncol* 6:172.

Gurney-Champion OJ, Dahele M, Mostafavi H, Slotman BJ, Verbakel WFAR (2013) Digital tomosynthesis for verifying spine position during radiotherapy: A phantom study. *Phys Med Biol* 58:5717–5733.

Gwak H-S, Youn S-M, Chang U, Lee DH, Cheon GJ, Rhee CH, Kim K, Kim H-J (2006) Usefulness of (18) F-fluorodeoxyglucose PET for radiosurgery planning and response monitoring in patients with recurrent spinal metastasis. *Minim Invasive Neurosurg MIN* 49:127–134.

Hopewell JW, Morris AD, Dixon-Brown A (1987) The influence of field size on the late tolerance of the rat spinal cord to single doses of x rays. *Br J Radiol* 60:1099–1108.

Kim S, Jin H, Yang H, Amdur RJ (2009) A study on target positioning error and its impact on dose variation in image-guided stereotactic body radiotherapy for the spine. *Int J Radiat Oncol Biol Phys* 73:1574–1579.

Kirkpatrick JP, van der Kogel AJ, Schultheiss TE (2010) Radiation dose-volume effects in the spinal cord. *Int J Radiat Oncol Biol Phys* 76:S42–S49.

Kuijper IT, Dahele M, Senan S, Verbakel WF (2010) Volumetric modulated arc therapy versus conventional intensity modulated radiation therapy for stereotactic spine radiotherapy: A planning study and early clinical data. *Radiother Oncol* 94:224–228.

Langer MP, Papiez L, Spirydovich S, Thai V (2005) The need for rotational margins in intensity-modulated radiotherapy and a new method for planning target volume design. *Int J Radiat Oncol Biol Phys* 63:1592–1603.

Li W, Sahgal A, Foote M, Millar B-A, Jaffray DA, Letourneau D (2012) Impact of immobilization on intrafraction motion for spine stereotactic body radiotherapy using cone beam computed tomography. *Int J Radiat Oncol Biol Phys* 84:520–526.

Loblaw DA, Perry J, Chambers A, Laperriere NJ (2005) Systematic review of the diagnosis and management of malignant extradural spinal cord compression: The cancer care ontario practice guidelines initiative's neuro-oncology disease site group. *J Clin Oncol* 23:2028–2037.

Lohr F, Debus J, Frank C, Herfarth K, Pastyr O, Rhein B, Bahner ML, Schlegel W, Wannenmacher M (1999) Noninvasive patient fixation for extracranial stereotactic radiotherapy. *Int J Radiat Oncol Biol Phys* 45:521–527.

Loughrey GJ, Collins CD, Todd SM, Brown NM, Johnson RJ (2000) Magnetic resonance imaging in the management of suspected spinal canal disease in patients with known malignancy. *Clin Radiol* 55:849–855.

Lutz S, Berk L, Chang E, Chow E, Hahn C, Hoskin P, Howell D et al. (2011) Palliative radiotherapy for bone metastases: An ASTRO evidence-based guideline. *Int J Radiat Oncol Biol Phys* 79:965–976.

Ma L, Sahgal A, Hossain S, Chuang C, Descovich M, Huang K, Gottschalk A, Larson DA (2009) Nonrandom intrafraction target motions and general strategy for correction of spine stereotactic body radiotherapy. *Int J Radiat Oncol Biol Phys* 75:1261–1265.

Mani VR., Rivazhagan DS (2013) Survey of medical image registration. *J Biomed Eng Technol* 1:8–25.

Matuszak MM, Yan D, Grills I, Martinez A (2010) Clinical applications of volumetric modulated arc therapy. *Int J Radiat Oncol Biol Phys* 77:608–616.

Nakazawa H, Mori Y, Komori M, Shibamoto Y, Tsugawa T, Kobayashi T, Hashizume C (2014) Validation of accuracy in image co-registration with computed tomography and magnetic resonance imaging in Gamma Knife radiosurgery. *J Radiat Res (Tokyo)* 55:924–933.

Nelson JW, Yoo DS, Sampson JH, Isaacs RE, Larrier NA, Marks LB, Yin FF, Wu QJ, Wang Z, Kirkpatrick JP (2009) Stereotactic body radiotherapy for lesions of the spine and paraspinal regions. *Int J Radiat Oncol Biol Phys* 73:1369–1375.

Pan H, Simpson DR, Mell LK, Mundt AJ, Lawson JD (2011) A survey of stereotactic body radiotherapy use in the United States. *Cancer* 117:4566–4572.

Patel VB, Wegner RE, Heron DE, Flickinger JC, Gerszten P, Burton SA (2012) Comparison of whole versus partial vertebral body stereotactic body radiation therapy for spinal metastases. *Technol Cancer Res Treat* 11:105–115.

Ryu S, Fang Yin F, Rock J, Zhu J, Chu A, Kagan E, Rogers L, Ajlouni M, Rosenblum M, Kim JH (2003) Image-guided and intensity-modulated radiosurgery for patients with spinal metastasis. *Cancer* 97:2013–2018.

Ryu S, Gerszten P, Yin F, Timmerman RD (n.d.) RTOG 0631 study protocol: Phase II/III study of image-guided radiosurgery/SBRT for localized spine metastasis. Available at: http://www.rtog.org/ClinicalTrials/ProtocolTable/StudyDetails.aspx?study=0631.

Ryu S, Jin JY, Jin R, Rock J, Ajlouni M, Movsas B, Rosenblum M, Kim JH (2007) Partial volume tolerance of the spinal cord and complications of single-dose radiosurgery. *Cancer* 109:628–636.

Ryu S, Rock J, Jain R, Lu M, Anderson J, Jin J-Y, Rosenblum M, Movsas B, Kim JH (2010) Radiosurgical decompression of metastatic epidural compression. *Cancer* 116:2250–2257.

Ryu S, Rock J, Rosenblum M, Kim JH (2004) Patterns of failure after single-dose radiosurgery for spinal metastasis. *J Neurosurg* 101(Suppl 3):402–405.

Sahgal A, Ames C, Chou D, Ma L, Huang K, Xu W, Chin C et al. (2009) Stereotactic body radiotherapy is effective salvage therapy for patients with prior radiation of spinal metastases. *Int J Radiat Oncol Biol Phys* 74:723–731.

Sahgal A, Bilsky M, Chang EL, Ma L, Yamada Y, Rhines LD, Letourneau D et al. (2011) Stereotactic body radiotherapy for spinal metastases: Current status, with a focus on its application in the postoperative patient. *J Neurosurg Spine* 14:151–166.

Sahgal A, Chou D, Ames C, Ma L, Chuang C, Lambom K, Huang K, Chin CT, Weinstein P, Larson D (2007) Proximity of spinous/paraspinous radiosurgery metastatic targets to the spinal cord versus risk of local failure. *Int J Radiat Oncol Biol Phys* 69:S243–S243.

Sahgal A, Larson DA, Chang EL (2008) Stereotactic body radiosurgery for spinal metastases: A critical review. *Int J Radiat Oncol Biol Phys* 71:652–665.

Sahgal A, Ma L, Fowler J, Weinberg V, Gibbs I, Gerszten PC, Ryu S et al. (2012a) Impact of dose hot spots on spinal cord tolerance following stereotactic body radiotherapy: A generalized biological effective dose analysis. *Technol Cancer Res Treat* 11:35–40.

Sahgal A, Ma L, Weinberg V, Gibbs IC, Chao S, Chang UK, Werner-Wasik M et al. (2012b) Reirradiation human spinal cord tolerance for stereotactic body radiotherapy. *Int J Radiat Oncol Biol Phys* 82:107–116.

Sahgal A, Weinberg V, Ma L, Chang E, Chao S, Muacevic A, Gorgulho A et al. (2013) Probabilities of radiation myelopathy specific to stereotactic body radiation therapy to guide safe practice. *Int J Radiat Oncol Biol Phys* 85:341–347.

Schultheiss TE, Kun LE, Ang KK, Stephens LC (1995) Radiation response of the central nervous system. *Int J Radiat Oncol Biol Phys* 31:1093–1112.

Stieb S, Lang S, Linsenmeier C, Graydon S, Riesterer O (2015) Safety of high-dose-rate stereotactic body radiotherapy. *Radiat Oncol Lond Engl* 10:27.

Stradiotti P, Curti A, Castellazzi G, Zerbi A (2009) Metal-related artifacts in instrumented spine. Techniques for reducing artifacts in CT and MRI: State of the art. *Eur Spine J* 18:102–108.

Thariat J, Castelli J, Chanalet S, Marcie S, Mammar H, Bondiau P-Y (2009) CyberKnife stereotactic radiotherapy for spinal tumors: Value of computed tomographic myelography in spinal cord delineation. *Neurosurgery* 64:A60–A66.

Tseng C-L, Sussman MS, Atenafu EG, Letourneau D, Ma L, Soliman H, Thibault I et al. (2015) Magnetic resonance imaging assessment of spinal cord and cauda equina motion in supine patients with spinal metastases planned for spine stereotactic body radiation therapy. *Int J Radiat Oncol Biol Phys* 91:995–1002.

van Luijk P, Bijl HP, Konings AW, van der Kogel AJ, Schippers JM (2005) Data on dose-volume effects in the rat spinal cord do not support existing NTCP models. *Int J Radiat Oncol Biol Phys* 61:892–900.

Vanel D (2004) MRI of bone metastases: The choice of the sequence. *Cancer Imaging* 4:30–35.

Verbakel WFAR, Gurney-Champion OJ, Slotman BJ, Dahele M (2015) Sub-millimeter spine position monitoring for stereotactic body radiotherapy using offline digital tomosynthesis. *Radiother Oncol J Eur Soc Ther Radiol Oncol* 115:223–228.

Verhey LJ (1995) Immobilizing and positioning patients for radiotherapy. *Semin Radiat Oncol* 5:100–114.

Wang H, Shiu A, Wang C, O'Daniel J, Mahajan A, Woo S, Liengsawangwong P, Mohan R, Chang EL (2008) Dosimetric effect of translational and rotational errors for patients undergoing image-guided stereotactic body radiotherapy for spinal metastases. *Int J Radiat Oncol Biol Phys* 71:1261–1271.

Wu QJ, Yoo S, Kirkpatrick JP, Thongphiew D, Yin FF (2009) Volumetric arc intensity-modulated therapy for spine body radiotherapy: Comparison with static intensity-modulated treatment. *Int J Radiat Oncol Biol Phys* 75:1596–1604.

10

Spine stereotactic body radiotherapy for the treatment of de novo spine metastasis

Ehsan H. Balagamwala, Jacob Miller, Lilyana Angelov, John H. Suh,
Simon S. Lo, Arjun Sahgal, Eric L. Chang, and Samuel T. Chao

Contents

10.1 INTRODUCTION

Skeletal metastases are the third most common site of metastasis, and up to 70% of all cancer patients may develop spine metastases during the natural course of their disease. Spine metastases most frequently present with back pain. As spine metastases progress, focal neurologic symptoms may develop due to epidural disease and/or cord compression. Approximately 10%–20% of patients with spine metastases will progress to develop symptomatic spinal cord compression (Fornasier and Horne, 1975; Grant et al., 1991; Finn et al., 2007). Typically, spine metastases are treated with a combination of conventionally fractionated external beam radiotherapy (EBRT) and/or surgery, with surgery often reserved for patients with mechanical spine instability or cord compression (Hartsell et al., 2005; Patchell et al., 2005; Lutz et al., 2011; Kaloostian et al., 2014). The biggest disadvantage of EBRT is the relative inability to spare organs at risk (OARs), most importantly the spinal cord. With recent improvements in systemic treatments, it is expected that the incidence of spine metastasis may rise necessitating novel treatments in order to spare toxicity.

The past decade has witnessed the development of and rapid adoption of spine stereotactic body radiotherapy (SBRT), which allows the delivery of high doses of radiation in a single or few fractions with the promise of improved palliation and local control. Radiosurgery was first described by Lars Leksell (1951) for the treatment of intracranial lesions, and since then, intracranial radiosurgery has become a cornerstone for the treatment of both benign and malignant intracranial diseases (Leksell, 1951; Guo et al., 2008; Suh, 2010; Murphy and Suh, 2011). With advancements in immobilization, image guidance, intensity modulation, and computerized treatment planning, extracranial radiosurgery has become a reality (Videtic and Stephans, 2010; Shin et al., 2011; Zaorsky et al., 2013). With SBRT, it is possible to deliver high radiotherapy doses to the tumor and its vasculature to overcome inherent radioresistance to fractionated radiotherapy thereby achieving superior local control and pain relief (Balagamwala et al., 2012b). The current understanding of the biology of hypofractionated radiotherapy was discussed in Chapter 1.

In this chapter, we discuss the indications, techniques, treatment planning, outcomes, as well as toxicities of spine SBRT.

10.2 INDICATIONS

Spine SBRT is typically performed in a single or limited number of fractions. Indications tend to be institution specific; however, patients with a long predicted life expectancy, good Karnofsky Performance Score (KPS), radioresistant histology, limited spine metastases, oligometastases, and well-controlled systemic disease are generally considered good candidates for spine SBRT. Spine SBRT is also a good treatment option in reirradiation of spine metastases, and this is discussed in more detail in Chapter 11. Relative contraindications include spinal cord compression, mechanical instability of the spine, active connective tissue disorder, and prior SBRT to the same level. In the setting of prior spine SBRT, fractionated SBRT may be considered. Some centers exclude radiosensitive histologies; however, in our institutional experience, select patients with traditionally radiosensitive histologies have done well with spine SBRT. A summary of indications and relative contraindications of spine SBRT is presented in Table 10.1. The American Society for Radiation Oncology (ASTRO) and American College of Radiology (ACR) have published guidelines for the treatment of spine metastases, with a section dedicated to the role of spine SBRT (Lutz et al., 2011; Lo et al., 2012).

10.3 TECHNICAL AND TREATMENT PLANNING CONSIDERATIONS

Spine SBRT is a resource-intensive treatment modality requiring extensive experience and expertise, as well as multidisciplinary involvement of radiation oncologists, medical physicists, neurosurgeons, and radiation therapists. Given the high dose per fraction delivered in spine SBRT and the proximity of the spinal cord, it is imperative to achieve accuracy of 1–2 mm (Sahgal et al., 2008). In order to perform spine SBRT safely and effectively, the following components are essential: a linear accelerator equipped with a multileaf

Table 10.1 Spine stereotactic body radiotherapy indications and contraindications

INDICATIONS	RELATIVE CONTRAINDICATIONS
Long predicted life expectancy	Poor expected survival
KPS ≥ 70	KPS 40–60
Radioresistant histology	Radiosensitive histology
≥3–5 mm separation from spinal cord	Spinal cord compression
Prior conventional EBRT	Prior spine SBRT to same level
Limited spine metastases	Multilevel or diffuse spine metastasis
Oligometastatic disease	Mechanical spine instability
Well-controlled systemic disease	Poorly controlled systemic disease
Post-separation surgery	Active connective tissue disease

collimator and onboard image guidance with cone beam CT (CBCT), a body immobilization system, and a sophisticated treatment planning system. Alternatively, a system such as CyberKnife (Accuray Inc., Sunnyvale, CA), which utilizes a linear accelerator mounted on a robotic arm, can be utilized. We will discuss each of these components in further detail.

10.3.1 IMMOBILIZATION

The utility of spine SBRT lies in maximizing dose distribution within the target volume with a steep dose gradient outside the target volume, thereby sparing the spinal cord. Therefore, spine SBRT requires a translational accuracy of <2 mm and a rotational accuracy of <2° (Chang et al., 2004; Lo et al., 2010). Although respiration does not significantly impact motion of spinal tumors, reproducibility of the spine setup especially in multifraction regimens poses a significant clinical challenge. Hamilton et al. (1995) have described an invasive rigid spine fixation device, similar to the head frame utilized in Gamma Knife radiosurgery; however, it is impractical for routine utilization. Therefore, most centers have utilized near rigid immobilization systems. Many centers utilize commercially available solutions, whereas other centers such as Memorial Sloan Kettering Cancer Center and the University of Heidelberg, Germany, have developed noninvasive in-house systems for near-rigid immobilization.

At the Cleveland Clinic, we perform CT-based simulation in the supine position and utilize a five-point head and neck mask for cervical and upper thoracic spinal lesions or the Elekta BodyFIX stereotactic body frame (Medical Intelligence, Schwabmünchen, Germany) for mid-to-low thoracic and lumbosacral spinal lesions. The BodyFIX system consists of a carbon fiber base plate, a whole-body vacuum cushion, a vacuum system, and a plastic fixation sheet (Figure 10.1). Hyde et al. reported their experience with the BodyFIX system in 42 consecutive patients. They found that using this system allowed for excellent precision: 90% of treatments had <1 mm translational error and 97% treatments had <1° rotational error. They also found that when they utilized a stricter threshold for repositioning the patient (1 vs 1.5 mm), the intrafraction translational motion was improved suggesting that accurate patient position prior to treatment delivery improves the overall precision of spine SBRT (Hyde et al., 2012).

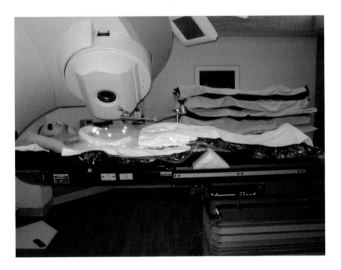

Figure 10.1 The Elekta BodyFIX stereotactic body frame (Medical Intelligence, Schwabmünchen, Germany), which consists of a carbon fiber base plate, whole-body vacuum cushion, vacuum system, and plastic fixation sheet for thoracic and lumbar lesions. (With kind permission from Springer Science+Business Media: *J. Radiat. Oncol.*, Stereotactic body radiotherapy for the treatment of spinal metastases, 1, 2012, 255–265, Balagamwala E.H., Cherian, S., Angelov, L., Suh, J.H., Djemil, T., Lo, S.S., Sahgal, A., Chang, E., Teh, B.S., Chao, S.T.)

10.3.2 TREATMENT PLANNING

10.3.2.1 Target volume delineation

The gross tumor volume (GTV) is defined as the radiographically visible tumor based on contrast-enhanced MRI. The clinical target volume (CTV) is defined as the margin applied to the GTV to account for microscopic disease in the vicinity of the GTV. The planning target volume (PTV) is formed by adding a margin to the CTV to account for daily patient setup errors. There are considerable differences between institutions regarding target volume definition as well as dose prescription (Guckenberger et al., 2011). Furthermore, there are differences between institutions on the imaging modalities used to delineate the target volume: some institutions rely on CT only, others rely on CT/MRI fusion, while others obtain CT or invasive myelogram.

Our practice involves obtaining a simulation CT as well as a high-definition (HD) MRI (1.5 mm slice thickness) of the region of interest. If there is instrumentation that significantly distorts the MRI images, a CT myelogram is performed to define the spinal cord and epidural disease. We subsequently fuse the simulation CT and the HD MRI (or CT myelogram if obtained) in MIM Maestro™ and MIMfusion® (MIM Software Inc., Cleveland, OH) or iPlan (BrainLab, AG). For a lesion within the vertebral body, the CTV is defined as the entire vertebral body +/− right or left pedicle depending on involvement. For a lesion involving only the posterior elements, the CTV may only include the spinous process and the lamina. Given the accuracy of immobilization as discussed earlier as well as image guidance (discussed below), we do not add a PTV. Our methodology is similar to the one utilized in the Radiation Therapy Oncology Group (RTOG) 0631 trial, which is a Phase II/III study of image-guided SBRT for localized spine metastasis.

10.3.2.2 Delineation of organs at risk (OARs) and dose limits

The principal OARs in spine SBRT are the spinal cord and/or the cauda equina because radiation myelopathy can be a devastating complication. This is especially a concern in patients treated for indolent tumors of the spine such as ependymoma, which has a long natural history. There is little consensus in the literature on the imaging modality used to delineate the neural structures or the definition of spinal cord and cauda equina (Balagamwala et al., 2012c). Furthermore, different institutions utilize different spinal cord and the cauda equina dose constraints.

We define the spinal cord and the cauda equina based on the HD MRI. The spinal cord is defined as the spinal cord at the vertebral level(s) being treated with a 4.5 mm cranial and caudal margin. Since the cauda equina is composed of nerve roots floating in the thecal sac, we define the cauda equina as the entire thecal sac at the vertebral level(s) being treated with a 4.5 mm cranial and cauda margin. We limit ≤10% of the contoured spinal cord to ≥10 Gy and limit the maximum point dose (0.03 cc) to <14 Gy. Since the cauda equina is composed of nerve roots with a potentially higher dose tolerance, we limit ≤10% of the contoured cauda equina to ≥12 Gy and limit the maximum point dose (0.03 cc) to <16 Gy.

Other OARs to consider during treatment planning include esophagus, kidneys, and bowel in select patients. Esophagus is an important OAR to consider especially for cervical and thoracic spine SBRT. Although no firm dose constraints for esophagus have been published, there have been reports of grade 3–4 esophageal toxicities (Yamada et al., 2008; Moulding et al., 2010). We attempt to limit the maximum dose to the esophagus to 16 Gy when possible. When planning treatment, it is also important to be mindful about avoiding entrance and exit dose through the kidney. This is especially important in patient with renal cell carcinoma who may already have undergone a nephrectomy and have limited renal reserve. Similarly, minimizing dose to the bowel can prevent acute nausea, vomiting, and diarrhea in patients undergoing lumbar spine SBRT.

10.3.2.3 Treatment dose and fractionation

Similar to the institutional differences described earlier, there is considerable institutional variation in terms of the optimal dose fractionation. Institutions utilize either a single-fraction regimen or a multifraction regimen. Single-fraction doses tend to range between 12 and 24 Gy with a recent trend toward delivering higher doses. Hypofractionated regimens include 25–30 Gy in 5 fractions, 24 Gy in 2–3 fractions, and 27 Gy in 3 fractions. As discussed below, there is evidence suggesting that

Table 10.2 **Treatment planning summary at the Cleveland Clinic**

TREATMENT PLANNING SUMMARY	
Immobilization	Cervical and upper thoracic spine: five-point head and neck mask
	Lower thoracic and lumbosacral spine: BodyFIX
Target volume delineation	GTV ± vertebral body ± right and left pedicles ± spinous process and lamina (depending on GTV location)
OAR delineation	*Spinal cord*: Spinal cord + 4.5 mm cranial and caudal extensions
	Cauda equina: Thecal sac + 4.5 mm cranial and caudal extensions
Prescription dose	16–18 Gy in 1 fraction
Image guidance	Cone-beam CT (Varian Edge, NovalisTX) ± orthogonal x-rays (NovalisTx)

≥20 Gy/fraction leads to a higher risk for development of vertebral compression fractures (VCF) (Sahgal et al., 2013a). Therefore, institutions that support a multifraction regimen argue that this approach allows the delivery of an equivalent biologically effective dose (BED) compared to a high-dose single-fraction regimen while minimizing the risk for VCF. However, there is no evidence yet suggestive of improved local control or improved toxicity profile when comparing multifraction regimens to single-fraction regimens utilizing <20 Gy. Therefore, we utilize single-fraction spine SBRT. Recently, we have dose escalated from 16 Gy in a single fraction to 18 Gy in a single fraction for select cases based on our institutional data suggesting a dose–response relationship for radioresistant histologies (unpublished). There certainly remains a need for high-quality evidence comparing the efficacy and side effect profile of single-fraction and multifraction regimens to guide future practice.

10.3.2.4 Image guidance for treatment delivery

Accurate treatment delivery is hinged upon accurate verification of patient position prior to and during the treatment. Incorporation of image guidance in radiotherapy has allowed for the meteoric rise of SBRT not only in spine radiotherapy but also other extracranial applications. Ma et al. (2009) identified a high incidence of translational variation >1 mm and rotational variation of >1° with treatment times >5 minutes. This study showed an important aspect of spine SBRT that is often underappreciated: fast treatment delivery is ideal. The majority of patients treated for spine metastases have significant pain associated with their disease, and therefore, positional changes are prevalent with long treatment durations. If long treatment duration is anticipated, an intrafraction break for image verification could be utilized, and any positional changes can be readjusted.

Image-guidance technologies include 2D techniques such as orthogonal x-rays as well as 3D techniques such as CBCT and megavoltage CT (MVCT). The major benefit of 3D techniques is the ability to visualize target volumes, OARs, as well as adjusting patient position based on soft tissues in addition to bone. MVCT may offer mitigation of metal artifact, however, at the expense of soft tissue resolution (Dahele et al., 2011). At the Cleveland Clinic, we utilize CBCT for every spine SBRT case and adjust patient position using a "6D" robotic couch. When patients are treated on our NovalisTX (BrainLAB AG, Feldkirchen, Germany), we verify patient position using orthogonal x-rays in addition to CBCT. A summary of our treatment methodology is demonstrated in Table 10.2.

10.4 SPINE SBRT OUTCOMES

Initial experience with spine SBRT was published when Hamilton and Lulu (1995) reported their experience with five patients using a rigid stereotactic frame and linear accelerator. However, due to the invasiveness of this rigid immobilization technique, spine SBRT was not rapidly adopted. Introduction of near-rigid, noninvasive immobilization (as discussed earlier) has led to significant increases in the utilization of spine SBRT.

In one of the earliest reports of spine SBRT, Benzil et al. (2004) described a series of 26 metastases in 22 patients. They reported that 94% of these patients experienced significant pain relief within 72 hours, which was durable for up to 3 months. Moreover, 63% of patients experienced improvement in neurologic deficits. This study set the foundation for the adoption of spine SBRT.

Since the early reports, many retrospective studies have been reported (select series are summarized in Table 10.3; Garg et al., 2012). Overall, spine SBRT has resulted in excellent radiographic and clinical control (>85%). In 2004, Gerszten et al. (2004) published their initial results of 125 spinal segments (115 patients) treated with CyberKnife. At 1 month follow-up, 94% of patients experienced pain relief, while 89% experienced control of neurologic deficits. Radiographic control was likewise excellent (96%). Gerszten and colleagues updated their initial results and reported their outcomes in 393 patients with 500 spinal segments treated with single-fraction spine SBRT. Median dose was 20 Gy (range 12.5–25 Gy) in a single fraction (Gerszten et al., 2007). The 1-year radiographic control was 90%, pain control was 86%, and 84% of patient experienced improvement of their neurologic deficits. Due to its large numbers, this study established spine SBRT as a safe and efficacious modality for the treatment of spine metastasis.

Ryu et al. evaluated their experience at Henry Ford in 49 patients with 61 spinal segments. They reported pain control of 85% and reported a relapse of pain at the treated spinal segment in 7%. Importantly, they reported an adjacent radiographic failure rate of 5% (Ryu et al., 2004). We also evaluated our risk for marginal failure, that is, tumor recurrence in one vertebral level above and below the treated spinal level (Koyfman et al., 2012). A total of 149 patients with 208 spinal segments were included in this study, and the rate of marginal failure was 12.5% and occurred a median time of 7.7 months. Patients who had paraspinal disease and those receiving <16 Gy were found to be at a higher risk for marginal failure. In 2011, Klish et al. published a prospective series of 65 spinal segments (58 patients), which were routinely irradiated in conjunction with adjacent segments. Eleven percent experienced failure in adjacent segments and at multiple other spinal segments, while only 3% of patients failed only in the adjacent segments. Given the high rate of failure outside of the irradiated field, these results suggested that routinely treating uninvolved adjacent segments with spine SBRT was unnecessary (Klish et al., 2011).

Chang et al. (2007) evaluated the patterns of failure in a prospective single institution series. This study included 63 patients with 74 segments, and the 1-year local control was 84% after a median follow-up of 21.3 months. Pattern of failure analysis demonstrated two primary mechanisms of failure: recurrence in the bone adjacent to the site of previous treatment and recurrence in the epidural space adjacent to the spinal cord. It is thought that epidural space failure is often due to the relative underdosing of the epidural disease in order to meet spinal cord constraints. Similarly, Patel et al. evaluated whole or partial vertebral body spine SBRT in a retrospective series of 154 segments (117 patients). Patients treated with the whole

Table 10.3 **Select outcomes for single-fraction spine stereotactic body radiotherapy**

SERIES (YEAR)	STUDY DESIGN	# PATIENTS (# SEGMENTS)	MEDIAN DOSE (RANGE) (Gy)	RADIOGRAPHIC CONTROL	PAIN CONTROL
Ryu et al. (2004)	Retrospective	49 (61)	14 (10–16)	96%	93%
Gerszten et al. (2007)	Retrospective	393 (500)	20 (12.5–25)	89%	86%
Yamada et al. (2008)	Retrospective	93 (103)	24 (18–24)	90%	NR
Garg et al. (2012)	Prospective	61 (63)	18 (16–24)	88%	NR
Ryu et al. (RTOG 0631) (2014)	Prospective (Phase II)	44 (55)	16	NR	NR

NR, not reported.

vertebral body approach achieved a superior radiographic local control (89% vs 71%, $p = 0.029$) and a lower retreatment rate (11% vs 19%, $p = 0.285$), albeit not statistically significant, compared to patients treated with the partial vertebral body approach (Patel et al., 2012).

Although many retrospective institutional studies have shown excellent radiographic and pain control rates, few studies have specifically evaluated radiation dose–response effects. Yamada and colleagues reported their initial experience with single-fraction high-dose spine SBRT in 93 patients with 103 spinal segments. They reported that although tumor histology was not a significant predictor of local control, dose >23 Gy was associated with better local control compared to dose <23 Gy (95% vs 80%, $p = 0.03$) (Yamada et al., 2008). They also evaluated this dose–response relationship in the post–spinal cord decompression and spinal instrumentation setting. In this small study of 21 patients, the authors found that 3 out of 5 patients (60%) who underwent low-dose radiosurgery experienced local failure compared to 1 out of 16 patients (6%) who underwent high-dose radiosurgery. They estimated that the 1-year cumulative incidence of local failure was 6.3% for the high-dose (24 Gy) group and 20% for the low-dose (<24 Gy) group ($p = 0.0175$) (Moulding et al., 2010). The same group also evaluated whether a dose–response relationship exists for the treatment of paraspinal disease in the recurrent setting. Among 97 treatments, those patients treated with 30 Gy total dose had a significantly lower risk of developing local failure compared to those treated with 20 Gy (HR 0.51, $p = 0.04$). Other factors such as tumor size or histology were not associated with local failure (Damast et al., 2011).

Chao et al. also analyzed our experience in single-fraction spine SBRT to evaluate dosimetric factors correlated with local control. A total of 189 patients with 256 spinal segments were included. Median prescription dose was 15 Gy (range, 8–16 Gy). We found that the presence of epidural disease, multilevel spinal disease, and lung cancer histology were associated with radiographic failure (Chao et al., 2012a). When we restricted the analysis to traditionally radiosensitive histologies (i.e., nonrenal cell, nonmelanoma), we found that lung cancer histology was associated with increased risk of radiographic failure. We also found that higher total dose and maximum dose to the target volume were associated with improved pain control (Balagamwala et al., 2012d). These data show that a dose–response relationship exists for radiosensitive histologies and suggests that there may be a role for dose escalation, especially for radioresistant histologies. Therefore, we have modified our institutional practice and currently treat all our patients with 18 Gy in 1 fraction as long as OAR constraints are met. It is important to note that although dose escalation leads to superior local control, the risk for development of VCF also increases, as discussed in detail in the succeeding text. Also, it may be suggestive of a benefit to a multifraction approach to increase the dose but keep the dose per fraction low to minimize risk of VCF.

Due to the ability to deliver a high BED with spine SBRT, it is most often utilized for the treatment of metastases from radioresistant tumors such as renal cell carcinoma. Nguyen et al. reported one of the first such series of 48 patients with 55 spinal metastases with a median follow-up time of 13.1 months. The actuarial 1-year progression-free survival was 82.1%. The complete pain response rate at 1 and 12 months post-SBRT was an impressive 44% and 52% (compared to baseline rate of 23%) (Nguyen et al., 2010). We also evaluated our series of 57 patients with 88 treated spinal segments treated with single-fraction spine SBRT. Median time to radiographic failure and pain progression was 26.5 and 26.0 months, respectively. The median time to pain relief was 0.9 months. In our series, the rate of VCF was 14% (Balagamwala et al., 2012a). In a more recent analysis, Thibault and colleagues evaluated 37 patients with 71 spinal segments noted a vertebral compression risk of 16% after spine SBRT for renal cell carcinoma metastasis (2014). Interestingly, Sohn and colleagues performed a matched-pair analysis comparing spine SBRT versus conventionally fractionated radiotherapy for spinal metastases from renal cell carcinoma. They found that more SBRT patients had complete or partial pain responses (however, the difference was not statistically different). Furthermore, they found that the progression-free survival was significantly higher for the SBRT patients ($p = 0.01$). There were no differences in toxicity (Sohn et al., 2014). These results are summarized in Table 10.4.

Chao and colleagues performed a recursive partitioning analysis (RPA) for patients undergoing spine SBRT at their institution. They evaluated 174 patients who underwent single-fraction spine SBRT with a median follow-up of 8.9 months. Histologies were divided into favorable (breast and prostate), radioresistant (renal cell, melanoma, sarcoma), and others (all other histologies). Median overall survival for favorable

Table 10.4 Outcome of spine stereotactic body radiotherapy in renal cell carcinoma

SERIES (YEAR)	STUDY DESIGN	# PATIENTS (# SEGMENTS)	MEDIAN DOSE (RANGE) (Gy)	MEDIAN # OF FRACTIONS	RADIOGRAPHIC CONTROL (%)	PAIN CONTROL
Nguyen et al. (2010)	Retrospective	48 (55)	27 (24–30)	3	78	75%
Balagamwala et al. (2012a)	Retrospective	57 (88)	15 (8–16)	1	71	68%
Thibault et al. (2014)	Prospective	37 (71)	24 (18–30)	2	83	NR
Sohn et al. (2014)	Retrospective	13 (31)	38 (mean)	4	86	100%

histologies was 14 months, 11.2 months for radioresistant histologies, and 7.2 months for other histologies ($p = 0.02$). RPA analysis resulted in three classes ($p < 0.01$). Class 1 was defined as time from primary disease (TPD) of >30 months, and KPS > 70. Class 2 was defined as TPD > 30 months and KPS ≤ 70, or TPD ≤ 30 months and age < 70 years. Class 3 was defined as TPD ≤ 30 months and age ≥ 70 years. Median overall survival was 21.1 months for Class 1, 8.7 months for Class 2, and 2.4 months for Class 3. This RPA analysis identified a subgroup of patients who may benefit the most from spine SBRT (Chao et al., 2012b).

An important consideration when evaluating treatment options for palliation is quality of life (QOL). Although clinical trials in recent years are incorporating QOL measures, much of the retrospective evidence does not include QOL measures. Degen et al. reported the earliest QOL outcomes after spine SBRT using the Short Form Health Survey (SF-12) at regular treatment intervals. No significant differences in QOL were observed up to 24 months after SBRT, suggesting that SBRT preserves patient QOL (Degen et al., 2005). RTOG 0631, the ongoing phase II/III trial comparing conventional EBRT (8 Gy in 1 fraction) to spine SBRT (16 or 18 Gy in 1 fraction), seeks not only to establish the safety and compare the efficacy of SBRT to conventional radiotherapy but also aims to establish the impact of radiotherapy on the QOL (Ryu et al., 2014). The phase II results of 44 patients were recently published, which showed a feasibility success rate of 74% in a rigorous quality-controlled setting (Ryu et al., 2014). The phase III component of the trial is currently accruing and clinicians are encouraged to enroll on this important trial.

10.5 TOXICITIES

10.5.1 PAIN FLARE

The goal of palliative radiotherapy for spine metastases has traditionally been short-term pain relief. With the advent of spine SBRT and the delivery of ablative radiotherapeutic doses, we are not only able to achieve adequate palliation but also hope to achieve excellent local control. With hypofractionated, high-dose radiotherapy, there is concern for the development of pain flare, which is a temporary worsening of bone pain at the treated site. Pain flare usually occurs within 1–2 weeks following radiotherapy and responds readily to steroids. With conventional radiotherapy, the incidence of pain flare has ranged between 16% and 41% (Chow et al., 2005; Loblaw et al., 2007; Hird et al., 2009). Recently, we have gained a better understanding of the incidence of pain flare in patients treated with spine SBRT.

Chiang et al. (2013) performed a prospective observational study in 41 patients undergoing multifraction spine SBRT to total doses of 24–35 Gy in 2–5 fractions. They reported an incidence of 68.3%, and pain flare was most commonly observed on day 1 after SBRT (29%). Majority of patients were successfully rescued with dexamethasone. Higher KPS and cervical or lumbar spine locations were associated with higher incidence of pain flare. Given the high incidence of pain flare, they have initiated a prophylactic dexamethasone protocol in all their patients undergoing spine SBRT (Khan et al., 2015).

Pan et al. retrospectively evaluated patients enrolled on institutional spine SBRT phase I/II clinical trials at MD Anderson Cancer Center and found an incidence of 23%. Median time to pain flare was 5 days, and multifraction SBRT was associated with higher incidence of pain flare (Pan et al., 2014). In our institutional experience of single-fraction SBRT (14–16 Gy), the incidence of pain flare has been approximately 15%, and we do not routinely premedicate with dexamethasone (Jung et al., 2013). We will reevaluate this as we go to higher doses and use multifraction SBRT.

10.5.2 VERTEBRAL COMPRESSION FRACTURE

With the delivery of high doses per fraction in spine SBRT, a major concern for late toxicity is the occurrence of VCF. Rose et al. evaluated their experience of high-dose (18–24 Gy in 1 fraction) spine SBRT and reported a vertebral fracture incidence of 39%. They found that lytic lesions involving >40% of the vertebral body as well as lesions caudal to T10 were 6.8 and 4.6 times, respectively, more likely to develop VCF (Rose et al., 2009). Similarly, Boehling et al. retrospectively evaluated patients treated on phase I/II trials with spine SBRT to doses of 18–30 Gy in 1–5 fractions. They reported a fracture incidence of 20%. In their series, age >55, a preexisting fracture and baseline pain were associated with increased risk for developing VCF (Boehling et al., 2012). Cunha et al. utilized the Spinal Instability Neoplastic Scoring (SINS) system to perform a more rigorous analysis of the risk factors predisposing for the development of VCF in patients treated with spine SBRT. They retrospectively evaluated 90 patients with 167 spinal segments and found that the incidence of compression fractures was 11%. Of the fractures, 63% were de novo, whereas 37% were fracture progression. Their analysis demonstrated that alignment, lytic lesions, lung and hepatocellular primary histologies, and dose >20 Gy per fraction were significant predictors of vertebral fractures (Cunha et al., 2012).

Sahgal et al. performed a multi-institutional retrospective study of 252 patients with 410 treated spinal segments and evaluated the risk of VCF. This study also utilized the SINS criteria. After a median follow-up of 11.5 months, the incidence of VCF was 14% and the median time to fracture was 2.46 months. Of the fractures, 47% were new fractures, and 53% were progression of preexisting fractures. Multivariate analysis demonstrated dose per fraction (greatest risk for ≥24 vs 20 Gy to 23 vs ≤19 Gy), baseline VCF, lytic tumor, and spine deformity were predictive of VCF (Sahgal et al., 2013a). We evaluated our experience in 348 patients with 507 treatments and found a VCF incidence of 15%. Multivariate analysis showed preexisting VCF and baseline pain were significant predictors for the development of VCF post-spine SBRT (Balagamwala et al., 2013).

10.5.3 NEUROLOGIC TOXICITY

Radiation myelopathy is the most feared complication of spine radiotherapy. Radiation myelopathy is a late toxicity of spine SBRT and rarely occurs <6 months after treatment and almost always presents within 3 years after treatment (Abbatucci et al., 1978). The risk of radiation myelopathy from spine SBRT has been estimated to be <1% (Kirkpatrick et al., 2010).

Sahgal et al. (2010) performed a multi-institutional retrospective analysis of 5 patients who developed myelopathy after spine SBRT and compared it to 19 patients without myelopathy. Patients who developed myelopathy received the following maximum point doses to the thecal sac: 10.6 Gy/13.1 Gy/14.8 Gy in 1 fraction, 25.6 Gy in 2 fractions, and 30.9 Gy in 3 fractions. When compared to those patients that did not develop myelopathy, the analysis showed that maximum dose up to 10 Gy in 1 fraction to the thecal sac is safe.

In those patients who have undergone previous conventionally fractionated radiotherapy, the dose tolerance of the spinal cord maybe different. This was evaluated by Sahgal et al. (2012) in five patients with radiation myelopathy. They reported that the risk of radiation myelopathy is very low when spine SBRT is delivered >5 months after conventional radiotherapy; normalized BED (nBED) to the thecal sac is 20–25 Gy provided that the total nBED does not exceed 70 Gy and the SBRT thecal sac nBED is not >50% of the total nBED. More recently, Sahgal et al. reported the first logistic regression model yielding estimates of radiation myelopathy from spine SBRT. They estimated that the risk of myelopathy is <5% when limiting maximum thecal sac dose to 12.4 Gy in 1 fraction, 17.0 Gy in 2 fractions, 20.3 Gy in 3 fractions, 23.0 Gy in 4 fractions, and 25.3 Gy in 5 fractions (Sahgal et al., 2013b).

10.6 CONTROVERSIES

Over the past decade, we have gained a great deal of understanding regarding the efficacy of and the toxicities associated with spine SBRT in the upfront as well as reirradiation setting. RTOG 0631, a prospective, randomized phase III trial comparing single-fraction spine SBRT with single-fraction conventional radiotherapy, will be an important addition to the developing literature. Data from multiple institutions show that spine SBRT is very efficacious, with radiographic and pain control rates of >85%. However, the major differences among institutions pertain to the technical nuances of spine SBRT including delineation of target volume and OARs, dose prescription, and most importantly single-fraction versus multifraction SBRT. Comparative data regarding the safety and efficacy of single-fraction versus multifraction SBRT do not exist, and current practice is based upon institutional preference.

10.7 FUTURE DIRECTIONS

Spine SBRT has established itself as the standard of care in the reirradiation setting, and the groundwork for its role in the upfront setting has been laid based upon the numerous studies showing its safety and efficacy. Although early experience shows that spine SBRT can be utilized in the setting of cord compression with good results, further work needs to be done to establish its role in the medically inoperable setting (Ryu et al., 2010; Suh et al., 2012). Furthermore, the role of spine SBRT after vertebral augmentation has not been clearly defined, and the jury is out on single-fraction versus multifraction spine SBRT. Multiple institutions have small prospective trials currently open evaluating the role of spine SBRT with concurrent immunologic therapy. As systemic chemotherapy continues to improve and patients with metastatic disease continue to have longer life expectancy, the role of repeat spine SBRT will have to be established.

CHECKLIST: KEY POINTS FOR CLINICAL PRACTICE

✓	ACTIVITY	SOME CONSIDERATIONS
	Patient selection	*Is the patient appropriate for SBRT?* • Patient has limited burden of metastatic disease • Other sites of metastases well controlled on systemic agents • At least 2 mm between epidural disease and the spinal cord • KPS ≥ 60
	SBRT vs EBRT	*Is the lesion amenable to SBRT?* • Nonlymphoma/nonmyeloma histology • Expected survival >6 months • ≤3 contiguous vertebral bodies involved
	Simulation	*Immobilization* • Supine • *Cervical and upper thoracic lesions*: Five-point aquaplastic mask • *Thoracic and lumbosacral lesions*: Stereotactic body immobilization system (Linac) or Alpha Cradle (CyberKnife) *Imaging* • CT simulation • High-definition planning MRI with T1 and STIR sequences • CT myelogram if patient unable to undergo MRI • MRI to simulation CT fusion of the region of interest

(Continued)

✔	ACTIVITY	SOME CONSIDERATIONS

	Treatment planning	*Contours*
		• CTV per published guidelines, dependent on extent of spine involvement
		• Spinal cord/cauda equina contour includes two slices above and below the vertebral level(s) of interest
		Treatment planning
		• Inverse treatment planning
		• CTV = PTV
		• Dose
		  • 16–18 Gy in 1 fraction, 24 Gy in 2 fractions, 27 Gy in 3 fractions or 30 Gy in 5 fractions to PTV
		• Coverage
		  • Achieve PTV coverage 90%
		  • May accept lower coverage in epidural space to achieve cord tolerance
		• Dose constraints
		  • Dependent upon dose fractionation

	Treatment delivery	*Imaging* • Patients are first set up conventionally using room lasers and manual shifts • Orthogonal kV matching to relevant bony anatomy • Cone-beam CT to match to tumor (CTV or PTV) and spinal cord • Treatment couch with 6 degrees of freedom to correct translational and rotational shifts if available • Repeat CBCT for patient movement or long treatment time interval (10–15 minutes) • Utilize volumetric modulated arc therapy and flattening filter-free mode whenever feasible to reduce treatment duration
	Outcomes	*Local control* • Excellent (>85%) radiographic as well as symptomatic local control rates
	Toxicity	*Acute* • Risk of pain flare ranges from 15% to 50% and is adequately treated with short course of steroids • Esophagitis may also occur and is treated with supportive management *Chronic* • >20 Gy/fraction leads to significant increase in risk for the development of vertebral compression fractures • Risk for spinal cord myelopathy is very low (<1%) so long as cord tolerance is not exceeded

REFERENCES

Abbatucci JS, Delozier T, Quint R, Roussel A, Brune D (1978) Radiation myelopathy of the cervical spinal cord: Time, dose and volume factors. *Int J Radiat Oncol Biol Phys* 4:239–248.

Balagamwala EH, Angelov L, Koyfman SA, Suh JH, Reddy CA, Djemil T, Hunter GK, Xia P, Chao ST (2012a) Single-fraction stereotactic body radiotherapy for spinal metastases from renal cell carcinoma. *J Neurosurg Spine* 17:556–564.

Balagamwala EH, Chao ST, Suh JH (2012b) Principles of radiobiology of stereotactic radiosurgery and clinical applications in the central nervous system. *Technol Cancer Res Treat* 11:3–13.

Balagamwala EH, Cherian S, Angelov L, Suh JH, Djemil T, Lo SS, Sahgal A, Chang E, Teh BS, Chao ST (2012c) Stereotactic body radiotherapy for the treatment of spinal metastases. *J Radiat Oncol* 1:255–265.

Balagamwala EH, Jung DL, Angelov L, Suh JH, Reddy CA, Djemil T, Magnelli A, Soeder S, Chao ST (2013) Incidence and risk factors for vertebral compression fractures from spine stereotactic body radiation therapy: Results of a large institutional series. *Int J Radiat Oncol Biol Phys* 87:S89.

Balagamwala EH, Suh JH, Reddy CA, Angelov L, Djemil T, Magnelli A, Soeder S, Chao ST (2012d) Higher dose spine stereotactic body radiation therapy is associated with improved pain control in radiosensitive histologies. *Int J Radiat Oncol Biol Phys* 84:S632–S633.

Benzil DL, Saboori M, Mogilner AY, Rocchio R, Moorthy CR (2004) Safety and efficacy of stereotactic radiosurgery for tumors of the spine. *J Neurosurg* 101(Suppl 3):413–418.

Boehling NS, Grosshans DR, Allen PK, McAleer MF, Burton AW, Azeem S, Rhines LD, Chang EL (2012) Vertebral compression fracture risk after stereotactic body radiotherapy for spinal metastases. *J Neurosurg Spine* 16:379–386.

Chang EL, Shiu AS, Lii M-F, Rhines LD, Mendel E, Mahajan A, Weinberg JS et al. (2004) Phase I clinical evaluation of near-simultaneous computed tomographic image-guided stereotactic body radiotherapy for spinal metastases. *Int J Radiat Oncol Biol Phys* 59:1288–1294.

Chang EL, Shiu AS, Mendel E, Mathews LA, Mahajan A, Allen PK, Weinberg JS et al. (2007) Phase I/II study of stereotactic body radiotherapy for spinal metastasis and its pattern of failure. *J Neurosurg Spine* 7:151–160.

Chao ST, Balagamwala EH, Reddy CA, Angelov L, Djemil T, Magnelli A, Soeder S, Suh JH (2012a) Spine stereotactic body radiation therapy outcomes correlated to dosimetric factors. *Int J Radiat Oncol Biol Phys* 84:S212.

Chao ST, Koyfman SA, Woody N, Angelov L, Soeder SL, Reddy CA, Rybicki LA, Djemil T, Suh JH (2012b) Recursive partitioning analysis index is predictive for overall survival in patients undergoing spine stereotactic body radiation therapy for spinal metastases. *Int J Radiat Oncol Biol Phys* 82:1738–1743.

Chiang A, Zeng L, Zhang L, Lochray F, Korol R, Loblaw A, Chow E, Sahgal A (2013) Pain flare is a common adverse event in steroid-naïve patients after spine stereotactic body radiation therapy: A prospective clinical trial. *Int J Radiat Oncol Biol Phys* 86:638–642.

Chow E, Ling A, Davis L, Panzarella T, Danjoux C (2005) Pain flare following external beam radiotherapy and meaningful change in pain scores in the treatment of bone metastases. *Radiother Oncol* 75:64–69.

Cunha MVR, Al-Omair A, Atenafu EG, Masucci GL, Letourneau D, Korol R, Yu E et al. (2012) Vertebral compression fracture (VCF) after spine stereotactic body radiation therapy (SBRT): Analysis of predictive factors. *Int J Radiat Oncol Biol Phys* 84:e343–e349.

Dahele M, Zindler JD, Sanchez E, Verbakel WF, Kuijer JPA, Slotman BJ, Senan S (2011) Imaging for stereotactic spine radiotherapy: Clinical considerations. *Int J Radiat Oncol Biol Phys* 81:321–330.

Damast S, Wright J, Bilsky M, Hsu M, Zhang Z, Lovelock M, Cox B, Zatcky J, Yamada Y (2011) Impact of dose on local failure rates after image-guided reirradiation of recurrent paraspinal metastases. *Int J Radiat Oncol Biol Phys* 81:819–826.

Degen JW, Gagnon GJ, Voyadzis J-M, McRae DA, Lunsden M, Dieterich S, Molzahn I, Henderson FC (2005) CyberKnife stereotactic radiosurgical treatment of spinal tumors for pain control and quality of life. *J Neurosurg Spine* 2:540–549.

Finn MA, Vrionis FD, Schmidt MH (2007) Spinal radiosurgery for metastatic disease of the spine. *Cancer Control J Moffitt Cancer Cent* 14:405–411.

Fornasier VL, Horne JG (1975) Metastases to the vertebral column. *Cancer* 36:590–594.

Garg AK, Shiu AS, Yang J, Wang X-S, Allen P, Brown BW, Grossman P et al. (2012) Phase 1/2 trial of single-session stereotactic body radiotherapy for previously unirradiated spinal metastases. *Cancer* 118:5069–5077.

Gerszten PC, Burton SA, Ozhasoglu C, Welch WC (2007) Radiosurgery for spinal metastases: Clinical experience in 500 cases from a single institution. *Spine* 32:193–199.

Gerszten PC, Ozhasoglu C, Burton SA, Vogel WJ, Atkins BA, Kalnicki S, Welch WC (2004) CyberKnife frameless stereotactic radiosurgery for spinal lesions: Clinical experience in 125 cases. *Neurosurgery* 55:89–98; discussion 98–99.

Grant R, Papadopoulos SM, Greenberg HS (1991) Metastatic epidural spinal cord compression. *Neurol Clin* 9:825–841.

Guckenberger M, Sweeney RA, Flickinger JC, Gerszten PC, Kersh R, Sheehan J, Sahgal A (2011) Clinical practice of image-guided spine radiosurgery—Results from an international research consortium. *Radiat Oncol Lond Engl* 6:172.

Guo S, Chao ST, Reuther AM, Barnett GH, Suh JH (2008) Review of the treatment of trigeminal neuralgia with gamma knife radiosurgery. *Stereotact Funct Neurosurg* 86:135–146.

Hamilton AJ, Lulu BA (1995) A prototype device for linear accelerator-based extracranial radiosurgery. *Acta Neurochir Suppl* 63:40–43.

Hamilton AJ, Lulu BA, Fosmire H, Stea B, Cassady JR (1995) Preliminary clinical experience with linear accelerator-based spinal stereotactic radiosurgery. *Neurosurgery* 36:311–319.

Hartsell WF, Scott CB, Bruner DW, Scarantino CW, Ivker RA, Roach M, Suh JH et al. (2005) Randomized trial of short- versus long-course radiotherapy for palliation of painful bone metastases. *J Natl Cancer Inst* 97:798–804.

Hird A, Chow E, Zhang L, Wong R, Wu J, Sinclair E, Danjoux C, Tsao M, Barnes E, Loblaw A (2009) Determining the incidence of pain flare following palliative radiotherapy for symptomatic bone metastases: Results From three Canadian Cancer Centers. *Int J Radiat Oncol Biol Phys* 75:193–197.

Hyde D, Lochray F, Korol R, Davidson M, Wong CS, Ma L, Sahgal A (2012) Spine stereotactic body radiotherapy utilizing cone-beam CT image-guidance with a robotic couch: Intrafraction motion analysis accounting for all six degrees of freedom. *Int J Radiat Oncol Biol Phys* 82:e555–e562.

Jung DL, Balagamwala EH, Angelov L, Suh JH, Reddy CA, Djemil T, Magnelli A, Soeder S, Chao ST (2013) Incidence and risk factors for pain flare following spine radiosurgery. *Int J Radiat Oncol Biol Phys* 87:S568–S569.

Kaloostian PE, Yurter A, Zadnik PL, Sciubba DM, Gokaslan ZL (2014) Current paradigms for metastatic spinal disease: An evidence-based review. *Ann Surg Oncol* 21:248–262.

Khan L, Chiang A, Zhang L, Thibault I, Bedard G, Wong E, Loblaw A et al. (2015) Prophylactic dexamethasone effectively reduces the incidence of pain flare following spine stereotactic body radiotherapy (SBRT): A prospective observational study. *Support Care Cancer* 23:2937–2943.

Kirkpatrick JP, van der Kogel AJ, Schultheiss TE (2010) Radiation dose-volume effects in the spinal cord. *Int J Radiat Oncol Biol Phys* 76:S42–S49.

Klish DS, Grossman P, Allen PK, Rhines LD, Chang EL (2011) Irradiation of spinal metastases: Should we continue to include one uninvolved vertebral body above and below in the radiation field? *Int J Radiat Oncol Biol Phys* 81:1495–1499.

Koyfman SA, Djemil T, Burdick MJ, Woody N, Balagamwala EH, Reddy CA, Angelov L, Suh JH, Chao ST (2012) Marginal recurrence requiring salvage radiotherapy after stereotactic body radiotherapy for spinal metastases. *Int J Radiat Oncol Biol Phys* 83:297–302.

Leksell L (1951) The stereotaxic method and radiosurgery of the brain. *Acta Chir Scand* 102:316–319.

Loblaw DA, Wu JSY, Kirkbride P, Panzarella T, Smith K, Aslanidis J, Warde P (2007) Pain flare in patients with bone metastases after palliative radiotherapy—A nested randomized control trial. *Support Care Cancer* 15:451–455.

Lo SS, Lutz S, Chang EL, Galanopoulos N, Howell DD, Kim EY, Konski AA et al. (2012) ACR Appropriateness Criteria® spinal bone metastases. American College of Radiology. Available at: https://acsearch.acr.org/docs/71097/Narrative/ [Accessed March 21, 2015].

Lo SS, Sahgal A, Wang JZ, Mayr NA, Sloan A, Mendel E, Chang EL (2010) Stereotactic body radiation therapy for spinal metastases. *Discov Med* 9:289–296.

Lutz S, Berk L, Chang E, Chow E, Hahn C, Hoskin P, Howell D et al. (2011) Palliative radiotherapy for bone metastases: An ASTRO evidence-based guideline. *Int J Radiat Oncol Biol Phys* 79:965–976.

Ma L, Sahgal A, Hossain S, Chuang C, Descovich M, Huang K, Gottschalk A, Larson DA (2009) Nonrandom intrafraction target motions and general strategy for correction of spine stereotactic body radiotherapy. *Int J Radiat Oncol Biol Phys* 75:1261–1265.

Moulding HD, Elder JB, Lis E, Lovelock DM, Zhang Z, Yamada Y, Bilsky MH (2010) Local disease control after decompressive surgery and adjuvant high-dose single-fraction radiosurgery for spine metastases. *J Neurosurg Spine* 13:87–93.

Murphy ES, Suh JH (2011) Radiotherapy for vestibular schwannomas: A critical review. *Int J Radiat Oncol Biol Phys* 79:985–997.

Nguyen Q-N, Shiu AS, Rhines LD, Wang H, Allen PK, Wang XS, Chang EL (2010) Management of spinal metastases from renal cell carcinoma using stereotactic body radiotherapy. *Int J Radiat Oncol Biol Phys* 76:1185–1192.

Pan HY, Allen PK, Wang XS, Chang EL, Rhines LD, Tatsui CE, Amini B et al. (2014) Incidence and predictive factors of pain flare after spine stereotactic body radiation therapy: Secondary analysis of phase 1/2 trials. *Int J Radiat Oncol Biol Phys* 90:870–876.

Patchell RA, Tibbs PA, Regine WF, Payne R, Saris S, Kryscio RJ, Mohiuddin M, Young B (2005) Direct decompressive surgical resection in the treatment of spinal cord compression caused by metastatic cancer: A randomised trial. *Lancet* 366:643–648.

Patel VB, Wegner RE, Heron DE, Flickinger JC, Gerszten P, Burton SA (2012) Comparison of whole versus partial vertebral body stereotactic body radiation therapy for spinal metastases. *Technol Cancer Res Treat* 11:105–115.

Rose PS, Laufer I, Boland PJ, Hanover A, Bilsky MH, Yamada J, Lis E (2009) Risk of fracture after single fraction image-guided intensity-modulated radiation therapy to spinal metastases. *J Clin Oncol* 27:5075–5079.

Ryu S, Pugh SL, Gerszten PC, Yin F-F, Timmerman RD, Hitchcock YJ, Movsas B et al. (2014) RTOG 0631 phase 2/3 study of image guided stereotactic radiosurgery for localized (1–3) spine metastases: Phase 2 results. *Pract Radiat Oncol* 4:76–81.

Ryu S, Rock J, Jain R, Lu M, Anderson J, Jin J-Y, Rosenblum M, Movsas B, Kim JH (2010) Radiosurgical decompression of metastatic epidural compression. *Cancer* 116:2250–2257.

Ryu S, Rock J, Rosenblum M, Kim JH (2004) Patterns of failure after single-dose radiosurgery for spinal metastasis. *J Neurosurg* 101(Suppl 3):402–405.

Sahgal A, Atenafu EG, Chao S, Al-Omair A, Boehling N, Balagamwala EH, Cunha M et al. (2013a) Vertebral compression fracture after spine stereotactic body radiotherapy: A multi-institutional analysis with a focus on radiation dose and the spinal instability neoplastic score. *J Clin Oncol* 31:3426–3431.

Sahgal A, Larson DA, Chang EL (2008) Stereotactic body radiosurgery for spinal metastases: A critical review. *Int J Radiat Oncol Biol Phys* 71:652–665.

Sahgal A, Ma L, Gibbs I, Gerszten PC, Ryu S, Soltys S, Weinberg V et al. (2010) Spinal cord tolerance for stereotactic body radiotherapy. *Int J Radiat Oncol Biol Phys* 77:548–553.

Sahgal A, Ma L, Weinberg V, Gibbs IC, Chao S, Chang U-K, Werner-Wasik M et al. (2012) Reirradiation human spinal cord tolerance for stereotactic body radiotherapy. *Int J Radiat Oncol Biol Phys* 82:107–116.

Sahgal A, Weinberg V, Ma L, Chang E, Chao S, Muacevic A, Gorgulho A et al. (2013b) Probabilities of radiation myelopathy specific to stereotactic body radiation therapy to guide safe practice. *Int J Radiat Oncol Biol Phys* 85:341–347.

Shin JH, Chao ST, Angelov L (2011) Stereotactic radiosurgery for spinal metastases: Update on treatment strategies. *J Neurosurg Sci* 55:197–209.

Sohn S, Chung CK, Sohn MJ, Chang U-K, Kim SH, Kim J, Park E (2014) Stereotactic radiosurgery compared with external radiation therapy as a primary treatment in spine metastasis from renal cell carcinoma: A multicenter, matched-pair study. *J Neurooncol* 119:121–128.

Suh JH (2010) Stereotactic radiosurgery for the management of brain metastases. *N Engl J Med* 362:1119–1127.

Suh JH, Balagamwala EH, Reddy CA, Angelov L, Djemil T, Magnelli A, Soeder S, Chao ST (2012) The use of spine stereotactic body radiation therapy for the treatment of spinal cord compression. *Int J Radiat Oncol Biol Phys* 84:S631.

Thibault I, Al-Omair A, Masucci GL, Masson-Côté L, Lochray F, Korol R, Cheng L et al. (2014) Spine stereotactic body radiotherapy for renal cell cancer spinal metastases: Analysis of outcomes and risk of vertebral compression fracture. *J Neurosurg Spine* 21:711–718.

Videtic GMM, Stephans KL (2010) The role of stereotactic body radiotherapy in the management of non-small cell lung cancer: An emerging standard for the medically inoperable patient? *Curr Oncol Rep* 12:235–241.

Yamada Y, Bilsky MH, Lovelock DM, Venkatraman ES, Toner S, Johnson J, Zatcky J, Zelefsky MJ, Fuks Z (2008) High-dose, single-fraction image-guided intensity-modulated radiotherapy for metastatic spinal lesions. *Int J Radiat Oncol Biol Phys* 71:484–490.

Zaorsky NG, Harrison AS, Trabulsi EJ, Gomella LG, Showalter TN, Hurwitz MD, Dicker AP, Den RB (2013) Evolution of advanced technologies in prostate cancer radiotherapy. *Nat Rev Urol* 10:565–579.

11 Image-guided hypo-fractionated stereotactic radiotherapy for reirradiation of spinal metastases

Nicholas Trakul, Sukhjeet S. Batth, and Eric L. Chang

Contents

11.1 INTRODUCTION

The spine is a common site of metastatic spread for tumors arising from solid cancers. Spinal metastases can cause significant pain and disability. Metastatic disease progression in the spine can lead to spinal cord compression and paralysis. Early palliative intervention is preferred to preserve neurological function and prevent any neurologic complications from occurring. Radiotherapy for spinal metastases is the treatment of choice when surgery is not indicated, or for patients who are medically inoperable. Historically, patients were treated with short courses of fractionated external beam radiation therapy (EBRT), which has been shown to be equivalent in terms of function and overall survival when compared to longer courses utilizing higher doses (Rades et al., 2009). The overall rate of local control of spinal tumors using EBRT is approximately 70% at 1 year and is even lower for so-called mass-type tumors (Mizumoto et al., 2011). However, several trends are worth noting. Recent advances in systemic therapies have improved expected survival for many cancers. There is also increasing recognition of the so-called oligometastatic state (Weichselbaum and Hellman, 2011). This subgroup of patients can often outlive the length of time that conventional short-course radiation would be expected to provide adequate pain and tumor control. Due to these trends, the clinical scenario in which the clinician is faced with locally recurrent cancer within the previously irradiated spine is becoming increasingly common. The spine is the second most commonly treated site with stereotactic body radiation therapy (SBRT) according to a recent survey (Pan et al., 2011). The ability to deliver high radiation doses with increased conformality provides an opportunity to effectively control tumors that have progressed despite prior irradiation with very low risk of spinal cord injury. Reirradiation using SBRT comprises a minority subset of spinal SBRT, which is still a relatively nascent field, and there are limited clinical data for this specific scenario. This chapter will seek to address the optimal treatment strategy for this clinical situation. SBRT, also known as stereotactic ablative radiotherapy, has been well described in this setting. In this chapter, we will also summarize the literature on retreatment of spinal metastases with SBRT and outline our institutional treatment procedure and technique.

11.2 INDICATIONS

Patients with malignant spinal cord compression symptoms should be evaluated by a multidisciplinary team and managed according to a well-defined algorithm such as the NOMS criteria (Laufer et al., 2013). Even patients without neurologic symptoms can benefit from formal multidisciplinary evaluation, including specialists from disciplines in spine neurosurgery, neuroradiation oncology, medical oncology, interventional radiology, and pain management specialists when appropriate. Patients with moderate to severe neurologic symptoms as a result of oncologic spinal cord compression or due to mechanical instability are usually best served by surgical intervention, if possible. Similarly, patients with a high degree of canal compromise (>25%–33%) or a short interval between the prior course of radiation (less than 6 months) may be less than optimal radiosurgery candidates. Tumors that extend over two contiguous vertebral bodies or are greater than 6 cm in length should be treated with caution. While treatment of multiple noncontiguous vertebral levels is technically feasible, one needs to carefully weigh the practicality and prognosis of patients with multilevel involvement (Ryu et al., 2015).

11.3 TREATMENT

11.3.1 IMMOBILIZATION AND IMAGING

Patient immobilization is of critical importance in all SBRT treatments, and the spine is no exception. High dose gradients around the spinal cord could lead to disastrous consequences if there is significant intrafraction motion. At our institution, linac-based treatments utilizing volumetric modulated arc therapy (VMAT) are most commonly used. For this treatment, immobilization is achieved using a semirigid vacuum-assisted body cushion such as the BodyFIX system (Elekta AB, Stockholm, Sweden). Even with the use of systems that provide real-time tracking such as CyberKnife, immobilization is important to avoid shifts between imaging cycles, but a simpler Vac-Lok or Alpha Cradle system may be utilized (Lo et al., 2010).

A treatment planning CT, with or without contrast, is obtained with 1–2 mm slices through the area of interest. A contrast-enhanced MRI of the spine area of interest is also obtained for treatment planning and fused to the treatment planning CT when feasible. This should be performed as close to the time of treatment as possible, and preferably in a similar position, although patient positioning may be limited by MRI bore size. It has been noted that the spinal cord itself can shift up to 2 mm based on position alone (van Mourik et al., 2014). Accurate assessment of the extent of disease on spinal MRI, or robust CT/MRI fusion, is of critical importance as significant contouring information is contained within the MRI.

After surgery, metallic artifact from spinal hardware instrumentation frequently limits the visibility of the spinal cord and epidural tumor on both MRI and CT. In this instance, a CT myelogram is highly recommended (Sahgal et al., 2011). Myelography does not assist with artifact masking of residual tumor or "at-risk" areas in the postsurgery spine. Verified fusion of the preoperative MRI with the planning CT and close communication by involvement of the operating neurosurgeon is important to accurately delineate the target volume in these cases.

11.3.2 TARGET VOLUME AND ORGAN AT RISK CONTOURING

Target volumes are typically defined on the treatment planning CT with the aid of fused MRI sequences obtained as described earlier. Gross tumor volume (GTV) is typically defined as gross disease seen on T1- and T2-weighted MRI sequences. Historically, there has been a great deal of variability regarding the delineation of target volumes in the spine from institution to institution. Currently, we define clinical treatment volume (CTV) as typically defined according to standards outlined in the Radiation Therapy Oncology Group (RTOG) 0631 protocol and the international standard, as published by the International Spine Radiosurgery Consortium (Cox et al., 2012; Ryu et al., 2014). A further expansion to form a planning treatment volume (PTV) is often done, typically 1.0–2.0 mm, depending on the institution, type of immobilization, and pretreatment imaging used.

Accurate contouring of the spinal cord is of critical importance, as it is the primary dose-limiting structure. This is important in all spinal SBRT cases, but it is of critical importance in the reirradiation setting, where commonly used dose constraints for the spinal cord are often exceeded and the margin for error is presumed to be small. Typically the treatment planning MRI fused to the treatment planning CT is used to delineate the cord. In the postoperative setting, hardware artifacts often make accurate delineation of the spinal cord difficult to impossible, and it is our preference to use a CT myelogram in this setting. Typically, either the thecal sac (which is usually equivalent to a 1.5 mm expansion around the cord) or a uniform 2 mm expansion is then used to create a spinal cord planning risk volume (PRV), in order to account for uncertainties in positioning and organ movement. Relevant at-risk organs near the spinal level to be treated are also contoured, that is, the kidney, esophagus, heart, and liver, with appropriate expansions to account for uncertainty to form PRVs as indicated by each particular organ.

11.3.3 TREATMENT PLANNING AND DELIVERY

In reirradiation, it is of critical importance to know the prior dose delivered to the spinal cord. It cannot be simply assumed that the spinal cord uniformly received the prescribed dose, even when using opposed fields. In many cases, the cord may have received doses above the prescribed dose, depending on the field arrangement and weighting. Many patients may have been treated emergently using a hand-calculated plan. In this case the relevant treatment parameters should be obtained and the fields reconstructed using the departmental computerized treatment planning system in order to accurately model prior dose to the spinal cord. Any uncertainty regarding prior dose should lead one to err on the side of caution, as the results of an inadvertent cord overdose could be significant.

SBRT planning for a previously irradiated cord is much the same as would occur if no prior radiation had been done. Treatment can be delivered as intensity-modulated radiotherapy using multiple step-and-shoot static fields typically ranging from seven to nine based on spine class solutions, or VMAT, which has the advantage of decreased treatment delivery times, and potentially reducing the possibility of intrafraction patient motion from occurring.

Prior reports of spinal SBRT have used a wide variety of dose/fractionation schemes, and there is little consensus as to the optimal dose and fraction number. Early studies have demonstrated the superiority of 30 Gy in 5 fractions over 20 Gy in 5 fractions, although this is generally not considered to be within the ablative range (Damast et al., 2011). In the dose range considered to be ablative, there has been no documented difference in responses between 27 Gy in 3 fractions and 30 Gy in 5 fractions (Garg et al., 2011). These are the two most commonly used dose/fractionation schemes in our institutional practice. While dosimetric coverage of the target volumes to 95% or better is routine for other anatomic sites, this level of coverage is often difficult or impossible to achieve, given the stricter cord constraints used in the retreatment setting. Coverage of as low as 70% may be adequate, if spinal cord limits and other organs at risk (OARs) dictate. Contouring of the GTV is often helpful to evaluate plans when coverage of the CTV and PTV are suboptimal. A representative treatment plan is demonstrated in Figures 11.1 through 11.4.

Constraints to the spinal cord PRV are dictated by the prior dose received by the spinal cord. This has been outlined in a report comparing doses received to the spinal cord in patients who subsequently developed radiation myelopathy. In summary, it was found that the risk of myelopathy was extremely low if the total dose, normalized to 2 Gy fractions, was kept to a BED < 70 Gy, and that the SBRT component comprised less than 50% of the normalized dose (Sahgal et al., 2012). The constraints suggested in this report are currently used in our practice and a table summarizing the constraints on the spinal cord for various prior dose and fraction schemes is contained within that publication. A notable exception would be patients treated with single-fraction EBRT as the initial treatment. Limited data in regard to development of myelopathy in this specific scenario should engender caution, and one should always err on the conservative side when discounting prior dose to the spinal cord in this setting.

Other OAR constraints within the treatment plan should be used according to institutional and published American College of Radiology (ACR), or institutional guidelines (Potters et al., 2010).

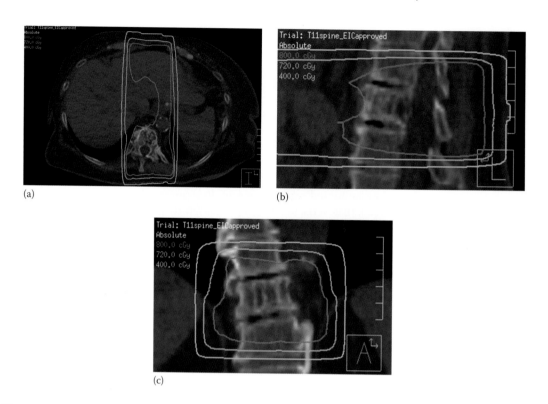

Figure 11.1 Representative images in the axial (a), sagittal (b), and coronal (c) planes from a conventional plan delivering 8 Gy in 1 fraction with anterior–posterior and posterior–anterior beams for a breast cancer patient with a painful T11 vertebral body metastasis. Isodose lines correspond to 8 Gy or 100% (red), 7.2 Gy or 90% (orange), and 4 Gy or 50% (teal).

Treatment delivery is performed using either pretreatment alignment via onboard kilovoltage (kV) imaging to match to bony anatomy followed by cone-beam CT or CT on rails. The CyberKnife system will adjust based on the alignment to bony anatomy using stereoscopic kV imaging in real time while radiation dose is being delivered. Adjustments are made with the goal of having the target and spinal cord volumes aligned within a preset tolerance (preferably <1 mm) in all directions, including rotation (<1°), which may be adjusted automatically or manually, depending on the capabilities of the treatment table and delivery system. Once treatment is initiated, realignment should be performed as needed, either after a preset elapsed time or if there is suspicion that there has been movement of the target.

11.4 OUTCOMES

Overall the outcomes from spinal metastasis reirradiation have been promising. Outcomes from relevant trials are summarized in Table 11.1. Interpretation of tumor control varies from study to study, and it is uncommon for suspected failures to be confirmed by biopsy, such as is commonly done in lung or other body sites. Local control rates range from 66% to 92%, with the majority of studies reporting control rates greater than 75%. In a series from University of California, San Francisco (UCSF), no significant difference was noted in failure rates between SBRT targets in a previously irradiated field and those that had not received prior radiation (Sahgal et al., 2009). Other outcomes besides local control are somewhat better defined and may be more clinically relevant in SBRT. Pain control using primary spinal SBRT is excellent, approaching 90% and comparing very favorably with standard EBRT (Ryu et al., 2008). Pain control in retreatment series appears to be less than that reported for primary spinal SBRT and is reported in the range of 65%–85% (Masucci et al., 2011). It is currently unclear and not possible to discern whether this represents a significant difference reflecting that pain control is more difficult to achieve in patients undergoing a second course of radiation with SBRT.

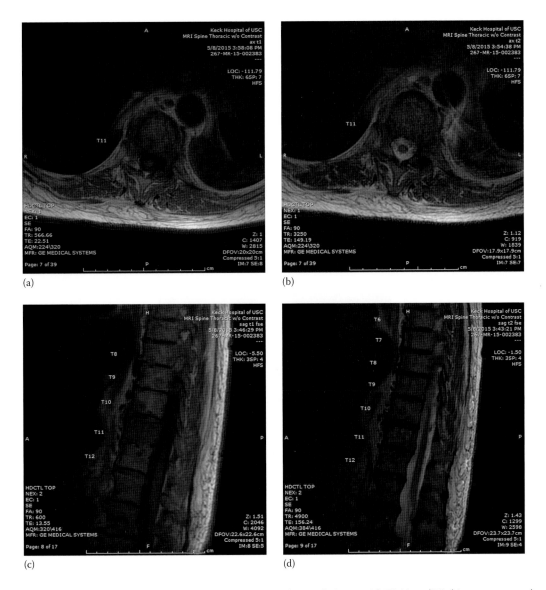

(a)   (b)

(c)   (d)

Figure 11.2 Representative magnetic resonance images in the axial planes with T1 (a) and T2 (b) sequences and sagittal planes with T1 (c) and T2 (d) sequences demonstrating progression of a T11 vertebral body metastasis 3 months after treatment with 8 Gy in 1 fraction.

The most common site of failure in patients retreated with SBRT is within the epidural space. This has been noted in a number of the published reports mentioned earlier (Sahgal et al., 2009; Choi et al., 2010; Garg et al., 2011). Failure was correlated with increasing proximity of the tumor to the thecal sac. This is expected, given the need to constrain the dose to the spinal cord, which is even more stringent in the reirradiation setting. It may also be caused by inadequate coverage of the "at-risk" epidural space in order to reduce the dose given circumferentially around the spinal cord. Perhaps better understanding of which areas of the epidural space are at highest risk for microscopic extension and utilization of standardized contouring guidelines will increase our understanding of epidural patterns of failures and how they can be minimized. Other common sites of failure include paraspinal tissue and segments of the spinal column not included within the treatment volume, such as the pedicles and lamina of the posterior elements (Masucci et al., 2011).

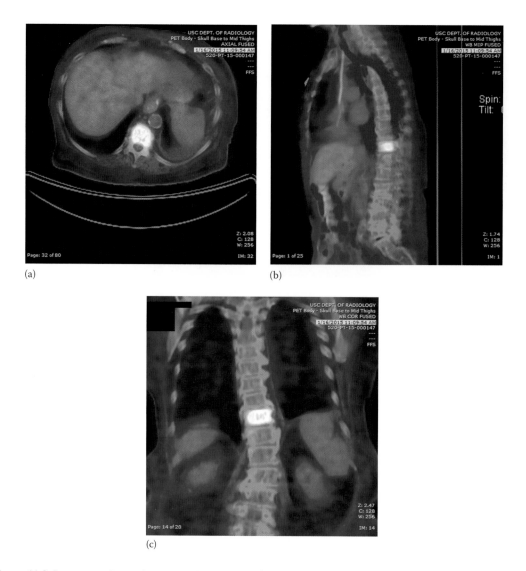

Figure 11.3 Representative positron-emission tomography/computed tomography images in the axial (a), sagittal (b), and coronal (c) planes demonstrating progression of a T11 vertebral body metastasis 3 months after treatment with 8 Gy in 1 fraction.

11.5 TOXICITY

Fortunately, radiation myelopathy secondary to spinal SBRT has been documented as a rare event in the literature. However, its rare incidence makes it difficult to accurately model toxicity. In the reirradiation setting, this is further complicated by the heterogeneity in dose fractionation schedules used in the treatment of spinal metastases. Animal models have been historically used to investigate the tolerance of the spinal cord to radiation. Ang et al. reirradiated monkey spinal cords after administering an initial treatment of 44 Gy. Using two different conventionally fractionated retreatment doses, it was found that even in a conservative model, 61% recovery of the cord could be assumed after 3 years, and in more optimistic models, higher rates of recovery could occur (Ang et al., 2001). Another study used pigs whose cords had been previously irradiated to 30 Gy in 10 fractions. After 1 year, an inhomogeneous ablative dose of radiation was delivered in a single fraction lateral to the spinal cord. No increase in myelopathy was seen in pigs that had been previously irradiated when compared to a control group, suggesting that significant, perhaps even complete cord recovery had occurred at 1 year (Medin et al., 2012). Data from clinical studies

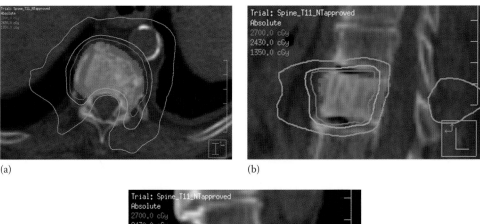

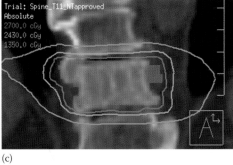

Figure 11.4 Representative images in the axial (a), sagittal (b), and coronal (c) planes from a stereotactic ablative radiotherapy plan delivering 27 Gy in 3 fractions with volumetric modulated arc therapy technique to a previously treated T11 vertebral body metastasis. Isodose lines correspond to 27 Gy or 100% (red), 24.3 Gy or 90% (orange), and 13.5 Gy or 50% (green).

are limited by small patient numbers but have informed practice. Neider et al., who used clinical data from patients treated with multiple courses of EBRT, found that significant spinal cord recovery did occur and that cumulative doses of 70–75 Gy (in 2 Gy fractionated equivalent doses) could be delivered safely (Nieder et al., 2006). The previously mentioned report by Sahgal et al. looked at reported cases of myelopathy in patients treated with SBRT. The analysis contained five patients, an indication of the rarity of the myelopathy. The results were summarized earlier, but the safety of a cumulative BED of 70 Gy correlates well with the data from Nieder. A reanalysis using the generalized linear-quadratic model found a similar result, with no myelopathy seen under 70 Gy (Huang et al., 2013).

Other organs previously exposed to radiation during a prior course of radiation can also receive significant radiation dose from spinal SBRT in the reirradiation setting. However, little is known about whether this has resulted in increased rates of toxicity. The esophagus can be another major organ injured by SBRT, and a report examining esophageal toxicity when treating spine and lung tumors urged conservatism in the setting of prior radiation and concurrent chemotherapy (Abelson et al., 2012). Overall, toxicity to other organs is a rare event in spinal SBRT, with only a few cases reported of overall mild toxicity (Chawla et al., 2013).

11.6 CONTROVERSIES

Reirradiation of any tumor has traditionally been controversial, due to the prevailing thought that any failure represents radioresistant biology and that additional radiation treatments would be futile. The radiobiology of doses in the ablative range, such as those used with SBRT is now understood to be very different from EBRT (Brown et al., 2014). Although these differences are still poorly understood, as the clinical literature on reirradiation demonstrates, local control rates associated with ablative doses are comparable whether or not prior radiation has been used. Other concerns regarding the use of SBRT primarily involve the use of a highly labor and cost-intensive treatment in patients whose life expectancy is

Table 11.1 Published outcomes for reirradiation of spinal tumors using stereotactic body radiation therapy

STUDY	NUMBER OF PATIENTS/ TOTAL TREATMENTS	MEDIAN DOSE FIRST RT COURSE (Gy)	MEDIAN INTERVAL (MONTHS)	REIRRADIATION TD/FRACTION NUMBER	MEDIAN ACCUMULATED DOSE TO SPINAL CORD (EQD2)	PLANNING	SETUP/ IMAGING	FOLLOW-UP (MONTHS)	MYELOPATHY	LOCAL CONTROL	PAIN CONTROL
Chang et al. (2012)	49/54	39.2	25	27 Gy/3	83.4 Gy	CyberKnife	kV tracking	17.3	0%	80.8% at 1 year	80.8% at 1 year
Choi et al. (2010)	42/51	40	19	20 Gy/2	76 Gy	CyberKnife	kV tracking	7	$n = 1$, G4	73% at 1 year	65%
Damast et al. (2011)	95/97	30	N/A	20–30 Gy/5	54.3 Gy	IMRT	CBCT or daily ports	12.1	0%	74% at 1 year (30 Gy/5)	77%
Garg et al. (2011)	59/63	30	N/A	27–30 Gy/3–5	N/A	IMRT	CBCT or daily CT on rails	13	$n = 2$, G3 peripheral nerve injury	76% at 1 year	N/A
Mahadevan et al. (2011)	60/81	30	20	24–30 Gy/3–5	N/A	CyberKnife	kV tracking	12	$n = 3$, persistent radicular pain; $n = 1$, lower extremity weakness	93% at f/u	65% at 1 month
Mahan et al. (2005)	8/8	30	N/A	30 Gy/15	48 Gy	Tomotherapy	Daily MVCT	15.2	0%	100% at f/u	100% at f/u
Milker-Zabel et al. (2003)	18/19	38	17.7	39.6 Gy	N/A	ss-IMRT	Stereotactic	12.3	0%	94.7% at 1 year	81.3%
Sahgal et al. (2009)	25/37	36	11	24 Gy/3	N/A	CyberKnife	kV tracking	7	0%	90%	N/A
Sterzing et al. (2010)	36/36	36.3	17.5	34.8 Gy/11	46.5 Gy	Tomotherapy	Daily MVCT	7.5	0%	76% at 1 year	N/A

Note: RT, radiotherapy; TD, total dose; N/A, not available; G, grade; CBCT, cone-beam computed tomography; kV, kilovolt; IMRT, intensity-modulated radiotherapy; MVCT, megavoltage computed tomography; ss-IMRT, single-shot intensity-modulated radiotherapy; f/u, follow-up.

presumed to be limited. However, cost-effective studies have shown SBRT to be a viable option in patients with pancreatic cancer, which carries a similarly poor prognosis (Murphy et al., 2012). The use of SBRT in cases of early metastatic epidural spinal cord compression with minimal motor dysfunction is an emerging indication requiring significant institutional resources to do in a timely fashion and considered controversial by some for both initial and retreatments. However, several groups are actively investigating the feasibility of this, and as more outcomes are published, it may be increasingly used (Ryu et al., 2010).

11.7 FUTURE DIRECTIONS

Future studies will seek to optimize the dose and fractionation to be used in SBRT. As more toxicity data are accumulated, we will hopefully be better able to define the maximal dose that can be delivered safely to the spinal cord in both the de novo and the reirradiation setting based on prospective studies involving systematic steps toward spinal cord constraint relaxation. The utility of combining SBRT and cement augmentation procedures is under investigation (Gerszten et al., 2005). It may be possible to combine SBRT with radiosensitizing agents in the hope of improving therapeutic efficacy. This is an area of active investigation in other SBRT sites, and its potential use in spinal SBRT is intriguing, assuming that it can be done safely. There is recent interest in the combination of brain radiosurgery and immunomodulators in the CNS (Patel et al., 2015). It appears inevitable that such immunomodulators will be evaluated in conjunction with spinal SBRT in the near future. On the opposite side of the spectrum, work is being undertaken to better understand the mechanism of myelopathy as there are agents that may prevent it (Wong et al., 2015). Looking to the future, as the numbers of patients reirradiated with spinal SBRT increases, a new problem may arise as to how to deal with spinal SBRT retreatment failures and whether reirradiation with a second course of SBRT to the same site is feasible and safe.

CHECKLIST: KEY POINTS FOR CLINICAL PRACTICE

✓	ACTIVITY	SOME CONSIDERATIONS
	Patient selection	*Is the patient appropriate for SBRT?* • Does not have high-grade spinal cord compression or is not a surgical candidate • KPS > 50 • No recent cytotoxic chemotherapy or radiotherapy • Single to few levels of vertebral bodies involved • No more than 2–3 contiguous vertebral bodies involved
	SBRT vs EBRT	*Is the lesion amenable to SBRT?* • Nonradiosensitive histology (e.g., lymphoma, multiple myeloma) • Expected survival >6 months
	Simulation	*Immobilization* • Supine with stereotactic body immobilization system (Linac) or Alpha Cradle (CyberKnife) *Imaging* • CT scan ± intrathecal contrast • MRI with gadolinium enhancement. T1/T2 sequences with 1–2 mm • Verified CT to MRI fusion over area of tumor involvement
	Treatment planning	*Contours* • GTV should include intact or postoperative residual contrast enhancing tumor as defined on T1/T2 • CTV per published guidelines, dependent on extent of spine involvement • PTV expansion 1–2 mm if necessary

(Continued)

✓	ACTIVITY	SOME CONSIDERATIONS
		Treatment planning • Inverse treatment planning • Dose • 27 Gy in 3 fractions or 30 Gy in 5 fractions to PTV • Coverage • PTV coverage 70%–95% • Dose constraints • Spinal cord: Based on prior cord dose (Sahgal et al., 2012)
	Treatment delivery	*Imaging* • Patients are first set up conventionally using room lasers and manual shifts • Orthogonal kV matching to relevant bony anatomy • CBCT to match to tumor (CTV or PTV) and the spinal cord • Treatment couch with six degrees of freedom to correct translational and rotational shifts if available • Repeat CBCT for patient movement or long treatment time interval (10–15 minutes)

REFERENCES

Abelson JA, Murphy JD, Loo BW, Chang DT, Daly ME, Wiegner EA, Hancock S et al. (October 2012) Esophageal tolerance to high-dose stereotactic ablative radiotherapy. *Dis Esophagus* 25(7):623–629.

Ang KK, Jiang GL, Feng Y, Stephens LC, Tucker SL, Price RE (July 1, 2001) Extent and kinetics of recovery of occult spinal cord injury. *Int J Radiat Oncol Biol Phys* 50(4):1013–1020.

Brown JM, Carlson DJ, Brenner DJ (February 1, 2014) The tumor radiobiology of SRS and SBRT: Are more than the 5 Rs involved? *Int J Radiat Oncol Biol Phys* 88(2):254–262.

Chang U-K, Cho W-I, Kim M-S, Cho CK, Lee DH, Rhee CH (May 2012) Local tumor control after retreatment of spinal metastasis using stereotactic body radiotherapy; comparison with initial treatment group. *Acta Oncol Stockh Swed* 51(5):589–595.

Chawla S, Schell MC, Milano MT (December 2013) Stereotactic body radiation for the spine: A review. *Am J Clin Oncol* 36(6):630–636.

Choi CYH, Adler JR, Gibbs IC, Chang SD, Jackson PS, Minn AY, Lieberson RE, Soltys SG (October 1, 2010) Stereotactic radiosurgery for treatment of spinal metastases recurring in close proximity to previously irradiated spinal cord. *Int J Radiat Oncol Biol Phys* 78(2):499–506.

Cox BW, Spratt DE, Lovelock M, Bilsky MH, Lis E, Ryu S, Sheehan J et al. (August 1, 2012) International spine radiosurgery consortium consensus guidelines for target volume definition in spinal stereotactic radiosurgery. *Int J Radiat Oncol Biol Phys* 83(5):e597–e605.

Damast S, Wright J, Bilsky M, Hsu M, Zhang Z, Lovelock M, Cox B, Zatcky J, Yamada Y (November 1, 2011) Impact of dose on local failure rates after image-guided reirradiation of recurrent paraspinal metastases. *Int J Radiat Oncol Biol Phys* 81(3):819–826.

Garg AK, Wang X-S, Shiu AS, Allen P, Yang J, McAleer MF, Azeem S, Rhines LD, Chang EL (August 1, 2011) Prospective evaluation of spinal reirradiation by using stereotactic body radiation therapy: The University of Texas MD Anderson Cancer Center experience. *Cancer* 117(15):3509–3516.

Gerszten PC, Germanwala A, Burton SA, Welch WC, Ozhasoglu C, Vogel WJ (October 2005) Combination kyphoplasty and spinal radiosurgery: A new treatment paradigm for pathological fractures. *J Neurosurg Spine* 3(4):296–301.

Huang Z, Mayr NA, Yuh WT, Wang JZ, Lo SS (June 2013) Reirradiation with stereotactic body radiotherapy: Analysis of human spinal cord tolerance using the generalized linear-quadratic model. *Future Oncol Lond Engl* 9(6):879–8l87.

Laufer I, Rubin DG, Lis E, Cox BW, Stubblefield MD, Yamada Y, Bilsky MH (June 2013) The NOMS framework: Approach to the treatment of spinal metastatic tumors. *Oncologist* 18(6):744–751.

Lo SS, Sahgal A, Wang JZ, Mayr NA, Sloan A, Mendel E, Chang EL (April 2010) Stereotactic body radiation therapy for spinal metastases. *Discov Med* 9(47):289–296.

Mahadevan A, Floyd S, Wong E, Jeyapalan S, Groff M, Kasper E (December 1, 2011) Stereotactic body radiotherapy reirradiation for recurrent epidural spinal metastases. *Int J Radiat Oncol Biol Phys* 81(5):1500–1505.

Mahan SL, Ramsey CR, Scaperoth DD, Chase DJ, Byrne TE (December 1, 2005) Evaluation of image-guided helical tomotherapy for the retreatment of spinal metastasis. *Int J Radiat Oncol Biol Phys* 63(5):1576–1583.

Masucci GL, Yu E, Ma L, Chang EL, Letourneau D, Lo S, Leung E et al. (December 2011) Stereotactic body radiotherapy is an effective treatment in reirradiating spinal metastases: Current status and practical considerations for safe practice. *Expert Rev Anticancer Ther* 11(12):1923–33.

Medin PM, Foster RD, van der Kogel AJ, Sayre JW, McBride WH, Solberg TD (July 1, 2012) Spinal cord tolerance to reirradiation with single-fraction radiosurgery: A swine model. *Int J Radiat Oncol Biol Phys* 83(3):1031–1037.

Milker-Zabel S, Zabel A, Thilmann C, Schlegel W, Wannenmacher M, Debus J (January 1, 2003) Clinical results of retreatment of vertebral bone metastases by stereotactic conformal radiotherapy and intensity-modulated radiotherapy. *Int J Radiat Oncol Biol Phys* 55(1):162–167.

Mizumoto M, Harada H, Asakura H, Hashimoto T, Furutani K, Hashii H, Murata H et al. (January 1, 2011) Radiotherapy for patients with metastases to the spinal column: A review of 603 patients at Shizuoka Cancer Center Hospital. *Int J Radiat Oncol Biol Phys* 79(1):208–213.

Murphy JD, Chang DT, Abelson J, Daly ME, Yeung HN, Nelson LM, Koong AC (February 15, 2012) Cost-effectiveness of modern radiotherapy techniques in locally advanced pancreatic cancer. *Cancer* 118(4):1119–1129.

Nieder C, Grosu AL, Andratschke NH, Molls M (December 1, 2006) Update of human spinal cord reirradiation tolerance based on additional data from 38 patients. *Int J Radiat Oncol Biol Phys* 66(5):1446–1449.

Pan H, Simpson DR, Mell LK, Mundt AJ, Lawson JD (October 1, 2011) A survey of stereotactic body radiotherapy use in the United States. *Cancer* 117(19):4566–4572.

Patel KR, Shoukat S, Oliver DE, Chowdhary M, Rizzo M, Lawson DH, Khosa F, Liu Y, Khan MK (May 16, 2015) Ipilimumab and stereotactic radiosurgery versus stereotactic radiosurgery alone for newly diagnosed melanoma brain metastases. *Am J Clin Oncol* [E-pub].

Potters L, Kavanagh B, Galvin JM, Hevezi JM, Janjan NA, Larson DA, Mehta MP et al. (February 1, 2010) American Society for Therapeutic Radiology and Oncology (ASTRO) and American College of Radiology (ACR) practice guideline for the performance of stereotactic body radiation therapy. *Int J Radiat Oncol Biol Phys* 76(2):326–332.

Rades D, Lange M, Veninga T, Rudat V, Bajrovic A, Stalpers LJA, Dunst J, Schild SE (January 1, 2009) Preliminary results of spinal cord compression recurrence evaluation (score-1) study comparing short-course versus long-course radiotherapy for local control of malignant epidural spinal cord compression. *Int J Radiat Oncol Biol Phys* 73(1):228–234.

Ryu S, Jin R, Jin J-Y, Chen Q, Rock J, Anderson J, Movsas B (March 2008) Pain control by image-guided radiosurgery for solitary spinal metastasis. *J Pain Symptom Manage* 35(3):292–298.

Ryu S, Pugh SL, Gerszten PC, Yin F-F, Timmerman RD, Hitchcock YJ, Movsas B et al. (March 2014) RTOG 0631 phase 2/3 study of image guided stereotactic radiosurgery for localized (1–3) spine metastases: Phase 2 results. *Pract Radiat Oncol* 4(2):76–81.

Ryu S, Rock J, Jain R, Lu M, Anderson J, Jin J-Y, Rosenblum M, Movsas B, Kim JH (May 1, 2010) Radiosurgical decompression of metastatic epidural compression. *Cancer* 116(9):2250–2257.

Ryu S, Yoon H, Stessin A, Gutman F, Rosiello A, Davis R (March 2015) Contemporary treatment with radiosurgery for spine metastasis and spinal cord compression in 2015. *Radiat Oncol J* 33(1):1–11.

Sahgal A, Ames C, Chou D, Ma L, Huang K, Xu W, Chin C et al. (July 1, 2009) Stereotactic body radiotherapy is effective salvage therapy for patients with prior radiation of spinal metastases. *Int J Radiat Oncol Biol Phys* 74(3):723–731.

Sahgal A, Bilsky M, Chang EL, Ma L, Yamada Y, Rhines LD, Létourneau D et al. (February 2011) Stereotactic body radiotherapy for spinal metastases: Current status, with a focus on its application in the postoperative patient. *J Neurosurg Spine* 14(2):151–166.

Sahgal A, Ma L, Weinberg V, Gibbs IC, Chao S, Chang U-K, Werner-Wasik M et al. (January 1, 2012) Reirradiation human spinal cord tolerance for stereotactic body radiotherapy. *Int J Radiat Oncol Biol Phys* 82(1):107–116.

Sterzing F, Hauswald H, Uhl M, Herm H, Wiener A, Herfarth K, Debus J, Krempie R (August 15, 2010) Spinal cord sparing reirradiation with helical tomotherapy. *Cancer* 116(16):3961–3968.

van Mourik AM, Sonke J-J, Vijlbrief T, Dewit L, Damen EM, Remeijer P, van der Heide UA (November 2014) Reproducibility of the MRI-defined spinal cord position in stereotactic radiotherapy for spinal oligometastases. *Radiother Oncol* 113(2):230–234.

Weichselbaum RR, Hellman S (June 2011) Oligometastases revisited. *Nat Rev Clin Oncol* 8(6):378–382.

Wong CS, Fehlings MG, Sahgal A (March 24, 2015) Pathobiology of radiation myelopathy and strategies to mitigate injury. *Spinal Cord* 53, 574–580.

12 IG-HSRT for benign tumors of the spine

Peter C. Gerszten and John C. Flickinger

Contents

12.1 INTRODUCTION

Benign tumors of the spine represent a wide variety of histologies that occur within the intradural space, as well as epidural, paraspinal, and vertebral body locations. The primary treatment option for most benign spinal neoplasms is open surgical resection. The safety and effectiveness of such surgery has been clearly documented (Seppala et al., 1995a,b; Klekamp and Samii, 1998; Gezen et al., 2000; Conti et al., 2004; Parsa et al., 2004; Dodd et al., 2006). The majority of spinal meningiomas, schwannomas, and neurofibromas are noninfiltrative and can be completely and safely resected using microsurgical techniques (McCormick, 1996a; Kondziolka et al., 1998; Asazuma et al., 2004; Parsa et al., 2004). When complete tumor removal is achieved, recurrence is unlikely (Roux et al., 1996; Cohen-Gadol et al., 2003; Conti et al., 2004; Parsa et al., 2004).

In certain circumstances, however, some patients are less than ideal candidates for open standard surgical resection because of age, medical comorbidities, the recurrent nature of their tumor, or anatomical location of the lesion (Dodd et al., 2006). Tumors that have recurred after open surgical resection may make safe surgical resection challenging or not possible. Multiple benign spinal tumors, as are common in familial neurocutaneous disorders, may be a pattern of spinal pathology better suited for a less invasive radiosurgical option. It is in such clinical circumstances that radiosurgery might serve as an important treatment option for these patients.

Radiation therapy has been used for the treatment of numerous benign diseases since the discovery of the therapeutic potential of ionizing radiation (Lukacs et al., 1978; Solan and Kramer, 1985; Klumpar et al., 1994; Seegenschmiedt et al., 1994; Keilholz et al., 1996). Although similar benign tumor histologies can be found in and around the spine, the initial radiosurgical instruments were frame based and unable to treat such extracranial targets (Dodd et al., 2006). The more recent emergence of frameless image-guided radiosurgery allows for the treatment of such tumors throughout the body (Adler et al., 1997; Ryu et al., 2001; Gerszten and Bilsky, 2006). Stereotactic radiosurgery for the treatment of a variety of benign intracranial lesions has become widely accepted with excellent long-term outcomes and minimal toxicity (Steiner et al., 1974; Lunsford et al., 1991; Flickinger et al., 1996; Kondziolka et al., 2003; Sachdev et al., 2011). Kondziolka et al. presented a long-term control rate of 93% after the use of the Gamma Knife (Elekta AB, Stockholm, Sweden) to treat 1045 intracranial benign meningiomas (2008). In a subset of 488 patients for which serial imaging was obtained, 215 tumors regressed and 256 were unchanged for a control rate of 97% (Murovic and Chang, 2014).

The development of frameless image-guided radiosurgery allows for the ability to treat such benign tumors throughout the body (Adler et al., 1997; Ryu et al., 2001; Dodd et al., 2006; Gerszten and Bilsky, 2006; Gerszten et al., 2012b). The current evidence supporting the use of extracranial radiosurgery for malignant spine tumors is considerable (Hitchcock et al., 1989; Colombo et al., 1994; Hamilton et al., 1995; Pirzkall et al., 2000; Chang and Adler, 2001; Ryu et al., 2001, 2003, 2004; Medin et al., 2002; Sperduto et al., 2002; Milker-Zabel et al., 2003; Benzil et al., 2004; Bilsky et al., 2004; Chang et al., 2004; De Salles et al., 2004; Gerszten and Welch, 2004, 2007; Rock et al., 2004; Yin et al., 2004; Degen et al., 2005). There is much less experience regarding the use of radiosurgery for the treatment of benign tumors of the spine (Adler et al., 1997; Ryu et al., 2001; Gerszten and Bilsky, 2006; Murovic and Charles Cho, 2010; Gerszten et al., 2012b). Radiosurgery has only recently become part of the multimodality management of benign spinal tumors.

12.2 INDICATIONS AND SPECIAL CONSIDERATIONS FOR BENIGN TUMORS

Open surgical extirpation remains the primary treatment option for benign spinal tumors if feasible. The majority of spinal meningiomas, schwannomas, and neurofibromas are noninfiltrative and can be completely and safely resected using microsurgical techniques (McCormick, 1996a,b; Asazuma et al., 2004; Parsa et al., 2004). However, the limitations of surgical options for some patients with benign tumors of the spine make radiosurgery an attractive alternative. Such limitations include medical comorbidities that preclude open surgery, recurrent tumors, anatomical constraints such as tumor location relative to the spinal cord, or patients with familial phakomatoses.

The current clinical indications for benign spine tumor radiosurgery include tumors located in surgically difficult regions of the spine, recurrent benign spinal tumors after prior surgical resection, and benign spine tumors in patients who have significant medical comorbidities that preclude open surgery. Relative contraindications for radiosurgery for benign spine tumors include tumors without well-defined margins, tumors with significant spinal cord compression resulting in acute neurologic symptoms, and tumors that can easily be resected with conventional surgical techniques (Gibbs et al., 2015).

In the most recently published series from our institution, 55% of cases demonstrated recurrence after prior open surgical resection (Gerszten et al., 2012b). In our previously published experience, only 23% of patients had undergone prior open surgical resection (Gerszten et al., 2008). This may reflect an increase in awareness among clinicians in the use of radiosurgery as a successful alternative to repeat open surgery in the event of tumor recurrence. However, as we have become more comfortable in the successful clinical outcome of radiosurgery, our more recent experience demonstrates that radiosurgery is still more commonly used as a primary treatment modality.

In addition to treating poor surgical candidates and recurrent or residual tumor after surgery, spinal radiosurgery is most appropriate for well-circumscribed lesions associated with minimal spinal cord or nerve root compromise and no biomechanical instability (Gerszten, 2007). Relative contraindications to spinal radiosurgery include evidence of overt spinal instability, neurologic deficit resulting from bony compression, or previous radiation treatment to spinal cord tolerance dose. A theoretical advantage of utilizing spinal radiosurgery as the frontline management for spinal tumors is the possibility that such treatment may act as prophylaxis to future spinal instability or neural element compression, obviating the need for extensive spinal surgery and instrumentation. Moreover, early conformal radiosurgery may obviate the need for large-field external beam radiation, which is known to suppress bone marrow function. Tumor shrinkage and complete obliteration aside, the minimally invasive technique of spinal radiosurgery may develop into an effective palliative strategy solely through local tumor control. Finally, the ability to perform spinal radiosurgery in the outpatient setting is an advantage that may spare patients with spinal tumors both time and the morbidities associated with hospitalization.

12.3 TECHNICAL AND DOSE CONSIDERATIONS

12.3.1 TARGET DELINEATION

The accurate geometric visualization of the tumor target on imaging is essential for radiosurgery contouring and treatment planning. Proper target delineation and contouring is indispensable to a safe and effective radiosurgery treatment. Although benign intradural, extramedullary tumors of the spine are often conspicuous by their homogeneous contrast enhancement, the large and often irregular shape of spinal neoplasms makes contouring a major challenge. The diagnostic imaging modality of choice for benign spinal tumors is magnetic resonance imaging (MRI). However, optimum radiosurgery treatment planning today rests on the modality of computed tomography (CT). Modern commercial radiosurgery systems use CT-based imaging for planning and delivery, and benign extramedullary spinal tumors are typically well visualized with postcontrast CT imaging. Most systems today also permit image fusion between MRI and CT images. Such image fusion may improve the target definition for spinal tumors, especially when such neoplasms exhibit heterogeneous contrast enhancement.

MRI-CT image fusion in the spine is usually more challenging than that required for intracranial radiosurgery because it is more dependent upon the technical aspects of image acquisition and patient positioning during image acquisition. The quality of MR spinal image fusion often requires that the patient's imaging position closely matches the intended treatment position. Issues of spatial distortion need to be considered if MRI is used directly for planning. Because signal intensities of MR images do not reflect a direct relationship with electron densities, unless attenuation coefficients are manually assigned to the region of interest, spatial distortion limits the accuracy of using MRI directly in radiosurgery planning (Gibbs et al., 2015). The ability to identify tumors on CT, together with the probability of generating an adequate image fusion with MRI imaging, is key to defining the radiosurgery target. Most benign tumors of the spine enhance brightly and have well-defined margins, making target delineation rather straightforward. Because virtually all extramedullary spinal tumors show some degree of contrast enhancement, postcontrast CT is sometimes used directly to define the target. CT myelography is an alternative imaging strategy which can provide superior tumor and spinal cord visualization in some intradural cases.

At our institution, patients are immobilized with the BodyFix (total body bag, Medical Intelligence®) when treatment sites are below T6; otherwise, a head and shoulder mask with S-board (CIVCO, Kalona, IA) is used. The target volume concept employed in all cases is the gross tumor volume (GTV) as seen on enhanced imaging. A clinical target volume (CTV) is not employed for these benign tumors. Our experience has shown that accurate contouring of benign spine tumors is nearly impossible without the use of MR imaging fusion. High-resolution sequence MR imaging often improves resolution degraded by instrumentation. In such cases of instrumentation, we have attempted to resolve questions of tumor definition with CT myelography. Titanium implants are preferred over stainless steel in order to decrease imaging artifact. In some cases when radiosurgery is already anticipated prior to open surgery, such as the case for large "dumbbell" foraminal tumors, spinal instrumentation is only placed on the *contralateral* side of the tumor in order to allow for maximum tumor definition for radiosurgery.

The International Commission on Radiation Units and Measurements has formalized the contouring process by defining three different volumes: The GTV represents the unambiguous radiologic confines of the target neoplasm, the clinical target volume (CTV) includes nearby anatomic regions such as the vertebral body where microscopic tumor extension is anticipated, and the planned target volume (PTV) is an augmented form of the CTV that corrects for movement and accuracy of treatment delivery. Since benign spinal tumors are usually well circumscribed and do not exhibit metastasis, the CTV will more closely approximate the GTV.

12.3.2 DOSE SELECTION

The goal of spinal radiosurgery for benign spinal tumors is to deliver a clinically significant radiation dose to the tumor via a plan that respects the radiation dose tolerance limits of the nearby spinal cord, cauda equina, and surrounding organs such as the intestines, esophagus, kidneys, larynx, and liver. By virtue of their origin along the dura and spinal nerve roots, extramedullary spinal tumors can significantly impinge upon the spinal cord or cauda equina, making contouring difficult (Gibbs et al., 2015). The degree of impingement of the tumor on the spinal cord may prevent the generation of a suitable radiosurgery treatment plan. Similar to radiosurgery doses prescribed for intracranial tumors, spine radiosurgery doses generally range from 12 to 20 Gy in a single fraction; doses as high as 30 Gy have been delivered when the treatment was hypofractionated in up to five sessions (Table 12.1). While spinal radiosurgery often is delivered using a single radiation fraction technique, the formal definition includes hypofractionated dosing plans with a maximum of five treatment sessions. Doses as high as 30 Gy have been reported in the setting of hypofractionation, and even higher total doses have been administered when one considers those patients who received conventional radiotherapy prior to spinal radiosurgery. The linear quadratic equation remains the most accepted mathematical model of cell kill secondary to ionizing radiation (Yamada et al., 2007). The equation has been used to compare various dose fractionation schedules.

The spinal cord is one of the most radiosensitive structures considered in radiosurgery treatment plans. Radiation myelopathy may occur in a delayed fashion, and the spinal cord tolerance to single fraction radiation has not been carefully determined with long-term follow-up studies. Nevertheless, some data point to a less than 5% probability of myelopathy at 5 years when the cord receives a 60 Gy dose using standard fractionation (Yamada et al., 2007). Most centers have tailored plans to ensure that the spinal cord is exposed to no more than 20 Gy during fractionated therapy (Gerszten et al., 2006). Borrowing again from the intracranial radiosurgery experience, some physicians have fashioned a cord avoidance strategy based on dosage tolerance data for the optic nerves and chiasm. In contrast to patients with metastatic cancer, patients receiving radiosurgery for benign paraspinal neoplasms are expected to survive longer with higher functional status. Consequently, it is prudent that planning for benign tumors err on underestimating spinal cord tolerance in case an unrecognized form of radiation myelopathy may manifest over decades. Whether prior radiosurgery will sensitize the spinal cord and cauda equina to degenerative insults in aging patients is currently not known.

In our institutional experience with benign spine tumors, the mean maximum dose received by the GTV is 16 Gy (range 12–24 Gy) delivered in a single fraction in the majority of cases. In a few cases in which the tumor was found to be intimately associated with the spinal cord with distortion of the spinal cord itself, the prescribed dose to the GTV was delivered in 3 fractions. The mean lowest dose received by the GTV was 12 Gy (range 8–16 Gy). The GTV ranged from 0.37 to 94.5 cm^3 (mean 13.7 cm^3; median 5.9 cm^3).

A planning target volume (PTV) expansion is sometimes employed at our institution in order to account for targeting inaccuracies. In the majority of our cases (85%), a PTV expansion of 2 mm was employed. The remaining PTV expansion ranged from 0 to 3 mm. The PTV prescription dose was as a rule 2 Gy less than the prescription dose to the GTV. The mean number of beams used to deliver the radiosurgery treatment was 10 (median 9, range 7–14 beams).

A planning target expansion of usually 2 mm is employed for all tumors of the neural foramen at the level of the spinal cord, as well as tumors of the cauda equina and paraspinal locations. However, in instances in which the tumor is intimately associated with the spinal cord itself, no such planned target expansion is used. For cases in which the spinal cord itself is deformed by the tumor within the spinal canal, the radiosurgery is delivered in three separate sessions.

Table 12.1 Series of radiosurgery for benign spine tumors

SERIES	MENINGIOMA	SCHWANNOMA	NEUROFIBROMA	MEAN AGE (YEARS)	N	INDICATION	DOSE PER FRACTIONS	LENGTH OF F/U (MONTHS)	OUTCOME
Dodd et al. (2006)	16	30	9	46	51	51% postsurgical recurrent/ residual	16–30 Gy/1–5 fractions	25	96% stable/ decreased 3 repeat surgery 1 progression 1 new myelopathy
Chopra et al. (2005)	0	0	1	12	1	Residual	12.5 Gy/1 fraction	20	Stable
De Salles et al. (2004)	1	1	1	62	3	Not reported	12–15 Gy/1 fraction	6	Stable
Benzil et al. (2004)	1	2	0	61	3	Not reported	5–50.4 Gy/ variable	Not reported	Rapid pain relief
Sahgal et al. (2007)	2	0	11	58	13	10 postsurgical 2 primary therapy	21 Gy/3 fractions	25	2 with progression (both NF1)
Gerszten et al. (2008)	13	35	25	44	73	30 pain 18 postsurgical 14 primary therapy 9 neuro deficit	12–20 Gy/1 fraction	37	73% pain improvement 100% stable/ decreased 3 new myelopathy

(*Continued*)

Table 12.1 (*Continued*) Series of radiosurgery for benign spine tumors

SERIES	MENINGIOMA	SCHWANNOMA	NEUROFIBROMA	MEAN AGE (YEARS)	N	INDICATION	DOSE PER FRACTIONS	LENGTH OF F/U (MONTHS)	OUTCOME
Gerszten et al. (2012a)	10	16	14	52	45	24 primary 21 postsurgical	12–24 Gy	32	100% control
Sachdev et al. (2011)	32	47	24	53	87	Surgery contraindicated	14–30 Gy/1–5 fractions	33	99% stable or decreased 1 progressed with new myelopathy 6 persistent symptoms
Selch et al. (2009)	NA	25 nerve sheath tumors	NA	61	25	12 pain 22 neurologic deficit	12–15 Gy/1 fraction	6	100% stable

12.4 OUTCOMES DATA

One of the first reports of the treatment of a benign spinal tumor, a hemangioblastoma, was reported by Chang et al. (1998). In 2001, the feasibility of image-guided spine radiosurgery for benign tumors was established when researchers at Stanford University reported the first clinical experience, which included two spinal schwannomas and one spinal meningioma (Ryu et al., 2001; Gibbs et al., 2015). Despite this relatively early use of stereotactic radiotherapy for benign spinal tumors, there has been a relative paucity of reports detailing clinical outcomes compared to spine malignancies. One issue that limits outcomes reporting for benign spine radiosurgery is that the evaluation of radiosurgery for these lesions requires longer follow-up to confirm durable safety and efficacy than for malignant tumors of the spine.

Nevertheless, there is a growing amount of experience to date with clinical outcomes after spine radiosurgery for benign tumors (Gerszten et al., 2003; Benzil et al., 2004; De Salles et al., 2004; Chopra et al., 2005; Dodd et al., 2006; Gibbs et al., 2015). Similar to the application of radiosurgery for intracranial pathology, spine radiosurgery has been primarily embraced as an adjunctive treatment technique for tumor recurrence or residual tumor after open surgery; tumor progression after failure of conventional irradiation treatment, in patients who have significant medical comorbidities that preclude open surgery; or as "salvage" therapy when further conventional irradiation or surgery is not appropriate.

Given their pathological similarities, it has been speculated that benign spinal lesions would be equally responsive to radiosurgery as their intracranial counterparts (Dodd et al., 2006). It is for this reason that centers experienced with radiosurgery have explored this developing technology to treat benign spine tumors. The extramedullary intradural spinal neoplasms treated with radiosurgery have primarily included meningiomas, schwannomas, and neurofibromas. Ryu et al. (2001) published the first clinical cohort description of radiosurgery being used to treat such lesions. In the initial experience from Stanford using the CyberKnife for 15 benign spinal tumors, no tumor progression was reported with a follow-up period of 12 months.

The largest published series to date for benign intradural extramedullary spinal tumors from Dodd et al. (2006) reported the results after radiosurgery treatment of 55 such tumors (30 schwannomas, 9 neurofibromas, and 16 meningiomas). Total treatment doses ranged from 16 to 30 Gy delivered in 1–5 fractions to tumor volumes that varied from 0.136 to 24.6 cm^3. Less than 1 year after radiosurgery, three patients (one meningioma, one schwannoma, and one neurofibroma) required open surgical resection of their tumor because of persistent or worsening symptoms. Only one of these three lesions was larger radiographically. Twenty-eight of the fifty-five lesions had greater than 24 months follow-up. The mean follow-up in this group was 36 months. All lesions in this group were either stable (61%) or smaller (39%). No tumor in this group increased in size. The following sections, subdivided by histopathology, summarize available clinical outcome data for benign spinal tumor radiosurgery.

12.5 SPINAL MENINGIOMAS

Spinal meningiomas are arachnoid cap cell-derived tumors of the fifth to seventh decade that have a female predominance and occur mainly in the thoracic region. They arise from cells of the meningeal coverings of the central nervous system and occur more frequently within the brain than the spine, with a 5:1 ratio. Gross total surgical resection optimizes outcome (Peker et al., 2005). Spinal meningiomas, in general, have a more favorable prognosis relative to their intracranial counterparts. In a study of histological and microarray data of meningiomas, Sayaguès et al. (2006) determined that spinal meningiomas were most commonly associated with lower proliferative rates and more indolent histologies (psammomatous, transitional variants) and showed characteristic genetic and genomic differences compared with intracranial meningiomas. Radiosurgery has become a well-accepted treatment option for the management of intracranial meningioma and schwannoma (Chang and Adler, 2001; Kondziolka et al., 2003). A more recent publication of over 1000 meningiomas revealed a 97% tumor control rate with a follow-up of up to 18 years (Kondziolka et al., 2008). Other studies have shown similar efficacy and safety of radiosurgery for meningiomas (Chang and Adler, 1997; Lee et al., 2002; Kondziolka et al., 2003).

Dodd et al. (2006) reported 16 treated spinal meningiomas (mean dose 20 Gy, mean tumor volume 2.4 cm^3, mean follow-up 27 months) and demonstrated radiologic stabilization in 67% and radiologic tumor decrease in 33% of the 15 who had radiographic follow-up. Only one patient required open surgery and another sustained the only complication reported in the series. Seventy percent of the meningiomas treated in this series were symptomatically stable or improved. Most patients experienced an improvement in pain and strength with radiosurgery (Dodd et al., 2006).

From our institution's published series, 13 spinal meningiomas were treated using a single-fraction technique (mean dose 21 Gy, mean tumor volume 4.9 cm^3). Eleven of thirteen patients had radiosurgery as an adjunctive treatment for residual or recurrent tumor following open surgical resection. Radiographic tumor control was demonstrated in all cases with a median follow-up of 17 months (Gerszten et al., 2008). Of the 11 patients who had undergone previous open surgical resection with residual or recurrent tumors, none demonstrated radiographic tumor progression on subsequent serial imaging after the radiosurgery treatment. Radiographic tumor control was also demonstrated for those two patients in whom radiosurgery was used as a primary treatment modality with median follow-up imaging at 14 months.

In the series of Sachdev et al. (2011) from Stanford University, all 32 meningiomas with radiographic follow-up were controlled at a mean follow-up of 33 months (range, 6–87) for all benign tumors. At last follow-up, 47% of meningiomas were stable, while 53% had decreased in volume. In a smaller series, Sahgal et al. (2007) reported treating two spine meningioma cases with radiosurgery (mean dose 23 Gy delivered in 2 fractions, mean tumor volume 1.6 cm^3) with no evidence of radiographic tumor progression. Benzil et al. (2004) and De Salles et al. (2004) have also reported an experience with treating spine meningiomas with good long-term radiographic follow-up (Chang and Adler, 1997).

12.6 SPINAL SCHWANNOMAS

Nerve sheath tumors are comprised of schwannomas and neurofibromas. Schwannoma is the most common spinal tumor and has no proclivity for spinal region or gender (Seppala et al., 1995b). Nerve sheath tumors are comprised of schwannomas and neurofibromas. Schwannomas account for nearly one-third of primary spinal tumors, whereas neurofibromas constitute only 3.5% (Klekamp and Samii, 1998; Gezen et al., 2000; Conti et al., 2004; Parsa et al., 2004). Spinal schwannomas typically arise from the posterolaterally placed dorsal nerve root. Given their posterior position relative to the spinal cord or cauda equina, their removal via a laminectomy approach is usually straightforward. Patients with nerve sheath tumors typically present with local pain, radiating pain, and/or paraparesis and a relatively long duration of symptoms varying from 6 weeks to over 5 years (Figure 12.1).

Although many reports in the literature describe these tumors collectively, there are sufficient differences between the two tumor types to warrant a separate discussion for each. Schwannomas arise most commonly in the dorsal nerve root, are more commonly completely intradural (>80%), and are generally amenable to complete resection (Conti et al., 2004; Gibbs et al., 2015). On the contrary, neurofibromas arise more commonly in the ventral nerve root, are predisposed to multiple tumors by a strong association with NF1, and present with both intradural and extradural components in 66% of cases. These tumors are also uniquely different with respect to predisposing genetic defects. The merlin/schwannomin gene on chromosome 22 is associated with schwannomas in NF2, whereas the neurofibromin gene on chromosome 17 is associated with NF1 and neurofibromas. NF2 is an autosomal dominant genetic disorder that predisposes patients to developing multiple central and peripheral nervous system tumors. Schwannomas are the most common spinal tumor that occurs in these patients. Though sporadic occurrences of schwannomas unrelated to NF2 are not uncommon, those tumors associated with NF2 are more aggressive and recur more often after treatment. In a retrospective review of 87 patients with spinal nerve sheath tumors removed by surgery, 17 of whom had NF2-associated schwannomas, all NF2-related tumors recurred by 9 years compared with a 10-year recurrence rate of 28% in tumors not associated with NF2 (Klekamp and Samii, 1998). Factors strongly predictive of recurrence after surgery were partial resection, prior recurrence, NF2, and advanced age (Klekamp and Samii, 1998).

There is significant experience with radiosurgery for the treatment of intracranial schwannomas. The long-term rate of growth control after radiosurgery for vestibular schwannoma has been documented as

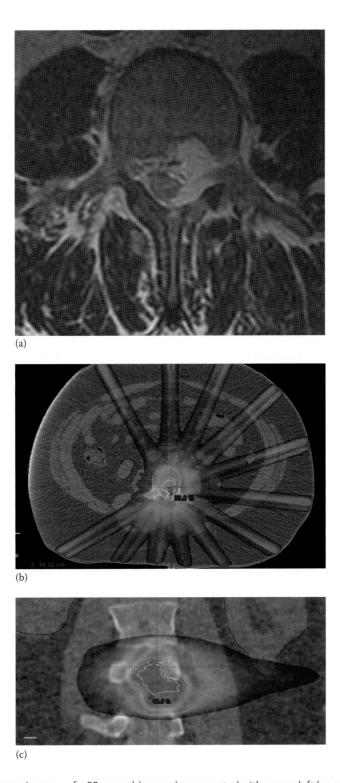

(a)

(b)

(c)

Figure 12.1 A representative case of a 33-year-old man who presented with severe left leg pain and a left L3 *de novo* nerve sheath tumor. He was neurological intact but had significant pain. Gadolinium-enhanced axial magnetic resonance imaging demonstrates the tumor filling the spinal canal and eroding the vertebral body (a). Axial and coronal projections of the isodose lines of the treatment plan (b, c). The tumor was treated with a prescribed dose of 16 Gy in a single fraction to a volume of 7.3 cm^3, with a cauda equina D$_{max}$ of 12 Gy, using 12 intensity-modulated radiation therapy beams. Near-complete pain relief was achieved within 1 month.

95%–98% (Kondziolka et al., 1998; Prasad et al., 2000). This successful tumor control should be similar for spinal schwannomas as well. From the Stanford series, Dodd et al. reported on 30 tumors (mean dose 19 Gy, mean tumor volume 5.7 cm³, mean follow-up 26 months) and all but one patient had radiographic tumor control after radiosurgery. One-third of patients reported improvement in pain, weakness, or sensation, but 18% had a clinical decline after treatment (Dodd et al., 2006). Forty percent of patients in this series had spinal schwannomas in the setting of NF2. Forty-one percent of these patients were treated for recurrent or residual tumor after surgery. At a mean follow-up of 33 months (range, 6–87) for all benign spine tumors, 51% of schwannomas were stable, while 47% had decreased in volume.

From our institution, 35 schwannomas were treated with spine radiosurgery (mean dose 22 Gy, mean tumor volume 11.0 cm³). Fourteen of seventeen patients (82%) described significant pain improvement for whom pain was the primary indication for radiosurgery. Radiographic tumor control was demonstrated in six of seven patients (86%) for which radiosurgery was used as the primary treatment. A total of three patients went on to undergo upon surgical resection for new or persistent neurological deficits (Gerszten et al., 2008). Benzil et al. (2004) and De Salles et al. (2004) have also reported an experience with treating spine schwannomas with good long-term radiographic follow-up and rapid pain relief.

Selch et al. (2009) retrospectively reviewed 20 patients with 25 nerve sheath tumors. Four patients had NF1 and four had NF2. Histopathology was available in seven patients after subtotal tumor removal 2–36 months prior to radiosurgery (four schwannomas, three neurofibromas). These patients underwent radiosurgery because of clinical and imaging evidence of tumor regrowth or persistent symptoms. Of the remaining 18 lesions, presumptive histopathology in nine was established after removal of peripheral nerve sheath tumors elsewhere in the patient (five neurofibromas, four schwannomas). Nine tumors without histopathologic confirmation were treated based on symptoms and imaging consistent with nerve sheath tumor. At a median follow-up of 12 months, there were no local failures. Tumor size remained stable in 18 cases, and 28% demonstrated more than 2 mm reduction in tumor size.

12.7 SPINAL NEUROFIBROMAS

Neurofibromas are benign nerve sheath tumors that may arise from either peripheral or spinal nerve roots. Neurofibromas of the spine are often multiple, predominate in the cervical region, and are commonly associated with NF1. Neurofibromas are less common than schwannomas, constituting only 3.5% of primary spinal tumors. Approximately 2% of patients with NFl will develop 2% symptomatic spinal tumors. Multiple spinal tumors are not uncommon (Seppala et al., 1995a,b). As with other nerve sheath tumors, patients present with pain and paraparesis. Two-thirds of neurofibromas occur in the cervical spine. These tumors grow both in the intradural as well as extradural spaces. Surgical extirpation of these tumors usually requires sectioning of the originating nerve root in order to completely resect the lesion.

Sahgal et al. (2007) reported a series of 11 treated neurofibromas (mean dose 21 Gy delivered in 3 fractions, mean tumor volume 6.0 cm³). Radiographic control was documented in nine patients. Three patients had NF1, and two of these suffered progression. In the published series from Stanford, nine neurofibromas in seven patients with NF1 (mean dose 10.6 Gy, mean tumor volume 4.3 cm³) were treated with radiosurgery, and tumor stabilization on imaging was documented in six of seven (86%) patients. After a mean follow-up of 20 months, half of the patients described an improvement in symptoms after radiosurgery and half of the patients documented a worsening in pain, weakness, or numbness at their last follow-up (Dodd et al., 2006). However, the tumors were radiographically stable in all patients. The authors caution that the role of radiosurgery for neurofibromas remains unclear, particularly considering that a significant number of the NF1 patients were myelopathic at presentation. They further state that the most realistic and attainable goal of neurofibroma treatment in myelopathic patients is tumor control without significant expectations for symptomatic improvement (Dodd et al., 2006).

Our institution has published an initial experience with 25 neurofibroma cases together with 35 schwannoma cases, and 13 meningioma cases (mean dose 21.3 Gy, mean tumor volume 12.6 cm³) (Gerszten et al., 2008). Similar to the Stanford experience, no patient has had evidence of radiographic tumor progression on follow-up. Twenty-one of these patients had NF1 and nine had NF2. Radiosurgery ameliorated discomfort in 8 of 13 patients (61.5%) treated for pain. All patients who saw no improvement in pain had NF1. These findings echo outcomes from the Stanford series that found pain control in NF1-associated spinal neurofibromas to respond less well to radiosurgery (Dodd et al., 2006). Poorer microsurgical results for neurofibromas have also been observed in patients with NF1 (Seppala et al., 1995b). The multiplicity of neurofibromas in NF1 may be partially to blame as this factor makes identifying the symptomatic neurofibroma in need of treatment more difficult. Furthermore, given that many of the patients with neurofibromas have multiple lesions along their spine, it can often be difficult to determine whether symptom progression is due to the treated lesion or from any of the other neurofibroma lesions within the spine (Roux et al., 1996). Moreover, the infiltrating nature of neurofibromas, in contrast to the other benign extramedullary intradural spinal tumors, may engender more irreversible neural damage and increase the susceptibility of the native nerve root to injury from both microsurgical and radiosurgical treatments. Finally, future genomic investigations may reveal that intrinsic genetic differences in NF1-associated neurofibromas predispose to a weaker radiobiologic response.

In the Sachdev et al. (2011) series, at a mean follow-up of 33 months (range, 6–87) for all benign tumors, 82% of the neurofibromas were stable, while 18% had decreased in volume. Considering pain as a separate component, 17% reported improvement, 50% reported minimal change, and 33% reported worsening. Given these results, the role of radiosurgery for neurofibromas remains unclear, particularly considering that a significant number of the NF1 patients were myelopathic at presentation. The poor clinical responses seen in this study appear to mimic the finding by Seppala et al. (1995a) that only 1 of 15 patients who were alive at long-term follow-up after surgery reported complete freedom from symptoms. It is likely that the most realistic and attainable goal of neurofibroma treatment in myelopathic patients is tumor control without significant expectations for symptomatic improvement. Furthermore, given that many of the patients with neurofibromas had multiple lesions along their spine, it can often be difficult to determine whether symptom progression was due to the treated lesion or from any of the other neurofibroma lesions within the spine (Murovic and Charles Cho, 2010; Gibbs et al., 2015).

12.8 SPINAL HEMANGIOBLASTOMAS

Hemangioblastomas are vascular tumors that are intramedullary in location and situated near the pial surface. Hemangioblastomas are the most common tumors associated with von Hippel–Lindau (VHL) disease. However, these tumors also can occur sporadically. Chang et al. (2011) evaluated 30 benign spinal tumors in 20 patients of which 8 tumors were hemangioblastomas. Three tumors were associated with VHL disease. All tumors were treated with 25.8 Gy as the mean stereotactic radiosurgery dosage, which was given in 1 fraction. The follow-up period was 50 months with a range between 23 and 72 months. Pain was not evaluated in this series. Of six patients who were asymptomatic prior to radiosurgery treatment, all (100%) remained asymptomatic after radiosurgery. In one patient who had gait difficulties, the gait improved. In another patient with unspecified neurological deficits, the neurological examination remained stable. Using imaging evaluation, six tumors regressed, and each one stabilized and progressed, respectively. No complications were mentioned to have occurred in any patients in this publication.

Moss et al. (2009) evaluated 92 cranial and spinal hemangioblastomas in 31 patients that were treated with radiosurgery. Of these patients, 17 tumors were located in the spinal cord. The median follow-up time for the spinal hemangioblastoma patients was 37 months with a mean of 37.2 months. The tumor dosages ranged from 20 to 25 Gy and were delivered in 1–3 fractions. Target volumes ranged from 0.06 to 2.65 cm^3. Fifteen of the sixteen hemangioblastomas either remained stable (nine) or improved (six). Only one patient's tumor increased in size during the follow-up period. Importantly, in this series of 16 hemangioblastomas managed with radiosurgery, no patients developed spinal cord radiation toxicity. Radiosurgery seems to be a safe and effective management strategy for spinal hemangioblastomas, especially in the setting of VHL.

12.9 TOXICITIES

For spine radiosurgery, the spinal cord and cauda equina are the organs at risk that most frequently limit the prescribed target dose. Spinal cord injury is arguably the most feared complication in radiotherapy and has historically limited the aggressiveness of spinal tumor treatment, whether benign or malignant

(Gibbs et al., 2009). There has been considerable attention given to attempted determinations of the human and animal radiation tolerance of spinal cord and cauda equina to stereotactic body radiotherapy (Ryu et al., 2003; Bijl et al., 2005; Sahgal et al., 2007, 2012) Complications from the radiosurgical treatment of benign spinal tumors are fortunately rare.

Dodd et al. (2006) reported the first published case of radiation-induced myelopathy after radiosurgery for benign spinal tumors. This case involved a 29-year-old woman with a cervicothoracic spinal meningioma who developed myelopathic symptoms 8 months after radiosurgery to a dose of 24 Gy over three sessions. It was felt that the relatively large volume of spinal cord (1.7 cm^3) irradiated above 18 Gy (over three sessions of 6 Gy) may have been the contributing factor. In their dose–volume analysis, the irradiated volume of spinal cord in this patient represented an outlier compared with other patients in the series.

In the most recent update from the Stanford experience of 103 tumors in 87 patients with a mean follow-up of 33 months, only a single patient in the entire series developed transient radiation myelitis 9 months after treatment (Sachdev et al., 2011). This patient had a C7-T2 recurrent (previously debulked) meningioma with no previous radiation to the area. The 7.6 cm^3 tumor was treated to 24 Gy in 3 fractions with an intratumoral maximum dose of 34.3 Gy. The volume of spinal cord irradiated included 4.7 cm^3 over 8 Gy and 0.1 cm^3 over 27 Gy. The maximum spinal cord dosage was 29.9 Gy. The patient developed posterior column dysfunction during the course of the myelitis but became neurologically stable after intervention with corticosteroids. The tumor volume had decreased at the time of intervention and the tumor remained radiographically controlled at last follow-up. Edema seen on initial imaging resolved over time and was replaced by some evidence of myelomalacia in the region.

In a series of 19 benign spine tumors from the University of California San Francisco, no late toxicity, including myelopathy, occurred (Sahgal et al., 2007). Previous studies have suggested that the risk of radiation myelopathy is a function of total dose, fraction size, length of spinal cord exposed, and duration of treatment (Rampling and Symonds, 1998; Isaacson, 2000). The benign spinal neoplasms that have thus far been treated with radiosurgery represent a heterogeneous group of neoplasms, some of which have had no prior treatment and some which have been treated with surgery and/or external beam radiation. Although already small, the risk of radiation myelopathy with radiosurgical treatment of benign spinal tumors may be further minimized once physicians understand how spinal cord radiation tolerance is influenced by previous treatments such a microsurgery.

In our initial series, of the total of 73 benign tumors, two schwannoma patients and one meningioma patient developed radiosurgery-related cord injury manifesting as a Brown-Sequard syndrome, 5–13 months posttreatment (Gerszten et al., 2008). These three patients were treated with a combination of steroids, vitamin E, and gabapentin; one patient underwent hyperbaric oxygen therapy. A commonality to all three cases was prior open surgical resections of the tumors, which may have caused a predisposition of the spinal cord to radiation injury.

In our most recent series using cone-beam CT image guidance and newer devices for radiosurgery delivery, no subacute or long-term spinal cord or cauda equina toxicity at a median follow-up of 32 months occurred (Gerszten et al., 2012a). Forty-five consecutive benign spine tumors were treated using the Elekta Synergy S 6-MV linear with a beam modulator and cone-beam CT image guidance technology for target localization. The mean maximum dose received by the gross tumor volume (GTV) was 16 Gy (range 12–24 Gy) delivered in a single fraction in 39 cases. The mean lowest dose received by the GTV was 12 Gy (range 8–16 Gy). The GTV ranged from 0.37 to 94.5 cm^3 (mean 13.7 cm^3). In the majority of cases, a planning target volume expansion of 2 mm was used. The technique used at our institution may serve as an important reference for treating benign spine tumors, with a high safety profile. Our prescribed dose to the target volume for benign tumors has decreased over time as we have become more comfortable with the successful long-term radiographic control of these lower doses and reinforced as well by the absence of radiation-induced toxicity.

12.10 CONTROVERSIES

Patients with benign lesions of the spine have prolonged life expectancies compared to their malignant counterparts. Therefore, the potential for delayed radiation myelopathy is of special concern when radiosurgery is considered. In addition, benign spine tumors have their own unique presentation, relationship to the spinal cord, and radiobiologic response to radiosurgery, any of which could represent unique challenges to the safe and effective application of radiosurgical ablation (Dodd et al., 2006). Because the life expectancy of most of these patients is considered to be normal and because radiation injury can take years to manifest (Dodd et al., 2006), there is more controversy regarding radiosurgery for the management of benign tumors of the spine. Especially in patients with greater survival and longer life span due to benign spinal tumor histology, the potential for delayed radiation myelopathy must be entertained. The low radiation tolerance of the spinal cord is the primary limiting factor when dosing therapeutic ionizing radiation for spinal tumors. Conventional external beam radiotherapy lacks the precision to deliver large single-fraction doses to benign spinal tumors that are in close proximity to the spinal cord. In fact, the radiosensitivity of the spinal cord has often required that the treatment dose is far below the optimal therapeutic dose (Faul and Flickinger, 1995; Loblaw and Laperriere, 1998; Ryu et al., 2001).

Spinal radiosurgery aims to deliver a highly conformal radiation dose to the treatment volume, which should increase the likelihood of successful tumor control and minimize the risk of spinal cord injury (Ryu et al., 2001, 2003; Benzil et al., 2004; Bilsky et al., 2004). The intradural nature of benign spinal tumors engenders a close proximity between tumor and spinal cord or cauda equina, an anatomical relationship that may influence the risk of neural toxicity. Moreover, since benign spinal tumors are prone to late recurrence and the late toxicity of radiation to the spinal cord may take years to develop, the evaluation of radiosurgical treatment in terms of efficacy, safety, and durability will necessitate longer follow-up than that which has been granted to metastatic spine tumors (Gerszten et al., 2002). Finally, since benign spinal tumors are rarely life threatening or destabilizing to the spinal column, treatment with high-dose ionizing radiation or by any other means creates more controversy than that seen in the context of malignant spinal tumors; morbidity from any intervention, whether immediate or delayed, becomes less acceptable in this usually younger, healthier patient population (Gerszten, 2011).

12.11 FUTURE DIRECTIONS

Radiosurgery for benign spinal tumors is feasible and safe. Delivering highly conformal tumoricidal doses of radiation to spinal tumors is an enormous technical hurdle that has recently been surmounted. Spinal radiosurgery is arguably more complex than intracranial radiosurgery and poses unique challenges. Preliminary studies have demonstrated that spine radiosurgery can afford patients with benign spinal tumors a high probability of symptomatic relief and excellent long-term radiographic tumor control. However, longer-term follow-up data are necessary to assess the durability of radiosurgical treatment. If further studies show a greater tolerance of the spinal cord to certain radiosurgical doses, a dose escalation might achieve greater obliteration rates while maintaining an acceptable risk. The optimum dose–fraction regimens required for lesion ablation and the dose tolerance limit of the normal spinal cord and cauda equina remain to be determined. The initial experience is very positive and should encourage future studies to help poise radiosurgery as a less invasive treatment option for benign spinal tumors, similar to the current role of intracranial radiosurgery. Its role in patients with neurofibromatosis will also need to be further defined.

Further clinical experience with radiosurgery for benign tumors of the spine also opens the opportunity for the investigation of the use of radiosurgery for nonneoplastic extracranial indications. Given our current knowledge base regarding the accuracy and conformality of current delivery technologies, combined with our understanding of dose tolerance for the spinal cord and cauda equina and neuroanatomical targets, the next logical step is for the development of "functional" spine radiosurgery. Such benign disease processes might include pain syndromes such as failed back surgery syndrome, reflex sympathetic dystrophy, facet-mediated pain, as well as other conditions such as hyperhydrosis and spinal cord injury. While perhaps there has been a trend to lower doses for benign spine tumors with an emphasis on safety, there is no reason that extremely high and highly conformal doses similar to those used for trigeminal neuralgia might not also be employed for the functional ablation of extracranial targets.

CHECKLIST: KEY POINTS FOR CLINICAL PRACTICE

✓	ACTIVITY	SOME CONSIDERATIONS
	Patient selection	*Is the patient appropriate for SBRT?* • The lesion is not amenable to open surgical resection • The lesion has recurred after prior open surgical resection • The patient has significant medical comorbidities that preclude open surgery • Absence of significant spinal cord compression or other acute neurologic symptoms
	SBRT vs external beam radiation therapy (EBRT)	*Is the lesion amenable to SBRT?* • The lesion has well-defined margins suitable for contouring and target delineation • Expected survival >6 months
	Simulation	*Immobilization* • Supine with appropriate stereotactic body immobilization system *Imaging* • MRI with gadolinium enhancement. T1/T2 sequences with 1–2 mm slide thickness • Possible use of CT scan ± intrathecal contrast for spinal cord delineation • Verified CT to MRI fusion over area of tumor involvement

	Treatment planning	*Contours*
		• GTV should include all portions of the tumor without a margin
		• CTV is not employed for these benign tumors
		• PTV expansion 1–2 mm if necessary
		• No PTV employed if tumor causes direct compression of the spinal cord
		Treatment planning
		• Inverse treatment planning
		• Dose
		  • 12–18 Gy in a single fraction if possible
		  • Fractionation employed only in cases of possible spinal cord overdose
		• Coverage
		  • GTV coverage 80%–95%
		• Dose constraints
		  • Spinal cord, 10–12 Gy maximum point dose

	Treatment delivery	*Imaging* • Patients are first set up conventionally using room lasers and manual shifts • Orthogonal kV matching to relevant bony anatomy • Cone-beam CT to match to tumor (GTV or PTV) • Treatment couch with 6 degrees of freedom to correct translational and rotational shifts if available

REFERENCES

Adler J Jr, Chang S, Murphy M, Doty J, Geis P, Hancock S (1997) The Cyberknife: A frameless robotic system for radiosurgery. *Stereotact Funct Neurosurg* 69:124–128.

Asazuma T, Toyama Y, Maruiwa H, Fujimura Y, Hirabayashi K (2004) Surgical strategy for cervical dumbbell tumors based on a three-dimensional classification. *Spine* 29:E10–E14.

Benzil DL, Saboori M, Mogilner AY, Rochio R, Moorthy CR (2004) Safety and efficacy of stereotactic radiosurgery for tumors of the spine. *J Neurosurg* 101:413–418.

Bijl H, van Luijk P, Coppes R, Schippers J, Konings A, van Der Kogel A (2005) Regional differences in radiosensitivity across the rat cervical spinal cord. *Int J Radiat Oncol Biol Phys* 61:543–551.

Bilsky M, Yamada Y, Yenice K et al. (2004) Intensity-modulated stereotactic radiotherapy of paraspinal tumors: A preliminary report. *Neurosurgery* 54:823.

Chang E, Shiu A, Lii M-F et al. (2004) Phase I clinical evaluation of near-simultaneous computed tomographic image-guided stereotactic body radiotherapy for spinal metastases. *Int J Radiat Oncol Biol Phys* 59:1288–1294.

Chang S, Adler J Jr (1997) Treatment of cranial base meningiomas with linear accelerator radiosurgery. *Neurosurgery* 41:1019–1025.

Chang S, Adler J (2001) Current status and optimal use of radiosurgery. *Oncology* 15:209–221.

Chang S, Murphy M, Geis P, Martin D, Hancock S, Doty J, Adler J Jr (1998) Clinical experience with image-guided robotic radiosurgery (the Cyberknife) in the treatment of brain and spinal cord tumors. *Neurol Med Chir* 38:780–783.

Chang U, Rhee C, Youn S, Lee D, Park S (2011) Radiosurgery using the Cyberknife for benign spinal tumors: Korea Cancer Center Hospital experience. *J Neurooncol* 101:91–99.

Chopra R, Morris C, Friedman W, Mendenhall W (2005) Radiotherapy and radiosurgery for benign neurofibromas. *Am J Clin Oncol* 28:317–320.

Cohen-Gadol A, Zikel O, Koch C et al. (2003) Spinal meningiomas in patients younger than 50 years of age: A 21-year experience. *J Neurosurg Spine* 98:258.

Colombo F, Pozza F, Chierego G, Casentini L, De Luca G, Francescon P (1994) Linear accelerator radiosurgery of cerebral arteriovenous malformations: An update. *Neurosurgery* 34:14–20.

Conti P, Pansini G, Mouchaty H, Capuano C, Conti R (2004) Spinal neurinomas: Retrospective analysis and long-term outcome of 179 consecutively operated cases and review of the literature. *Surg Neurol* 61:34–44.

Degen J, Gagnon G, Voyadzis J, McRae D, Lunsden M, Dieterich S, Molzahn I, Henderson F (2005) CyberKnife stereotactic radiosurgical treatment of spinal tumors for pain control and quality of life. *J Neurosurg Spine* 2:540–549.

De Salles AA, Pedroso A, Medin P, Agazaryan N, Solberg T, Cabatan-Awang C, Epinosa DM, Ford J, Selch MT (2004) Spinal lesions treated with Novalis shaped beam intensity modulated radiosurgery and stereotactic radiotherapy. *J Neurosurg* 101:435–440.

Dodd RL, Ryu MR, Kammerdsupaphon P, Gibbs IC, Chang J, Steven D, Adler J, John R (2006) CyberKnife radiosurgery for benign intradural extramedullary spinal tumors. *Neurosurgery* 58:674–685.

Faul CM, Flickinger JC (1995) The use of radiation in the management of spinal metastases. *J Neurooncol* 23:149–161.

Flickinger JC, Pollock BE, Kondziolka D (1996) A dose-response analysis of arteriovenous malformation obliteration after radiosurgery. *Int J Radiat Oncol Biol Phys* 36:873–879.

Gerszten P (2007) The role of minimally invasive techniques in the management of spine tumors: Percutaneous bone cement augmentation, radiosurgery and microendoscopic approaches. *Orthop Clin North Am* 38:441–450.

Gerszten P (2011) Radiosurgery for benign spine tumors and vascular malformations. In: Winn HR (ed.), *Youmans Neurological Surgery*, Vol. 3, pp. 2686–2692. Philadelphia, PA: Elsevier Saunders.

Gerszten P, Burton S, Ozhasoglu C, McCue K, Quinn A (2008) Radiosurgery for benign intradural spinal tumors. *Neurosurgery* 62:887–895.

Gerszten P, Chen S, Quadar M, Xu Y, Novotny J Jr, Flickinger J (2012a) Radiosurgery for benign tumors of the spine using the Synergy S with cone beam CT image guidance. *J Neurosurg* 117:197–202.

Gerszten P, Ozhasoglu C, Burton S, Kalnicki S, Welch WC (2002) Feasibility of frameless single-fraction stereotactic radiosurgery for spinal lesions. *Neurosurg Focus* 13:1–6.

Gerszten P, Quader M, Novotny J Jr, Flickinger J (2012b) Radiosurgery for benign tumors of the spine: Clinical experience and current trends. *Technol Cancer Res Treat* 11:133–139.

Gerszten P, Welch W (2004) CyberKnife radiosurgery for metastatic spine tumors. *Neurosurg Clin North Am* 15:491.

Gerszten P, Welch W (2007) Combined percutaneous transpedicular tumor debulking and kyphoplasty for pathological compression fractures. *J Neurosurg Spine* 6:92–95.

Gerszten PC, Bilsky MH (2006) Spine radiosurgery. *Contemp Neurosurg* 28:1–8.

Gerszten PC, Burton SA, Ozhasoglu C, Vogel WJ, Quinn AE, Welch WC (2006). Radiosurgery for the management of spinal metastases. In: Kondziolka D (ed.), *Radiosurgery*, Vol. 6, pp. 199–210. Basel, Switzerland: Karger.

Gerszten PC, Ozhasoglu C, Burton S, Vogel WJ, Atkins BA, Kalnicki S (2003) CyberKnife frameless single-fraction stereotactic radiosurgery for benign tumors of the spine. *Neurosurg Focus* 14:1–5.

Gezen F, Kahraman S, Canakci Z, Beduk A (2000) Review of 36 cases of spinal cord meningioma. *Spine* 27:727–731.

Gibbs I, Chang S, Dodd R, Adler J (2015) Radiosurgery for benign extramedullary tumors of the spine. In: Gerszten P and Ryu S (eds.), *Spine Radiosurgery*, pp. 164–169. New York: Thieme.

Gibbs I, Patil C, Gerszten P, Adler J Jr, Burton S (2009) Delayed radiation-induced myelopathy after spinal radiosurgery. *Neurosurgery* 64:A67–A72.

Hamilton A, Lulu B, Fosmire H, Stea B, Cassady J (1995) Preliminary clinical experience with linear accelerator-based spinal stereotactic radiosurgery. *Neurosurgery* 36:311–319.

Hitchcock E, Kitchen G, Dalton E, Pope B (1989) Stereotactic LINAC radiosurgery. *Br J Neurosurg* 3:305–312.

Isaacson S (2000). Radiation therapy and the management of intramedullary spinal cord tumors. *J Neurooncol* 47:231–238.

Keilholz L, Seegenschmiedt M, Sauer R (1996) Radiotherapy for prevention of disease progression in early-stage Dupuytren's contracture: Initial and long-term results. *Int J Radiat Oncol Biol Phys* 36:891–897.

Klekamp J, Samii M (1998) Surgery of spinal nerve sheath tumors with special reference to neurofibromatosis. *Neurosurgery* 42:279–290.

Klumpar D, Murray J, Ancher M (1994) Keloids treated with excision followed by radiation therapy. *J Am Acad Dermatol* 31:225–231.

Kondziolka D, Lunsford L, McLaughlin M (1998) Long-term outcomes after radiosurgery for acoustic neuromas. *N Engl J Med* 339:1426–1433.

Kondziolka D, Mathieu D, Lunsford L, Martin J, Madhok R, Niranjan A, Flickinger J (2008) Radiosurgery as definitive management of intracranial meningiomas. *Neurosurgery* 62:53–58.

Kondziolka D, Nathoo N, Flickinger JC (2003) Long term results after radiosurgery for benign intracranial tumors. *Neurosurgery* 53:815–821; discussion 821–822.

Lee J, Niranjan A, McInerney J, Kondziolka D, Flickinger JC, Lunsford LD (2002) Stereotactic radiosurgery providing long-term tumor control of cavernous sinus meningiomas. *J Neurosurg* 97:65–72.

Loblaw DA, Laperriere NJ (1998) Emergency treatment of malignant extradural spinal cord compression: An evidence-based guideline. *J Clin Oncol* 16:1613–1624.

Lukacs S, Braun-Falco O, Goldschmidt H (1978) Radiotherapy of benign dermatoses: Indications, practice, and results. *J Dermatol Surg Oncol* 4:620–625.

Lunsford LD, Kondziolka D, Flickinger JC (1991) Stereotactic radiosurgery for arteriovenous malformations of the brain. *J Neurosurg* 75:512–524.

McCormick P (1996a) Surgical management of dumbbell and paraspinal tumors of the thoracic and lumbar spine. *Neurosurgery* 38:67–74.

McCormick P (1996b) Surgical management of dumbbell tumors of the cervical spine. *Neurosurgery* 38:294–300.

Medin P, Solberg T, DeSalles A (2002) Investigations of a minimally invasive method for treatment of spinal malignancies with LINAC stereotactic radiation therapy: Accuracy and animal studies. *Int J Radiat Oncol Biol Phys* 52:1111–1122.

Milker-Zabel S, Zabel A, Thilmann C, Schlegel W, Wannemacher M, Debus J (2003) Clinical results of retreatment of vertebral bone metastases by stereotactic conformal radiotherapy and intensity-modulated radiotherapy. *Int J Radiat Oncol Biol Phys* 55:162–167.

Moss J, Choi C, Adler J Jr, Soltys SG, Gibbs IC, Chang S (2009) Stereotactic radiosurgical treatment of cranial and spinal hemangioblastomas. *Neurosurgery* 65:79–85.

Murovic J, Chang S (2014) Treatment of benign spinal tumors with radiosurgery. In: Sheehan J and Gerszten P (eds.), *Controversies in Stereotactic Radiosurgery Best Evidence Recommendations*, pp. 240–246. New York: Thieme.

Murovic J, Charles Cho S (2010) Surgical strategies for managing foraminal nerve sheath tumors: The emerging role of CyberKnife ablation. *Eur Spine J* 19:242–256.

Parsa A, Lee J, Parney I, Weinstein P, McCormick M, Ames C (2004) Spinal cord and intradructal-extraparenchymal spinal tumors: Current best care practices and strategies. *J Neurooncol* 69:219–318.

Peker S, Cerci A, Ozgen S, Isik N, Kalelioglu M, Pamir M (2005) Spinal meningiomas: Evaluation of 41 patients. *J Neurosurg Sci* 49:7–11.

Pirzkall A, Lohr F, Rhein B, Höss A, Schlegel W, Wannenmacher M, Debus J (2000) Conformal radiotherapy of challenging paraspinal tumors using a multiple arc segment technique. *Int J Radiat Oncol Biol Phys* 48:1197–1204.

Prasad D, Steiner M, Steiner L (2000) Gamma surgery for vestibular schwannoma. *J Neurosurg* 92:745–759.

Rampling R, Symonds P (1998) Radiation myelopathy. *Curr Opin Neurol* 11:627–632.

Rock J, Ryu S, Yin F (2004) Novalis radiosurgery for metastatic spine tumors. *Neurosurg Clin North Am* 15:503.

Roux F, Nataf F, Pinaudeau M, Borne G, Devaux B, Meder J (1996) Intraspinal meningiomas: Review of 54 cases with discussion of poor prognosis factors and modern therapeutic management. *Surg Neurol* 46:458–463.

Ryu S, Chang S, Kim D, Murphy M, Quynh-Thu L, Martin D, Adler J (2001) Image-guided hypo-fractionated stereotactic radiosurgery to spinal lesions. *Neurosurgery* 49:838–846.

Ryu S, Fang Yin F, Rock J, Zhu J, Chu A, Kagan E, Rogers L, Ajlouni M, Rosenblum M, Kim J (2003) Image-guided and intensity-modulated radiosurgery for patients with spinal metastasis. *Cancer* 97:2013–2018.

Ryu S, Rock J, Rosenblum M, Kim J (2004) Patterns of failure after single-dose radiosurgery for spinal metastasis. *J Neurosurg* 101(Suppl 3):402–405.

Sachdev S, Dodd R, Chang S, Soltys S, Adler J, Luxton G, Choi C, Tupper L, Gibbs I (2011) Stereotactic radiosurgery yields long-term control for benign intradural, extramedullary spinal tumors. *Neurosurgery* 69:533–539.

Sahgal A, Chou D, Ames C et al. (2007) Image-guided robotic stereotactic body radiotherapy for benign spinal tumors: The University of California San Francisco preliminary experience. *Technol Cancer Res Treat* 6:595–604.

Sahgal A, Ma L, Weinberg V et al. (2012) Reirradiation human spinal cord tolerance for stereotactic body radiotherapy. *Int J Radiat Oncol Biol Phys* 82:107–116.

Sayagués J, Tabernero M, Maíllo A et al. (2006) Microarray-based analysis of spinal versus intracranial meningiomas: Different clinical, biological, and genetic characteristics associated with distinct patterns of gene expression. *J Neuropathol Exp Neurol* 65:445–454.

Seegenschmiedt M, Martus P, Goldmann A, Wölfel R, Keilholz L, Sauer R (1994) Preoperative versus postoperative radiotherapy for prevention of heterotopic ossification (HO): First results of a randomized trial in high-risk patients. *Int J Radiat Oncol Biol Phys* 30:63–73.

Selch M, Lin K, Agazaryan N, Tenn S, Gorgulho A, DeMarco J, DeSalles A (2009) Initial clinical experience with image-guided linear accelerator-based spinal radiosurgery for treatment of benign. *Surg Neurol* 72:668–674.

Seppala M, Haltia M, Sankila R, Jaaskelainen J, Heiskanen O (1995a) Long-term outcome after removal of spinal neurofibroma. *J Neurosurg* 82:572–577.

Seppala M, Haltia M, Sankila R, Jaaskelainen J, Heiskanen O (1995b) Long term outcome after removal of spinal schwannoma: A clinicopathological study of 187 cases. *J Neurosurg* 83:621–626.

Solan M, Kramer S (1985) The role of radiation therapy in the management of intracranial meningiomas. *Int J Radiat Oncol Biol Phys* 11:675–677.

Sperduto P, Scott C, Andrews D (2002) Stereotactic radiosurgery with whole brain radiation therapy improves survival in patients with brain metastases: Report of radiation therapy oncology group phase III study. *Int J Radiat Oncol Biol Phys* 54:3a.

Steiner L, Leksell L, Forster D (1974) Stereotactic radiosurgery in intracranial arterio-venous malformations. *Acta Neurochir* Suppl 21:195–209.

Yamada Y, Lovelock D, Bilsky M (2007) A review of image-guided intensity-modulated radiotherapy for spinal tumors. *Neurosurgery* 61:226–235.

Yin F, Ryu S, Ajlouni M, Yan H, Jin J, Lee S, Kim J, Rock J, Rosenblum M, Kim J (2004) Image-guided procedures for intensity-modulated spinal radiosurgery. Technical note. *J Neurosurg* 101(Suppl 3):419–424.

Postoperative spine IG-HSRT outcomes

Ariel E. Marciscano and Kristin J. Redmond

Contents

13.1 INTRODUCTION

Spinal metastatic disease is common among patients with cancer and has traditionally been associated with a poor prognosis. Approximately 40% of cancer patients are affected by spinal metastases during their disease course (Klimo and Schmidt, 2004), and autopsy series have estimated that 30%–70% of patients with known malignancy have evidence of spinal metastatic disease on postmortem examination (Fornasier and Horne, 1975; Wong et al., 1990; Harrington, 1993).

The incidence of spinal metastases is anticipated to increase as systemic therapies continue to improve and prolong survival of patients with metastatic disease. As such, a shift toward definitive rather than palliative management of spinal metastases has become increasingly important. In general, management of spinal metastatic disease consists of radiation therapy, surgery, and systemic therapy, as well as combined modality approaches such as surgery followed by postoperative radiation therapy. The focus of this chapter is on postoperative spine image-guided hypofractionated stereotactic radiation therapy (IG-HSRT) with an emphasis upon treatment considerations and technique, clinical outcomes, and future directions.

13.1.1 HISTORICAL PERSPECTIVE

Surgery and radiation therapy represent the mainstay treatments for spinal metastases. Traditionally, conventionally fractionated low-dose radiotherapy alone has been used for patients with oncologic/nonmechanical pain with minimal epidural disease. Conventionally fractionated palliative radiotherapy is the primary treatment option for patients with widely metastatic spinal disease that cannot be addressed with surgical resection, patients that are medically unfit for surgery, and among patients with limited life expectancy. Patients with spinal instability, malignant epidural spinal cord compression (MESCC), and/or mechanical pain resulting from pathologic fracture are generally best addressed with up-front surgery.

A landmark trial by Patchell et al. (2005) established the standard of care in the management of MESCC by demonstrating the superiority of surgical decompression followed by postoperative low-dose radiotherapy over low-dose radiotherapy alone. This multi-institutional prospective study randomized 101 patients to immediate direct circumferential decompressive surgery plus postoperative radiotherapy or up-front radiation therapy alone (Patchell et al., 2005). Radiation dose, fractionation, and delivery were identical in both arms with 30 Gy in 10 fractions delivered with anterior–posterior/posterior–anterior technique to a treatment field encompassing a vertebral body above and below the level of involvement (enrollment restricted MESCC to a single area of involvement). There was a significantly higher proportion of patients in the surgical arm that were ambulatory following treatment, and those treated in the surgical arm retained the ability to walk for a significantly longer period of time than those treated in the radiation therapy alone. Among the patients that were nonambulatory prior to treatment, those treated with surgery and postoperative radiation therapy were more likely to regain the ability to walk. Additionally, patients treated with surgery and postoperative radiation therapy had a significant benefit with regard to maintenance of continence, motor strength by the American Spinal Injury Association (ASIA) score, functional ability by Frankel score, and survival. Although criticized for small sample size and lack of standardization of surgical technique, this study was influential in establishing the role of surgery in the management of MESCC as other series had previously failed to demonstrate the benefit of surgery alone or in combination with radiotherapy over radiotherapy alone (Gilbert et al., 1978; Yong et al., 1980; Findlay, 1984). Furthermore, it gave credence to the concept that aggressive intervention combining surgery and postoperative radiotherapy is able to improve patient outcomes in the appropriate clinical scenario.

13.1.2 RECENT ADVANCES

Improvements in surgical technique and spinal instrumentation over the past two decades have expanded the armamentarium against spinal metastatic disease and MESCC (Sciubba and Gokaslan, 2006). An emphasis upon minimally invasive approaches, for decompression and tumor resection, that focus on removal of only the compressive epidural tumor or bone fragment has served to lessen perioperative morbidity, improve surgical outcomes, and quicken recovery, thus reducing the interval to administration of adjuvant therapies. Indeed, innovative new approaches such as minimal access spinal surgery (MASS) and minimally invasive separation surgery represent the significant advances in this growing field where traditionally fewer options such as posterior laminectomy and major invasive decompressive surgeries were available (Massicotte et al., 2012; Sharan et al., 2014). Management of spinal instability has also matured with percutaneous vertebroplasty and kyphoplasty allowing for minimally invasive stabilization. Anterior and posterior stabilization approaches have accordingly evolved when spinal stabilization and fixation is warranted after invasive decompressive surgeries (Sciubba et al., 2010).

Radiotherapy technique and administration has similarly evolved at a rapid pace. Historically, the concerns and limitations of radiation therapy largely revolved around the low tolerance of the spinal cord and significant potential for spinal cord injury (SCI) and/or radiation-induced myelopathy. As such, low-dose conventionally fractionated radiotherapy has been utilized as a palliative treatment for temporary control of oncologic pain and neurological symptoms. This is a particularly salient issue among patients with favorable histologies and longer life expectancy as well as patients with radioresistant histologies where low-dose conventional radiotherapy has limited success at achieving durable local tumor control (Greenberg et al., 1980; Maranzano and Latini, 1995). Klekamp and Samii reported on their experience with surgery followed by postoperative low-dose conventionally fractionated radiotherapy for spinal metastases. The overall local recurrence rate was approximately 70% at 1 year. On multivariate analysis, preoperative ambulatory status, favorable histology, complete tumor resection, and low number of involved vertebral levels were identified as independent predictors of a lower rate of recurrence (Klekamp and Samii, 1998). Although limited by the single-institution and retrospective nature of this work, this report alludes to the poor long-term local control rates achieved with low-dose conventional radiotherapy and the significant potential for disease progression and recurrence even after treatment with a combined modality approach.

Long-term, high-quality evidence exists supporting the safe and efficacious use of single-fraction radiosurgery (SRS) and HSRT in the management of intracranial benign and malignant tumors. More recently, this technology has been applied to several extracranial sites, including the spine. Although long-term data are only beginning to emerge, short-term reports suggest excellent local control and acceptable toxicity. Advances in image guidance, precise spine and body immobilization, multileaf collimation, robotic technologies, and intensity-modulated treatment planning have afforded radiation oncologists the ability to deliver highly conformal ablative doses of radiation with steep dose gradients thereby sparing nearby critical structures, such as the spinal cord. With regard to spinal metastases, IG-HSRT has several advantages compared to conventionally fractionated radiotherapy: (1) highly precise and accurate beam delivery allows for a higher biologically effective dose (BED) to be delivered and (2) smaller treatment volumes and sharp dose falloff allow for sparing of critical organs at risk (OAR) such as the spinal cord, esophagus, bowel, and kidneys. Given the potential for spine IG-HSRT to overcome shortcomings associated with low-dose conventional radiotherapy, there have been significant efforts to evaluate spine IG-HSRT in the definitive setting and, more recently, in the postoperative setting with encouraging results.

With the complementary evolution of both surgical and IG-HSRT technologies, there has been a clear paradigm shift from the historic goal of short-term palliation, toward offering definitive management with long-term local control as the primary goal of therapy. With respect to postoperative spine management, IG-HSRT can serve as a potent adjuvant treatment to enhance local control. Fundamentally, if a patient undergoes a major spinal procedure, it is intuitive that the adjuvant therapy be equally as aggressive, hence the role of IG-HSRT. Surgery is also now being reconceptualized as neoadjuvant to definitive postoperative IG-HSRT. By focusing the surgical goals on epidural tumor resection and spinal cord decompression using minimally invasive techniques, with or without stabilization, the morbidity of surgery can be further minimized with delivery of tumoricidal doses with IG-HSRT as the primary therapy. This approach will also increase the utilization of spine surgery as currently it is thought of as a highly selective treatment for highly selected patients with single-level MESCC.

13.2 INDICATIONS

As the available treatment options continue to expand, it is of critical importance to better understand which patients will benefit from up-front surgical management versus those that should be addressed with radiotherapy alone. Ultimately, optimal treatment selection requires a multidisciplinary approach with consideration of (1) patient-specific factors such as neurologic symptoms, etiology of pain, medical comorbidities, surgical contraindications, systemic disease burden, and prognosis, as well as (2) disease-specific factors, including spinal instability, degree of epidural spinal cord compression (ESCC), relative tumor radiosensitivity, and prior radiation treatment.

Several groups have sought to develop patient stratification schema to guide management. The Spine Oncology Study Group developed a novel comprehensive classification system to diagnose neoplastic spinal instability based upon patient symptoms and radiographic criteria (Fisher et al., 2010). The objective of the spinal instability neoplastic score (SINS) criteria is to help identify patients who may benefit from surgical intervention and stabilization. The SINS is a qualitative score assigned based on the importance of several factors including pain, location, spinal alignment, presence of vertebral compression fracture (VCF), type of lesion, and posterolateral involvement of spinal elements (e.g., as shown in Table 13.1). By SINS, patients with a total score ranging 0–6 are considered stable, those scoring 7–12 are considered potentially unstable, and those with a total score ≥13 are considered unstable, and surgical management is highly recommended. The SINS has been validated as a reliable, reproducible tool among radiologists, radiation oncologists, and surgeons (Fourney et al., 2011; Fisher et al., 2014a,b). The ASIA has utilized the

Table 13.1 Spinal instability neoplastic score (SINS)

PATIENT-SPECIFIC	SPINE-SPECIFIC	TUMOR-SPECIFIC	TOTAL SINS SCORE
Pain (max score = 3)	**Location (max score = 3)**	**Type of lesion (max score = 2)**	**Stable = 0–6**
Mechanical pain (3)	Junctional: occiput–C2, C7–T2, T11–L1, L5–S1 (3)	Osteolytic (3)	**Potentially unstable = 7–12**
Occasional pain, nonmechanical (1)	Mobile: C3–C6, L2–L4 (2)	Mixed (1)	**Unstable ≥ 13**
No pain (0)	Semi-rigid: T3–T10 (1)	Osteosclerotic (0)	
	Rigid: S2–S5 (0)	**Posterolateral involvement of spinal elements (max score =3)**	
	Spinal alignment (max score = 4)	Bilateral (3)	
	Subluxation/translation (4)	Unilateral (1)	
	Kyphosis/scoliosis (2)	None (0)	
	Normal (0)		
	Presence of vertebral compression fracture (max score = 3)		
	≥50% collapse (3)		
	<50% collapse (2)		
	No collapse with ≥50% involvement (1)		
	None (0)		

Source: Reproduced from Fisher, C.G. et al., *Spine* (1976), 35(22), E1221, 2010. With permission.

ASIA impairment scale (AIS) to classify the degree of SCI (Kirshblum et al., 2011; Kirshblum and Waring, 2014). In the context of MESCC, surgeons and oncologists have used this scale to help stratify patients and the need for surgical intervention. In general, surgical resection should be considered in patients presenting with AIS grades A–D without improvement after high-dose corticosteroid administration, especially if the neurologic impairment developed over the preceding 48 hours (e.g., as shown in Table 13.2).

Bilsky and colleagues developed and validated a six-point grading system to define ESCC (Bilsky et al., 2010). Prior iterations have used myelography and MRI to define the degree of CSF space impingement and spinal cord compression to assist surgical decision making (Bilsky et al., 2001). The advent of spine IG-HSRT necessitated a more detailed system to specify the degree of thecal sac impingement as this information is important in assessing the safety and feasibility of a definitive IG-HSRT approach. In general, patients with low-grade ESCC (ESCC scale grades 0–1b) have sufficient distance at the spinal cord–tumor interface to permit safe IG-HSRT administration (e.g., as shown in Figure 13.1). On the ESCC scale, grade 0 indicates bone-only disease; grade 1a indicates epidural impingement without deformation of the thecal sac; and grade 1b indicates deformation of the thecal sac without spinal cord abutment. Grade 1c disease (deformation of the thecal sac with spinal cord abutment but without cord compression) is an intermediate between consideration as low-grade versus high-grade ESCC, and management is dependent upon multidisciplinary communication between the surgeon and radiation oncologist as well as other clinical factors. High-grade ESCC consists of grade 2 and grade 3 disease, which often warrants separation surgery with epidural tumor resection in order to achieve spinal cord decompression and sufficient separation between the spinal cord–tumor interface to permit safe

Table 13.2 ASIA impairment scale (AIS)

A = Complete	No sensory or motor function is preserved in the sacral segments S4–S5.
B = Sensory incomplete	Sensory but no motor function is preserved below the neurological level and includes the sacral segments S4–S5 (light touch, pin prick at S4–S5, or deep anal pressure) and no motor function is preserved more than three levels below the motor level on either side of the body.
C = Motor incomplete	Motor function is preserved below the neurological level[a] and more than half of the key muscle functions below the single neurological level of injury have a muscle grade <3.
D = Motor incomplete	Motor function is preserved below the neurological level[a] and at least half of the key muscle functions below the neurological level of injury have a muscle grade of ≥3.
E = Normal	If sensation and motor function are tested and graded as normal in all segments, and the patient had prior deficits, then the AIS grade is E. Someone without an initial spinal cord injury does not receive and AIS grade.

Source: Reproduced from Kirshblum, S.C. et al., *J. Spinal Cord Med.*, 34(6), 535, 2011. With permission.

Note: When assessing the extent of motor sparing below the level for distinguishing between AIS B and C, the motor level on each side is used, whereas to differentiate between AIS C and D the single neurological level is used.

[a] For an individual to receive grade C or D (motor incomplete), they must have either (1) voluntary anal sphincter contraction or (2) sacral sensory sparing with sparing of motor function more than three levels below the motor level for that side of the body.

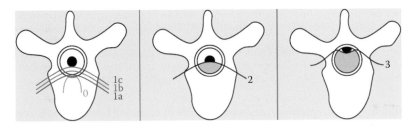

Figure 13.1 Schematic representation of the six-point epidural spinal cord compression scale. A grade of 0 indicates bone-only disease; 1a, epidural impingement, without deformation of the thecal sac; 1b, deformation of the thecal sac, without spinal cord abutment; 1c, deformation of the thecal sac with spinal cord abutment, but without cord compression; 2, spinal cord compression, but with CSF visible around the cord; and 3, spinal cord compression, no CSF visible around the cord. (Adapted from Bilsky, M.H. et al., *J. Neurosurg. Spine*, 13(3), 324, 2010. With permission.)

postoperative IG-HSRT (Moussazadeh et al., 2014). On the ESCC scale, grade 2 disease denotes evidence of spinal cord compression with CSF visible around the spinal cord, and grade 3 disease denotes spinal cord compression without CSF visible around the cord.

Memorial Sloan Kettering Cancer Center (MSKCC) has adopted the NOMS framework as a decision-making algorithm for patients with metastatic spinal cancers (Laufer et al., 2013). The acronym, NOMS, depicts the neurologic, oncologic, mechanical, and systemic considerations that guide patient stratification and treatment selection (e.g., as shown in Figure 13.2). In general, patients with low-grade ESCC and without myelopathy do not require decompressive surgery. Radiosensitive histologies may be addressed with conventionally fractionated low-dose radiotherapy, and radioresistant histologies are treated with IG-HSRT. Patients presenting with high-grade ESCC and/or myelopathy should be considered for surgical intervention with stabilization and/or decompression. Radioresistant histologies may benefit from decompressive or separation surgery prior to delivery of IG-HSRT; however, it is also reasonable to manage patients with radiosensitive histologies with conventionally fractionated low-dose radiotherapy alone. Patients that are unable to tolerate surgery with high-grade ESCC and/or myelopathy should be managed

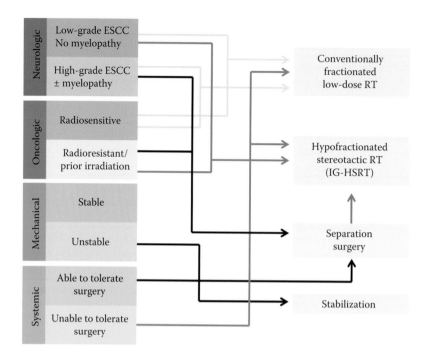

Figure 13.2 The NOMS (neurologic, oncologic, mechanical, and systemic) decision framework. By the NOMS algorithm, patients with radiosensitive histologies are treated with conventionally fractionated low-dose RT, regardless of myelopathy or degree of epidural spinal cord compression (ESCC). Patients with radioresistant histologies can be treated with up-front separation surgery followed by image-guided hypofractionated stereotactic radiation therapy (IG-HSRT) or definitive IG-HSRT alone, depending on the presence of myelopathy or degree of ESCC. Patients with mechanical spinal instability should be considered for spinal stabilization procedures. Patients unfit for surgery should be treated with conventionally fractionated low-dose RT or IG-HSRT depending on the degree of ESCC, presence of myelopathy, safety, and relative radiosensitivity. (Adapted from Laufer, I. et al., *Oncologist*, 18(6), 744, 2013a. With permission. Copyright Clearance Center, Inc.)

in a palliative fashion with administration of conventionally fractionated low-dose radiotherapy, as the spinal cord–tumor interface is often insufficient to safely deliver IG-HSRT. Of note, any patient with evidence of mechanical spinal instability should be considered for up-front stabilization via a percutaneous or open approach. While the NOMS framework is a useful tool for patient stratification, there are alternatives management strategies available for patients with spinal metastatic disease. For example, the application of spine IG-HSRT may be considered for radiosensitive histologies to enhance local control in the definitive setting. Additionally, patients presenting with high-grade ESCC or myelopathy are often considered for stabilization/decompression followed by IG-HSRT regardless of histologic radiosensitivity.

Additional considerations for selecting between surgical and nonsurgical approaches are prior irradiation, need for pathologic diagnosis, and older age (Chi et al., 2009). Table 13.3 summarizes the absolute and relative indications for surgical intervention in the patient with metastatic spinal disease.

Table 13.3 **Absolute and relative indications for surgical intervention**

ABSOLUTE INDICATIONS	RELATIVE INDICATIONS
SINS ≥ 13 (unstable)	ESCC grade 2 or 3
ASIA grade A–D, due to ESCC, especially if not responsive to high-dose corticosteroids and onset <48 hours	Oligometastatic disease
	Recurrence following prior radiotherapy
	Need for pathologic diagnosis
	Intractable pain from mechanical instability

13.3 TREATMENT PLANNING

13.3.1 SIMULATION

Safe and accurate delivery of postoperative spine IG-HSRT requires the construction of a highly reproducible setup with rigid body immobilization at time of CT simulation. Proper immobilization reduces the potential for intrafraction and interfraction shifts and position changes, which is particularly relevant for treatment systems that do not utilize real-time imaging (Sahgal et al., 2011). The location of the spinal lesion dictates the type of immobilization device, as outlined in Table 13.4.

Once the patient is immobilized with a reproducible setup, a thin-slice (≤2 mm) CT simulation scan with and without intravenous iodinated contrast is performed. Patients with contraindications to CT contrast should forego contrast administration. Subsequently, a thin-slice MRI is acquired with a region of interest that encompasses the target area and at least one vertebral body above and below the target. If there are no contraindications to MRI and/or gadolinium contrast, an MRI simulation scan with and without contrast administration should be performed. At a minimum, the following MRI sequences should be acquired: (1) T1-weighted axial images with and without gadolinium for tumor delineation and (2) T2-weighted or short tau inversion recovery (STIR) axial images for visualization of spinal cord and further tumor delineation. Sagittal reconstruction images and other MRI sequences may assist fusion and treatment planning and, therefore, should be considered. Additionally, incorporation of preoperative MRI axial images is critical to understanding the initial extent of disease and therefore clinical target volume (CTV) delineation. In situations where metallic artifact from spinal instrumentation obscures the delineation of critical neural structures (spinal cord, thecal sac) on MRI, a CT myelogram should be obtained. Following completion of simulation scans, the following datasets are fused within the treatment planning software: (1) CT simulation, (2) MRI simulation, (3) preoperative MRI, and (4) CT myelogram (if performed).

13.3.2 TARGET DELINEATION

The CTV encompasses all areas of gross disease on preoperative MRI imaging, the postoperative residual, and sites of potential microscopic disease spread. As a general rule, the involved regions of the spine and the immediately adjacent segments are believed to be at risk and are included in the CTV; guidelines for bony CTV delineation are shown in Table 13.5 (Cox et al., 2012). In addition, a firm understanding of the surgical approach as documented in the operative report through discussions with the surgeon is critical. The surgical incision and scar are generally not included within the CTV unless considered to be at high risk of subclinical involvement or recurrence. Likewise, spinal hardware and instrumentation is generally not included within the CTV unless there is increased concern for subclinical involvement. Although, often unavoidable due to the initial or postoperative extent of disease, efforts should be taken to limit circumferential or "donut" treatment fields when possible as this can impact subsequent treatment planning and the achievement of OAR constraints and target coverage. The planning target volume (PTV) includes the CTV with a geometric expansion of 1.0–2.0 mm. In situations where the PTV margin extends into the spinal cord or nearby OARs, a planning organ-at-risk volume (PRV) is generated to account for residual setup error and intrafractional organ motion for patients and is subtracted from the PTV. PRV delineation

Table 13.4 Reasonable spine IG-HSRT immobilization strategies for treatment planning

SPINAL LOCATION	IMMOBILIZATION
Cervical and upper-thoracic (C1–T3)	Long thermoplastic cranial mask with mold cushioning for support
Mid-thoracic and lumbar (T4–L3)	Alpha Cradle® and wing board set up with arms positioned above the head with appropriate headrest support and memory foam
Inferior (L4–sacrum)	Hevezi pad or Vac Lok® immobilization with the patient's arms positioned at their side

Table 13.5 **Guidelines for spine IG-HSRT bony CTV delineation**

GTV INVOLVEMENT	CTV DESCRIPTION
Any portion of the vertebral body	Include the entire vertebral body
Lateralized within the vertebral body	Include the entire vertebral body and the ipsilateral pedicle/transverse process
Diffusely involves the vertebral body	Include the entire vertebral body and the bilateral pedicle/transverse processes
GTV involves vertebral body and unilateral pedicle	Include entire vertebral body, pedicle, ipsilateral transverse process, and ipsilateral lamina
GTV involves vertebral body and bilateral pedicles/transverse processes	Include entire vertebral body, bilateral pedicles/transverse processes, and bilateral laminae
GTV involves unilateral pedicle	Include pedicle, ipsilateral transverse process, and ipsilateral lamina ± vertebral body
GTV involves unilateral lamina	Include lamina, ipsilateral pedicle/transverse process and spinous process
GTV involves spinous process	Include entire spinous process and bilateral laminae

Source: Reproduced from Cox, B.W. et al. *Int. J. Radiat. Oncol. Biol. Phys.* 83(5), e597, 2012. With permission.

is discussed in Section 13.3.3. A representative postoperative spine IG-HSRT treatment plan is shown in Figure 13.3.

13.3.3 ORGANS AT RISK AND NORMAL TISSUE CONSTRAINTS

The location of the postoperative target dictates which nearby OARs should be contoured. All OARs must be contoured at a minimum of one vertebral body superior and inferior to the PTV and to a greater extent for nonisocentric planning systems where hot spots may be found remotely from the target. The critical neural structures, spinal cord and/or the cauda equina, are delineated using T2-variant MRI or CT myelogram in cases of significant metal artifact from spinal instrumentation, and a 0–2 mm expansion is used as the spinal cord PRV for planning purposes. The cauda equina OAR structure is defined as the thecal sac without an expansion. To further characterize spinal cord tolerance in the context IG-HRST, Sahgal and colleagues analyzed the risk of radiation myelopathy (RM) among patients previously treated with conventionally fractionated radiotherapy and reirradiated with IG-HSRT (Sahgal et al., 2012a). Example spinal cord constraints for various fractionation schemes are shown in Table 13.6. Further studies have helped to define safe practice guidelines by developing a percentage probability of RM based upon point max doses to the thecal sac as referenced in Table 13.7 (Sahgal et al., 2010, 2012b; Masucci et al., 2011). In addition to the use of point max as a spinal cord dose constraint parameter, the maximum dose per small volume (0.1–0.2 cc) is also utilized. Reasonable spine IG-HSRT dose constraints for other critical structures based on American Association of Physicist in Medicine Task Group 101 recommendations are outlined in Table 13.8 (Benedict et al., 2010).

13.3.4 DOSE SELECTION

The optimal dose and fractionation schema for postoperative spine IG-HSRT remains undefined. Dose prescription is often a complex decision that is contingent upon multiple considerations including nearby OARs, target volume, extent of epidural disease, number of involved vertebral levels, and prior irradiation (Sahgal et al., 2011). The existing literature for postoperative spine IG-HSRT is highly variable with regard to dose selection and fractionation making it difficult to offer generalized recommendations. Current data suggest that higher dose per fraction delivered over less fractions may enhance local control, while other series have contradictory findings. Moreover, lower BED and hypofractionated regimens have been associated with a lower risk of radiation-associated toxicities (Cunha et al., 2012; Al-Omair et al., 2013a;

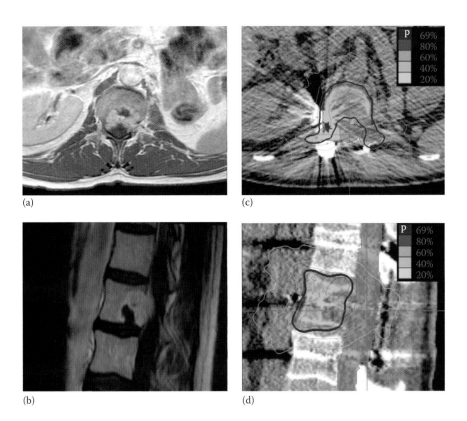

Figure 13.3 Representative postoperative image-guide hypofractionated stereotactic radiation therapy (IG-HSRT) treatment plan. This patient with metastatic clear cell renal cell carcinoma developed lower back pain. MRI demonstrated an L1 vertebral body lesion with bilateral pedicle involvement and extension into the anterior epidural space with likely impingement of the L2 nerve root and mild left neuroforaminal narrowing. (a) Preoperative axial T1-weighted MRI. (b) Preoperative sagittal T1-weighted MRI. The patient underwent L1 corpectomy, left L1 laminectomy, left T12–L1, and L1–L2 facetectomy. Posterior stabilization with instrumented allograft arthrodesis from T12–L2 was performed. (c, d) Noncircumferential postoperative IG-HSRT plan taking into account initial bony involvement and epidural disease. The PTV is denoted in red, and thecal sac organ at risk is denoted in green. Isodose line color scheme: prescription (blue), 80% (violet), 60% (orange), 40% (maize), and 20% (yellow).

Sahgal et al., 2013b). There is likely a delicate balance of dose escalation and hypofractionation needed to achieve optimal local control and to minimize toxicities. The aforementioned controversies and uncertainties outline the need for future randomized prospective studies. In general, 85%–95% target coverage by 90%–100% of the prescription dose is acceptable; however, dose and fractionation selection is ultimately contingent upon the clinical context and physician preference. PTV coverage should always be compromised in order to meet OAR constraints. Reasonable fractionation regiments include 18–20 Gy in 1 fraction, 24 Gy in 2 fractions, 27–30 Gy in 3 fractions, and 30–40 Gy in 5 fractions.

13.3.5 TREATMENT DELIVERY

In general, spine IG-HSRT is delivered by either multileaf collimator (MLC) linear accelerator (LINAC) treatment units or CyberKnife technology (Adler et al., 1997, 1999; Ryu et al., 2001). Key differences between treatment delivery systems have been previously described and are beyond the scope of this chapter (Sahgal et al., 2011). MLC-based LINAC treatment delivery entails the use of 7–11 coplanar beams with 5–15 MLC apertures or segments per beam. Intensity-modulated beam delivery can be delivered via "step-and-shoot" technique with a fixed gantry position or dynamic MLC approach. Additionally, CT image guidance and/ or stereoscopic kilovoltage x-rays allow for treatment setup verification and interfraction shifts; however, there is no intrafractional imaging during the course of treatment. Alternatively, the CyberKnife technology is

Table 13.6 Reasonable reirradiation IG-HSRT doses to the thecal sac Pmax following common initial conventional radiotherapy regimens

PRIOR CONVENTIONAL RADIOTHERAPY DOSE (nBED)	IG-HSRT DOSE TO THECAL SAC Pmax				
	1 FRACTION	2 FRACTIONS (Gy)	3 FRACTIONS (Gy)	4 FRACTIONS (Gy)	5 FRACTIONS
0 Gy	10 Gy	14.5	17.5	20	22 Gy
					25 Gy to <0.1 cc
20 Gy in 5 fractions (30 Gy$_{2/2}$)	9 Gy	12.2	14.5	16.2	18 Gy
30 Gy in 10 fractions (30 Gy$_{2/2}$)	9 Gy	12.2	14.5	16.2	18 Gy
37.5 Gy in 15 fractions (42 Gy$_{2/2}$)	9 Gy	12.2	14.5	16.2	18 Gy
40 Gy in 20 fractions (40 Gy$_{2/2}$)	n/a	12.2	14.5	16.2	18 Gy
45 Gy in 25 fractions (43 Gy$_{2/2}$)	n/a	12.2	14.5	16.2	18 Gy
50 Gy in 25 fractions (50 Gy$_{2/2}$)	n/a	11	12.5	14	15.5 Gy

Source: Reproduced from Sahgal, A. et al., *Int. J. Radiat. Oncol. Biol. Phys.*, 82(1), 107, 2012a. With permission.
Abbreviations: nBED, normalized biologically effective dose; Pmax, point max.

Table 13.7 Predicted % probability of radiation myelopathy (RM) by Pmax volume absolute doses in Gy for IG-HSRT

% RISK OF RM	Pmax LIMIT (Gy)				
	1 FRACTION	2 FRACTIONS	3 FRACTIONS	4 FRACTIONS	5 FRACTIONS
1% probability	9.2	12.5	14.8	16.7	18.2
2% probability	10.7	14.6	17.4	19.6	21.5
3% probability	11.5	15.7	18.8	21.2	23.1
4% probability	12.0	16.4	19.6	22.2	24.4
5% probability	12.4	17.0	20.3	23.0	25.3

Source: Reproduced from Sahgal, A. et al., *Int. J. Radiat. Oncol. Biol. Phys.*, 85(2), 341, 2012b. With permission.
Abbreviations: Pmax, point max to thecal sac.

nonisocentric, and treatment is delivered via a mobile robotic arm that is mounted with a compact LINAC. Highly conformal dose delivery is achieved by using a large number beam angles (~100–200) with circular collimators of varying size to maximize target coverage and spare OARs. A distinct difference between technologies is that CyberKnife uses near real-time imaging during treatment to make intrafractional adjustments (Sahgal et al., 2011).

13.3.6 SPECIAL CONSIDERATIONS

Treatment planning and delivery in the postoperative spine setting poses several unique challenges. Target delineation and visualization of residual tumor is complicated by imaging artifact created by hardware (Pekmezci et al., 2006; Sahgal et al., 2011). Postsurgical blood products and debris are often difficult to differentiate from residual tumor, which introduces uncertainty into contouring. As previously mentioned, a CT myelogram may be helpful for spinal cord delineation in this setting. An additional challenge is

Table 13.8 Reasonable organ at risk (OAR) dosimetric constraints for spine IG-HSRT

OAR		MAX POINT DOSE (Gy)[a]		
		1 FRACTION	3 FRACTIONS	5 FRACTIONS
Esophagus		15.4	25.2	35
Heart/pericardium		22	30	38
Great vessels		37	45	53
Trachea		20.2	30	40
Stomach		12.4	22.2	32
Bowel		12.4	22.2	32
Skin[b]		26	33	39.5
	Max critical volume above threshold (cc)[c]	Threshold dose (Gy)		
Combined kidneys	200	8.4	16	17.5
Combined lung	1000	7.4	12.4	13.5
Liver	700	9.1	19.2	21

Source: Adapted from Benedict, S.H. et al., *Med. Phys.*, 39(1), 63, 2010.
[a] Max point dose defined as <0.035 cc.
[b] Unless intentionally included as CTV.
[c] For parallel tissues, the volume-dose constraints are based on critical minimum volume of tissue that should receive a dose equal to or less than the indicated threshold dose.

the impact of spinal instrumentation and surgical hardware upon radiation dosimetry. The amount of artifact and image distortion increases as the amount of hardware increases for both CT and MR imaging (Sahgal et al., 2011). High Z (atomic number) material such as titanium can result in significant dose perturbation. Dosimetric modeling studies on the effect of hardware on spine IG-HSRT reported that in relation to the prescribed dose, the dose in front of the hardware was approximately 6% higher due to electron back scatter, while it was approximately 7% lower beyond the hardware due to photon attenuation (Wang et al., 2013). Indeed, other studies have observed up to 13% attenuation with 6 MV photon beams where hardware material consisted of both titanium rods and pedicle screws (Liebross et al., 2002). These dosimetric inconsistencies are clinically relevant as it may result in hot spots within OARs or in underdosing of the target volume. It is important to communicate the composition of the surgical hardware to the radiation dosimetrists so that an appropriate CT number to electron density conversion and density override is used for treatment planning. Furthermore, utilizing multiple-beam arrangements may reduce the impact of high Z material dose perturbation (Wang et al., 2013).

13.4 OUTCOMES

13.4.1 REPORTED CLINICAL OUTCOMES IN THE LITERATURE

The literature to date evaluating postoperative spine IG-HSRT is limited but suggests promising clinical outcomes. There is currently no level 1 randomized data comparing conventional radiotherapy and IG-HSRT in the postoperative spine setting. The current data reporting clinical outcomes using postoperative spine IG-HSRT are summarized in Table 13.9.

Early reports by Gerszten et al. (2005) described outcomes of 26 patients with symptomatic pathological compression fracture treated with kyphoplasty-based closed fracture reduction followed by SRS with doses ranging from 16 to 20 Gy. With a median follow-up of 16 months (range, 11–24 months), 92% of patients reported long-term improvement in back pain, and there were no neurological symptoms attributable to radiosurgery. Given concerns regarding kyphoplasty alone, the same group from the University of Pittsburgh Medical Center subsequently evaluated minimally invasive surgery involving tumor debulking

Table 13.9 Clinical studies evaluating postoperative spine IG-HSRT

STUDY (AUTHOR/ YEAR)	NO. PATIENTS	HSRT CHARACTERISTICS						CLINICAL OUTCOMES						
		SURGICAL TECHNIQUE	MEDIAN TOTAL DOSE (Gy/NO. FRACTIONS)	TOTAL DOSE (Gy)/NO. FRACTIONS	nBED (GY$_{2/2}$)[a]	CTV DESCRIPTION	MEDIAN FOLLOW-UP (MONTHS)	LOCAL CONTROL (LAST FOLLOW-UP)	LOCAL CONTROL (1-YEAR)	PATTERNS OF LOCAL FAILURE	MEDIAN SURVIVAL (MONTHS)	PAIN CONTROL	TOXICITY
Gerszten/ 2005	n = 26	Kyphoplasty	18 Gy/1 Fx	16–20 Gy/1 Fx	72–110 Gy	Entire VB and any adjacent tumor extension	16 (11–124)	92%	NR	NR	NR	92% long-term improvement in pain by VAS	No RT-induced SCI
Rock/2006	n = 18	Variable, open surgery	12 Gy/1 Fx	6–16 Gy/1 Fx	12–72 Gy	Involved VB and anterior 1/3 of pedicles, including paravertebral/ epidural soft tissue component	7 (4–36)	94%	NR	NR	NR	33% with complete pain relief	6% possible RT-induced toxicity 92% with pre-treatment deficits stable/ improved
Gerszten/ 2009	n = 11	Percutaneous transpedicular corpectomy with closed fracture reduction	19 Gy/1 Fx	16–22.5 Gy/ 1 Fx	72–138 Gy	Entire VB and any adjacent tumor extension	11 (7–44)	100%	NR	NR	NR	100% long-term improvement in back pain	No RT-induced SCI
Moulding/ 2010	n = 21	Posterolateral decompression and instrumentation	24 Gy/1 Fx	18–24 Gy/1 Fx	90–156 Gy	Preoperative gross tumor volume, including entire VB, accounting for microscopic tumor	10.3	81%	90.5%	NR	10.2	NR	Acute neuritis (n = 1), grade 4 esophagitis (n = 1)

(*Continued*)

Table 13.9 (Continued) Clinical studies evaluating postoperative spine IG-HSRT

| STUDY (AUTHOR/YEAR) | NO. PATIENTS | SURGICAL TECHNIQUE | HSRT CHARACTERISTICS | | | | CLINICAL OUTCOMES | | | | | | |
			MEDIAN TOTAL DOSE (Gy)/NO. FRACTIONS	TOTAL DOSE (Gy)/NO. FRACTIONS	nBED (GY$_{2/2}$)[a]	CTV DESCRIPTION	MEDIAN FOLLOW-UP (MONTHS)	LOCAL CONTROL (LAST FOLLOW-UP)	LOCAL CONTROL (1-YEAR)	PATTERNS OF LOCAL FAILURE	MEDIAN SURVIVAL (MONTHS)	PAIN CONTROL	TOXICITY
Massicotte/ 2012	n = 10	Minimal access spine surgery (MASS)	24 Gy/2 Fx	20 Gy/1 Fx (20%); 18–24 Gy/2–3 Fx (70%); 35 Gy/5 Fx (10%)	110 Gy; 36–84 Gy; 80 Gy	Radiographically visible tumor, potential areas at risk of microscopic disease, and ipsilateral trajectory of tube	13 (3–18)	70%	NR	67% epidural, 33% "in-field" within VB	NR	80% with improvement of pain by VAS	Acute pain flare (n = 2), no reports of RT myelopathy
Laufer/ 2013	n = 186	Separation surgery with circumferential epidural resection	24 Gy/1 Fx; 27 Gy/3 Fx; 30 Gy/5–6 Fx	24 Gy/1 Fx (22%); 24–30 Gy/3 Fx (20%); 18–36 Gy/5–6 Fx (58%)	156 Gy; 60–90 Gy; 24–72 Gy	Preoperative gross tumor volume, including entire VB, accounting for microscopic tumor	7.6 (1–66.4)	NR	83.6%	NR	54% alive at last follow-up	NR	NR
Al-Omair/ 2013a	n = 80	Stabilization alone OR decompression ± stabilization	24 Gy/2 Fx	18–26 Gy/1–2 Fx (44%); 18–40 Gy/3–5 Fx (56%)	50–156 Gy; 36–100 Gy	Circumferential "donut" (90%)	8.3	74%	84%	71% epidural, 5% posterior elements, 24% "mixed" in-field bony + epidural	64%	NR	Acute pain flare (n = 7); no RT myelopathy

Abbreviations: IG-HSRT, image-guided hypofractionated stereotactic radiotherapy; Gy, Gray; No., number; CTV, clinical target volume; nBED, normalized biologically effective dose; Fx, fractions; VB, vertebral body; NR, not reported; VAS, 10-point visual analog scale of pain; RT, radiation therapy; SCI, spinal cord injury.

[a] nBED using α/β ratio = 2, delivered in 2 Gy per fraction.

and kyphoplasty followed by adjuvant SRS (Gerszten and Monaco, 2009). In this study, 11 patients with pain secondary to pathological compression fracture with moderate (20%–50%) spinal canal compromise underwent transpedicular coblation corpectomy combined with closed fracture reduction and fixation and were then treated with conformal radiosurgery with doses ranging from 16 to 22.5 Gy delivered over one session. Similarly, all patients experienced some degree of improvement in back pain after treatment by 10-point visual analog scale. There were no documented acute radiation toxicities nor new neurological deficits during the median follow-up period of 11 months (range, 7–44 months).

Using comparatively modest radiosurgical doses, Rock et al. (2006) retrospectively reviewed a series of 18 patients treated with a variety of surgical approaches followed by SRS. The operative procedure varied based upon the extent of osseous involvement and need for stabilization after decompression/resection. The radiosurgery dose ranged between 6 and 16 Gy. Among the patients with neurological deficits preceding therapy, 62% of these patients demonstrated improvement in neurological status, and 30% had stable neurological deficits during the median follow-up period of 7 months. One patient demonstrated neurological deterioration after radiosurgery that was attributed to rapid tumor progression. Although pain control was not comprehensively reported, 33% of patients experienced complete resolution of pretreatment pain.

The MSKCC experience with decompressive surgery and adjuvant radiosurgery was initially described by Moulding et al. In this study, 21 patients were treated with posterolateral decompression and fixation followed by high-dose SRS of 18–24 Gy. They reported an excellent 1 year local control rate of 90.5% and a median survival of 10.2 months after IG-HSRT (Moulding et al., 2010). In general, toxicities were considered acceptable with a single self-limited episode of acute neuritis immediately following radiosurgery, three cases of grade 2 esophagitis, and a single case of grade 4 esophagitis requiring surgical intervention. Subsequently, Laufer and colleagues reported the largest series of postoperative spine IG-HSRT cases to date with an overall median follow-up of 7.6 months. In this retrospective analysis, 186 patients underwent separation surgery consisting of posterolateral laminectomy and circumferential epidural tumor resection followed by adjuvant hypofractionated or single-fraction stereotactic radiosurgery (Laufer et al., 2013). IG-HSRT was delivered as either single-fraction treatment of 24 Gy, high-dose hypofractionated treatment of 24–30 Gy delivered over 3 fractions, or low-dose hypofractionated treatment of 18–36 Gy delivered over 5–6 fractions. The cumulative incidence of local progression was 16.4% at 1 year. On multivariate analysis, a statistically significant increase in local control using high-dose hypofractionated treatment compared with low-dose hypofractionated regimens was noted; however, no significant difference in local control was observed between single-fraction and high-dose hypofractionated treatments.

MASS was explored as a neoadjuvant therapy to spine stereotactic radiosurgery by Massicotte et al. at the University of Toronto. Ten consecutive patients with mechanical pain due to single-level metastatic spine involvement with variable degrees of epidural disease (Bilsky grade 1a–3) were treated with MASS followed by stereotactic radiosurgery that was delivered over a range of doses and fractionation schedules (18–35 Gy over 1–5 fractions; Massicotte et al., 2012). With a median follow-up of 13 months, the local control rate was favorable at 70%, and 80% of patients reported some degree of improvement in pain by 10-point visual analog scale. With regards to toxicity, there were no reports of RM, and two patients experienced transient acute pain flares within a week of treatment completion. Of note, an improvement in disability and quality of life following decompression and radiosurgery was also observed.

More recently, Al-Omair et al. (2013a) reported on a series of 80 patients that underwent postoperative spine IG-HSRT with a median follow-up of 8.3 months. Surgical technique was heterogeneous ranging from stabilization alone to epidural disease resection/decompression with or without stabilization. Procedure selection was largely driven by preoperative epidural disease (55% of patients with high-grade Bilsky 2–3 epidural disease) extent and instability. Similarly, high-dose single-fraction and hypofractionated radiosurgery (18–26 Gy over 1–2 fractions) as well low-dose hypofractionated radiosurgery (18–40 Gy over 3–5 fractions) treatments were utilized; the median total dose was 24 Gy over 2 fractions. Local control rate at 1 year was 84% and overall survival rate at 1 year was 64%. Isolated progression within the epidural space was the most common pattern of local failure. The toxicity analysis identified seven patients that experienced transient acute pain flare events, similar to previous reports (Moulding et al., 2010; Massicotte et al., 2012; Chiang et al., 2013), and there were no reported cases of RM.

13.4.2 PREDICTORS OF LOCAL CONTROL

Local control rates with postoperative spine IG-HSRT based upon imaging and pain control are consistently superior to those reported following conventional radiotherapy. Acknowledging differences in definitions of local control, median follow-up, technique, dose, and fractionation, local control rates of 70%–100% (84%–90.5% at 1 year) have been achieved with postoperative spine IG-HSRT. By way of comparison, Klekamp and Samii reported local control rates of 40.1% at 6 months and 30.7% at 1 year with postoperative low-dose conventional radiotherapy (Klekamp and Samii, 1998).

There may be several factors that impact local control in postoperative spine IG-HSRT. Dose and fractionation have been evaluated in several studies. In the Moulding et al. series, a cumulative 1 year local control rate of 90.5% was observed. However, a cumulative local failure rate of 60% among patients undergoing low-dose IG-HSRT prompted further investigation (Moulding et al., 2010). Stratifying local failure status by IG-HSRT dose, there was a significant difference in risk of local failure among patients treated with 24 Gy (6.3%) and those treated with 18 or 21 Gy (20%). As mentioned earlier, Laufer et al. reported a significant improvement in local control among patients treated with high-dose hypofractionated regimens (24–30 Gy in 3 fractions) compared to low-dose hypofractionated regimens (18–36 Gy in 5–6 fractions). However, no differences were observed between high-dose single-fraction and high-dose hypofractionated regimens (Laufer et al., 2013). Similarly, Al-Omair et al. identified high-dose regimens of 18–26 Gy in 1–2 fractions as a significant predictor of local control when compared to low-dose regimens (18–40 Gy in 3–5 fractions). Again, there were no differences in local control among patients treated with high-dose IG-HSRT when comparing single-fraction and hypofractionated regimens.

The extent of postoperative residual epidural disease also plays a role, as one series identified ESCC grade 0–1 disease (e.g., as shown in Figure 13.1) as a significant predictor of local control. Further subgroup analysis of patients with preoperative ESSC grade 2–3 disease demonstrated that successful decompression to postoperative ESCC grade 0–1 disease resulted in a significantly improved local control rate compared to patients with postoperative ESCC grade 2 disease (Al-Omair et al., 2013a).

13.4.3 PATTERNS OF FAILURE

Emerging data suggest that patterns of failure following postoperative spine SBRT are similar to those following SBRT for intact vertebral bodies. Specifically, local failure within the epidural space appears to be the most common (Massicotte et al., 2012; Al-Omair et al., 2013a) likely because of the decreased dose at the spinal cord–tumor interface required to achieve OAR constraints. In a recent study, 21 patients experienced local failure of which 71% developed isolated progression within the epidural space, 5% developed failure in the posterior elements alone when outside of the CTV, and 24% developed a mixed pattern of infield bony and/or epidural space progression. Approximately 90% of patients in this study were treated with circumferential "donut" fields, which may explain this predominant pattern of failure (Al-Omair et al., 2013a). The role of intraoperative phosphorous-32 brachytherapy plaque placement for residual or progressive epidural disease is being evaluated as a therapeutic modality to address local failure within the epidural space (Tong et al., 2014).

13.4.4 SINGLE-FRACTION VERSUS MULTIFRACTION HSRT

Postoperative spine IG-HSRT is in its nascency as a therapeutic modality, and there is no prospective randomized data to compare outcomes with single-fraction versus multifraction technique. Recent work by Al-Omair et al. (2013a) and Laufer et al. suggests that higher-dose regimens are associated with increased local control; however, there does not appear to be a difference when comparing single-fraction and multifraction high-dose IG-HSRT regimens. This alludes to the probability that cumulative BED is more predictive of local control than fractionation schema. It should be noted that hypofractionation should be considered in situation where single-fraction IG-HSRT is not able to achieve OAR or target coverage parameters.

13.5 TOXICITIES

Toxicities associated with postoperative spine IG-HSRT mostly mirror those following IG-HSRT for intact vertebral bodies and relate to injury of the spinal cord resulting in RM, injury to nearby OARs (e.g., esophageal stricture), and VCF.

Although exceedingly uncommon, RM is the most serious toxicity associated with spine IG-HSRT as it can result in irreversible paralysis, permanent neurological deficits, or death. Adherence to appropriate spinal cord dose constraints results in a very low number of RM events. The current evidence suggests that high doses to small volumes of the spinal cord are predictive of RM and therefore point max (Pmax) is an important dosimetric parameter. Sahgal and colleagues developed a predictive probability model of RM based upon Pmax to the thecal sac, which is referenced in Table 13.7 (Sahgal et al., 2012b). Furthermore, dose constraints for thecal sac Pmax for reirradiation with IG-HSRT have been previously described in Section 13.3.3 and suggests that reirradiation is reasonably safe if cumulative normalized BED (nBED) does not exceed 70 $Gy_{2/2}$ (e.g., as shown in Table 13.6; Sahgal et al., 2012a). There have been few cases of RM documented in the spine IG-HSRT literature (Ryu et al., 2007; Gibbs et al., 2009). Among the postoperative IG-HSRT series, no cases of RM have been reported, and Rock et al. described one case of progressive neurological decline that was ultimately attributed to tumor progression as opposed to RM (Rock et al., 2006).

The presence of spinal instrumentation likely reduces the risk for VCF, but it remains a possibility. Extrapolating from the intact spine IG-HSRT literature and accounting for differences in dose and fractionation, the risk of VCF is approximately 11%–39% (Sahgal et al., 2013a). Factors such as lower location within the spine, lytic vertebral body disease, kyphotic/scoliotic deformity, high dose per fraction, age greater than 55 years, and baseline VCF confer an increased risk of VCF (Rose et al., 2009; Boehling et al., 2012; Cunha et al., 2012; Sahgal et al., 2013a,b). Stabilization alone does not offset radiation-induced collagen damage and osteoradionecrosis, and therefore postoperative spine IG-HSRT patients remain at risk for VCF (Sahgal et al., 2013b). The Al-Omair series reported a total of nine VCF events, five of which were de novo and four progressing from preexisting fractures. This corresponded to a crude VCF rate of 11%, which is consistent with the existing literature and supports that surgical stabilization alone does decrease risk of VCF related to osteoradionecrosis (Al-Omair et al., 2013a,b).

Acute pain flare is a transient common adverse reaction that occurs within the first 10 days after spine IG-HSRT (Chiang et al., 2013). It is usually responsive to steroid administration and dexamethasone prophylaxis is a consideration to reduce to incidence and/or severity of acute pain flare. Other spine IG-HSRT-related toxicities include damage to nearby OARs such as the esophagus, bowel, and kidneys, and care should be taken to minimize dose to these critical structure in order to reduce risk of acute and long-term side effects.

There are also risks that are unique to the postoperative setting, specifically wound-healing complications and hardware failure. Importantly, preliminary data suggest that the rates of hardware failure may be approximately 1%–2%, which is similar to the rate following surgery alone (Al-Omair et al., 2013a; Amankulor et al., 2014). These limited data suggest that postoperative spine IG-HSRT does not increase the risk of hardware failure.

13.6 CONTROVERSIES

Important questions regarding optimal dose and fractionation, target volumes, and dose constraints remain unanswered in the postoperative spine IG-HSRT setting. Early data suggest that higher nBED regimens may result in higher local control rates; however, it remains unclear if tumoricidal doses delivered in single-fraction or multifraction impact local control. Furthermore, dose escalation can increase the incidence of serious late toxicities, such as RM and VCF, which were significantly less common in the pre-spine IG-HSRT era. Discerning the balance between dose escalation and minimizing toxicity are critical for optimizing patient outcomes. Insufficient distance between the spinal cord–tumor interface is not an uncommon scenario that requires a compromise in either target coverage or prescription dose to meet normal tissue constraints. The superiority of compromising coverage with a higher prescription dose over improved coverage with a lower prescription dose is unknown and merits further investigation. Additionally, there are no specific consensus guidelines for target delineation in the postoperative setting. The need to include the entire preoperative extent of disease or only residual and high-risk areas is another area of uncertainty. Finally, it is unknown whether more conservative dose constraints for OARs should be adopted in consideration of surgical manipulation and disruption of spinal vasculature as this may be associated with an increased incidence of RM and other toxicities, although initial studies do not suggest an increased risk (Sahgal et al., 2011).

13.7 FUTURE DIRECTIONS

Prospective trials will be critical in gaining a better understanding of the nuances of postoperative spine IG-HSRT. While the existing literature is promising from a local control, pain control, and toxicity perspective, it remains inadequate at present. As IG-HSRT and surgical technique continually evolve, it will be important to further understand which patients will maximally benefit from open decompressive surgeries versus those which can be managed with minimally invasive procedures followed by IG-HSRT. Minimally invasive approaches have the advantage of reducing surgical morbidity and improving patient quality of life while maximizing local control. Future investigations will further bolster the concept of surgical intervention as a neoadjuvant modality to postoperative spine IG-HSRT.

CHECKLIST: KEY POINTS FOR CLINICAL PRACTICE

KEY POINTS FOR POSTOPERATIVE SPINE IG-HSRT

- Postoperative spine IG-HSRT is an emerging treatment modality for the management of spinal metastases.
- Current literature suggests excellent local control rates (70%–100%) can be achieved with postoperative spine IG-HSRT. While no level 1 randomized evidence exists, results are consistently superior to conventional radiotherapy.
- Surgical hardware introduces complexities in target delineation and treatment planning. Strict adherence to normal tissue constraints can reduce the risk of RM and other toxicities.

✓	ACTIVITY	SOME CONSIDERATIONS
	Patient selection	*Is the patient appropriate for surgery followed by HSRT?* • Medically fit for surgery • Patients with spinal instability (SINS ≥13) • Neurological symptoms due to MESCC (Bilksy grade 2–3) • Recurrent disease following prior radiotherapy • Oligometastatic disease and good KPS
	SRS vs HSRT	*Is the lesion amenable to single-fraction SRS?* • In general, HSRT should be considered in situations where SRS is not able to achieve OAR constraints while maintaining target coverage • Some prefer HSRT and not single fraction and is an area of controversy
	Simulation	*Immobilization* • Supine, positioning, and immobilization device are dependent on location of spinal lesion • For C1–T3, consider long thermoplastic mask *or* custom head mold and open face mask • For T4–L3, consider wing board setup with arms positioned above the head • For L4–sacrum, consider Hevezi pad or Vac Lok

(Continued)

✓	ACTIVITY	SOME CONSIDERATIONS
		Imaging • Both CT scan and MRI both with/without contrast should be performed in treatment position • MRI should be thin-slice (≤2 mm slices), T1 ± gadolinium sequence for tumor delineation, and T2/STIR sequence for spinal cord and tumor delineation • Use preoperative MRI to assist CTV delineation • CT myelogram for spinal cord delineation, if there is significant metallic artifact from instrumentation

	Treatment planning	*Contours*
		• GTV: gross residual disease as delineated on postoperative imaging
		• CTV includes GTV plus
		  • Entire anatomic compartment of preoperative disease plus immediately adjacent spinal segment
		  • Do not need to include surgical hardware or operative scar unless clinically indicated
		• PTV is a geometric expansion of ~1–2 mm from the CTV
		• Contour all OARs at least one vertebral body superior/inferior to PTV and to a further extent for nonisocentric planning due to potential for remote hot spots
		• Consider subtracting PRVs (planning organ at risk volume) from PTV
		Treatment planning
		• Isocentric or nonisocentric
		• Must account for location of instrumentation in selection of beam angles and use a density override in dose calculations
		• Dose (variable depending on clinical scenario)
		  • 18–20 Gy in 1 fraction
		  • 24 Gy in 2 fractions
		  • 27–30 Gy in 3 fractions
		  • 30–40 Gy in 5 fractions
		• Coverage
		  • 85%–95% target coverage by 90%–100% of prescription dose is generally acceptable
		  • Compromise PTV coverage to meet OAR constraints (or consider hypofractionation)

	Treatment delivery	*Technology/imaging*
		• Generally, the two types of treatment delivery systems are
		  1. MLC-based LINAC
		    • 7–11 coplanar beams
		    • 5–15 MLC apertures or segments per beam
		    • "Step-and-shoot" technique with fixed gantry position or dynamic MLC
		  2. CyberKnife
		    • Nonisocentric
		    • Mobile robotic arm mounted with compact LINAC
		    • Large number of beam angles (~100–200)
		    • Circular collimators of varying size to maximize target coverage and spare OARs

(Continued)

✓	ACTIVITY	SOME CONSIDERATIONS
		• Cone-beam CT and/or stereoscopic kV x-rays for treatment setup verification and interfraction shifts • CyberKnife permits near real-time imaging during treatment to make intrafractional adjustments *Concurrent systemic therapy* • Caution should be utilized to limit concurrent administration of targeted agents and systemic therapy unless clinically essential

REFERENCES

Adler JR, Jr, Chang SD, Murphy MJ, Doty J, Geis P, Hancock SL (1997) The Cyberknife: A frameless robotic system for radiosurgery. *Stereotact Funct Neurosurg* 69:124–128.

Adler JR, Jr, Murphy MJ, Chang SD, Hancock SL (1999) Image-guided robotic radiosurgery. *Neurosurgery* 44:1299–1307.

Al-Omair A, Masucci L, Masson-Cote L, Campbell M, Atenafu EG, Parent A, Letourneau D et al. (2013a) Surgical resection of epidural disease improves local control following postoperative spine stereotactic body radiotherapy. *Neuro-Oncology* 15(10):1413–1419.

Al-Omair A, Smith R, Kiehl TR, Lao L, Yu E, Massicotte EM, Keith J, Fehlings MG, Sahgal A (2013b) Radiation-induced vertebral compression fracture following spine stereotactic radiosurgery: Clinicopathological correlation. *J Neurosurg Spine* 18(5):430–435.

Amankulor NM, Xu R, Iorgulescu JB, Chapman T, Reiner AS, Riedel E, Lis E, Yamada Y, Bilsky M, Laufer I (2014) The incidence and pattern of hardware failure after separation surgery in patients with spinal metastatic tumors. *J Neurosurg Spine* 14(9):1850–1859.

Benedict SH, Yenice KM, Followill D, Galvin JM, Hinson W, Kavanagh B, Keall P et al. (2010) Stereotactic body radiation therapy: The report of AAPM Task Group 101. *Med Phys* 39(1):563.

Bilsky MH, Boland PJ, Panageas KS, Woodruff JM, Brennan MF, Healey JH (2001) Intralesional resection of primary and metastatic sarcoma involving the spine: Outcome analysis of 59 patients. *Neurosurgery* 49(6):1277–1286.

Bilsky MH, Laufer I, Fourney DR, Groff M, Schmidt MH, Varga PP, Vrionis FD, Yamada Y, Gerszten PC, Kuklo TR (2010) Reliability analysis of the epidural spinal cord compression scale. *J Neurosurg Spine* 13(3):324–328.

Boehling NS, Grosshans DR, Allen PK, McAleer MF, Burton AW, Azeem S, Rhines LD, Chang EL (2012) Vertebral compression fracture risk after stereotactic body radiotherapy for spinal metastases. *J Neurosurg Spine* 16(4):379–386.

Chi JH, Gokaslan Z, McCormick P, Tibbs PA, Kryscio RJ, Patchell RA (2009) Selecting treatment for patients with malignant epidural spinal cord compression—Does age matter?: Results from a randomized clinical trial. *Spine* (1976) 34(5):431–435.

Chiang A, Zeng L, Zhang L, Lochray F, Korol R, Loblaw A, Chow E, Sahgal A (2013) Pain flare is a common adverse events in steroid-naïve patients after spine stereotactic body radiation therapy: A prospective clinical trial. *Int J Radiat Oncol Biol Phys* 86(4):638–642.

Cox BW, Spratt DE, Lovelock M, Bilsky MH, Lis E, Ryu S, Sheehan J (2012) International Spine Radiosurgery Consortium consensus guidelines for target volume definition in spinal stereotactic radiosurgery. *Int J Radiat Oncol Biol Phys* 83(5):e597–e605.

Cunha MV, Al-Omair A, Atenafu EG, Masucci GL, Letourneau D, Korol R, Yu E et al. (2012) Vertebral compression fracture (VCF) after spine stereotactic body radiation therapy (SBRT): Analysis of predictive factors. *Int J Radiat Oncol Biol Phys* 84(3):e343–e349.

Findlay GF (1984) Adverse effects of the management of malignant spinal cord compression. *J Neurol Neurosurg Psychiatry* 47(8):761–768.

Fisher CG, DiPaola CP, Ryken TC, Bilsky MH, Shaffrey CI, Berven SH, Harrop JS et al. (2010) A novel classification system for spinal instability in neoplastic disease: An evidence based-approach and expert consensus from the Spine Oncology Study Group. *Spine* (1976) 35(22):E1221–E1229.

Fisher CG, Schouten R, Versteeg AL, Boriani S, Varga PP, Rhines LD, Kawahara N et al. (2014b) Reliability of the Spinal Instability Neoplastic Score (SINS) among radiation oncologists: An assessment of instability secondary to spinal metastases. *Radiat Oncol* 9:69.

Fisher CG, Versteeg AL, Schouten R, Boriani S, Varga PP, Rhines LD, Heran MK et al. (2014a) Reliability of the spinal instability neoplastic scale among radiologists: An assessment of instability secondary to spinal metastases. *Am J Roentgenol* 203(4):869–874.

Fornasier VL, Horne JG (1975) Metastases to the vertebral column. *Cancer* 36(2):590–594.

Fourney DR, Frangou EM, Ryken TC, Dipaola CP, Shaffrey CI, Berven SH, Bilsky MH et al. (2011) Spinal instability neoplastic score: An analysis of reliability and validity from the spine oncology study group. *J Clin Oncol* 29(22):3072–3077.

Gerszten PC, Germanwala A, Burton SA, Welch WC, Ozhasoglu C, Vogel WJ (2005) Combination kyphoplasty and spinal radiosurgery: A new treatment paradigm for pathological fractures. *J Neurosurg Spine* 3:296–301.

Gerszten PC, Monaco EA, III (2009) Complete percutaneous treatment of vertebral body tumors causing spinal canal compromise using transpedicular cavitation, cement augmentation, and radiosurgical technique. *Neurosurg Focus* 27(6):E9.

Gibbs IC, Patil C, Gerszten PC, Adler JR, Jr., Burton SA (2009) Delayed radiation-induced myelopathy after spinal radiosurgery. *Neurosurgery* 64(2 Suppl):A67–A72.

Gilbert RW, Kim JH, Posner JB (1978) Epidural spinal cord compression from metastatic tumor: Diagnosis and treatment. *Ann Neurol* 3(1):40–51.

Greenberg HS, Kim JH, Posner JB (1980) Epidural spinal cord compression from metastatic tumor: Results from a new treatment protocol. *Ann Neurol* 8:361–366.

Harrington KD (1993) Metastatic tumors of the spine: Diagnosis and treatment. *J Am Acad Orthop Surg* 1:76–86.

Kirshblum S, Waring W, III (2014) Updates for the international standards for neurological classification of spinal cord injury. *Phys Med Rehabil Clin North Am* 25(3):505–517.

Kirshblum SC, Burns SP, Biering-Sorensen F, Donovan W, Graves DE, Jha A, Johansen M et al. (2011) International standards for neurological classification of spinal cord injury (revised 2011). *J Spinal Cord Med* 34(6):535–546.

Klekamp J, Samii M (1998) Surgical results for spinal metastases. *Acta Neurochir* 140:957–967.

Klimo P, Jr., Schmidt MH (2004) Surgical management of spinal metastases. *Oncologist* 9(2):188–196.

Laufer I, Iorgulescu JB, Chapman T, Lis E, Shi W, Zhang Z, Cox BW, Yamada Y, Bilsky MH (2013b) Local disease control for spinal metastases following "separation surgery" and adjuvant hypofractionated or high-dose single-fraction stereotactic radiosurgery: Outcome analysis in 186 patients. *J Neurosurg Spine* 18(3):207–214.

Laufer I, Rubin DG, Lis E, Cox BW, Stubblefield MD, Yamada Y, Bilsky MH (2013a) The NOMS framework: Approach to treatment of spinal metastatic tumors. *Oncologist* 18(6):744–751.

Liebross RH, Starkschall G, Wong PF, Horton J, Gokaslan ZL, Komaki R (2002) The effect of titanium stabilization rods on spinal cord radiation dose. *Med Dosim* 27(1):21–24.

Maranzano E, Latini P (1995) Effectiveness of radiation therapy without surgery in metastatic spinal cord compression: Final results from a prospective trial. *Int J Radiat Biol Phys* 32:959–967.

Massicotte E, Foote M, Reddy R, Sahgal A (2012) Minimal access spine surgery (MASS) for decompression and stabilization performed as an out-patient procedure for metastatic spinal tumors followed by spine stereotactic body radiotherapy (SBRT): First report of technique and preliminary outcomes. *Technol Cancer Res Treat* 11(1):15–25.

Masucci GL, Yu E, Ma L, Chang EL, Letourneau D, Lo S, Leung E et al. (2011) Stereotactic body radiotherapy is an effective treatment in re-irradiating spinal metastases: Current status and practical considerations for safe practice. *Expert Rev Anticancer Ther* 11(12):1923–1933.

Moulding HD, Elder JB, Lis E, Lovelock DM, Zhang Z, Yamada Y, Bilsky MH (2010) Local disease control after decompressive surgery and adjuvant high-dose single-fraction radiosurgery for spine metastases. *J Neurosurg Spine* 13(1):87–93.

Moussazadeh N, Laufer I, Yamada Y, Bilsky MH (2014) Separation surgery for spinal metastases: Effect of spinal radiosurgery on surgical treatment goals. *Cancer Control* 21(2):168–174.

Patchell RA, Tibbs PA, Regine WF, Payne R, Saris S, Kryscio RJ, Mohiuddin M, Young B (2005) Direct decompressive surgical resection in the treatment of spinal cord compression caused by metastatic cancer: A randomized trial. *Lancet* 366(9486):643–648.

Pekmezci M, Dirican B, Yapici B, Yazici M, Alanay A, Gürdalli S (2006) Spinal implants and radiation therapy: The effect of various configurations of titanium implant systems in a single-level vertebral metastasis model. *J Bone Joint Surg Am* 88(5):1093–1100.

Rock JP, Ryu S, Shukairy MS, Yin FF, Sharif A, Schreiber F, Abdulhak M, Kim JH, Rosenblum ML (2006) Postoperative radiosurgery for malignant spinal tumors. *Neurosurgery* 58(5):891–898.

Rose PS, Laufer I, Boland PJ, Hanover A, Bilsky MH, Yamada J, Lis E (2009) Risk of fracture after single fraction image-guided intensity-modulated radiation therapy to spinal metastases. *J Clin Oncol* 27(30):5075–5079.

Ryu S, Jin JY, Jin R, Rock J, Ajlouni M, Movsas B, Rosenblum M, Kim JH (2007) Partial volume tolerance of the spinal cord and complications of single-dose radiosurgery. *Cancer* 109(3):628–636.

Ryu SI, Chang SD, Kim DH, Murphy MJ, Le QT, Martin DP, Adler JR, Jr. (2001) Image-guided hypo-fractionated stereotactic radiosurgery to spinal lesions. *Neurosurgery* 49(4):838–846.

Sahgal A, Atenafu EG, Chao S, Al-Omair A, Boehling N, Balagamwala EH, Cunha M et al. (2013b) Vertebral compression fracture after spine stereotactic body radiotherapy: A multi-institutional analysis with a focus on radiation dose and spinal instability neoplastic score. *J Clin Oncol* 31(27):3426–3431.

Sahgal A, Bilsky M, Chang EL, Ma L, Yamada Y, Rhines LD, Létourneau D et al. (2011) Stereotactic body radiotherapy for spinal metastases: Current status, with a focus on its application in the post-operative patient. *J Neurosurg Spine* 14:151–166.

Sahgal A, Ma L, Gibbs I, Gerszten PC, Ryu S, Soltys S, Weinberg V et al. (2010) Spinal cord tolerance for stereotactic body radiotherapy. *Int J Radiat Oncol Biol Phys* 77:548–553.

Sahgal A, Ma L, Weinberg V, Gibbs IC, Chao S, Chang UK, Werner-Wasik M et al. (2012a) Reirradiation human spinal cord tolerance for stereotactic body radiotherapy. *Int J Radiat Oncol Biol Phys* 82(1):107–116.

Sahgal A, Weinberg V, Ma L, Chang E, Chao S, Muacevic A, Gorgulho A et al. (2012b) Probabilities of radiation myelopathy specific to stereotactic body radiation therapy to guide safe practice. *Int J Radiat Oncol Biol Phys* 85(2):341–347.

Sahgal A, Whyne CM, Ma L, Larson DA, Fehlings MG (2013a) Vertebral compression fracture after stereotactic body radiotherapy for spinal metastases. *Lancet Oncol* 14(8):e310–e320.

Sciubba DM, Gokaslan ZL (2006) Diagnosis and management of metastatic spine disease. *Surg Oncol* 15(3):141–151.

Sciubba DM, Petteys RJ, Dekutoski MB, Fisher CG, Fehlings MG, Ondra SL, Rhines LD, Gokaslan ZL (2010) Diagnosis and management of metastatic spine disease. A review. *J Neurosurg Spine* 13(1):94–108.

Sharan AD, Szulc A, Krystal J, Yassari R, Laufer I, Bilsky MH (2014) The integration of radiosurgery for the treatment of patients with metastatic spine diseases. *J Am Acad Orthop Surg* 22(7):447–454.

Tong WY, Folkert MR, Greenfield JP, Yamada Y, Wolden SL (2014) Intraoperative phosphorous-32 brachytherapy for multiply recurrent high-risk epidural neuroblastoma. *J Neurosurg Pediatr* 13(4):388–392.

Wang X, Yang JN, Li X, Tailor R, Vassilliev O, Brown P, Rhines L, Chang E (2013) Effect of spine hardware on small spinal stereotactic radiosurgery dosimetry. *Phys Med Biol* 58(19):6733–6747.

Wong DA, Fornasier VL, MacNab I (1990) Spinal metastases: The obvious, the occult, and the impostors. *Spine* 15(1):1–4.

Yong RF, Post EM, King GA (1980) Treatment of spinal epidural metastases. Randomized prospective comparison of laminectomy and radiotherapy. *J Neurosurg* 53(6):741–748.

14 Postoperative cavity image-guided stereotactic radiotherapy outcomes

Mary Frances McAleer and Paul D. Brown

Contents

14.1 INTRODUCTION

For decades, whole-brain radiotherapy (WBRT) has been considered the standard of care therapy following resection of brain metastases based on positive results of prospective study by Patchell et al. (1998), which has more recently been confirmed by another trial conducted by the European Organisation for Research and Treatment of Cancer (Kocher et al., 2011). Addition of WBRT significantly improved local control (LC) in patients randomized to receive the adjuvant radiation compared with those followed by observation after surgical resection of one (Patchell et al., 1998) and up to two (Kocher et al., 2011) brain lesions in these trials. However, due to concerns of adverse effects of WBRT on patient neurocognitive function and quality of life (QOL) as well as delay of potentially life-prolonging systemic therapies (see, e.g., Soffietti et al., 2013), there is an emerging body of evidence in support of limited field, single-fraction or short hypofractionated courses of stereotactic radiotherapy to the at-risk intracranial surgical cavity for improved local tumor control for select patients with brain metastases. Utilization of such therapy has become more widely accepted, given the advances in neuroradiographic imaging, neurosurgical, and radiotherapeutic techniques. Here we present the current data and indications for postoperative image-guided single-fraction or hypofractionated stereotactic radiotherapy (IG-HSRT) for resected brain metastases.

14.2 INDICATIONS

Since there are limited prospective data on the utility of postoperative IG-HSRT for patients with resected intracranial metastases, indications for the use of this therapy must be considered relative and not absolute.

There is a contemporary, multi-institutional, Alliance cooperative group phase III trial comparing postsurgical stereotactic radiosurgery (SRS) to WBRT for resected brain metastases (North Central Cancer Treatment Group [NCCTG] N107C) that specifies the following criteria for adult subjects to be eligible to receive focal SRS to the surgical cavity with a 2 mm margin (http://www.cancer.gov/types/metastatic-cancer/research/postop-radiation-therapy-brain, last access November 30, 2015):

- A maximum of four brain metastases, with resection of at least one of these lesions
- Confirmed metastatic solid tumor, excluding germ cell, small cell, and lymphoma
- <3.0 cm maximum diameter of any unresected metastasis on postcontrast MRI
- <5.0 cm maximum diameter of resection cavity on postoperative imaging
- All lesions >5 mm from the optic chiasm and outside the brainstem
- Good performance status (Karnofsky performance status [KPS] > 60)
- Nonpregnant, nonnursing females
- No prior cranial RT
- No concurrent cytotoxic systemic therapy during postoperative SRS
- No leptomeningeal metastatic disease

14.3 APPROACH

Due to current dearth of prospective evidence to support the use of adjuvant IG-HSRT following resection of intracranial metastases, with rare exception, SRS to the resection cavity has been utilized at our institution only in the setting of an in-house, phase III, randomized controlled trial. Patients >3 years of age, with a KPS of at least 70, and with a total of up to three intracranial metastases from solid tumors (excluding small cell lung cancer, lymphoma, leukemia, or multiple myeloma), who have undergone gross total resection (GTR) of at least one lesion and have a surgical cavity ≤4 cm in maximum diameter based on postoperative MRI imaging review by the study neuroradiologist, are eligible for this trial. Eligible subjects are then randomized to receive either SRS or be followed with close imaging surveillance. Those subjects with more than one lesion are then treated with SRS to the unresected lesions having maximum diameter ≤3 cm using the dose regimen established by Radiation Therapy Oncology Group (RTOG) 90-05 (Shaw et al., 1996, 2000). Patients are stratified by histology (melanoma vs other), diameter of resected lesion (< or ≥3 cm), and number of metastases (1 vs 2–3).

Patients randomized to the SRS arm must undergo volumetric MRI of brain within 7 days of SRS, and the radiation procedure must be delivered within 3–30 days of surgical resection. The SRS treatment is frame based and may be delivered using either the Gamma Knife (GK) Perfexion unit (Elekta, Stockholm, Sweden) or a linear accelerator (LINAC) with arcs or mini-multileaf collimation. Real-time treatment planning is employed. The target is defined as the surgical cavity plus 1 mm margin, and the single-fraction dose to be delivered is defined based on the target volume: 16 Gy for 0–10 cubic centimeters (cc), 14 Gy for 10.1–15 cc, and 12 Gy for >15 cc. The maximum point dose to the optic apparatus is limited to <9 Gy, and 1.0 cc of brainstem may receive 12 Gy.

After treatment, patients are monitored every 6–9 weeks for 1 year, then every 3–4 months during year 2, then every 6 months thereafter with MRI of brain to assess LC. Secondary endpoints include overall survival (OS), distant brain failure (DBF), and treatment-related complications. The planned total accrual is 132 subjects, and this study is estimated to meet accrual by spring 2015. Results from this trial are still pending, as the study is still actively accruing patients. The treatment approach to administer adjuvant SRS to the surgical bed in the contemporary, phase III N107C multi-institutional, cooperative group trial specifies the following:

- *Technique:* GK or ≥4 MV x-rays using LINAC
  - Mandatory patient immobilization/localization system
- *Target:* MRI-defined surgical cavity +2 mm margin, excluding anatomic barriers
  - <5.0 cm maximum diameter of cavity
- *Dose:* Prescribed to highest isodose line covering cavity + margin
  - Based on cavity (without margin) volume (see Table 14.1)

Table 14.1 Prescription dose for postoperative stereotactic radiosurgery as defined by N107C trial

CAVITY VOLUME (cc)	DOSE (Gy)
<4.2	20
≥4.2 to <8.0	18
≥8.0 to <14.4	17
≥14.4 to <20	15
≥420 to <30	14
≥30 to <5.0 cm max diam	12

Abbreviations: cc, cubic cm; max diam, maximum diameter.

14.4 OUTCOME DATA

14.4.1 PROSPECTIVE TRIALS

In the published literature, there is currently only one prospective, single-institution, phase II trial from Memorial Sloan Kettering Cancer Center (MSKCC) that has reported results of LC using single-fraction SRS following resection of intracranial metastases (Brennan et al., 2014). Forty-nine evaluable patients with fifty surgical cavities were enrolled on this study. Patients had median age of 59 years (range, 23–81), median KPS of 90 (range, 70–100), and the majority (57%) had histologically confirmed diagnosis of non–small cell lung cancer (NSCLC), followed by 20% with breast cancer and 8% with melanoma. Sixty-five percent of patients had primary tumor site controlled, and 45% had extracranial metastatic disease. Ninety-eight percent of patients had single metastases, with median tumor diameter of 2.9 cm (range, 1.0–5.2), and GTR was reported in 92% of cases. Of the 49 evaluable patients, 39 received adjuvant SRS to a total of 40 surgical cavities, with median cavity diameter of 2.8 cm (range, 1.7–5.4) that included the surgical track. A 2 mm margin was added to each cavity, as identified on postcontrast MRI and CT head, and single-fraction SRS was delivered using RTOG 90-05 dose guidelines (Shaw et al., 1996, 2000) and LINAC delivery of 8–12 noncoplanar, static beams with micro-multileaf collimation a median of 31 days (range, 7–56) from surgery.

In this MSKCC prospective trial, 1 year local failure (LF) was identified in 15% of patients treated with postoperative SRS (Brennan et al., 2014). This percentage increased to 53% in those patients with cavities measuring at least 3 cm and having dural invasion. DBF at 1 year was seen in 44% at a median of 4.4 months (range 1.1–17.9), and median OS was 14.7 months (range 1–94.1). WBRT was used as salvage in approximately 2/3 of patients who had intracranial disease progression. The results of this study revealed comparable cavity LC as was observed in the WBRT arms of the prospective randomized controlled studies of adjuvant WBRT following surgery or SRS, and the rate of DBF was similar to the observation arms of these studies (Patchell et al., 1998; Kocher et al., 2011). Of note, however, OS in the MSKCC study was almost 4 months longer than in the older trials, and WBRT was delayed or omitted in the majority of these patients.

14.4.2 RETROSPECTIVE STUDIES

Since 2008, there have been numerous retrospective analyses of postoperative focal brain radiotherapy published, including 13 reports of outcomes using single-fraction SRS (Table 14.2), 7 with multifraction IG-HSRT regimens (Table 14.3), and another 7 that combined both single- and multiple-fraction treatments (Table 14.4). Four of these twenty-seven reports also compared the outcomes of patients receiving focal radiotherapy to those receiving WBRT following resection of intracranial metastases (Lindvall et al., 2009; Hwang et al., 2010; Al-Omair et al., 2013; Patel et al., 2014). The patient populations of these investigations were similar to that described for the prospective MSKCC trial. Specifically, the median age of patients was 55–65 years, virtually all had KPS of at least 70, and the most common tumor histologies were NSCLC, breast cancer, and melanoma. The majority of subjects

Table 14.2 Summary of studies using postoperative single-fraction stereotactic radiosurgery

STUDY	NO.	AGE (YEARS)	KPS	% HISTOLOGY	% ONE LESION	% GTR	CAVITY (cc)	MARGIN (mm)	DOSE (Gy)	% LC	% DBF	OS (MONTHS)	% NECROSIS	% LMD	% WBRT SALVAGE
MSKCC[a] (Brennan et al., 2014)	39	59 (23–81)	90 (70–100)	57 L 20 B 8 M	97	92	2.8[c] (1.7–5.4) +track	2 +track	18 (15–22)	85	44[d]	14.7 (1–94.1)	17.5	NR	65
Dartmouth (Hartford et al., 2013)	47	64 (24–85)	80 (50–100)	49 L 11 B 15 M	70	76	3.0[c] (1.3–4.6) preop	2	10 (8–20)	85.5[d]	56.2[d]	52.5%[d]	NR	NR	45
Tufts (Hwang et al., 2010)	25	59.5[b] (48–71)	NR	NR	NR	95	NR	NR	NR	100	28	15 (6.0–35.8)	NR	NR	NR
Sherbrooke (Iorio-Morin et al., 2014)	110	58 (37–84)	90 (50–100)	50 L 13 B 10 M	30	81	12 (0.6–43)	1	18 (10–20)	73[d]	54	11 (1.4–84)	0.9	11	28
Osaka (Iwai et al., 2008)	21	61 (41–80)	88[b] (70–100)	24 L 10 B	76	86	10.7 (3.4–26.9)	NR	17 (13–23)	76	48	20 postop	NR	24	NR
UVA (Jagannathan et al., 2009)	47	61 (37–88)	90 (60–100)	40 L 15 B 21 M	13	100	10.5 (1.75–35.45)	2–3	19 (6–22)	94	6	11 (7–36)	0	NR	28
Wake Forest (Jensen et al., 2011)	106	56.1 (22.6–88.0)	NR	47 L 14 B 10 M	57.5	96.4	8.0 (0.32–33.4)	0	17 (11–23)	80.3[d]	NR	10.9	3	7.5	37
Barrow (Kalani et al., 2010)	68	60[b] (28–89)	90 (40–100)	44 L 15 B 13 M	100	NR	10.35 (0.9–45.4)	1–3	15 (14–30)	79.5	39.7	13.2	NR	NR	NR
Allegheny (Karlovits et al., 2009)	52	61 (31–85)	NR	46 L 17 B	34	92.3	3.85 (0.08–22)	2	15 (8–18)	93	44	15	NR	NR	31
Wash U (Limbrick et al., 2009)	15	56.8 (41–85)	93% ≥70	40 L 27 B NR	80	(0.18–16.0)	NR	(16–24)	83GTR	73.3[d]‖	60	20 (5–68)	NR	NR	40
U Pitt (Luther et al., 2013)	120	58[b]	NR	40 L 21 B 16 M	NR	100	7.3 PTV	NR	16	85.8	40	NR	NR	NR	16

(Continued)

Table 14.2 (Continued) Summary of studies using postoperative single-fraction stereotactic radiosurgery

STUDY	NO.	AGE (YEARS)	KPS	% HISTOLOGY	% ONE LESION	% GTR	CAVITY (cc)	MARGIN (mm)	DOSE (Gy)	% LC	% DBF	OS (MONTHS)	% NECROSIS	% LMD	% WBRT SALVAGE
U Pitt & Sherbrooke (Mathieu et al., 2008)	40	59.5	80 (60–100)	40 L 10 B 20 M	67.5	80	9.1 (0.6–39.9)	1	16 (11–20)	73	54	13 (2–56)	0	NR	16
U Penn (Ojerholm et al., 2014b)	91	60 (22–82)	96% ≥70	43 L 13 B 14 M	57	82	9.2 (0.6–34.7)	0	16 (12–21)	82	64	22.3	7	14	33
Henry Ford (Robbins et al., 2012)	85	58 (38–83)	80 (60–100)	59 L 11 B 13 M	62	68	13.96	2–3	16 (12–20)	81.2	55	12.1	8	8	35

Note: Data presented as median (range) unless otherwise indicated.
a Phase 2 study.
b Mean (range).
c Size in cm.
d At 1 year.

Abbreviations: KPS, Karnofsky performance status; Histology L, non–small cell lung cancer; B, breast cancer; M, melanoma; GTR, gross total resection; cc, cubic cm; preop, preoperative; LC, local control; DBF, distant brain failure; OS, overall survival; postop, postoperative; LMD, leptomeningeal disease; WBRT, whole–brain radiotherapy; NR, not reported.

Table 14.3 Summary of studies using postoperative multiple-fraction stereotactic radiotherapy

STUDY	NO.	AGE (YEARS)	KPS	% HISTOLOGY	% ONE LESION	% GTR	CAVITY (cc)	MARGIN (mm)	DOSE (Gy)/FX	% LC	% DBF	OS (MONTHS)	% NECROSIS	% WBRT SALVAGE
Sunnybrook (Al-Omair et al., 2013)	20	70 (41–90)	100% ≥70	50 L / 15 B	NR	85	23.6ᵃ (3.1–42.1)	2	25–37.5/5	79ᵇ	NR	23.6	NR	NR
NYU (Connolly et al., 2013)	33	56.6 (27–82)	90 (70–90)	39 L / 27 B / 24 M	100	NR	3.3ᶜ (1.7–5.7) preop	10	40.05/15	85	39.3ᵇ	65.6%ᵇ	0	14.3
Emory (Eaton et al., 2013)	22ᵈ	58 (23–81)	88% ≥60	24 L / 29 B / 21 M	NR	50	24.5ᵃ (0.8–122.0)	2 (0–10)	21/3ᶜ (67%) 24/4ᵉ (14%) 30/5ᶜ (12%)	61ᵇ	71	Not reached	9.5	40
Umeå (Lindvall et al., 2009)	47	64.9ᶠ	87% >70	45 L / 21 B / 2 M	100	100ᵍ	6 (0.6–26) 74% <10 cc	NR	35–40/5	84	19	5	2.1	NR
Sapienza (Minniti et al., 2013)	101	57	80 (60–100)	22.8 L / 18.8 B / 27.8 M	100	100	29.5ᵃ (18.5–52.7)	3	27/3	92	53	17	9	24
Hannover (Steinmann et al., 2012)	33	58 (33–73)	100% ≥70	42 L / 27 B / 9 M	100	75	22.6ᵃ (4.9–93.6)	4	40/4 (67%) 35/7 (21%) 30/6 (12%)	73	47	20	NR	39
Beth Isreal Deaconess (Wang et al., 2012)	37	73% <65	97% ≥70	27 L / 24 B / 32 M	24	NR	28.8ᶠ (11.1–81.0)	2–3	24/3 fx	80ᵇ	20	5.5	2.9	14.3

Note: Data presented as median (range) unless otherwise indicated.
ᵃ Planning target volume.
ᵇ At 1 year.
ᶜ Size in cm.
ᵈ Patients with resected brain metastases (total 44 patients in study).
ᵉ Additional 1–1.5 Gy simultaneous integrated boost for subtotally resected lesions.
ᶠ Mean (range).
ᵍ Based on neurosurgeon's report.

Abbreviations: KPS, Karnofsky performance status; Histology L, non–small cell lung cancer; B, breast cancer; M, melanoma; GTR, gross total resection; cc, cubic cm; preop, preoperative; fx, fraction; LC, local control; DBF, distant brain failure; OS, overall survival; LMD, leptomeningeal disease; WBRT, whole-brain radiotherapy; NR, not reported.

Table 14.4 Summary of studies using both postoperative single- and multiple-fraction stereotactic radiotherapy

STUDY	NO.	AGE (YEARS)	KPS	% HISTOLOGY	% ONE LESION	% GTR	CAVITY (cc)	MARGIN (mm)	DOSE (Gy)/FX	% LC	% DBF	OS (MONTHS)	% NECROSIS	% LMD	% WBRT SALVAGE
Stanford (Choi et al., 2012b)	112	61 (18–86)	92% ≥70	43 L 16 B 16 M	63	90	8.5 (0.08–66.8)	0 (48%) 2 (52%)	12–30/1–5 86% 1 fx, 76% 3 fx	89.2	46[a]	17 (2–114)	3.5	NR	28
City of Hope (Do et al., 2009)	30	61.5 (40–93)	96.7% ≥70	47 L 20 B 20 M	NR	NR	NR	1[b] 2–3[c]	15–18/1 22–27.5/4–6	82	63	51%[a]	6.6	NR	47
Dana Farber (Kelly et al., 2012)	17	61.8 (38–81)	80 (70–100)	35 L 35 M	NR	94.4	3.49 (0.53–10.8)	0	18 (15–18)/1 (82%) 25/5 (12%) 30/10 (6%)	89	35	Not reached	NR	NR	41
U Pitt (Ling et al., 2015)	99	64 (39–81)	80 (60–100)	40 L 18 B 17 M	61	81	12.9[d] (0.6–51.1)	0–1[c]	15–21/1 (26%) 20–24/2 18–27/3 (56%) 24/4 20–28/5	72[a]	36[a]	12.7	9[g]	6	50
Emory (Patel et al., 2014)	96	56 (20–83)	96% ≥70	47 L	71	74	7.19 (0.90–35.70)	1	21 Gy ≤ 2 cm 18 Gy 2.1–3 cm 15 Gy 3.1–4 cm 3–5 fx >4 cm	83[a]	50	12.7	27	31[f]	14
Emory (Prabhu et al., 2012)	62	55 (20–75)	85% ≥70	41 L 11 B 23 M	71	81	8.5 (0.7–57)	≥1 (95%)	18 (15–24)/1 (86%) 3–4 fx (14%)	78[a]	49[a]	13.4 (9.3–17.5)	8	NR	26[a]
U Pitt (Rwigema et al., 2011)	77	63 (39–83)	80 (60–100)	43 L 14 B 12 M	85.7	NR	7.6 (1.1–59)	1	18 (12–27)/1–3	76.1[a]	53.3[a]	14.5 (1.6–51.4)	2.6	NR	26

Note: Data presented as median (range) unless otherwise indicated.

[a] At 1 year.
[b] Frame-based, preoperative lesion <3 cm.
[c] Mask-based, preoperative lesion ≥3 cm.
[d] Planning target volume.
[e] Physician preference.
[f] Compared with 13% incidence LMD in patients receiving WBRT.
[g] "Radiation injury".

Abbreviations: KPS, Karnofsky performance status; Histology L, non–small cell lung cancer; B, breast cancer; M, melanoma; GTR, gross total resection; cc, cubic cm; fx, fraction; LC, local control; DBF, distant brain failure; OS, overall survival; LMD, leptomeningeal disease; WBRT, whole-brain radiotherapy; NR, not reported.

in these studies had solitary brain metastases, and the extent of resection was considered gross total in approximately 70% of cases or more, with exception of one of the multifractionation studies where only half of patients had GTR (Eaton et al., 2013). While 2/3 of the subjects in the MSKCC trial were reported to have had control of their primary tumor, this percentage ranged from ~20% to 70% in the retrospective series, often reported as recursive partitioning analysis (RPA) class I (Gaspar et al., 1997, 2000). The percentage of patients with other extracranial disease was not reported for the majority of these analyses, but when reported, ranged from 21% to 65%.

The majority of patients in the retrospective series received their adjuvant treatment approximately 4 weeks from time of surgery. The surgical cavity size ranged from 3.5 to 17.5 cc in the studies that included singe-fraction SRS (Tables 14.2 and 14.4). The median volume treated in the MSKCC series would approximately equal 11.5 cc, converting the maximum target diameter to spherical volume. For the series that had only fractionated treatments, the target often included the setup margin and was reported as planning target volume. Consequently, the target size in these reports often exceeded 20 cc (Table 14.3). The margin added to the surgical cavity, identified by postcontrast MRI in the majority of the retrospective analyses, ranged from 0 to 3 mm for the investigations that included single-fraction SRS, up to 10 mm in the multifraction series.

There was considerable variability in technique and dose regimens utilized in the various retrospective series. The majority (10/13) of single-fraction-only studies were conducted using GK radiosurgery, whereas the analyses that included any patients treated with multiple fractions all utilized LINAC-based approaches, including CyberKnife and helical TomoTherapy. In the series that included only one-fraction adjuvant treatments (Table 14.2), the median marginal dose was 15–18 Gy, but the dose range was wider compared to the MSKCC trial (8–30 vs 15–22 Gy, respectively). The total dose delivered in the hypofractionated regimens was 21–40 Gy administered using various regimens of 3–15 fractions (Table 14.3). For the studies that combined single- and multiple-fraction courses, the median marginal dose was 10–30 Gy delivered in 1–5 fractions (Table 14.4). While the debate continues as to the applicability of the commonly accepted models to determine biologic equivalent dose of hypofractionated and single-fraction radiotherapy courses to standard 1.8–2 Gy per fraction treatments delivered over several weeks, Eaton et al. (2013) estimated that the most common hypofractionated regimens used would have comparable tumor control as 17.4–22.8 Gy single-fraction SRS.

Despite the lack of uniformity in approach to focal adjuvant treatment of resected brain metastases, the outcomes are remarkably similar across the spectrum of treatment employed. The mean LC rate for single-fraction cavity SRS studies was 83% (range, 73%–100%), 79% (range, 61%–92%) for multifraction IG-HSRT, and 81% (range, 72%–89%) for the cohorts that included both single and multiple fractions (Tables 14.2 through 14.4). Restricting the analysis to only that subset of studies that reported LC at 1 year, the mean result was 80% for the SRS group (n = 3; Jensen et al., 2011; Hartford et al., 2013; Iorio-Morin et al., 2014), 73% for the IG-HSRT group (n = 3; Wang et al., 2012; Al-Omair et al., 2013; Eaton et al., 2013), and 77% for the combination group (n = 4; Rwigema et al., 2011; Prabhu et al., 2012; Patel et al., 2014; Ling et al., 2015). While the 1 year LC appears almost 10% lower in the three multifraction studies than was reported for the entire MSKCC study population, the median tumor diameter in these three retrospective was >3 cm, a factor that was identified in the MSKCC trial as associated with decreased LC and worse prognosis (Brennan et al., 2014). The mean DBF rate for the retrospective single-fraction, multifraction, and combination cohorts was 47%, 45%, and 48%, respectively, which is comparable to 44% DBF reported at 1 year for the subjects in the MSKCC trial. The average median OS time was 14.9 months based on 11 of 13 retrospective studies of the one-fraction SRS patients (Table 14.2). This value was 14.2 months based on five of seven studies in the multifraction IG-HSRT, excluding one study where the OS at 1 year was 73%, and the median OS was not yet reached (Eaton et al., 2013; Table 14.3). Similarly, the average median OS was 14.4 months for subjects in four of seven of the combined single- and multifraction reports, and one report excluded with 93% 1 year OS and median OS not yet reached (Kelly et al., 2012; Table 14.4). These results are comparable to the median OS of 14.7 months reported in the MSKCC trial (Brennan et al., 2014) and superior to the <12 month median OS of all subjects in the prospective studies of adjuvant WBRT by Patchell et al. (1998) and Kocher et al. (2011). In those retrospective studies where salvage WBRT was reported, less than half of patients received this treatment (Tables 14.2 through 14.4), which compares favorably to the 65% of subjects in the MSKCC trial receiving this treatment.

As always, when comparing the results of any retrospective analysis to those of prospective clinical trials, the issue of selection bias in the retrospective cohorts must be taken into consideration. Nevertheless, it is encouraging that the outcomes of the many retrospective studies that employed various techniques of postoperative partial brain radiotherapy following resection of intracranial metastases are not inferior to the prospective trials investigating either the adjuvant focal brain radiation or WBRT approach. Given concerns of adverse effects of WBRT on neurocognitive function and functional independence, the observation that, where reported in these retrospective analyses, over 50% of subjects treated with SRS or IG-HSRT to the surgical cavity of resected brain metastases were spared this treatment.

14.5 TOXICITIES

Postoperative radiotherapy delivered to the resection cavity of a metastatic brain lesion has very little toxicity. The most commonly cited adverse event following this therapy is brain radionecrosis. In the only prospective phase II trial from MSKCC of single-fraction SRS following metastatectomy, radiation necrosis was reported in 17.5% of subjects (Brennan et al., 2014). The incidence of necrosis following single-fraction SRS in retrospective studies ranged from 0% to 8% (Table 14.2) and from approximately 3% to 30% in the studies that included multifraction IG-HSRT (Tables 14.3 and 14.4). While median volume of normal brain receiving 24 Gy was identified as the most significant factor associated with necrosis in one study of multifraction IG-HSRT (Minniti et al., 2013), most of the studies identifying necrosis following adjuvant focal tumor bed radiotherapy failed to identify specific risk factors for this treatment-associated injury (see, e.g., Choi et al., 2012a; Eaton et al., 2013). When reported, necrosis was most often managed by steroid administration and/or surgical resection (Jensen et al., 2011; Choi et al., 2012a; Robbins et al., 2012; Wang et al., 2012; Eaton et al., 2013; Gans et al., 2013; Patel et al., 2014; Ling et al., 2015).

Another adverse outcome often reported in analyses of postoperative SRS for resected brain metastases is the development of leptomeningeal disease (LMD). None of the subjects in the MSKCC phase II trial were found to have LMD (Brennan et al., 2014). However, LMD was noted in five of the retrospective single-fraction SRS studies, occurring at a mean incidence of 13% (range, 7.5%–24%, Table 14.2) and in 6%–31% of subjects in two of the seven combination single-/multiple-fraction reports (Table 14.4). In the one study with the highest occurrence of LMD following postoperative cavity radiation, this incidence was more than halved in subjects treated at the same institution with adjuvant WBRT (Patel et al., 2014). Although not universal among all of these investigations, the use of adjuvant SRS alone (Patel et al., 2014; Hsieh et al., 2015), breast cancer histology (Jensen et al., 2011; Atalar et al., 2013b; Ojerholm et al., 2014b), and infratentorial tumor location (Iwai et al., 2008; Jensen et al., 2011; Ojerholm et al., 2014b) have been identified as potential risk factors for LMD. Figure 14.1 demonstrates a case of LMD 5 months following postoperative SRS in a patient with breast cancer.

14.6 CONTROVERSIES

Given the current limited prospective data to guide clinicians as to the utility and optimal implementation of postoperative IG-HSRT in the management of patients with brain metastases, there remain many unanswered questions and topics for debate. A few of the outstanding controversial issues related to this adjuvant therapy are highlighted below.

14.6.1 PATIENT-RELATED ISSUES

As with any clinical therapeutic decision, the risk versus benefit of treatment must be carefully measured and weighed against anticipated patient survival, particularly in this setting of metastatic cancer to the brain. Thus, proper patient selection is critical. Extrapolating from the RPA (Gaspar et al., 1997, 2000) and the more recent graded prognostic assessment (GPA) index (Sperduto et al., 2008) that have been developed to

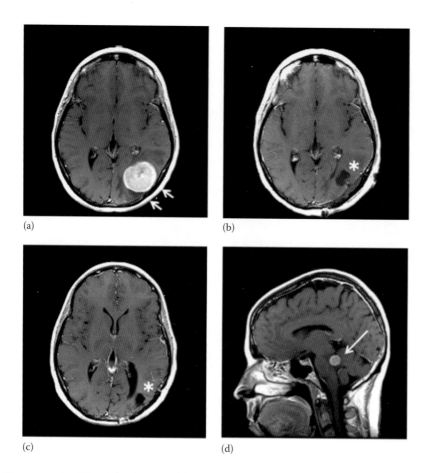

Figure 14.1 Representative T1-weighted postgadolinium MRI images of 38-year-old female with breast cancer. (a) Axial image demonstrating 4.3 cm left occipital metastasis abutting the dura (short arrows). (b) Status post gross total resection followed by 16 Gy single-fraction stereotactic radiosurgery to cavity + 1 mm margin on protocol (*cavity). (c) Five months after stereotactic radiosurgery, there is no evidence of local recurrence (*cavity), but (d) sagittal image shows leptomeningeal disease with metastasis in fourth ventricle (long arrow).

identify those patients with brain metastases likely to have the best survival, younger patients (<50–65 years) with excellent performance status (≥70), limited number of brain lesions (1–3), and controlled primary site of disease with no or limited extracranial metastases would be the ones anticipated to benefit most from aggressive therapy including surgical resection and postoperative SRS or IG-HSRT. While the sole prospective, and majority of the retrospective, study highlighted above included subjects with more favorable RPA/GPA scores, the precise optimal patient profile for adjuvant cavity radiotherapy is still unknown.

Related to the prognostic factor of number of brain metastases, another undefined patient-specific issue is, what is the maximal number of lesions (resected and not) that can be treated with postoperative SRS/IG-HSRT? Most patients in the analyses above had only one resected brain metastasis, although several studies included subjects with more than one resection cavity and/or up to a total of 10 intracranial lesions treated using the focal adjuvant radiation treatment (Jensen et al., 2011; Brennan et al., 2014).

14.6.2 SURGICAL ISSUES

The majority of lesions treated with postoperative SRS/IG-HSRT in the reported studies were considered to be GTR. How extent of resection was established in the various retrospective analyses was often not specified, and in one study, this determination was based on self-reporting by the neurosurgeon (Lindvall et al., 2009). Uniform criteria to measure completeness of resection are as yet not defined but likely necessary in the determination of appropriateness of adjuvant radiotherapy to the surgical cavity in these cases.

While the dose to the completely resected tumor bed is still indeterminate (see Section 14.6.4, below), the question of how to clinically manage subtotally resected (STR) brain metastases is another matter for debate. In two of the retrospective studies, a higher radiation dose was delivered to STR lesions than to the GTR ones (Robbins et al., 2012; Eaton et al., 2013). What the dose should be to residual tumor in the resection cavity and benefit of dose escalation in that circumstance are presently unknown.

Another ill-defined area likely to impact outcomes of patients treated with this approach is the skill of the neurosurgeon, as mentioned in the review of postoperative tumor bed radiosurgery by Roberge and Souhami (2010). Given anticipated wide variability on this front and no easily applied algorithm to normalize surgical outcomes, establishing a uniform system to report extent of resection is again underscored.

14.6.3 TUMOR-SPECIFIC ISSUES

While it is known that a wide variety of tumor histologies are able to metastasize to the brain, not all are amenable to resection and/or focal treatment options given predilection for microscopic dissemination (e.g., small cell lung cancer; Ojerholm et al., 2014a). Other tumor types are known to be resistant to standard fractionated radiotherapy and, therefore, may benefit more from ablative hypofractionated radiation treatment (e.g., melanoma or renal cell carcinoma; Chang et al., 2005). The most common diagnoses of patients in the prospective and retrospective studies of cavity SRS/IG-HSRT summarized above include NSCLC, breast cancer, and melanoma. Whether the same radiation dose delivered adjuvantly to the tumor bed is equally efficacious regardless of histology is not known.

Also unknown is the maximal lesion size that should be considered amenable to postoperative radiotherapy. When reported in the studies above, the median tumor size ranged from 2.7 to 3.9 cm, with largest tumor diameter measuring 6.9 cm in the analysis by Ojerholm et al. (2014b). The maximum tumor diameter treated in the dose escalation trial of SRS for intact, previously irradiated brain tumors (either primary or metastatic) was 4 cm (Shaw et al., 1996, 2000). Since the dose-limiting factor is the amount of "normal" brain receiving radiation, it can be reasoned that the dose-volume constraints would still apply in the postoperative setting, but without ample prospective data, this statement must be considered purely conjecture.

Another as-yet undefined matter related, in part, to the target size is the optimal timing of cavity SRS/IG-HRST following resection. Several investigators have reported on the dynamics of the surgical cavity volume over time. Jarvis et al. (Hartford et al., 2013) analyzed the volume change of 43 resection cavities in 41 patients from first postoperative day T1-weighted postcontrast MRI imaging until similar imaging prior to planned SRS over an average of 24 days (range, 2–104 days). The majority of tumor beds remained stable, but one-quarter collapsed and another third increased by >2 cc. In a separate analysis of 63 subjects with 68 resection cavities, Atalar et al. (2013a) observed no significant difference in immediate postoperative target volume within 3 days of surgery and that measured at time of SRS treatment planning up to 33 days following craniotomy. In this series, the largest tumors (>4.2 cc and diameter of 2 cm) were associated with the greatest reduction in tumor bed volume (median change −35%, $p < 0.001$), whereas those smaller than 4.2 cc had larger cavity volumes (median change +46%, $p = 0.001$). Figure 14.2 demonstrates a case of cavity "shrinkage" 1 month following resection of a brain metastasis. Neither of these studies reported on potential factors contributing to cavity shrinkage, which is an important consideration for patients with initial postoperative tumor bed volumes deemed too large for focal radiotherapy who may be found eligible for cavity SRS/IG-HRST on delayed repeat imaging. In a separate analysis of 37 patients with 39 resection cavities, Ahmed et al. (2014) identified presence of at least 15 mm vasogenic edema surrounding the tumor bed immediately following surgery to be predictive of at least a 10% reduction of cavity size when imaged within 30 days after resection. Besides cavity size dynamics, other issues surrounding the timing of adjuvant SRS/IG-HRST following surgery include the competing risk of adequate wound healing and that of tumor recurrence. Historically, postoperative radiotherapy is typically delayed for a period of 10–14 days to allow wound repair, with treatment to be

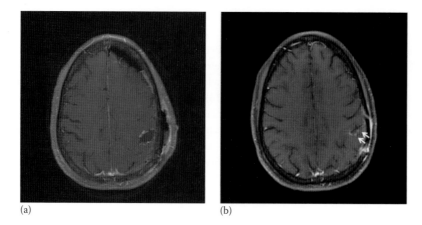

(a)   (b)

Figure 14.2 Representative T1-weighted postgadolinium MRI images of 48-year-old female with breast cancer. Axial image at (a) 24 hours and (b) 1 month following gross total resection of a left parietal metastasis. Note collapse of the cavity (short arrows) at the later postoperative time point.

initiated ideally within 4–6 weeks of surgery (see, e.g., Patchell et al., 1998; Kocher et al., 2011). The issue of tumor recurrence and timing of SRS following surgery was commented upon in the work by Jarvis et al. (2012), since 2 of the 13 subjects found to have increase in tumor bed size prior to SRS had clear evidence of disease regrowth and another two potential recurrence by 19 days (range, 4–76 days) after initial postoperative MRI. The current open Alliance NCCTG N107C cooperative group phase III trial comparing postsurgical SRS to WBRT for resected brain metastases has defined the time to adjuvant radiation treatment delivery to be less than 21 days from date of study registration, which must be within 28 days of postoperative brain imaging.

14.6.4 RADIOTHERAPY ISSUES

The dose regimens utilized in the prospective and many of the retrospective studies of cavity SRS/ IG-HSRT were based on RTOG 90-05 dose escalation trial of SRS for intact, previously irradiated brain tumors (either primary or metastatic) (Shaw et al., 1996, 2000). There are two distinctions that must be highlighted between SRS in the postoperative setting and the treatment used in RTOG 90-05: (1) altered blood supply leading to hypoxic area along the resection cavity and (2) radiation-naïve brain in patients receiving the cavity SRS/IG-HSRT treatment. Hypoxic tissue is well recognized as resistant to standard fractionated doses of radiation, and a recent *in silico* analysis of radiosurgery for brain metastases further suggests benefit of hypofractionation for large tumors as a means to overcome hypoxia (Toma-Dasu et al., 2014). Following resection of a brain metastasis, the blood vessels "feeding" the tumor are severed, thus rendering the potential residual microscopic malignant cells hypoxic. The optimal adjuvant radiation treatment approach in this setting, namely, single-fraction versus hypofractionated radiotherapy, is unknown, as underscored by the *in silico* study of unresected metastases.

The other factor that distinguishes radiation dose selection in the postoperative state from the dose-finding RTOG 90-05 study is that in the former case, the tumor has been removed and the target is a rind of "normal tissue," whereas in the latter case, the tumor was intact and previously irradiated and hence, arguably, radiation resistant. Because RTOG 90-05 defined the maximum safe tolerated radiosurgical dose to be delivered to the target (based on maximum diameter) in the setting of prior brain radiation, use of these established dose-volume criteria could be considered conservative for unirradiated brain tissue. Use of the RTOG 90-05 dose regimen has been criticized, for this reason, since LC following cavity SRS/IG-HSRT has been found to be only approximately 80% in the studies reporting outcomes using this technique, and since >90% of failures were identified as occurring within the prescription isodose line in the analysis of patients treated with postoperative SRS at Emory University by Prabhu et al. (2012) (same patient population updated in Patel et al. 2014) in Table 14.4.

Another area of great contention is the target definition for radiotherapy following resection of brain metastases. In the studies of adjuvant SRS/IG-HSRT tabulated above, the radiation target included a margin on the surgical cavity of 0–3 mm for single-fraction treatments and of up to 10 mm for the multifraction regimens. The addition of a 2 mm margin obviates the benefit of postoperative cavity shrinkage by increasing the target volume by 2%–200%, as was demonstrated by Atalar et al. (2013a). The need for, and extent of, margin will also vary depending upon the immobilization (frame based or frameless) and technique (GK or LINAC-based, use of intrafractional image guidance) employed to deliver the adjuvant radiotherapy. However, since the incidence of marginal failures is low (Prabhu et al., 2012; Eaton et al., 2013), the extent of marginal expansion of the cavity and consequent treatment of more "normal" brain must be carefully considered. Other debated issues related to target definition for cavity SRS include the need to cover the surgical track (yes: Brennan et al., 2014; no: Kelly et al., 2012; Minniti et al., 2013; Patel et al., 2014), and the optimal timing and sequence of imaging used to delineate the target (discussed in Section 14.6.3, above).

14.6.5 OTHER ISSUES

In addition to the myriad unresolved questions regarding optimal patient, target, dose, or technique of radiotherapy to be delivered adjuvantly following resection of solid brain metastases, there remain other outstanding issues as to the utility of cavity SRS/IG-HSRT. One such issue is the timing of cavity radiotherapy and systemic therapy. Despite being closed early due to poor accrual, the phase III trial by the RTOG (RTOG 0320) investigating potential OS benefit of adding temozolomide or erlotinib to WBRT and SRS for subjects with NSCLC and 1–3 brain metastases revealed statistically higher incidence of grade 3–5 toxicity in the concurrent systemic therapy arms (Sperduto et al., 2013). While the use of single-fraction or short-course hypofractionated radiotherapy has often been advocated to reduce delays in systemic chemo-, biologic, or immunotherapy for patients with metastatic disease, there are limited prospective data on the optimal interval from this type of radiation delivery (either before or after planned systemic therapy) in the setting of unresected brain metastases, and no data in the setting of postoperative SRS/IG-HSRT with respect to potential exacerbation of toxicity versus either therapy alone. Another advantage often cited to support the use of focal brain radiotherapy versus WBRT for subjects with metastatic brain disease is improved neurocognitive function and overall QOL (see, e.g., Chang et al., 2009). The effect of SRS to the postoperative resection cavity on neurocognition and patient functional status has not been reported in the published single prospective or multiple retrospective studies reviewed here. Finally, in this era of responsible medical care expenditures, the cost of SRS has been reported to exceed that of standard of care WBRT by as much as threefold in one of the retrospective studies of postoperative SRS for brain metastases (Kalani et al., 2010). The clinical and financial benefit of cavity SRS/IG-HRST vis-à-vis improved OS, QOL, and neurocognitive function, compared with WBRT, following resection of brain metastases remains to be determined in large, randomized controlled prospective studies.

14.7 FUTURE DIRECTIONS

By the time of publication of this book, the Alliance cooperative NCCTG N107C phase III trial is anticipated to have met accrual. The primary endpoints of this study are to evaluate OS and neurocognitive impairment at 6 months following randomization in subjects with 1–4 total brain metastases treated with surgery and cavity SRS versus WBRT. The secondary endpoints of this trial include assessment of QOL, time to CNS failure, duration of functional independence, long-term duration of neurocognitive status, posttreatment adverse events, LF at 6 months, time to LF, and differences in CNS failure patterns. Additional correlative analyses include investigation of radiation changes in limbic system, Apo E genotyping and determination if Apo E is a predictor of neurocognitive decline or protection, and determination if inflammatory markers or hormone, and growth factors are predictors of radiation-induced neurocognitive decline. The results of this study are eagerly anticipated, since they will help define the practice and utility of adjuvant SRS/IG-HSRT for patients with resected solid brain metastases.

CHECKLIST: KEY POINTS FOR CLINICAL PRACTICE

✓	ACTIVITY	SOME CONSIDERATIONS
	Patient selection	*Is the patient appropriate for postoperative HSRT?* • Due to limited prospective data, optimal patient characteristics unknown • Consider enrollment on clinical trial
	SRS vs HSRT	*Is the surgical cavity amenable to single-fraction SRS?* • Current dose-volume constraints based on extrapolation from studies of unresected brain metastases • Consider enrollment on clinical trial
	Simulation	*Immobilization* • Rigid frame versus thermoplastic mask *Imaging* • CT and/or MRI should be performed in treatment position • MRI should ideally include volumetric (1 mm) images, T1 sequence with IV contrast • Accurate image coregistration should be verified by clinician and used for contouring targets
	Treatment planning	*Contours* • Surgical cavity = clinical target volume • No consensus on margin, dose, fractionation, timing • Consider enrollment on clinical trial
	Treatment delivery	*Technique* • Both Gamma Knife and linear accelerator–based approaches acceptable • Patient localization system versus near- to real-time image guidance is mandatory *Concurrent systemic therapy* • Limited prospective data • Consider enrollment on clinical trial

REFERENCES

Ahmed S, Hamilton J, Colen R, Schellingerhout D, Vu T, Rao G, McAleer MF, Mahajan A (2014) Change in postsurgical cavity size within the first 30 days correlates with extent of surrounding edema: Consequences for postoperative radiosurgery. *J Comput Assist Tomogr* 38:457–460.

Al-Omair A, Soliman H, Xu W, Karotki A, Mainprize T, Phan N, Das S et al. (2013) Hypofractionated stereotactic radiotherapy in five daily fractions for post-operative surgical cavities in brain metastases patients with and without prior whole brain radiation. *Technol Cancer Res Treat* 12:493–499.

Atalar B, Choi CY, Harsh GR, 4th, Chang SD, Gibbs IC, Adler JR, Soltys SG (2013a) Cavity volume dynamics after resection of brain metastases and timing of postresection cavity stereotactic radiosurgery. *Neurosurgery* 72:180–185; discussion 185.

Atalar B, Modlin LA, Choi CY, Adler JR, Gibbs IC, Chang SD, Harsh GR, 4th et al. (2013b) Risk of leptomeningeal disease in patients treated with stereotactic radiosurgery targeting the postoperative resection cavity for brain metastases. *Int J Radiat Oncol Biol Phys* 87:713–718.

Brennan C, Yang TJ, Hilden P, Zhang Z, Chan K, Yamada Y, Chan TA et al. (2014) A phase 2 trial of stereotactic radiosurgery boost after surgical resection for brain metastases. *Int J Radiat Oncol Biol Phys* 88:130–136.

Chang EL, Selek U, Hassenbusch SJ, 3rd, Maor MH, Allen PK, Mahajan A, Sawaya R, Woo SY (2005) Outcome variation among "radioresistant" brain metastases treated with stereotactic radiosurgery. *Neurosurgery* 56:936–945.

Chang EL, Wefel JS, Hess KR, Allen PK, Lang FF, Kornguth DG, Arbuckle RB et al. (2009) Neurocognition in patients with brain metastases treated with radiosurgery or radiosurgery plus whole-brain irradiation: A randomised controlled trial. *Lancet Oncol* 10:1037–1044.

Choi CY, Chang SD, Gibbs IC, Adler JR, Harsh GR IV, Atalar B, Lieberson RE, Soltys SG (2012a) What is the optimal treatment of large brain metastases? An argument for a multidisciplinary approach. *Int J Radiat Oncol Biol Phys* 84:688–693.

Choi CY, Chang SD, Gibbs IC, Adler JR, Harsh GR IV, Lieberson RE, Soltys SG (2012b) Stereotactic radiosurgery of the postoperative resection cavity for brain metastases: Prospective evaluation of target margin on tumor control. *Int J Radiat Oncol Biol Phys* 84:336–342.

Connolly EP, Mathew M, Tam M, King JV, Kunnakkat SD, Parker EC, Golfinos JG, Gruber ML, Narayana A (2013) Involved field radiation therapy after surgical resection of solitary brain metastases–mature results. *Neuro Oncol* 15:589–594.

Do L, Pezner R, Radany E, Liu A, Staud C, Badie B (2009) Resection followed by stereotactic radiosurgery to resection cavity for intracranial metastases. *Int J Radiat Oncol Biol Phys* 73:486–491.

Eaton BR, Gebhardt B, Prabhu R, Shu HK, Curran WJ, Jr., Crocker I (2013) Hypofractionated radiosurgery for intact or resected brain metastases: Defining the optimal dose and fractionation. *Radiat Oncol* 8:135.

Gans JH, Raper DM, Shah AH, Bregy A, Heros D, Lally BE, Morcos JJ, Heros RC, Komotar RJ (2013) The role of radiosurgery to the tumor bed after resection of brain metastases. *Neurosurgery* 72:317–325; discussion 325–316.

Gaspar L, Scott C, Rotman M, Asbell S, Phillips T, Wasserman T, McKenna WG, Byhardt R (1997) Recursive partitioning analysis (RPA) of prognostic factors in three Radiation Therapy Oncology Group (RTOG) brain metastases trials. *Int J Radiat Oncol Biol Phys* 37:745–751.

Gaspar LE, Scott C, Murray K, Curran W (2000) Validation of the RTOG recursive partitioning analysis (RPA) classification for brain metastases. *Int J Radiat Oncol Biol Phys* 47:1001–1006.

Hartford AC, Paravati AJ, Spire WJ, Li Z, Jarvis LA, Fadul CE, Rhodes CH (2013) Postoperative stereotactic radiosurgery without whole-brain radiation therapy for brain metastases: Potential role of preoperative tumor size. *Int J Radiat Oncol Biol Phys* 85:650–655.

Hsieh J, Elson P, Otvos B, Rose J, Loftus C, Rahmathulla G, Angelov L, Barnett GH, Weil RJ, Vogelbaum MA (2015) Tumor progression in patients receiving adjuvant whole-brain radiotherapy vs localized radiotherapy after surgical resection of brain metastases. *Neurosurgery* 76:411–420.

Hwang SW, Abozed MM, Hale A, Eisenberg RL, Dvorak T, Yao K, Pfannl R et al. (2010) Adjuvant Gamma Knife radiosurgery following surgical resection of brain metastases: A 9-year retrospective cohort study. *J Neuro-Oncol* 98:77–82.

Iorio-Morin C, Masson-Cote L, Ezahr Y, Blanchard J, Ebacher A, Mathieu D (2014) Early Gamma Knife stereotactic radiosurgery to the tumor bed of resected brain metastasis for improved local control. *J Neurosurg* 121(Suppl):69–74.

Iwai Y, Yamanaka K, Yasui T (2008) Boost radiosurgery for treatment of brain metastases after surgical resections. *Surg Neurol* 69:181–186; discussion 186.

Jagannathan J, Yen CP, Ray DK, Schlesinger D, Oskouian RJ, Pouratian N, Shaffrey ME, Larner J, Sheehan JP (2009) Gamma Knife radiosurgery to the surgical cavity following resection of brain metastases. *J Neurosurg* 111:431–438.

Jarvis LA, Simmons NE, Bellerive M, Erkmen K, Eskey CJ, Gladstone DJ, Hug EB, Roberts DW, Hartford AC (2012) Tumor bed dynamics after surgical resection of brain metastases: Implications for postoperative radiosurgery. *Int J Radiat Oncol Biol Phys* 84:943–948.

Jensen CA, Chan MD, McCoy TP, Bourland JD, deGuzman AF, Ellis TL, Ekstrand KE et al. (2011) Cavity-directed radiosurgery as adjuvant therapy after resection of a brain metastasis. *J Neurosurg* 114:1585–1591.

Kalani MY, Filippidis AS, Kalani MA, Sanai N, Brachman D, McBride HL, Shetter AG, Smith KA (2010) Gamma Knife surgery combined with resection for treatment of a single brain metastasis: Preliminary results. *J Neurosurg* 113(Suppl):90–96.

Karlovits BJ, Quigley MR, Karlovits SM, Miller L, Johnson M, Gayou O, Fuhrer R (2009) Stereotactic radiosurgery boost to the resection bed for oligometastatic brain disease: Challenging the tradition of adjuvant whole-brain radiotherapy. *Neurosurg Focus* 27:E7.

Kelly PJ, Lin YB, Yu AY, Alexander BM, Hacker F, Marcus KJ, Weiss SE (2012) Stereotactic irradiation of the postoperative resection cavity for brain metastasis: A frameless linear accelerator-based case series and review of the technique. *Int J Radiat Oncol Biol Phys* 82:95–101.

Kocher M, Soffietti R, Abacioglu U, Villà S, Fauchon F, Baumert BG, Fariselli L et al. (2011) Adjuvant whole-brain radiotherapy versus observation after radiosurgery or surgical resection of one to three cerebral metastases: Results of the EORTC 22952-26001 study. *J Clin Oncol Off J Am Soc Clin Oncol* 29:134–141.

Limbrick DD, Jr., Lusis EA, Chicoine MR, Rich KM, Dacey RG, Dowling JL, Grubb RL et al. (2009) Combined surgical resection and stereotactic radiosurgery for treatment of cerebral metastases. *Surg Neurol* 71:280–288; disucssion 288–289.

Lindvall P, Bergstrom P, Lofroth PO, Tommy Bergenheim A (2009) A comparison between surgical resection in combination with WBRT or hypofractionated stereotactic irradiation in the treatment of solitary brain metastases. *Acta Neurochir* 151:1053–1059.

Ling DC, Vargo JA, Wegner RE, Flickinger JC, Burton SA, Engh J, Amankulor N, Quinn AE, Ozhasoglu C, Heron DE (2015) Postoperative stereotactic radiosurgery to the resection cavity for large brain metastases: Clinical outcomes, predictors of intracranial failure, and implications for optimal patient selection. *Neurosurgery* 76:150–157.

Luther N, Kondziolka D, Kano H, Mousavi SH, Engh JA, Niranjan A, Flickinger JC, Lunsford LD (2013) Predicting tumor control after resection bed radiosurgery of brain metastases. *Neurosurgery* 73:1001–1006; discussion 1006.

Mathieu D, Kondziolka D, Flickinger JC, Fortin D, Kenny B, Michaud K, Mongia S, Niranjan A, Lunsford LD (2008) Tumor bed radiosurgery after resection of cerebral metastases. *Neurosurgery* 62:817–824.

Minniti G, Esposito V, Clarke E, Scaringi C, Lanzetta G, Salvati M, Raco A, Bozzao A, Maurizi Enrici R (2013) Multidose stereotactic radiosurgery (9 Gy × 3) of the postoperative resection cavity for treatment of large brain metastases. *Int J Radiat Oncol Biol Phys* 86:623–629.

Ojerholm E, Alonso-Basanta M, Simone CB II (2014a) Stereotactic radiosurgery alone for small cell lung cancer: A neurocognitive benefit? *Radiat Oncol* 9:218.

Ojerholm E, Lee JY, Thawani JP, Miller D, O'Rourke DM, Dorsey JF, Geiger GA et al. (2014b) Stereotactic radiosurgery to the resection bed for intracranial metastases and risk of leptomeningeal carcinomatosis. *J Neurosurg* 121(Suppl):75–83.

Patchell RA, Tibbs PA, Regine WF, Dempsey RJ, Mohiuddin M, Kryscio RJ, Markesbery WR, Foon KA, Young B (1998) Postoperative radiotherapy in the treatment of single metastases to the brain: A randomized trial. *JAMA* 280:1485–1489.

Patel KR, Prabhu RS, Kandula S, Oliver DE, Kim S, Hadjipanayis C, Olson JJ et al. (2014) Intracranial control and radiographic changes with adjuvant radiation therapy for resected brain metastases: Whole brain radiotherapy versus stereotactic radiosurgery alone. *J Neurooncol* 120:657–663.

Prabhu R, Shu HK, Hadjipanayis C, Dhabaan A, Hall W, Raore B, Olson J, Curran W, Oyesiku N, Crocker I (2012) Current dosing paradigm for stereotactic radiosurgery alone after surgical resection of brain metastases needs to be optimized for improved local control. *Int J Radiat Oncol Biol Phys* 83:e61–e66.

Robbins JR, Ryu S, Kalkanis S, Cogan C, Rock J, Movsas B, Kim JH, Rosenblum M (2012) Radiosurgery to the surgical cavity as adjuvant therapy for resected brain metastasis. *Neurosurgery* 71:937–943.

Roberge D, Souhami L (2010) Tumor bed radiosurgery following resection of brain metastases: A review. *Technol Cancer Res Treat* 9:597–602.

Rwigema JC, Wegner RE, Mintz AH, Paravati AJ, Burton SA, Ozhasoglu C, Heron DE (2011) Stereotactic radiosurgery to the resection cavity of brain metastases: A retrospective analysis and literature review. *Stereot Funct Neuros* 89:329–337.

Shaw E, Scott C, Souhami L, Dinapoli R, Bahary JP, Kline R, Wharam M et al. (1996) Radiosurgery for the treatment of previously irradiated recurrent primary brain tumors and brain metastases: Initial report of radiation therapy oncology group protocol (90-05). *Int J Radiat Oncol Biol Phys* 34:647–654.

Shaw E, Scott C, Souhami L, Dinapoli R, Kline R, Loeffler J, Farnan N (2000) Single dose radiosurgical treatment of recurrent previously irradiated primary brain tumors and brain metastases: Final report of RTOG protocol 90-05. *Int J Radiat Oncol Biol Phys* 47:291–298.

Soffietti R, Kocher M, Abacioglu UM, Villa S, Fauchon F, Baumert BG, Fariselli L et al. (2013) A European Organisation for Research and Treatment of Cancer phase III trial of adjuvant whole-brain radiotherapy versus observation in patients with one to three brain metastases from solid tumors after surgical resection or radiosurgery: Quality-of-life results. *J Clin Oncol Off J Am Soc Clin Oncol* 31:65–72.

Sperduto PW, Berkey B, Gaspar LE, Mehta M, Curran W (2008) A new prognostic index and comparison to three other indices for patients with brain metastases: An analysis of 1,960 patients in the RTOG database. *Int J Radiat Oncol Biol Phys* 70:510–514.

Sperduto PW, Wang M, Robins HI, Schell MC, Werner-Wasik M, Komaki R, Souhami L et al. (2013) A phase 3 trial of whole brain radiation therapy and stereotactic radiosurgery alone versus WBRT and SRS with temozolomide or erlotinib for non-small cell lung cancer and 1 to 3 brain metastases: Radiation Therapy Oncology Group 0320. *Int J Radiat Oncol Biol Phys* 85:1312–1318.

Steinmann D, Maertens B, Janssen S, Werner M, Frühauf J, Nakamura M, Christiansen H, Bremer M (2012) Hypofractionated stereotactic radiotherapy (hfSRT) after tumour resection of a single brain metastasis: Report of a single-centre individualized treatment approach. *J Cancer Res Clin Oncol* 138:1523–1529.

Toma-Dasu I, Sandstrom H, Barsoum P, Dasu A (2014) To fractionate or not to fractionate? That is the question for the radiosurgery of hypoxic tumors. *J Neurosurg* 121(Suppl):110–115.

Wang CC, Floyd SR, Chang CH, Warnke PC, Chio CC, Kasper EM, Mahadevan A, Wong ET, Chen CC (2012) Cyberknife hypofractionated stereotactic radiosurgery (HSRS) of resection cavity after excision of large cerebral metastasis: Efficacy and safety of an 800 cGy × 3 daily fractions regimen. *J Neurooncol* 106:601–610.

15

Brain metastases image-guided hypofractionated radiation therapy: Rationale, approach, outcomes

John M. Boyle, Paul W. Sperduto, Steven J. Chmura,
Justus Adamson, John P. Kirkpatrick, and Joseph K. Salama

Contents

15.1 INTRODUCTION

Image-guided hypofractionated stereotactic radiation therapy (HSRT) is an emerging technique used primarily for the treatment of brain metastases. This technique employs noninvasive immobilization methods and modern imaging techniques at the time of treatment delivery for precise localization, allowing for the delivery of highly conformal radiation therapy. HSRT differs from commonly employed methods for delivering stereotactic radiosurgery (SRS) in the use of noninvasive immobilization and in the delivery of fractionated courses of radiation. In addition, HSRT aims to maximize the therapeutic ratio, permitting safe and efficacious treatment of large intracranial lesions or those in close proximity to critical normal tissues that would normally be unsuitable for single-fraction radiosurgery.

In this chapter, we briefly review the epidemiology of brain metastases to define the scope of the problem for which HSRT is well suited. Subsequently, we present the data and rationale for the use of SRS in this setting, followed by a discussion of the rationale for HSRT. Next, we present our institutional techniques for HSRT with a focus on clinical scenarios where HSRT may be beneficial. Finally, we will review the available data supporting the use of HSRT with a focus on treated metastasis control and normal tissue toxicity. The chapter closes with a brief discussion of controversies and future directions.

15.1.1 EPIDEMIOLOGY AND BACKGROUND

Brain metastases are a significant cause of morbidity and mortality among cancer patients. It is estimated that 20%–40% of cancer patients will develop brain metastases in the course of their disease, translating to an incidence of between 100,000 and 300,000 cases annually in the United States (Johnson and Young, 1996; Mehta et al., 2005). Evidence suggests that the incidence of brain metastases may be on the rise. This may be due to increased detection of clinically occult disease through staging magnetic resonance scans (MRI) (Sundermeyer et al., 2005). Alternatively, the increased incidence may be due to the increasing efficacy of systemic therapies unmasking the true incidence in a systemically controlled cancer population. For example, a number of retrospective series have a trend toward higher rates of brain metastases in women with breast cancer, in some cases associated with HER2 overexpression and treatment with trastuzumab (Crivellari et al., 2001; Slimane et al., 2004; Lin and Winer, 2007). In non–small cell lung cancer, the success of erlotinib for patients with epidermal growth factor receptor (EGFR) mutant tumors has led to a similar hypothesis (Patel et al., 2014).

A number of prognostic indices have been developed to predict survival of patients with newly diagnosed brain metastases. The Radiation Therapy Oncology Group's (RTOG) recursive partitioning analysis (RPA) was first published in 1997 and identified higher Karnofsky Performance Status (KPS), younger age, control of the primary tumor, and absence of extracranial metastases as prognostic factors for overall survival (OS) (Gaspar et al., 1997). This work was further developed, taking into account the number of brain metastases and control of the primary tumor (Sperduto et al., 2008, 2010, 2012). The resulting graded prognostic assessment (GPA) and the subsequent diagnosis-specific GPA provide a useful tool for predicting the prognosis of patients with brain metastases. These prognostic indices aid in the appropriate selection of patients where HSRT is an appropriate treatment strategy.

For oncologists treating patients with brain metastases, there is an increasing arsenal of therapeutic options including best supportive care, surgery, radiation therapy, and current clinical trials evaluating novel systemic therapies (NCT01622868, NCT02015117). Selecting the best option for the patient often involves weighing the potential toxicities and benefits of treatment, while minding the devastating implications of uncontrolled intracranial disease. Unfortunately, both aggressive metastasis-directed treatment and uncontrolled disease can cause a decline in neurocognition, performance status and even increase mortality (DeAngelis et al., 1989; Regine et al. 2001). Therefore, the optimal choice of therapy is best made on an individual level and tailored to the clinical scenario at hand. Thankfully, a growing body of level 1 evidence and consensus guidelines are aiding oncologists and patients in selecting an appropriate treatment strategy.

15.1.2 RANDOMIZED TRIALS AND RATIONALE FOR STEREOTACTIC RADIOSURGERY

Multiple randomized trials have been conducted, attempting to establish the appropriate roles of surgery and radiation therapy in management of brain metastases. Three early randomized controlled trials investigated the benefit of adding treated metastasis therapy in the form of surgery to whole-brain radiation therapy (WBRT) for patients with single-brain metastases (Patchell et al., 1990; Noordijk et al., 1994; Mintz et al., 1996). The first two of these trials found improvement in OS in the surgical arm (Patchell et al., 1990; Noordijk et al., 1994). The third trial by Mintz et al., while failing to establish a benefit for surgery, clarified the importance of systemic disease status. Indeed, it was found that a lack of control resulted in increased mortality (risk ratio 2:3; Mintz et al., 1996). The benefit of surgery for patients with controlled extracranial disease was also seen in the study by Noordijk et al. Median survival for patients in either arm with uncontrolled extracranial disease was 5 months. Conversely, survival was improved by 5 months for patients with controlled extracranial disease (median 12 vs 7 months). These studies established the benefit of aggressive metastasis-directed therapy with surgery in the setting of patients with good performance status and relatively stable extracranial disease.

Based in part on the results of the surgical trials, similar attempts have also been made to improve treated metastasis control by adding radiosurgery to WBRT. RTOG 9805 selected patients with one to three brain metastases (stratified a priori) and a KPS ≥ 70, and randomized to WBRT or WBRT followed by aggressive metastasis-directed therapy using radiosurgery (Andrews et al., 2004). The addition of radiosurgery to

WBRT failed to meet the primary endpoint of improved OS for the entire cohort, which was no different between arms. However on subset analysis, there was a survival benefit favoring SRS for patients with a single metastasis (4.9 vs 6.5 months). In addition, KPS was significantly improved for *all* patients receiving SRS, and steroid requirements were significantly reduced when compared with patients receiving WBRT alone. In summary, both surgical studies and subset analysis of the RTOG trials suggest that aggressive metastasis-directed therapy of a single-brain lesion improves survival. Furthermore, treatment of up to three metastases with radiosurgery following WBRT was safe and improved performance status.

The utility of WBRT following craniotomy for metastases was established by Patchell et al. (1998), investigating the role of WBRT following surgical resection of a single brain metastases. In patients undergoing surgery alone, there were substantially higher rates of both treated metastasis (46% vs 10%) and distant recurrence (37% vs 14%). Despite a dramatic reduction in the risk of in-brain recurrence with the addition of WBRT (70% vs 18%), this did not translate into an OS benefit. However, the percentage of deaths due to neurologic causes was reduced with the addition of WBRT (44% vs 14%), presumably due to a reduction in intracranial recurrence. In the modern MRI era, the European Organisation for Research and Treatment of Cancer (EORTC) conducted a similar study, addressing the benefit of immediate WBRT following surgical resection or radiosurgery. This trial randomized WBRT following intracranial metastasis-directed therapy for one to three metastases, allowing both complete surgical resection and SRS at the discretion of the treating surgeon/radiation oncologist (Kocher et al., 2011). The primary endpoint of the study, time to decline in WHO performance status >2, was no different with or without WBRT (median 9 vs 10 months). The secondary endpoint, OS, was not different (11 months). Similar to the study by Patchell et al., there was a significant reduction in intracranial progression with WBRT (78% vs 48%, $p < 0.02$), observed as a decrease in the rate of both treated metastasis and distant in-brain recurrence. Intriguingly, rates of 2-year treated metastasis progression were markedly lower with SRS as compared to surgery, either with WBRT (19% vs 37%) or without WBRT (31% vs 69%).

These studies demonstrate that the addition of WBRT immediately following ablative therapy with surgery or radiation results in decreased intracranial progression and neurocognitive death rates while failing to alter OS. The benefits of WBRT must be weighed against any toxicity of treatment. While WBRT is associated with mild acute symptoms, there is concern that long-term effects on cognitive and cerebellar function may be worse than with intracranial metastasis-directed therapy alone (Regine et al., 2001; Tallet et al., 2012). Thus, more contemporary studies have therefore sought to clarify the role WBRT in patients with brain metastases treated with SRS.

A Japanese study enrolled patients with good performance status (Eastern Cooperative Oncology Group [ECOG] ≥ 2) and up to four brain metastases and randomized to WBRT followed by SRS or SRS alone (Aoyama et al., 2006). Consistent with other trials, the primary endpoint of OS was not significantly different between treatment arms despite a reduction of in-brain recurrence with WBRT. The authors of the trial hypothesize that the efficacy of salvage therapy, which was more often used in patients treated initially with SRS alone, could explain the findings. A similar study randomized patents with up to three metastases to SRS followed by WBRT or SRS alone (Chang et al., 2009). The primary endpoint of the study was neurocognitive function, as measured by a standard battery of neurocognitive function tests. The study was stopped early by the institutions data and safety monitoring board when interim analysis suggested a high probability (96%) that the addition of WBRT resulted in a decline in learning and memory function (52%) at 4 months following treatment compared to those receiving SRS alone (24%). Interestingly, higher mortality was also seen in the WBRT arm (hazard ratio [HR] 2.47) suggesting that people dying are more likely to demonstrate reduced learning and memory. Furthermore, a recent pooled analysis was conducted of three trials comparing SRS with and without WBRT. It was found that for patients younger than 50, SRS alone led to improved survival compared to with the addition of WBRT (Sahgal et al., 2015). Together these data suggest that the omission of WBRT does not compromise survival and in fact may lead to a reduction in morbidity and potentially improved quality of life.

In summary, there are a number of treatment strategies for patients with brain metastases. Given the results of the above trials, SRS with or without WBRT has become widely adopted as a treatment modality for patients with a limited number of brain metastases as reflected by recently published guidelines (Tsao et al., 2012).

15.1.3 RATIONALE FOR HYPOFRACTIONATED STEREOTACTIC RADIOSURGERY

HSRT has been proposed as an alternative to single-fraction SRS. This technique uses a noninvasive stereotactic head frame system to deliver ablative doses of radiation to intracranial targets in 2–10 fractions with precision and conformity. There are several hypothetical advantages to this approach including (1) reduced toxicity of treatment, particularly for large metastases or those in eloquent regions of the brain, (2) improved control through delivery of total radiation doses with greater biological effectiveness, (3) increased patient comfort and compliance, and (4) use in radioresistant histologies, such as melanoma and renal cell carcinoma, in which WBRT is of unclear benefit and clear toxicity.

The initial trials of SRS employed doses of 15–24 Gy delivered in a single fraction to metastases of <4 cm in size. The doses selected in these studies were based on the results of the RTOG 90-05 (Shaw et al., 2000). Previously irradiated patients with recurrent brain tumors (primary or metastatic) measuring up to 4 cm were enrolled and underwent treatment with either Gamma Knife or linear accelerator–based SRS. The maximally tolerated dose based on clinically determined toxicity was deemed acceptable for tumors ≤20, 21–30, and 31–40 mm, with doses of 24, 18, and 15 Gy, respectively. The target volume included the contrast-enhancing volume on MRI or CT. The dose was prescribed to the 50%–90% isodose line with the entire target volume covered by the prescription isodose line without a margin. As a result of this trial, enrollment into the randomized trials of SRS was limited to tumors <4 cm. Subsequently, there is appropriate hesitation in treating larger tumors with single-fraction SRS, given that these doses were associated with clinically significant rates of grade 3 or higher chronic toxicity: 10% for metastases <2 cm treated to 24 Gy, 20% for metastases 21–30 mm treated with 18 Gy, and 14% for metastases 31–40 mm treated with 16 Gy. This initial trial demonstrated the maximum acceptable toxicity for a given size and dose. However, efficacy of treatment was not an evaluated endpoint.

There is a large body of work, primarily in benign lesions such as arteriovenous malformations (AVMs), investigating the dose–volumetric tolerance of normal brain tissue. A number of metrics have been found to be predictive of radiation injury, defined as neuroradiological changes and subsequent clinical side effects. For single-fraction SRS, parameters such as volume of tissue receiving 8 Gy (Flickinger et al., 1997), 10 Gy (Voges et al., 1996; Flickinger et al., 1997; Levegrun et al., 2004), or 12 Gy (Flickinger et al., 1997, 1998, 2000; Lawrence et al., 2010) have been evaluated. In a representative series, it was found that among 133 patients treated with SRS, 0% with a volume of brain receiving 10 Gy (V10 Gy) of <10 cc developed neuroradiological changes as opposed to a rate of 24% if V10 was ≥10 cc (Voges et al., 1996). Moreover, these radiographic changes were highly correlated with symptoms and neurological changes. These studies demonstrate a dose–volume relationship for single-fraction treatment in the absence of WBRT and with long-term follow-up.

Beyond dose–volume constraints, it has been found that the location of the treated lesions is predictive of radiation injury following SRS (Flickinger et al., 1992, 1998, 2000). In a series of 422 patients treated with SRS for AVM, of whom 85 experienced radiation injury, multivariate logistic regression analysis was used to identify correlations between lesion location and risk of injury (Flickinger et al., 2000). It was found that in addition to V12 Gy, location was predictive of radiation injury with the lowest risks for lesions in the cerebral and cerebellar hemispheres, followed by midbrain structures, and the highest risk in the brain stem. For example, it was found that a V12 Gy of 20 cc predicted an approximate 40% risk of radiation injury for a lesion in the basal ganglia as opposed to a risk of <5% for a lesion in the frontal lobe. These studies further refined the understanding of risk–benefit in patients undergoing SRS in the absence of WBRT.

There is a limited amount of data reporting the risk of brain stem toxicity with single-fraction SRS. Interpretation of these studies is complicated by the differing dose prescription and volumetric information reported. The largest study by Foote et al. (2001) analyzed outcomes among 149 patient treated with SRS for vestibular schwannomas using doses from 10 to 22.5 Gy. The reported outcomes were stratified by years of treatment as before 1994, when planning was largely CT based, and after 1994 when MRI was primarily used. For patients treated in the more contemporary cohort, the 2-year rates of facial and trigeminal neuropathies were 5% and 2%, respectively, as opposed to 29% and 7% in the older patient cohort. A multivariate analysis of risk factors found that a maximum dose (D_{max}) of

17.5 Gy and proximity of the tumor to the brain stem predicted for subsequent development of cranial nerve neuropathy. Available data suggest greater risks for higher V12 Gy, prescription doses >15 Gy, and volumes >4 cc (Spiegelmann et al., 2001; Kano et al., 2012). Based on this and other series, it seems that doses of 12.5–13 Gy is associated with a risk of <5% clinically observed toxicity.

The optic nerves and optic chiasm are critical structures that are often dose limiting for patient receiving single-fraction SRS. Most studies reporting on radiation-induced optic neuropathy report the maximum point dose. The majority of the published data support doses of <8 Gy as safe, but with increasing risk from 8 to 12 Gy (Leber et al., 1998; Stafford et al., 2003; Pollock et al., 2008). For doses >12 Gy, the risks are typically judged to be unacceptably high at >10% (Mayo et al., 2010).

When the limitations of dose, volume, and location are applied to metastatic lesions, the shortcomings of single-fraction SRS for all metastases are apparent. It has been found that large volume metastases, where dose is most limited by toxicity, SRS doses of <15–18 Gy are predictive of worse tumor control (Shiau et al., 1997; Mori et al., 1998; Vogelbaum et al., 2006; Yang et al., 2011). These limitations may also extend to lesions in critical structures of the brain.

A radiobiologic argument for HSRT, presented in work by Hall and Brenner (1993), highlights the benefits of a fractionated approach to SRS. It is recognized that many tumors contain a population of hypoxic and, therefore, radioresistant cells. Fractionated SRS takes advantage of the reoxygenation that occurs between treatments. Each dose of radiation will primarily kill the aerobic cell population. It has been documented that following each dose, the tumor will reestablish its original proportion of oxygenated cells, maximizing the effect of the subsequent dose. Additionally, fractionation takes advantage of the differing dose–response curves of early responding tissues (including tumors) and late responding tissues (such as brain). Radiobiological principles dictate that total dose, rather than fractional dose, most affects the amount of cell killing for early responding tissues. This is, to a degree, in opposition to late responding normal tissues that are more sensitive to changes in fractionation. Thus, more fractionated schedules will preferentially spare normal tissues and reduce late effects without compromising tumor control probability.

Employing the linear–quadratic model of cell kill and extrapolating from experience with low-dose rate brachytherapy for recurrent gliomas, Brenner et al. (1991) described alternative dose fractionations and their equivalent single-fraction dose. For example, alternative fractionation schemes equivalent to 18 Gy in a single fraction would be 27.4 Gy in 3 fractions or 32.6 Gy in 5 fractions. These doses have been employed in the setting of brain metastases, resulting in high rates of treated metastasis control and low rates of late toxicity (Manning et al., 2000). Further support for the safety of hypofractionated regimens comes from the use of HSRT for recurrent gliomas where 25 Gy in 5 fractions has been utilized (Landy et al., 1994; Cabrera et al., 2013). Concerns regarding the applicability of the linear–quadratic model to SRS and HSRT are discussed in Section 15.5.

15.2 RADIATION TECHNIQUE

The following section will detail the radiation technique employed at Duke University since 2008, having been employed in >400 patients. While other centers may employ slightly different hardware and software solutions, the basic technique will be similar for most centers using a linear accelerator to deliver HSRT.

15.2.1 SIMULATION

Unless contraindicated, all patients we plan to treat with HSRT at Duke are immobilized in a thermoplastic face mask that is either frameless or attaches to a rigid U-frame system (BrainLAB, Munich, Germany) as shown in Figure 15.1. Subsequently, all patients undergo CT imaging, and for those patients without contraindication to MRI, fine-cut (1 mm) contrast-enhanced spoiled gradient (3D SPGR) MR imaging is obtained. The MR and CT images are fused using iPlan treatment planning software (BrainLAB) and the target lesion contoured on the 3D SPGR images. For patients who are unable to undergo an MRI, a fine-cut CT image with contrast enhancement is obtained.

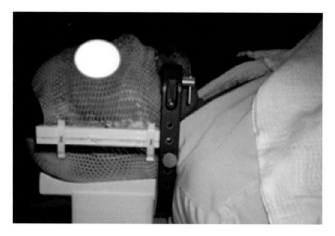

Figure 15.1 Example of frameless thermoplastic immobilization system.

15.2.2 CONTOURS

We define the gross tumor volume (GTV) as the contrast-enhancing lesion on the 3D SPGR MRI. A margin of 1 mm is applied to the GTV to generate a planning target volume (PTV), based on our prior analyses demonstrating equivalent treated metastasis control between 1 and 3 mm PTV expansions with a trend toward reduced rates of radionecrosis favoring the smaller expansion (Kirkpatrick et al., 2015). In the situation where only a fine-cut CT is available, we may utilize a slightly larger margin expansion (typically 2 mm), as it is more difficult to visualize the boundary between tumor and normal brain. Normal tissues contoured include the whole brain, optic nerves, optic chiasm, and brain stem delineated on fine-cut images.

15.2.3 TREATMENT PLANNING

At our institution, we employ a number of different treatment techniques, all utilizing linear accelerator SRS systems equipped with a high-definition multileaf collimator with 2.5 mm wide leaves in the proximal ±4 cm from the isocenter, and 5 mm wide leaves at >4 cm (Novalis Tx & Truebeam STX, Varian Medical Systems, Palo Alto, CA). Dynamic conformal arc therapy (DCAT) and volumetric modulated arc therapy (VMAT) both take advantage of rotational radiation therapy to deliver highly conformal treatment. DCAT is a technique where the aperture of the MLC leaves in the beams eye view dynamically conforms to the PTV with an added margin during gantry rotation. We typically utilize a 1–2 mm margin between the projection of the PTV and the MLC aperture in order to achieve a balance between dose conformity and falloff, and acceptable dose heterogeneity. A full description of the dosimetric trade-offs associated with choice of PTV to MLC margin is given by Zhao et al. (2014). VMAT consists of rotational arcs where the MLC leaf positions, dose, and dose rate of the linear accelerator are varied as a function of gantry angle; the combination of which is determined via an inverse optimization to achieve a desired dose distribution. DCAT is typically ideal for smaller or more uniformly shaped targets, while VMAT may be preferred for larger or irregularly shaped targets. Intensity-modulated radiation therapy (IMRT) employs nonrotational "static" beams where the MLC leaves are again modified via optimization to allow "dose painting." Similar to VMAT, IMRT is typically best suited for larger or nonuniform target volumes, especially when the concave surface of a target volume abuts a sensitive normal structure. In this instance, it may be desirable to employ static beams to tightly constrain the angles from which dose is delivered, allowing preferential sparing via a nonuniform dose distribution. At our institution, DCAT planning is performed using iPlan (BrainLAB), while VMAT and IMRT are planned using Eclipse (Varian).

Our standard dose fractionation for HSRT is 25 Gy delivered in 5 fractions on consecutive days, allowing breaks for weekends. Dose is prescribed to the 100% isodose line, allowing for up to a 110% hot spot. When using DCAT, hotspots of 120% may be acceptable. Our goal is for a minimum of 99% coverage of the PTV by the prescription dose. Conformity index is an additional metric of dose conformity that is calculated by dividing the volume of tissue receiving the prescription dose by the volume of the PTV. Our goal is to achieve a conformity index of <2.0, though this may not be feasible with DCAT when treating very small or highly irregularly shaped lesions.

One benefit of HSRT is that, in general, no hard dose constraints for normal tissues are necessary. We seek to minimize dose to the noninvolved normal tissues such as brain stem, optic nerves, and optic chiasm. In cases of prior treatment, whether surgical or radiation therapy, constraints may be needed as judged clinically applicable. We do not typically make dose reductions in the setting of prior WBRT, though we find it useful to calculate the biologically equivalent dose (BED) for the HSRT course and the previous treatment, particularly as regards the total BED versus the maximum tolerated BED observed in conventionally fractionated radiotherapy.

15.2.4 TREATMENT DELIVERY

The time interval between simulation and treatment delivery should optimally be less than 5–7 days. Immediately before each treatment, patients are first positioned on the table using the BrainLAB localization system, which consists of a target aperture position overlay (TAPO) box placed over the patient mask that delineates the isocenter position on each axis. All patients then undergo image guidance with cone-beam CT and kV orthogonal imaging while on the treatment table. Appropriate adjustment of the isocenter is made and imaging is repeated. This technique ensures a translation position deviation of less than 1 mm in any direction and <1.0 degrees of rotation (Ma et al., 2009). The final treatment position is verified in the room using the TAPO. Imaging is independently checked by both the planner and the physician who must agree on the accuracy of image guidance. For plans with high monitor units (MU), we often utilize either a 6X flattening filter free photon energy (Truebeam STX) or an SRS-specific 6X photon energy with a smaller flattening filter (Novalis Tx), allowing a dose rate of up to 1400 and 1000 MU per minute, respectively. Following the delivery of the final treatment, all patients are discharged on a short course of steroids (typically dexamethasone) with dose and duration guided by patient reported symptoms. For those patients already on steroids, no dose adjustments are made.

15.3 INDICATIONS

Single-fraction SRS remains a reasonable option for tumors that are small and in noneloquent areas of the brain. However, for larger tumors or tumors that are located within or in close proximity to critical areas of the brain, HSRT should be considered. Such critical areas include the optic chiasm, thalamus, basal ganglia, corpus callosum, or brain stem. In general, HSRT should be considered for lesions that are >4 cm in any axis or >3–4 cc in total volume. Included below are clinical scenarios where HSRT was deemed appropriate and employed.

Case 1 (Figure 15.2)
A 70-year-old woman presents with a 2-month history of progressive dizziness and disequilibrium. Her primary care physician ordered a CT of the head to rule out a stroke that revealed a hypodense region in the brain stem. A follow-up MRI revealed an enhancing lesion in the pons. A subsequent staging CT scan of the chest, abdomen, and pelvis showed nodularity in the right lung apex, but no clear lesion amenable to biopsy. Thus, a biopsy was performed of the lesion in the brain stem returning metastatic adenocarcinoma consistent with primary non–small cell lung cancer. Given a KPS of 90 and a low burden of disease, SRS was recommended. Given the location in the brain stem, HSRT was selected.

For this case of HSRT, a five dynamic conformal arc HSRT plan was selected. A dose of 2500 cGy was delivered in 5 fractions over 5 days. The dose was prescribed to the 100% isodose line yielding a maximum dose of 27.9 Gy and 99.2% coverage of the PTV with a conformity index of 1.4. A short course of dexamethasone was prescribed at discharge.

Case 2 (Figure 15.3)
A 65-year-old male with a history of T1bN0 squamous cell carcinoma of the left upper lobe underwent lobectomy 2 years ago. On routine follow-up, he complained of ataxia and multiple episodes of falling down. A brain MRI was ordered by his oncologist revealing a solitary brain metastasis in the medulla. The patient was restaged with positron emission tomography and CT scans (PET/CT) showing no evidence of

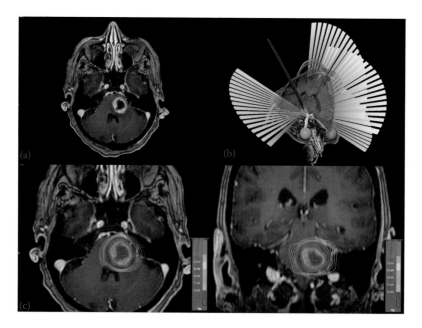

Figure 15.2 Case example of brain stem metastasis. Representative axial image from planning MRI with gross tumor volume (red contour) and planning target volume (PTV) (magenta contour) (a). Diagram demonstrating the orientation and degree of the five dynamic conformal arcs. Notice the avoidance of the optic nerves and optic chiasm (b). Axial image with isodose lines demonstrating steep dose gradient. Prescription isodose line (orange contour) and PTV (magenta volume) (c). Coronal image with isodose lines again demonstrating steep dose gradient and coverage of the PTV (d).

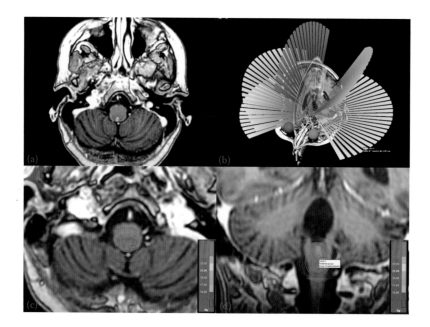

Figure 15.3 A second case example of brain stem metastasis. Representative axial image from planning MRI with gross tumor volume (red contour) and planning target volume (PTV) (magenta contour) (a). Diagram demonstrating the orientation and degree of rotation in the five dynamic conformal arcs. Notice the avoidance of the optic nerves and optic chiasm (b). Axial image with isodose lines demonstrating steep dose gradient. Prescription isodose line (orange contour) and PTV (magenta volume) (c). Coronal image with isodose lines again demonstrating steep dose gradient and good coverage of the PTV (d).

extracranial disease. Given his reasonable performance status and absence of extracranial disease, SRS was recommended. Due to the location in the brain stem, HSRT was selected.

For this case, a five dynamic conformal arc HSRT plan was selected. A dose of 2500 cGy in 5 fractions was delivered over 7 days. The dose was prescribed to the 100% isodose line yielding a maximum dose of 29.3 Gy with 99.5% coverage of the PTV, and a conformity index of 2.0. SRS was tolerated without difficulty, and the patient was discharged on a short course of dexamethasone.

Case 3 (Figure 15.4)

A 68-year-old male presented to his primary care physician complaints of memory loss and word-finding difficulties. He reported trouble recalling his children's and pets' names. An MRI of the brain was ordered revealing a 5.5 × 4 × 3.5 cm enhancing lesion involving the left splenium of the corpus callosum. He underwent staging PET/CT showing a mass in the right lower lobe and mediastinal adenopathy. He underwent bronchoscopy and biopsy of the lung mass revealing adenocarcinoma. He was staged clinically as T2N2M1 lung cancer. At the patient's request, the plan was to proceed with primary chemotherapy following treatment of the brain metastases. He was evaluated by neurosurgery and deemed inappropriate for surgical management. After discussing WBRT, SRS, and HSRT, the latter was selected. This approach was chosen to maximize the chance of treated metastasis control while minimizing the risk of treatment toxicity. This approach also prevented excessive delay prior to initiation of systemic therapy.

For this case, a three-arc VMAT HSRT plan was selected. A dose of 2500 cGy was delivered in 5 fractions on consecutive days. The prescription was made to the 100% isodose line yielding a maximum dose of 27.7 Gy and 99% coverage of the PTV and a conformity index of 99%. Therapy was tolerated well, and he was discharged to proceed with systemic therapy.

Case 4 (Figure 15.5)

A 42-year-old woman with a history of HER2+ pathologic T2N2a breast cancer was diagnosed 7 years ago. She was treated with mastectomy followed by adjuvant chemotherapy as well as adjuvant chest wall and nodal radiation therapy. Five years ago, she developed a chest wall recurrence for which she was

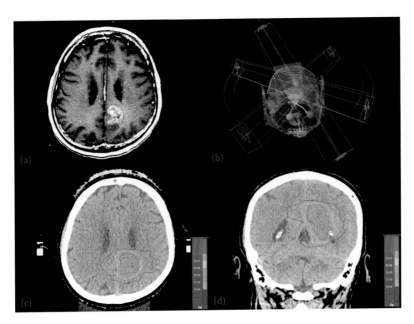

Figure 15.4 Case example of large metastasis. Representative axial image from planning MRI with gross tumor volume (GTV) (red contour) and planning target volume (PTV) (magenta contour) (a). Diagram demonstrating the orientation and degree of rotation of arcs in the five arc volumetric modulated arc therapy plan. These angles were selected to avoid the optic structures anteriorly and brain stem inferiorly (b). Axial image with isodose lines demonstrating steep dose gradient. Prescription isodose line (orange contour), GTV (red volume), and PTV (magenta contour) (c). Coronal image with isodose lines again demonstrating a steep dose gradient with good coverage of the GTV and PTV (d).

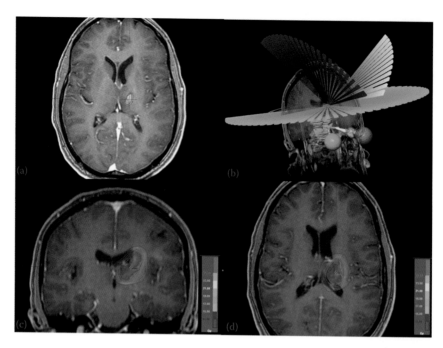

Figure 15.5 Case example of thalamic metastasis. Representative axial image from the planning MRI with gross tumor volume (red contour) and planning target volume (PTV) (magenta contour) (a). Diagram demonstrating the orientation and degree of the three arcs in the volumetric modulated arc therapy plan (b). Axial image with isodose lines demonstrating steep dose gradient. Prescription isodose line (orange contour) and PTV (magenta volume) (c). Coronal image with isodose lines again demonstrating steep dose gradient and good coverage of the PTV (d).

treated with reirradiation followed by trastuzumab chemotherapy. Four years ago, she developed new onset disequilibrium. An MRI of the brain was concerning for leptomeningeal disease. A spine MRI showed no spinal leptomeningeal disease. She underwent a course of WBRT to a dose of 30 Gy in 10 fractions. She presents now with new onset ataxia and occasional falls. A repeat brain MRI showed a solitary enhancing lesion in the left thalamus. She has no current evidence of extracranial disease. Treatment options including repeat WBRT, SRS, and HSRT were discussed with the patient. In an effort to maximize control while minimizing toxicity, HSRT was selected.

For this case, a three-arc VMAT HSRT plan was selected to deliver a dose of 2500 cGy over a 7-day period. A VMAT plan was selected given the abnormal shape of the target volume. In this instance, VMAT allows for high conformality despite the nonuniform volume. The prescription was made to the 100% isodose line yielding a maximum dose of 27.4 Gy with 99% coverage of the PTV and a conformity index of 1.2. Treatment was tolerated well, and she was discharged to continue systemic therapy.

Case 5 (Figure 15.6)

A 53-year-old male presented 1 year ago to the emergency room of an outside hospital with intense headaches associated with nausea and vomiting. A CT scan showed a hyperdense lesion in the left frontal lobe. An MRI showed an enhancing mass in the left frontal lobe measuring 4.5 × 4.0 × 3.5 cm. Staging studies showed a lesion in the left kidney. He underwent craniotomy and resection of a tumor with pathology returning metastatic renal cell carcinoma. He was then lost to follow-up and received no adjuvant therapy but, recently, established care at our institution. A repeat brain MRI showed an enhancing lesion posterior to the resection cavity involving the left caudate head and left lentiform nucleus. A separate enhancing lesion was seen at the lateral periphery of the resection cavity. He was evaluated by neurosurgery but deemed not to be a surgical candidate. Treatment options were discussed including WBRT and HSRT to the new enhancing lesions, as well as to the entire resection bed. Given the size of the projected treatment volume, as well as the left optic nerve, and the desire to maximize treated metastasis control, HSRT was selected.

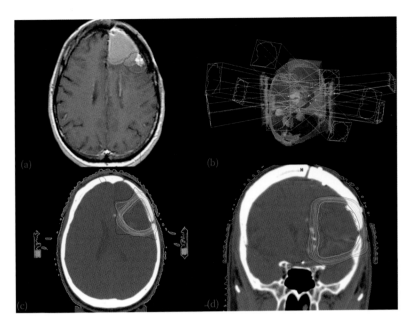

Figure 15.6 Case example of large target volume adjacent to critical normal tissues. Representative axial image from planning MRI with gross tumor volume (GTV) (red contour) and planning target volume (PTV) (magenta contour) (a). Diagram demonstrating the orientation and degree of rotation of arcs in the five field IMRT plan. These angles were selected to avoid the optic structures inferiorly (b). Axial image with isodose lines demonstrating steep dose gradient. Prescription isodose line (orange contour), GTV (red volume), and PTV (magenta contour) (c). Coronal image with isodose lines again demonstrating a steep dose gradient with good coverage of the GTV and PTV. Inferior to the PTV is the left optic nerve (yellow contour) (d).

For this case, a five-field IMRT HSRT plan was selected to deliver 2500 cGy in five consecutive daily treatments. The prescription was made to the 100% isodose line yielding a maximum dose of 27.5 Gy and 99% coverage of the PTV with a conformity index of 1.3. He received a single dose of dexamethasone and was discharged to the care of urologic oncology.

15.4 OUTCOMES DATA

There is a no level 1 evidence for HSRT with regard to toxicity or of treated metastasis control. Furthermore, there are limited data to compare the effectiveness of SRS as compared to HSRT, though available literature will be reviewed. However, there is a growing body of retrospective and prospective evidence to support the feasibility, safety, and efficacy of using HSRT (Manning et al., 2000; Aoyama et al., 2003; Ernst-Stecken et al., 2006; Kwon et al., 2009; Kim et al., 2011; Fokas et al., 2012; Minniti et al., 2014). These series will be discussed in more detail below, but universally show high rates of treated metastasis control and favorable rates of toxicity.

15.4.1 TREATED METASTASIS CONTROL

A review of the available literature demonstrates universally high rates of treated metastasis control with the use of HSRT. Reported rates are typically between 70% and 90% at 1 year (Table 15.1). Caution must be made when interpreting these rates and comparing them to outcomes in patients treated with SRS, where patients are more likely to have smaller metastases in less sensitive areas of the brain. When employing HSRT, a number of factors have been identified as predictive of treated metastasis control including tumor volume, tumor dimensions, use of concurrent chemotherapy, histology, and RPA class (Manning et al., 2000; Kwon et al., 2009; Fokas et al., 2012; Minniti et al., 2014). Interestingly, various analyses of treated metastasis control by dose and fractionation have yielded mixed results with some studies showing a dose–response (Aoyama et al., 2003), whereas others have found no such relationship (Fahrig et al., 2007; Kwon et al., 2009).

Table 15.1 Treated metastasis control in select series of hypofractionated stereotactic radiation therapy for brain metastases

SERIES	NO. OF PATIENTS/ METASTASES	PRIOR WBRT (DOSE/ FRACTIONS)	HSRT (DOSE/ FRACTIONS)	MEDIAN TUMOR VOLUME (cc)	LOCAL CONTROL (%)	TIME POINT	COMMENT
Manning et al. (2000)	32/57	30 Gy/10	18–36 Gy/3 Marginal dose	2.16	91	Crude Median 37 months follow-up	—
Aoyama et al. (2003)	87/159	None	35 Gy/4 Isocentric dose	3.3	81	1 year	Tumor volume >3 cc predictive of local failure
Ernst-Stecken et al. (2006)	51/72	None 40 Gy/20	35 Gy/5 30 Gy/5 Marginal dose	6	76	1 year	Phase II study, including large tumors or those in critical brain regions not amenable to SRS
Fahrig et al. (2007)	150/243	72/150 patients treated with 40 Gy/20	30–35 Gy/5 (n = 51) 40 Gy/10 (n = 36) 35 Gy/7 (n = 63) Marginal dose	Tumor volume not reported Median PTV 6.1	93	Crude	No difference in tumor response by dose fractionation

(Continued)

Table 15.1 (*Continued*) Treated metastasis control in select series of hypofractionated stereotactic radiation therapy for brain metastases

SERIES	NO. OF PATIENTS/ METASTASES	PRIOR WBRT (DOSE/ FRACTIONS)	HSRT (DOSE/ FRACTIONS)	MEDIAN TUMOR VOLUME (CC)	LOCAL CONTROL (%)	TIME POINT	COMMENT
Kwon et al. (2009)	27/52	45/52 metastases treated with 37.5 Gy/15	20–36 Gy/4–6; 25 Gy/5 most common	0.52	68.2	1 year	Smaller tumor dimension, smaller tumor volume, concurrent chemotherapy predictive of local control
Lindvall et al. (2009)	47/47	None	35–40 Gy/5; Marginal dose	6	84	Crude	—
Kim et al. (2011)	40/49	16/40 patients 30 Gy/10	30–42 Gy/6; Marginal dose; Median dose 36 Gy	Tumor volume not reported; Median PTV 5.0	69	1 year	—
Fokas et al. (2012)	122/not reported	None	35 Gy/7; 40 Gy/10	2.04; 5.93	75; 71	1 year	—
Minniti et al. (2014)	135/170	None	27 Gy/3 if ≤2 cn; 36 Gy/3 if >2 cm	10.1	72	1 year	Melanoma histology predictive of failure (HR 6.1)

An early series by Manning et al. (2000) included 57 metastases in 32 patients treated with WBRT (30 Gy in 10 fractions) followed by HSRT. Patients were treated with 3 fractions of 6–12 Gy, prescribed to the 80%–90% isodose line. With a median follow-up of 37 months in surviving patients, only two experienced treated metastasis progression. Of note, this series included relative small metastases (median volume 2.16 cm^3 corresponding to a diameter of 1–2 cm) and excluded patients with brain stem involvement or within 5 mm of the chiasm.

Consistent with a broader trend, Aoyama et al. explored the efficacy of HSRT without WBRT in an attempt to minimize toxicity. Patients with up to four metastases were included (Aoyama et al., 2003). The analysis included 159 metastases in 87 patients with a median tumor volume of 3.3 cc (approximately equivalent to 2 cm spherical tumor diameter). A dose of 35 Gy in 4 fractions was prescribed to the treatment isocenter. The acceptable minimum dose was 28 Gy, though 32 Gy was the median minimum dose. For lesions in the brain stem, dose was reduced by 10%–20%, whereas for targets <1 cc, the dose was increased by 10%–20%. Treated metastasis control was 85%, 81%, and 69% at 6 months, 1 year, and 2 years, respectively. A univariate analysis identified a tumor volume of >3 cc, a minimum dose of <32 Gy, and dose at the isocenter of <35 Gy as predictive of progression. Only tumor volume remained significant on multivariable analysis. Thirty patients developed metachronous brain metastasis, of whom 22 had sufficient performance status to receive additional therapy.

As early experiences of HSRT have demonstrated acceptable rates of treated metastasis control, interest has grown in the use of HSRT to treat lesions where SRS was seen as ineffective or overly risky. As discussed previously, such situations include large tumor or those involving sensitive structures of the brain. To this end, a German phase II study enrolled 51 patients with 72 brain metastases, which involved the brain stem, mesencephalon, basal ganglia, or capsula interna or with a sum tumor volume of >3 cc (Ernst-Stecken et al., 2006). Patients with extracranial disease controlled were treated with HSRT alone (35 Gy in 5 fractions to the 90% isodose line), and those without extracranial disease control were treated with WBRT followed by HSRT (30 Gy in 5 fractions to the 90% isodose line). This approach resulted in treated metastasis control of 89% and 76% at 6 months and 1 year, respectively, despite a median tumor volume of 6 cc. A total of 14 patients (27%) experienced neurologic symptoms largely (11/14) due to metachronous brain metastases. This seminal study demonstrates that even large tumors, or those in sensitive areas, can be effectively controlled utilizing HSRT, notably without employing WBRT.

Further support for these findings comes from a German study where HSRT was routinely employed in patients with tumors in the brain stem, mesencephalon, basal ganglia, or capsula interna, or with a total tumor volume of >3cc (Fahrig et al., 2007). A total of 243 brain metastases in 150 patients were treated at one of three treatment centers, which employed various dose fractionations schedules including 30–35 Gy in 5 fractions (n = 51), 40 Gy in 4 fractions (n = 36), or 35 Gy in 7 fractions (n = 63). Of note, 72 patients had prior WBRT though outcomes were not stratified to identify its effect. With a median follow-up of 28 months, the crude rate of treated metastasis control was 93% with no differences observed with the differing dose regimens.

At least three institutions have compared outcomes between patients treated with single-fraction SRS and HSRT. One such series from Korea included 130 metastases in 98 patients treated with SRS (n = 58) or HSRT (n = 40) (Kim et al., 2011). Prior WBRT had been employed in 12/58 and 16/40 patients treated with SRS and HSRT, respectively. The groups differed in size of the PTV, with median volumes of 2.21 cc for SRS as compared to 5.0 cc for HSRT (p = 0.02 for comparison). The SRS dose was 18–22 Gy in all patients. Patients treated with HSRT received 30–42 Gy (median 36 Gy) in 6 fractions daily. Despite the HSRT group having significantly larger tumor volumes, treated metastasis control between the two groups was no different at 1 year (71% vs 69%), and there was no statistical difference in toxicity. Further support for their findings comes from a German series (Fokas et al., 2012). As was the institutional policy, patients with lesions >3 cm or involvement of the brain stem, mesencephalon, basal ganglia, or capsula interna were treated with HSRT. In their analysis, which included 260 patients with one to three brain metastases treated with SRS (n = 138) or HSRT at 35 Gy in 7 fractions (n = 61) or 40 Gy in 10 fractions (n = 61), they found no difference in treated metastasis control (71%–75%). This was despite the respective differences in median tumor volumes of patients receiving SRS (0.87 cc), HSRT at 35 Gy in 7 fractions (2.04 cc), and HSRT at 40 Gy in 10 fractions (5.93 cc).

In sum, these data provided ample support for the efficacy of HSRT. The reported rates of treated metastasis control are high and consistent with those reported for brain metastases treated with SRS. Furthermore, these data demonstrate that HSRT maintains favorable control rates and toxicity despite being used near critical structures and in lesions with larger volumes. A discussion on the safety of HSRT is presented below.

15.4.2 TOXICITY

One hypothetical benefit of HSRT is the ability to treat larger lesions or those in critical areas of the brain without excess acute and late toxicity. As the vast majority of data is retrospective in nature, caution should again be taken when interpreting the data. That being said, the literature does indeed support the hypothesis. Patients treated with SRS are at risk of acute toxicity such as nausea, vomiting, seizures, and headaches, as well as late toxicity including alopecia, chronic headaches, visual or hearing impairment, as well as chronic seizures. Rates of acute toxicity are typically mild and consistently <5%, whereas rates of grade 3–4 late toxicity are consistently <10%. Table 15.2 is a summary of acute and late toxicities from our literature review.

In the study by Manning et al. (2000), there was no acute toxicity. Of the 32 patients, 4 (12.5%) experienced seizures up to 12 months following treatment. Aoyama et al. documented very low rates of both acute (4.6%) and late (2.7%) toxicity (Aoyama et al., 2003). It is worth noting that one patient who received a dose of 40 Gy experienced brain edema 2 weeks following therapy resulting in hemiparesis, not responsive to steroids. Both late toxicities were due to biopsy-proven radionecrosis. The phase II study by Ernest-Stecken et al. (2006), while including somewhat higher-risk patients, demonstrated a significantly higher rate of radionecrosis (34%). Of 72 lesions treated, 48 (61%) experienced increasing

Table 15.2 Toxicity in select series of hypofractionated stereotactic radiation therapy for brain metastases

SERIES	NUMBER OF PATIENTS	ACUTE TOXICITY[a] (%)	LATE TOXICITY[a] (%)	COMMENT
Manning et al. (2000)	32	0	12	Seizures from 3 weeks to 12 months post-HSRT
Aoyama et al. (2003)	87	4.6	2.7	Both late toxicities due to radionecrosis
Ernst-Stecken et al. (2006)	72	0	34 symptomatic necrosis	Radiographic necrosis correlated to V4 Gy per fraction (5 fractions)
Kwon et al. (2009)	27	0	14.8 Grade 2	Only one grade 3 toxicity (headache)
			3.7 Grade 3	
Fahrig et al. (2007)	150	0	10	Various dose fractionations
				No toxicities in patients treated with 40 Gy/10 fractions
Lindvall et al. (2009)	47	0	2.1	—
Kim et al. (2011)	40	0	0	—
Fokas et al. (2012)	122	0	1–2% Grade 3	One case of radionecrosis
Minniti et al. (2014)	135	Not reported	4% Grade 3–4	9% rate of radiographic radionecrosis, predicted by increasing V18 Gy and V21 Gy

[a] Grade included if reported.

edema (enlargement of T2-weighted signal) of which 25 (34%) required an increase in steroid medication. A test of correlation identified increasing V4 Gy per fraction to be predictive of radiographic necrosis, with a threshold of 23 cc (14% if <23 cc and 70% if >23 cc). The analysis of different dose fractionation schemes by Fahrig et al. (2007) found no acute or late toxicity in patients treated with 40 Gy in 10 fractions, suggesting that increased fractionation may reduce complications.

In their comparison of outcomes for patients treated with SRS and HSRT, Kim et al. (2011) showed a reduction in toxicity with the latter without compromising treated metastasis control. Despite significantly larger target volumes and higher rates of WBRT in the HSRT cohort (5.0 vs 2.2 cc), there was no grade ≥2 toxicity. In comparison, there was a 2% rate of grade 3, and a 7% rate of grade 4, toxicity in the SRS cohort. Similarly, Fokas et al. (2012) found higher rates of grade 1–3 toxicity in patients treated with SRS (14%) as compared to HSRT at 35 Gy in 7 fractions (6%) or HSRT at 40 Gy in 10 fractions (2%). As noted above, this is despite the significantly larger target volumes in the latter groups.

From the available retrospective data, it is apparent from the available data that HSRT can be delivered safely without excess acute and late toxicity. Again, this is despite the treatment of larger tumors and/or those in close approximation to sensitive normal tissues.

15.5 CONTROVERSIES AND FUTURE DIRECTIONS

While there is a growing body of retrospective literature analyzing the efficacy and safety of HSRT, there is much room for future investigations. Ideally, future studies would prospectively enroll patients onto protocols that employ HSRT allowing for prospective collection of data on disease control, neurocognitive function, and toxicity. These data could then be compared to prospectively gathered outcomes for patients treated with SRS, allowing clarification of the role of HSRT, particularly in regard to patient selection.

There is a paucity of data on the optimal dose fractionation schemes for different clinical scenarios, focusing both on disease control and normal tissue toxicity. First principles dictate that higher doses will always improve disease control. However, a much more nuanced understanding of the dose–response curve for different dose–fractionation schedules is needed for selecting an optimal schedule. As a correlate to this, similar models are needed to predict normal tissue damage. Much as has been done for single-fraction SRS, work should focus on establishing the dose–volume–location relationships for lesions treated with HSRT. In particular, data are needed for treatment of lesions in the brain stem or within close proximity to the optic nerves and optic chiasm.

Furthermore, it is imperative to further investigate the mechanisms of tissue damage when large doses of radiation are used. There is a great deal of controversy regarding the utility of the linear–quadratic model in the setting of SRS or HSRT (Kirkpatrick et al., 2008, 2009; Hanin and Zaider, 2010). Traditional understanding of radiation biology principle does not explain the excellent clinical outcomes when using these techniques. Standard principles of radiation biology dictate that cell killing occurs largely through DNA damage via strand breaks. Alternative mechanisms of cell killing have been proposed such as vascular damage (Kirkpatrick et al., 2008; Park et al., 2012; Song et al., 2013) or radiation-induced immune response (Finkelstein et al., 2011). In the absence of clinical data, preclinical models may be needed for mechanistic studies that will inform future clinical investigations.

CHECKLIST: KEY POINTS FOR CLINICAL PRACTICE

✓	ACTIVITY	SOME CONSIDERATIONS
	Patient selection	*Is the patient appropriate for SRS/HSRT?* • Consider surgery if mass effect or no tissue diagnosis • One to four metastases • KPS > 70 • Recent chemotherapy?

(Continued)

✓	ACTIVITY	SOME CONSIDERATIONS
	SRS vs HSRT	*Is the lesion amenable to single-fraction SRS?* • Is the lesion >4 cm in size or >3–4 cc? • Is the lesion in the brain stem or in cross proximity to the optic nerves or optic chiasm? • If yes, consider HSRT
	Simulation	*Immobilization* • Rigid frameless system with <1 mm of translational or <1.0 degree of rotation • Consider patient comfort and replicability of setup *Imaging* • Both CT scan and MRI should be performed in treatment position • MRI ideally should be fine cut (1 mm slices), T1 sequence with IV contrast • Accurate image fusion should be verified by clinician and used for contouring targets
	Treatment planning	*Contours* • GTV should include contrast-enhancing lesions • PTV created by using institutionally appropriate expansion (1–3 mm) • Organs at risk should be contoured on MRI *Treatment planning* • For uniform lesions, consider DCAT • For nonuniform lesions or those in close proximity to sensitive normal tissues, consider VMAT or IMRT • Evidence to support multiple-dose–fractionation schemes • Consider 25–35 Gy in 5 fractions; marginal dose prescription • >99% PTV coverage • Conformity index <2.0 • No firm dose constraints for normal tissues but should be kept as low as feasible • Exceptions in the setting or prior radiation or surgical manipulation
	Treatment delivery	*Imaging* • At a minimum, orthogonal kV imaging should be performed to verify isocenter prior to treatment delivery • If available, cone-beam CT should be performed to assess translational and rotational accuracy of setup • Reimage if shifts are made *HSRT discharge* • For large lesions or those near brain stem, consider a short course of low-dose dexamethasone • For patients already on steroids, dose adjustments should be symptom bases • Repeat MRI at 3 months post-HSRT

REFERENCES

Andrews DW, Scott CB, Sperduto PW, Flanders AE, Gaspar LE, Schell MC, Werner-Wasik M et al. (2004) Whole brain radiation therapy with or without stereotactic radiosurgery boost for patients with one to three brain metastases: Phase III results of the RTOG 9508 randomised trial. *Lancet* 363(9422):1665–1672.

Aoyama H, Shirato H, Onimaru R, Kagei K, Ikeda J, Ishii N, Sawamura Y, Miyasaka K (2003) Hypofractionated stereotactic radiotherapy alone without whole-brain irradiation for patients with solitary and oligo brain metastasis using noninvasive fixation of the skull. *Int J Radiat Oncol Biol Phys* 56(3):793–800.

Aoyama H, Shirato H, Tago M, Nakagawa K, Toyoda T, Hatano K, Kenjyo M et al. (2006) Stereotactic radiosurgery plus whole-brain radiation therapy vs stereotactic radiosurgery alone for treatment of brain metastases: A randomized controlled trial. *JAMA* 295(21):2483–2491.

Brenner DJ, Martel MK, Hall EJ (1991) Fractionated regimens for stereotactic radiotherapy of recurrent tumors in the brain. *Int J Radiat Oncol Biol Phys* 21(3):819–824.

Cabrera AR, Cuneo KC, Desjardins A, Sampson JH, McSherry F, Herndon JE, 2nd, Peters KB et al. (2013) Concurrent stereotactic radiosurgery and bevacizumab in recurrent malignant gliomas: A prospective trial. *Int J Radiat Oncol Biol Phys* 86(5):873–879.

Chang EL, Wefel JS, Hess KR, Allen PK, Lang FF, Kornguth DG, Arbuckle RB et al. (2009) Neurocognition in patients with brain metastases treated with radiosurgery or radiosurgery plus whole-brain irradiation: A randomised controlled trial. *Lancet Oncol* 10(11):1037–1044.

Crivellari D, Pagani O, Veronesi A, Lombardi D, Nole F, Thurlimann B, Hess D et al. (2001) High incidence of central nervous system involvement in patients with metastatic or locally advanced breast cancer treated with epirubicin and docetaxel. *Ann Oncol* 12(3):353–356.

DeAngelis LM, Delattre JY, Posner JB (1989) Radiation-induced dementia in patients cured of brain metastases. *Neurology* 39(6):789–796.

Ernst-Stecken A, Ganslandt O, Lambrecht U, Sauer R, Grabenbauer G (2006) Phase II trial of hypofractionated stereotactic radiotherapy for brain metastases: Results and toxicity. *Radiother Oncol* 81(1):18–24.

Fahrig A, Ganslandt O, Lambrecht U, Grabenbauer G, Kleinert G, Sauer R, Hamm K (2007) Hypofractionated stereotactic radiotherapy for brain metastases—Results from three different dose concepts. *Strahlenther Onkol* 183(11):625–630.

Finkelstein SE, Timmerman R, McBride WH, Schaue D, Hoffe SE, Mantz CA, Wilson GD (2011) The confluence of stereotactic ablative radiotherapy and tumor immunology. *Clin Dev Immunol* 2011:439752.

Flickinger JC, Kondziolka D, Lunsford LD, Kassam A, Phuong LK, Liscak R, Pollock B (2000) Development of a model to predict permanent symptomatic postradiosurgery injury for arteriovenous malformation patients. Arteriovenous Malformation Radiosurgery Study Group. *Int J Radiat Oncol Biol Phys* 46(5):1143–1148.

Flickinger JC, Kondziolka D, Maitz AH, Lunsford LD (1998) Analysis of neurological sequelae from radiosurgery of arteriovenous malformations: How location affects outcome. *Int J Radiat Oncol Biol Phys* 40(2):273–278.

Flickinger JC, Kondziolka D, Pollock BE, Maitz AH, Lunsford LD (1997) Complications from arteriovenous malformation radiosurgery: Multivariate analysis and risk modeling. *Int J Radiat Oncol Biol Phys* 38(3):485–490.

Flickinger JC, Lunsford LD, Kondziolka D, Maitz AH, Epstein AH, Simons SR, Wu A (1992) Radiosurgery and brain tolerance: An analysis of neurodiagnostic imaging changes after gamma knife radiosurgery for arteriovenous malformations. *Int J Radiat Oncol Biol Phys* 23(1):19–26.

Fokas E, Henzel M, Surber G, Kleinert G, Hamm K, Engenhart-Cabillic R (2012) Stereotactic radiosurgery and fractionated stereotactic radiotherapy: Comparison of efficacy and toxicity in 260 patients with brain metastases. *J Neuro-Oncol* 109(1):91–98.

Foote KD, Friedman WA, Buatti JM, Meeks SL, Bova FJ, Kubilis PS (2001) Analysis of risk factors associated with radiosurgery for vestibular schwannoma. *J Neurosurg* 95(3):440–449.

Gaspar L, Scott C, Rotman M, Asbell S, Phillips T, Wasserman T, McKenna WG, Byhardt R (1997) Recursive partitioning analysis (RPA) of prognostic factors in three Radiation Therapy Oncology Group (RTOG) brain metastases trials. *Int J Radiat Oncol Biol Phys* 37(4):745–751.

Hall EJ, Brenner DJ (1993) The radiobiology of radiosurgery: Rationale for different treatment regimes for AVMs and malignancies. *Int J Radiat Oncol Biol Phys* 25(2):381–385.

Hanin LG, Zaider M (2010) Cell-survival probability at large doses: An alternative to the linear-quadratic model. *Phys Med Biol* 55(16):4687–4702.

Johnson JD, Young B (1996) Demographics of brain metastasis. *Neurosurg Clin North Am* 7(3):337–344.

Kano H, Kondziolka D, Flickinger JC, Yang HC, Flannery TJ, Niranjan A, Novotny J, Jr., Lunsford LD (2012) Stereotactic radiosurgery for arteriovenous malformations, Part 5: Management of brainstem arteriovenous malformations. *J Neurosurg* 116(1):44–53.

Kim YJ, Cho KH, Kim JY, Lim YK, Min HS, Lee SH, Kim HJ, Gwak HS, Yoo H, Lee SH (2011) Single-dose versus fractionated stereotactic radiotherapy for brain metastases. *Int J Radiat Oncol Biol Phys* 81(2):483–489.

Kirkpatrick JP, Brenner DJ, Orton CG (2009) Point/Counterpoint. The linear-quadratic model is inappropriate to model high dose per fraction effects in radiosurgery. *Med Phys* 36(8):3381–3384.

Kirkpatrick JP, Meyer JJ, Marks LB (2008) The linear-quadratic model is inappropriate to model high dose per fraction effects in radiosurgery. *Semin Radiat Oncol* 18(4):240–243.

Kirkpatrick JP, Wang Z, Sampson JH, McSherry F, Herndon JE 2nd, Allen KJ, Duffy E et al. (2015) Defining the optimal planning target volume in image-guided stereotactic radiosurgery of brain metastases: Results of a randomized trial. *Int J Radiat Oncol Biol Phys* 91(1):100–108.

Kocher M, Soffietti R, Abacioglu U, Villa S, Fauchon F, Baumert BG, Fariselli L et al. (2011) Adjuvant whole-brain radiotherapy versus observation after radiosurgery or surgical resection of one to three cerebral metastases: Results of the EORTC 22952-26001 study. *J Clin Oncol* 29(2):134–141.

Kwon AK, Dibiase SJ, Wang B, Hughes SL, Milcarek B, Zhu Y (2009) Hypofractionated stereotactic radiotherapy for the treatment of brain metastases. *Cancer* 115(4):890–898.

Landy HJ, Schwade JG, Houdek PV, Markoe AM, Feun L (1994) Long-term follow-up of gliomas treated with fractionated stereotactic irradiation. *Acta Neurochir Suppl* 62:67–71.

Lawrence YR, Li XA, el Naqa I, Hahn CA, Marks LB, Merchant TE, Dicker AP (2010) Radiation dose-volume effects in the brain. *Int J Radiat Oncol Biol Phys* 76(3 Suppl):S20–S27.

Leber KA, Bergloff J, Pendl G (1998) Dose-response tolerance of the visual pathways and cranial nerves of the cavernous sinus to stereotactic radiosurgery. *J Neurosurg* 88(1):43–50.

Levegrun S, Hof H, Essig M, Schlegel W, Debus J (2004) Radiation-induced changes of brain tissue after radiosurgery in patients with arteriovenous malformations: Correlation with dose distribution parameters. *Int J Radiat Oncol Biol Phys* 59(3):796–808.

Lin NU, Winer EP (2007) Brain metastases: The HER2 paradigm. *Clin Cancer Res* 13(6):1648–1655.

Lindvall P, Bergstrom P, Lofroth PO, and Tommy Bergenheim A (2009) A comparison between surgical resection in combination with WBRT or hypofractionated stereotactic irradiation in the treatment of solitary brain metastases. *Acta Neurochir* 151(9):1053–1059.

Ma J, Chang Z, Wang Z, Jackie Wu Q, Kirkpatrick JP, Yin FF (2009) ExacTrac X-ray 6 degree-of-freedom image-guidance for intracranial non-invasive stereotactic radiotherapy: Comparison with kilo-voltage cone-beam CT. *Radiother Oncol* 93(3):602–608.

Manning MA, Cardinale RM, Benedict SH, Kavanagh BD, Zwicker RD, Amir C, Broaddus WC (2000) Hypofractionated stereotactic radiotherapy as an alternative to radiosurgery for the treatment of patients with brain metastases. *Int J Radiat Oncol Biol Phys* 47(3):603–608.

Mayo C, Martel MK, Marks LB, Flickinger J, Nam J, Kirkpatrick J (2010) Radiation dose-volume effects of optic nerves and chiasm. *Int J Radiat Oncol Biol Phys* 76(3 Suppl):S28–S35.

Mehta MP, Tsao MN, Whelan TJ, Morris DE, Hayman JA, Flickinger JC, Mills M, Rogers CL, Souhami CL (2005) The American Society for Therapeutic Radiology and Oncology (ASTRO) evidence-based review of the role of radiosurgery for brain metastases. *Int J Radiat Oncol Biol Phys* 63(1):37–46.

Minniti G, D'Angelillo RM, Scaringi C, Trodella LE, Clarke E, Matteucci P, Osti MF, Ramella S, Enrici RM, Trodella L (2014) Fractionated stereotactic radiosurgery for patients with brain metastases. *J Neuro-Oncol* 117(2):295–301.

Mintz AH, Kestle J, Rathbone MP, Gaspar L, Hugenholtz H, Fisher B, Duncan G, Skingley P, Foster G, Levine M (1996) A randomized trial to assess the efficacy of surgery in addition to radiotherapy in patients with a single cerebral metastasis. *Cancer* 78(7):1470–1476.

Mori Y, Kondziolka D, Flickinger JC, Kirkwood JM, Agarwala S, Lunsford LD (1998) Stereotactic radiosurgery for cerebral metastatic melanoma: Factors affecting local disease control and survival. *Int J Radiat Oncol Biol Phys* 42(3):581–589.

Noordijk EM, Vecht CJ, Haaxma-Reiche H, Padberg GW, Voormolen JH, Hoekstra FH, Tans JT et al. (1994) The choice of treatment of single brain metastasis should be based on extracranial tumor activity and age. *Int J Radiat Oncol Biol Phys* 29(4):711–717.

Park HJ, Griffin RJ, Hui S, Levitt SH, Song CW (2012) Radiation-induced vascular damage in tumors: implications of vascular damage in ablative hypofractionated radiotherapy (SBRT and SRS). *Radiat Res* 177(3):311–327.

Patchell RA, Tibbs PA, Regine WF, Dempsey RJ, Mohiuddin M, Kryscio RJ, Markesbery WR, Foon KA, Young B (1998) Postoperative radiotherapy in the treatment of single metastases to the brain: A randomized trial. *JAMA* 280(17):1485–1489.

Patchell RA, Tibbs PA, Walsh JW, Dempsey RJ, Maruyama Y, Kryscio RJ, Markesbery WR, Macdonald JS, Young B (1990) A randomized trial of surgery in the treatment of single metastases to the brain. *N Engl J Med* 322(8):494–500.

Patel S, Rimner A, Foster A, Zhang M, Woo KM, Yu HA, Riely GJ, Wu AJ (2014) Risk of brain metastasis in EGFR-mutant NSCLC treated with erlotinib: A role for prophylactic cranial irradiation? *Int J Radiat Oncol Biol Phys* 90(1s):S643–S644.

Pollock BE, Cochran J, Natt N, Brown PD, Erickson D, Link MJ, Garces YI, Foote RL, Stafford SL, Schomberg PJ (2008) Gamma knife radiosurgery for patients with nonfunctioning pituitary adenomas: Results from a 15-year experience. *Int J Radiat Oncol Biol Phys* 70(5):1325–1329.

Regine WF, Scott C, Murray K, Curran W (2001) Neurocognitive outcome in brain metastases patients treated with accelerated-fractionation vs. accelerated-hyperfractionated radiotherapy: An analysis from Radiation Therapy Oncology Group Study 91-04. *Int J Radiat Oncol Biol Phys* 51(3):711–717.

Sahgal A, Aoyama H, Kocher M, Neupane B, Collette S, Tago M, Shaw P, Beyene J, Chang EL (2015) Phase 3 trials of stereotactic radiosurgery with or without whole-brain radiation therapy for 1 to 4 brain metastases: Individual patient data meta-analysis. *Int J Radiat Oncol Biol Phys* 91(4):710–717.

Shaw E, Scott C, Souhami L, Dinapoli R, Kline R, Loeffler J, Farnan N (2000) Single dose radiosurgical treatment of recurrent previously irradiated primary brain tumors and brain metastases: Final report of RTOG protocol 90-05. *Int J Radiat Oncol Biol Phys* 47(2):291–298.

Shiau CY, Sneed PK, Shu HK, Lamborn KR, McDermott MW, Chang S, Nowak P et al. (1997) Radiosurgery for brain metastases: Relationship of dose and pattern of enhancement to local control. *Int J Radiat Oncol Biol Phys* 37(2):375–383.

Slimane K, Andre F, Delaloge S, Dunant A, Perez A, Grenier J, Massard C, Spielmann M (2004) Risk factors for brain relapse in patients with metastatic breast cancer. *Ann Oncol* 15(11):1640–1644.

Song CW, Cho LC, Yuan J, Dusenbery KE, Griffin RJ, Levitt SH (2013) Radiobiology of stereotactic body radiation therapy/stereotactic radiosurgery and the linear-quadratic model. *Int J Radiat Oncol Biol Phys* 87(1):18–19.

Sperduto PW, Berkey B, Gaspar LE, Mehta M, Curran W (2008) A new prognostic index and comparison to three other indices for patients with brain metastases: An analysis of 1,960 patients in the RTOG database. *Int J Radiat Oncol Biol Phys* 70(2):510–514.

Sperduto PW, Chao ST, Sneed PK, Luo X, Suh J, Roberge D, Bhatt A et al. (2010) Diagnosis-specific prognostic factors, indexes, and treatment outcomes for patients with newly diagnosed brain metastases: A multi-institutional analysis of 4,259 patients. *Int J Radiat Oncol Biol Phys* 77(3):655–661.

Sperduto PW, Kased N, Roberge D, Xu Z, Shanley R, Luo X, Sneed PK et al. (2012) Summary report on the graded prognostic assessment: An accurate and facile diagnosis-specific tool to estimate survival for patients with brain metastases. *J Clin Oncol* 30(4):419–425.

Spiegelmann R, Lidar Z, Gofman J, Alezra D, Hadani M, Pfeffer R (2001) Linear accelerator radiosurgery for vestibular schwannoma. *J Neurosurg* 94(1):7–13.

Stafford SL, Pollock BE, Leavitt JA, Foote RL, Brown PD, Link MJ, Gorman DA, Schomberg PJ (2003) A study on the radiation tolerance of the optic nerves and chiasm after stereotactic radiosurgery. *Int J Radiat Oncol Biol Phys* 55(5):1177–1181.

Sundermeyer ML, Meropol NJ, Rogatko A, Wang H, Cohen SJ (2005) Changing patterns of bone and brain metastases in patients with colorectal cancer. *Clin Colorectal Cancer* 5(2):108–113.

Tallet AV, Azria D, Barlesi F, Spano JP, Carpentier AF, Goncalves A, Metellus P (2012) Neurocognitive function impairment after whole brain radiotherapy for brain metastases: Actual assessment. *Radiat Oncol* 7:77.

Tsao MN, Rades D, Wirth A, Lo SS, Danielson BL, Gaspar LE, Sperduto PW et al. (2012) Radiotherapeutic and surgical management for newly diagnosed brain metastasis(es): An American Society for Radiation Oncology evidence-based guideline. *Pract Radiat Oncol* 2(3):210–225.

Vogelbaum MA, Angelov L, Lee SY, Li L, Barnett GH, Suh JH (2006) Local control of brain metastases by stereotactic radiosurgery in relation to dose to the tumor margin. *J Neurosurg* 104(6):907–912.

Voges J, Treuer H, Sturm V, Buchner C, Lehrke R, Kocher M, Staar S, Kuchta J, Muller RP (1996) Risk analysis of linear accelerator radiosurgery. *Int J Radiat Oncol Biol Phys* 36(5):1055–1063.

Yang HC, Kano H, Lunsford LD, Niranjan A, Flickinger JC, Kondziolka D (2011) What factors predict the response of larger brain metastases to radiosurgery? *Neurosurgery* 68(3):682–690; discussion 690.

Zhao B, Jin, J-Y, Wen N, Huang Y, Siddiqui MS, Chetty IJ, Ryu S (2014) Prescription to 50–75% isodose line may be optimum for linear accelerator based radiosurgery of cranial lesions. *J Radiosurg SBRT* 3(2):139.

16 Image-guided hypofractionated stereotactic whole-brain radiotherapy and simultaneous integrated boost for brain metastases

Alan Nichol

Contents

16.1 INTRODUCTION

The Radiation Therapy Oncology Group (RTOG) 9508 study showed that whole-brain radiotherapy (WBRT) followed by a stereotactic radiosurgery (SRS) boost, compared to WBRT alone, resulted in better survival for patients with solitary brain metastases and for patients with good prognosis (Andrews et al., 2004; Sperduto et al., 2014). The development of helical tomotherapy and volumetric modulated arc therapy (VMAT) led to an interest in using these technologies to deliver the WBRT and boost the metastases simultaneously (Bauman et al., 2007; Gutierrez et al., 2007; Hsu et al., 2009). This technique for treating the whole brain with a concurrent boost to the brain metastases is often referred to as WBRT and simultaneous integrated boost (SIB). This chapter will review the issue of using WBRT for selected patients, summarize publications about the various WBRT and SIB prescriptions, and describe the delivery of WBRT and SIB in 5 fractions.

WBRT has been shown to reduce the risk of local and distant recurrences of brain metastases when combined with SRS compared to SRS alone (Aoyama et al., 2006; Chang et al., 2009; Kocher et al., 2011; Sahgal et al., 2015). According to the Cochrane Database of Systematic Review of WBRT+SRS versus SRS alone, WBRT decreased the relative risk of any intracranial disease progression at 1 year by 53% (HR 0.47, 95%CI 0.34–0.66, $p < 0.0001$), but there was no significant difference in overall survival (HR 1.11, 95%CI 0.83–1.48, $p = 0.47$) (Soon et al., 2014). The randomized clinical trials also showed that focal treatment was associated with less alopecia, less fatigue, less cognitive impairment, and better quality of life as compared to treatment including WBRT, although the quality of the evidence was judged to be low (Patchell et al., 1998; Aoyama et al., 2006, 2007; Roos et al., 2006; Chang et al., 2009; Kocher et al., 2011; McDuff et al., 2013; Soffietti et al., 2013).

The best evidence for the neurocognitive effects of WBRT comes from studies of prophylactic cranial irradiation (PCI) because they isolate the effects of WBRT from those of palliative chemotherapy and progression or recurrence of brain metastases (McDuff et al., 2013). Gondi et al. reported on a combined analysis of the RTOG 0212 and 0214 studies. Of the patients that completed the baseline cognitive testing, 410 received PCI and 173 were observed (Gondi et al., 2013). This analysis showed that PCI was significantly associated with declines in self-reported cognitive functioning ($p < 0.0001$) and Hopkins Verbal Learning Test–Recall ($p = 0.002$) at 6 and 12 months. The RTOG 0212 study compared two PCI prescriptions (Wolfson et al., 2011). Significant declines on one or more cognitive tests had an incidence of 62% in the 25 Gy arm and 85% in the 36 Gy arm. Age was shown to predispose patients to a higher risk of new neurocognitive impairment with 36 Gy, compared to 25 Gy PCI in the RTOG 0212 trial (Wolfson et al., 2011). In a multivariable model, age over 60 predicted for a decline in the Hopkins verbal learning test-delayed recall at 12 months in Gondi's RTOG 0212 and 0214 analysis (Gondi et al., 2013).

WBRT diminishes neurocognitive function, but so too does poor control of brain metastases (Regine et al., 2001; Meyers et al., 2004; Aoyama et al., 2007; Li et al., 2007). Regine et al. reported a significant drop in Mini-Mental Status Examination (MMSE) scores in patients whose brain metastases were not controlled by WBRT (Regine et al., 2001). Progression of treated metastases was shown to diminish neurocognitive function in an exploratory analysis of a randomized controlled trial of WBRT with or without motexafin gadolinium. Meyers et al. (2004) demonstrated that patients with partial responses were more likely than patients with progressive disease to have improvement of their cognition on a battery of eight tests. Li et al. (2007) assessed volumetric response of brain metastases treated with WBRT and found that better volumetric response correlated with better neurocognitive function. The median time to cognitive impairment was longer in good responders than in poor responders on all eight cognitive tests used in the study ($p = 0.008$). In their randomized controlled trial, Aoyama et al. observed that 3-point declines in MMSE were delayed from 7.6 months in the SRS alone arm to 16.5 months in the WBRT+SRS arm ($p = 0.05$), which they attributed to better local and distant control of brain metastases, most evident in the first 2 years after treatment (Aoyama et al., 2007).

The American Society for Radiation Oncology issued a Choosing Wisely® statement: "Don't routinely add whole brain radiotherapy to stereotactic radiosurgery for limited brain metastases" because of the absence of a survival benefit with WBRT and its known side effects (American Society of Radiation Oncology, 2014). The Choosing Wisely® campaign is intended to encourage more detailed conversations between physicians and patients about treatment options. The conversation about WBRT and SIB should include a frank discussion about the patient's prognosis and the likelihood of side effects from WBRT, compared to the likelihood of complications that could arise from intracranial relapse. There are a number of characteristics of patients that predispose them to a high risk of new brain metastases and need for salvage treatment, both of which can be reduced with adjuvant WBRT. Given its known toxicities, the use of WBRT must be selective (Abe and Aoyama, 2012; Mehta, 2015; Sahgal, 2015). The following section presents a discussion about the patients for whom the risk/benefit ratio may favor WBRT.

16.2 INDICATIONS FOR WBRT

Selected patients at high risk of new and early new metastases may be suited to adjuvant WBRT. Rodrigues et al. (2014) and Ayala-Peacock et al. (2014) have published nomograms that identify the patients at high risk of intracranial relapse. These tools can be employed to make individualized treatment recommendations to patients about adjuvant WBRT regarding their personal risk of new metastases and the reduction in the incidence of relapse and in the need for salvage therapies.

Rodrigues et al. (2014) studied the incidence of new brain metastases at 1 year in a cohort of patients treated with SRS alone for one to three metastases. Their univariate analysis showed that younger patients with smaller gross tumor volumes (GTV), two to three brain metastases, and better performance status were more likely to have new metastases at 1 year. The odds ratio for new metastases was 0.78 (95%CI, 0.63–0.95) for each increase in age of 10 years, indicating that older patients had a lower risk of new metastases. Worse performance status also conferred a lower risk of new metastases: World Health Organization Performance Status (WHO-PS) = 0 versus 1, odds ratio = 0.50 (95%CI, 0.29–0.87), and 2 versus 3, odds ratio = 0.24 (95%CI, 0.11–0.56). On multivariable analysis, patients with larger GTVs also had a lower risk of new metastases: odds ratio = 0.66 (95%CI, 0.44–0.99). Patients with two to three lesions had a higher risk of new metastases than patients with only one lesion: odds ratio = 2.27 (95%CI, 1.37–3.75). Thus, the risk of new metastases at 1 year generally increased with predictors of longer survival, presumably because the incidence of new brain metastases competes with the risk of death from extracranial disease. Their recursive partitioning analysis identified a high-risk subgroup of patients with WHO-PS = 0 and two to three metastases, with a cumulative incidence of new metastases at 1 year of almost 70%, and an intermediate risk group with a cumulative incidence of new metastases of almost 50%.

Ayala-Peacock et al. (2014) performed a similar retrospective study on 464 patients treated with SRS alone for up to 13 brain metastases. The risk of new brain metastases was significantly increased with a higher baseline number of brain metastases, progressive extracranial disease, detection of new metastases on the SRS planning MRI, and high-risk histologies: melanoma and Her2-negative breast cancer. Although salvage WBRT was only required in 30% of the patients before death, there were some patients that had a high risk of needing early WBRT. For example, salvage WBRT was required at a median of 3.3 months for patients with melanoma and at a median of 3.0 months for patients with poorly differentiated lung cancer.

Patients' long-term risk of dementia from whole brain radiotherapy rises with better prognosis. Considering their dementia risk and risk of new metastases suggests that patients with a life expectancy of 6–12 months and a high risk of new metastases are good candidates for WBRT with SRS or WBRT and SIB.

16.3 STUDIES OF WBRT AND SIB

There are several radiotherapy systems that can deliver WBRT and SIB. Helical tomotherapy has been used in both planning studies and clinical trials (Bauman et al., 2007; Gutierrez et al., 2007; Sterzing et al., 2009; Edwards et al., 2010; Rodrigues et al., 2011). VMAT can also deliver WBRT and SIB (Hsu et al., 2009; Lagerwaard et al., 2009; Weber et al., 2011; Awad et al., 2013; Oehlke et al., 2015). Reports of clinical experience with WBRT and SIB are summarized in Table 16.1.

Sterzing et al. (2009) retreated two patients with WBRT and SIB using helical tomotherapy for relapsed brain metastases following prior 40 Gy in 20 fractions of WBRT. One patient had 8 new metastases and the other had 11 new metastases. The prescriptions were 15 Gy in 10 fractions to the whole brain and 30 Gy in 10 fractions to a 2 mm planning target volume (PTV) on the brain metastases. The authors reported local control in both patients at 6 and 12 months of follow-up.

Edwards et al. (2010) treated 11 patients with WBRT and SIB using helical tomotherapy for one to four brain metastases measuring up to 8 cm in diameter. The prescriptions were 30 Gy in 10 fractions to the whole brain and 40 Gy in 10 fractions to the metastases. The whole-brain PTV was an expansion of the brain by 3 mm. The brain metastases were treated with a 0 mm GTV–PTV margin. All patients received corticosteroids during the treatment. The lesions all exhibited 1 month local responses and the patients experienced no complications.

Lagerwaard et al. (2009) treated three patients with WBRT and SIB using VMAT for one to three new brain metastases. The prescriptions employed were 20 Gy in 5 fractions to the whole-brain PTV and 40 Gy in 5 fractions to the metastasis PTVs. Both the brain and the GTVs were expanded by a 2 mm margin to create the PTVs. The patients were positioned using cone-beam CT (CBCT). All head rotations over 0.8° were corrected clinically by repositioning the patients and repeating the CBCT. All translational corrections were applied. A follow-up publication reported on 50 patients treated using a robotic couch capable of correcting patient positioning in 6 degrees of freedom. The crude rate of radionecrosis reported in this subgroup was 6% (3/50) (Rodrigues et al., 2012b).

Rodrigues et al. (2011) reported a phase I trial of 30 Gy/10 WBRT and increasing SIB doses from 35 to 60 Gy in 5 Gy increments to the brain metastasis GTVs with no margin. Dose-limiting toxicity was not observed at the final dose level of 60 Gy. A phase II study of the final dose level from the phase I study

Table 16.1 Summary of whole-brain radiotherapy and simultaneous integrated boost results

STUDY	NUMBER OF SUBJECTS	NUMBER OF METASTASES MEDIAN (RANGE)	SIZE OF METASTASES	RESULT	ENDPOINT	MEDIAN SURVIVAL (MONTHS)	WBRT DOSE (Gy)/ FRACTIONS	SIB DOSE (Gy)/ FRACTIONS
Lagerwaard (Lagerwaard et al., 2009)	3	3 (1–3)	1.5–25.8 cm³	NR	NR	NR	20/5	40/5
Sterzing (Sterzing et al., 2009)	2	8–11	NR	NR	Reirradiation	NR	15/10	30/10
Awad (Awad et al., 2013)[a]	30	2 (1–8)	Mean 6.9 cm³	81%	3.5 month LC	9.4	28.6–37.5/15	50–70.8/15
Edwards (Edwards et al., 2010)	11	(1–4)	2.5–8 cm	100%	3 month LC	>4	30/10	40/10
Rodrigues Phase I (Rodrigues et al., 2011)	48	<3 cm	<3 cm	66%	PFS	5.3	30/10	35–60/10
Rodrigues and Lagerwaard (Rodrigues et al., 2012b)[b]	120	2 (1–6)	<3 cm	~67%[c]	1 year	5.9	30/10 20/5	35–60/10 40/5
Weber (Weber et al., 2011)	29	(1–4)	<40 cm³	78%	6 month PFS	>6	30/10	40/10
Oehlke (Oehlke et al., 2015)[a]	20	5 (2–13)	0.78 ± 1.17 cm³	~64%[c]	1 year local control	16.5	30/12	51/12
Kim (Kim et al., 2015)[a]	11	4 (2–15)	<10.1 cm³	67%	1 year brain control	14.5	25–28/10–14	40–48/10–14 to GTV alone
BC Cancer Agency (Nichol et al., 2015)	60	3 (1–10)	Diameter <3 cm	90%	3 month freedom from progression	10.1	20/5	47.5/5 to GTVs alone; 38/5 to 2 mm PTVs

a Hippocampal avoidance.

b Combined cohorts—including all 48 patients from Rodrigues' phase I study.

c Estimated from a figure.

Abbreviations: NR, not reported; LC, local control; PFS, progression-free survival; GTV, gross tumor volume; PTV, planning target volume.

has been conducted (Rodrigues et al., 2012a). An early publication about this 60 Gy/10 SIB prescription reported no radionecrosis (Rodrigues et al., 2012b).

Weber et al. (2011) treated 29 patients with 1–4 brain metastases with 30 Gy in 10 fractions to the whole brain and 40 Gy in 10 fractions to the metastases using VMAT. The report did not mention margins. The 6-month overall survival was 55%.

Awad et al. (2013) reported on 30 patients treated with VMAT using a variety of planning techniques and fractionation schedules: 17 received hippocampal avoidance (HA)-WBRT and SIB and 5 received WBRT and SIB. Twenty-six of the patients had melanoma. The whole brain and the metastases were expanded by 2 mm margins to create PTVs. The hippocampus was contoured according to the RTOG 0933 study guidelines (Gondi et al., 2010a, 2014) and was expanded to create an HA volume. The median dose to the whole brain was 30 Gy in 15 fractions and to the brain metastases was 50 Gy in 15 fractions. Overall survival for the 22 patients treated with WBRT and SIB was not separately reported, but it was 9.4 months for the whole cohort.

Oehlke et al. (2015) reported on the first 20 patients treated with HA-WBRT and SIB using VMAT in a single-institution study. The whole-brain prescription was 30 Gy in 12 fractions, and the boost prescription was 51 Gy in 12 fractions. The hippocampus was contoured according to the RTOG 0933 study guidelines (Gondi et al., 2010a, 2014). The whole-brain PTV was an expansion of the brain parenchyma by 3 mm. The PTV for brain metastases was an expansion of the GTVs by 1 mm. The treatment was well tolerated. Grade 1 and 2 treatment-related toxicities were commonly observed, but only one patient developed radionecrosis requiring surgery (grade 4 toxicity). The median intracranial progression-free survival was 9.2 months and the median overall survival was 16.6 months.

Kim et al. (2015) reported on 11 patients with metastatic lung cancer treated with HA-WBRT and SIB using helical tomotherapy for 70 brain metastases (Kim et al., 2015). They delivered 25–28 Gy to the whole brain and 40–48 Gy to metastases with no margin in 10–14 fractions. The mean dose to the hippocampus was 13.7 Gy. The intracranial control rate at 1 year was 67%.

At the BC Cancer Agency, a phase 2 study with 60 patients was conducted using VMAT to deliver 20 Gy/5 WBRT and 47.5 Gy/5 SIB to brain metastases without a margin (Nichol et al., 2015). The median survival was 10.1 months. Freedom from in-brain progression was 90% at 3 months. The crude incidence of symptomatic radionecrosis was 12% in the 60 patients. However, the incidence of radionecrosis with this boost prescription was unacceptable for metastases located in the thalamus and basal ganglia (Flickinger et al., 1998; Wegner et al., 2011). The SIB prescription of 47.5 Gy/5 to metastases with no margin that was used in our study is not recommended for metastases in the brain stem, thalamus, and basal ganglia.

16.4 CHOOSING A PRESCRIPTION FOR WBRT AND SIB

A 5-fraction course of WBRT and SIB was selected at our center because it was more convenient for patients than longer courses and it could be delivered during the second week of a 3-week cycle of chemotherapy. The WBRT prescription of 20 Gy in 5 fractions has a long history of use for treating brain metastases with local control results that are slightly lower than prescriptions like 30 Gy in 10 fractions, 37.5 Gy in 15 fractions, and 40 Gy in 20 fractions (Table 16.2) (Tsao et al., 2012). However, when focal high-dose treatment is delivered to the GTVs, the WBRT is adjuvant therapy, intended only to control

Table 16.2 Equivalent doses in 2 Gy fractions (EQD2) for whole-brain radiotherapy prescriptions

WHOLE-BRAIN RADIOTHERAPY DOSE (Gy)	FRACTIONS	EQD2 (Gy_3)	EQD2 (Gy_{10})
20	5	28	23
25	10	28	26
30	10	36	33
37.5	15	41	39
40	20	40	40

Table 16.3 Equivalent doses in 2 Gy fractions (EQD2) for hypofractionated simultaneous integrated boosts, compared to the whole-brain radiotherapy + stereotactic radiosurgery dosing schema of the Radiation Therapy Oncology Group 9508 study

INDICATIONS	BOOST (Gy)	FRACTIONS	EQD2 (Gy_3)	EQD2 (Gy_{10})
Brain stem metastases[a]	12	1	67	56
Brain stem metastases and brain metastases PTVs treated with translational setup correction (2 mm GTV–PTV margins)	35	5	70	50
	50	10	80	63
Brain metastases 3–4 cm[a]	15	1	95	70
Brain metastases 2–3 cm[a]	18	1	117	81
Brain metastases treated with translational and rotational setup correction (no GTV–PTV margin)	45	5	108	71
	60	10	108	80
Brain metastases <2 cm[a]	24	1	171	107

[a] With 37.5 Gy in 15 fractions WBRT.

subclinical microscopic disease. In this setting, we regard doses like those used for PCI as being sufficient to manage the risk of subclinical disease in the brain. Based on calculations of biologically equivalent doses, the WBRT prescription of 20 Gy/5 would be expected to have similar effectiveness and late side effects to the PCI prescription of 25 Gy/10.

The WBRT+SRS arm of the RTOG 9508 study employed a WBRT prescription of 37.5 Gy/15 and SRS prescribing based on the diameter of metastases: 24 Gy for <2 cm, 18 Gy for 2–3 cm, and 15 Gy for 3–4 cm (Andrews et al., 2004). Table 16.3 presents the biologically equivalent doses from RTOG 9508 for comparison with possible 5-fraction and 10-fraction boost prescriptions. There are a number of studies of 5- to 10-fraction schedules of stereotactic radiotherapy that can inform the choice of an SIB prescription (Lindvall et al., 2005; Aoki et al., 2006; Ernst-Stecken et al., 2006; Fahrig et al., 2007; Narayana et al., 2007; Tomita et al., 2008; Kwon et al., 2009; Scorsetti et al., 2009; Chen et al., 2011; Fokas et al., 2012; Jiang et al., 2012; Martens et al., 2012; Ogura et al., 2012; De Potter et al., 2013; Eaton et al., 2013; Ahmed et al., 2014; Murai et al., 2014; Rajakesari et al., 2014). In addition, there are investigations of the biological equivalence of various boost schedules for hypofractionated stereotactic radiotherapy (Yuan et al., 2008; Wiggenraad et al., 2011).

All of the phase 3 trials of WBRT+SRS or SRS alone for patients with brain metastases excluded patients with brain stem metastases. There are a number of retrospective reports that describe treatment of brain stem metastases with SRS (Kased et al., 2008; Samblas et al., 2009; Koyfman et al., 2010; Valery et al., 2010; Leeman et al., 2012; Kilburn et al., 2014; Hsu et al., 2015). For patients with multiple brain metastases, the likelihood of one of them being deep and within eloquent brain increases with the number of metastases at baseline, so it is important to have a safe and effective SIB prescription for these metastases (Leeman et al., 2012). Kilburn et al. (2014) reported that the risk of toxicity rises when treating metastases with volumes over 1 cm³ using 18 Gy to the GTV with no margin. We employ a prescription of 45 Gy to metastases with no margin when treating with translational and rotational setup correction and 35 Gy to a 2 mm PTV margin on metastases when treating with translational setup correction (Table 16.4).

16.5 SIMULATION

Patients require highly reproducible immobilization for WBRT and SIB. The treatment can be delivered in about 12 min, and the chosen mask system should ensure intrafraction stability for this treatment duration. At our center, we used a thick thermoplastic mask with a patient-specific mouthpiece attached inside the shell. If only translational image-guided setup correction is available, the use of a mouthpiece offers the advantage of reducing rotation (Dhabaan et al., 2012; Dincoglan et al., 2012; Theelen et al., 2012).

The planning CT scan must be performed with high resolution in both the axial plane and the craniocaudal direction. To optimize contouring of small structures like the chiasm and metastases, the field of view needs to be as small as possible. To optimize dose calculations, the field of view needs to

Table 16.4 Dose constraints and typical volumetric modulated arc therapy planning priorities for 20 Gy/5 whole-brain radiotherapy and 45 Gy/5 simultaneous integrated boost to gross tumor volumes or 20 Gy/5 whole-brain radiotherapy and 35 Gy/5 simultaneous integrated boost to planning target volumes

STRUCTURE	DOSE CONSTRAINTS		
	LIMIT	VOLUME (%)	DOSE (Gy)
Anterior chambers (no margin)	Upper	<1	10
Cochleas and middle ears (no margin)	Upper	<1	10
Retinas (no margin)	Upper	<1	25
Optics_PRV (2 mm margin)	Upper	<1	25
Spinal cord_PRV (2 mm margin)	Upper	<1	25
Whole-brain planning target volume (2 mm margin)	Upper	<5	25
	Lower	98	19
Brain metastasis GTVs (no margin) translational and rotational setup correction	Upper	0	55
	Lower	98	45
Or			
Brain metastasis PTVs (2 mm margin) translational setup correction	Upper	0	45
	Lower	98	35

be large enough to include the entire immobilization device. A 35 cm field of view covers most head immobilization devices and offers 0.7 mm resolution in the axial plane. The CT reconstruction resolution in the craniocaudal direction must be ≤1 mm for targeting metastases with high-dose radiotherapy.

The MRI sequence used for radiotherapy planning should be a gadolinium-enhanced T1, high-resolution 3D acquisition with axial and craniocaudal resolution of ≤1 mm to permit accurate segmentation of the metastases in three dimensions. T1 with gadolinium sequences sometimes exhibit high-signal artifacts that can resemble metastases, often near blood vessels. Thus, in addition to the high-resolution 3D sequence, it is important to acquire another T1 with gadolinium sequence in a different plane to avoid misinterpretation of the radiotherapy planning sequence and the inadvertent targeting of MRI artifacts. In addition, coregistrations of CT and MRI are imperfect: MRIs can have geometric distortions and artifact due to movement of the patients during acquisition. At our center, we use contrast enhancement for the planning CT scan because it is the fiducial image set on which the dose calculations are performed and the setup correction is executed. It is reassuring to have another contrast-enhanced imaging modality to verify the presence of metastases and to confirm the accuracy of the MRI-CT fusion in the regions of the high-dose boosts. CT imaging with contrast also helps to prevent the targeting of MRI artifacts.

16.6 STRUCTURE AND TARGET DELINEATION

Organs at risk must be contoured to ensure that the dose they receive is controlled in the planning optimization. Organs at risk include anterior chambers (lenses), retinas, brain stem, and spinal cord. In addition, the optic nerves and chiasm should be contoured as a single structure to avoid gaps that can be left between separate chiasm and optic nerve structures. Figure 16.1 illustrates the contours of a patient who was treated in our study. The whole-brain contour can be automatically generated by the planning software. If present, skull metastases can be added to the whole-brain clinical target volume to ensure that they receive the whole-brain prescription dose. This whole-brain and skull metastasis clinical target volume can be expanded by 2 mm to create a whole-brain PTV. The GTVs (or PTVs) of the metastases should be Boolean subtracted from the whole-brain PTV, so it can be used for dose–volume histogram calculations on the "normal" (or nontarget) brain.

Studies comparing different GTV to PTV margin widths for SRS showed that wider margins increase the risk of radionecrosis. In a study by Nataf et al. (2008), the incidence of radionecrosis was 20% for a 2 mm margin in contrast to 7% for a 0 mm margin. This study suggested that the PTV margin employed

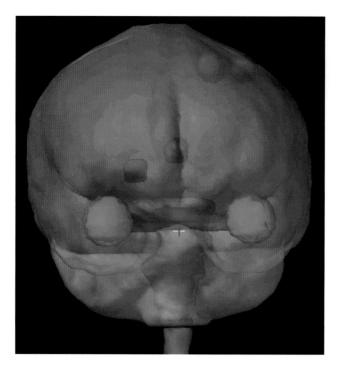

Figure 16.1 Contouring for WBRT and SIB. The Anterior Chambers (green) are the anterior 1/3 of the globes and the retinas (orange) are the posterior 2/3 of the globes. The Optics_PRV is the optic chiasm and optic nerves expanded by 2 mm (brown). The brainstem is magenta. The Spinal_PRV is the spinal cord expanded by 2 mm (cyan). The 2-mm thick Shells (blue) surround the GTVs (not visible). The whole brain PTV (yellow) is the brain expanded by 2 mm.

for SIB likely has a clinically significant effect on the risk of radionecrosis. The simplest prescription is to PTV = GTV because the multiple GTVs do not need margins added to create PTVs, but it is important to select a dose and a margin that are appropriate for the accuracy of the treatment planning process and the setup correction that is employed. With single-isocenter, multiple-target SIB, a 1° rotation around the plan isocenter could cause a 2 mm displacement of metastases near the skull (Lagerwaard et al., 2009; Peng et al., 2011). Thus, for treatment delivered with only translation setup correction, prescribing to a PTV defined by a 2 mm margin from the GTV is recommended (Table 16.4).

16.7 RADIOTHERAPY PLANNING

At our center, WBRT and SIB was delivered on Varian linear accelerators with multi-leaf collimators that had 5 mm central leaves using two 360° coplanar arcs at a dose rate of 600 monitor units/min, with the collimator set at 45°–30° for the clockwise arc and 315°–330° for the counterclockwise arc. For 5-fraction WBRT and SIB, dose constraints for the organs at risk and target doses are found in Table 16.4. This table illustrates two possible SIB prescriptions: either to GTVs with no margin or to PTVs with a margin on GTVs. The dose–volume histogram of a plan with five brain metastases is illustrated in Figure 16.2.

Dose profiles through SIB volumes for a patient on our study are illustrated in Figure 16.3. They show that, when VMAT is delivered with two coplanar axial arcs, the penumbra width around the high-dose target volumes is wider in the axial plane than in the craniocaudal direction. Noncoplanar beam arrangements can provide more uniform penumbras in three dimensions by reducing the penumbra width in the axial plane. However, noncoplanar arcs send a moderate exit dose through the whole body. In addition, they prolong treatment times considerably and diminish setup accuracy unless a patient-tracking system is employed at nonzero couch angles. They are very useful for reducing the dose to normal brain when planning VMAT SRS to metastases alone but rarely offer an advantage for WBRT and SIB, where the whole brain is a

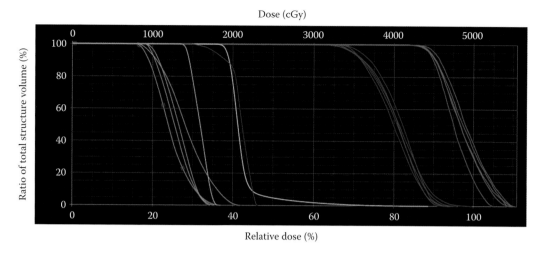

Figure 16.2 Dose–volume histograms for 20 Gy/5 WBRT and 45 Gy/5 SIB: GTVs, red; shells, blue; whole brain PTV, yellow; optics PRV, salmon; retinas, orange; anterior chambers, green; spinal cord PRV, cyan.

target volume. For patients with multiple brain metastases in the same axial plane, the narrower penumbra in the axial plane, which can be achieved with noncoplanar arcs, may be important to diminish the likelihood of bridges of high dose forming between the boost volumes. There are other strategies for eliminating high-dose bridges between adjacent metastases, including the use of multiple shells around the target volumes and the segmentation of avoidance structures between the target volumes (Thomas et al., 2014).

16.8 TREATMENT DELIVERY

The BC Cancer Agency phase 2 study of WBRT and SIB used daily online translational image-guided setup correction with orthogonal kV imaging. However, we now use CBCT for daily setup correction because of improvements in CBCT image reconstruction times. Automated 6-degree-of-freedom matching can detect smaller rotations than the radiation therapists can perceive with orthogonal kV imaging. Even when using a mask with a mouthpiece, rotations of over 1°, particularly in the sagittal plane, are sometimes detected by CBCT. A 6-degree-of-freedom robotic couch to correct for both translations and rotations is ideal for delivery of WBRT and SIB but is not required.

16.9 TOXICITY

An oral corticosteroid dose of dexamethasone 4 mg is recommended daily 1 hour before each fraction and in the mornings on weekend days during a 5-fraction course of WBRT and SIB. An oral antinauseant is also recommended 1 hour before each fraction. In our experience, headache and nausea are rare with the use of these prescriptions. For patients who are not taking dexamethasone before treatment, dexamethasone is stopped on the last day of WBRT and SIB. For patients who are taking dexamethasone before treatment, individualized tapering instructions are provided for after the treatment. The acute side effects of WBRT and SIB are the same as would be expected from WBRT alone.

Alopecia occurred after WBRT and SIB with a similar time course as the alopecia seen after conventional WBRT, so VMAT delivery did not reduce acute alopecia. This observation was similar to the observations of De Puysseleyr et al. who reported that alopecia, assessed at 1 month, was not diminished in the 10 patients treated in their study (De Puysseleyr et al., 2014). However, the hair grew back better at the vertex and occiput of the scalp than after WBRT with open beams. Thus, the use of IGRT setup for a 2 mm PTV margin around the brain can prevent the formation of the traditional permanent regions of alopecia caused by tangential dose in open beams.

In our study, acute and subacute serous otitis media occurred in several patients. The middle ears were not contoured as organs at risk. Because they are located between the inferior temporal lobes and the posterior

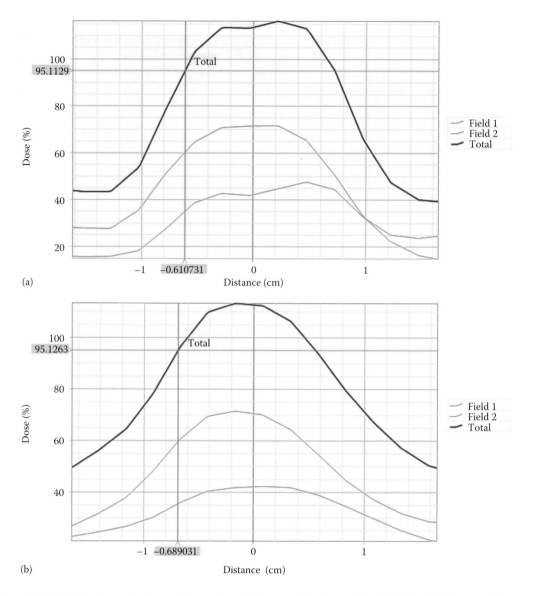

Figure 16.3 (a) Anterior–posterior and (b) craniocaudal dose profiles through the WBRT and SIB boost volume delivered with VMAT using clockwise and anticlockwise coplanar axial arcs. The penumbra is narrower in the craniocaudal direction than in the axial plane.

fossa, the VMAT dose optimization delivered the full WBRT prescription dose to the middle ears. The likelihood of serous otitis media might have been reduced by contouring the middle ear and Eustachian tubes as an organ at risk and limiting the dose using VMAT (Wang et al., 2009). The cochleas could have been spared at the same time by including them in "L Ear" and "R Ear" structures. For the middle ear and cochlea, recommended threshold doses are not known, so a dose that is as low as reasonably achievable below the WBRT dose would be a suitable goal (Bhandare et al., 2010; Theunissen et al., 2014).

16.10 CONTROVERSIES

There is ongoing debate about the relative cost of treatment with SRS alone, as compared to WBRT+SRS. Patients treated with SRS alone require many more SRS treatments for salvage because of their elevated risk of new brain metastases. With WBRT+SRS, up-front WBRT in 10–15 fractions is delivered to every patient, whereas with SRS alone, WBRT for salvage was required in only about 30% of patients (Ayala-Peacock et al., 2014). A study by Hall et al. (2014) showed that SRS alone was more cost effective than WBRT+SRS when long courses of WBRT were employed. It remains to be seen whether 5-fraction or 10-fraction WBRT and SIB might be more cost effective than either SRS alone or WBRT+SRS, because it reduces the duration of up-front treatment compared to WBRT+SRS and the need for salvage therapies compared to SRS alone.

16.11 FUTURE DIRECTIONS

A clinical trial comparing WBRT plus salvage SRS, if needed, WBRT and SIB, and SRS/stereotactic radiotherapy alone plus salvage WBRT, if needed, for patients with 2–10 metastases would help clarify the risks and benefits of these three approaches for patients with multiple brain metastases and could be designed to shed light on the controversy about the risks and benefits of WBRT. The 5-fraction and 10-fraction WBRT and SIB regimens could also be compared in a trial to determine which one offers the best balance of disease control and risk of radionecrosis.

HA-WBRT can be delivered with helical tomotherapy or VMAT (Gutierrez et al., 2007; Hsu et al., 2009; Awad et al., 2013; Oehlke et al., 2015). Following promising results of the RTOG 0933 study (Gondi et al., 2014), which showed less neurocognitive decline with HA-WBRT compared to historical controls, a randomized study (NRG CC001, clinicaltrials.gov identifier: NCT02360215) was designed to determine whether hippocampal sparing during WBRT reduces the subacute and long-term memory impairment that may occur after conventional WBRT. The results of this NRG Oncology study will determine whether HA offers clinical benefit. HA-WBRT and SIB has been shown to be technically and clinically feasible (Awad et al., 2013; Kim et al., 2015; Oehlke et al., 2015). The NRG CC001 study will determine whether HA-WBRT and SIB should be offered routinely.

CHECKLIST: KEY POINTS FOR CLINICAL PRACTICE

✓	ACTIVITY	SOME CONSIDERATIONS
	Patient selection	*Is the patient appropriate for WBRT and SIB?* • Up to 10 brain metastases • Largest metastasis <3 cm diameter • Patient with good enough prognosis to justify a boost to brain metastases • Patient with high enough risk of new brain metastases to justify use of WBRT
	Simulation	• Immobilization using a head shell designed to prevent rotation • CT and MRI imaging, both with 1 mm resolution in three dimensions • CT with contrast in the immobilization device • MRI with contrast for fusion with CT • Verify the accuracy of the CT-MRI fusion
	Treatment planning	• Metastases segmented on MRI and verified on CT • Whole brain segmented on CT with automated tools, if available • Organs at risk segmented on MRI and verified on CT • Inverse-planned volumetric modulated arc therapy using two coplanar arcs • With 6-degree-of-freedom setup correction, prescribe 45 Gy in 5 fractions to brain metastases with no margin and employ shells around the metastases to control dose falloff. Prescribe 35 Gy/5 to brain stem metastases • With 3-degree-of-freedom setup correction, add margins of 2 mm to make PTVs, employ shells around the PTVs to control dose falloff, add margins of 2 mm to make PRVs, and prescribe 35 Gy in 5 fractions to brain and brain stem metastasis PTVs

(Continued)

✓	ACTIVITY	SOME CONSIDERATIONS
	Treatment delivery	*Imaging* • Corticosteroids and antinauseants daily 1 hour before treatment • Clinical setup using lasers in head shell • For translational setup correction, cone-beam CT is used to assess rotation in head shell; if head rotation is >1°, correct • Position head clinically, repeat cone-beam CT to verify correction of head rotation, and apply translational setup correction • For translational and rotational setup correction, obtain cone-beam CT and apply the translational and rotational setup correction • No systemic cancer therapy for 1 week before, or until 1 week after, WBRT and SIB

REFERENCES

Abe E, Aoyama H (2012) The role of whole brain radiation therapy for the management of brain metastases in the era of stereotactic radiosurgery. *Curr Oncol Rep* 14:79–84.

Ahmed KA, Sarangkasiri S, Chinnaiyan P, Sahebjam S, Yu HH, Etame AB, Rao NG (2014) Outcomes following hypofractionated stereotactic radiotherapy in the management of brain metastases. *Am J Clin Oncol* 2014 Apr 21. [Epub ahead of print].

American Society of Radiation Oncology (September 14, 2014) ASTRO releases second list of five radiation oncology treatments to question, as part of national *choosing wisely*® campaign, 2015.

Andrews DW, Scott CB, Sperduto PW, Flanders AE, Gaspar LE, Schell MC, Werner-Wasik M et al. (2004) Whole brain radiation therapy with or without stereotactic radiosurgery boost for patients with one to three brain metastases: Phase III results of the RTOG 9508 randomised trial. *Lancet* 363:1665–1672.

Aoki M, Abe Y, Hatayama Y, Kondo H, Basaki K (2006) Clinical outcome of hypofractionated conventional conformation radiotherapy for patients with single and no more than three metastatic brain tumors, with noninvasive fixation of the skull without whole brain irradiation. *Int J Radiat Oncol Biol Phys* 64:414–418.

Aoyama H, Shirato H, Tago M, Nakagawa K, Toyoda T, Hatano K, Kenjyo M et al. (2006) Stereotactic radiosurgery plus whole-brain radiation therapy vs stereotactic radiosurgery alone for treatment of brain metastases: A randomized controlled trial. *JAMA* 295:2483–2491.

Aoyama H, Tago M, Kato N, Toyoda T, Kenjyo M, Hirota S, Shioura H et al. (2007) Neurocognitive function of patients with brain metastasis who received either whole brain radiotherapy plus stereotactic radiosurgery or radiosurgery alone. *Int J Radiat Oncol Biol Phys* 68:1388–1395.

Awad R, Fogarty G, Hong A, Kelly P, Ng D, Santos D, Haydu L (2013) Hippocampal avoidance with volumetric modulated arc therapy in melanoma brain metastases—The first australian experience. *Radiat Oncol* 8:62.

Ayala-Peacock DN, Peiffer AM, Lucas JT, Isom S, Kuremsky JG, Urbanic JJ, Bourland JD (2014) A nomogram for predicting distant brain failure in patients treated with gamma knife stereotactic radiosurgery without whole brain radiotherapy. *Neuro Oncol* 16:1283–1288.

Bauman G, Yartsev S, Fisher B, Kron T, Laperriere N, Heydarian M, VanDyk J (2007) Simultaneous infield boost with helical tomotherapy for patients with 1 to 3 brain metastases. *Am J Clin Oncol* 30:38–44.

Bhandare N, Jackson A, Eisbruch A, Pan CC, Flickinger JC, Antonelli P, Mendenhall WM (2010) Radiation therapy and hearing loss. *Int J Radiat Oncol Biol Phys* 76:S50–S57.

Chang EL, Wefel JS, Hess KR, Allen PK, Lang FF, Kornguth DG, Arbuckle RB et al. (2009) Neurocognition in patients with brain metastases treated with radiosurgery or radiosurgery plus whole-brain irradiation: A randomised controlled trial. *Lancet Oncol* 10:1037–1044.

Chen XJ, Xiao JP, Li XP, Jiang XS, Zhang Y, Xu YJ, Dai JR, Li YX (2011) Risk factors of distant brain failure for patients with newly diagnosed brain metastases treated with stereotactic radiotherapy alone. *Radiat Oncol* 6:175.

De Potter B, De Meerleer G, De Neve W, Boterberg T, Speleers B, Ost P (2013) Hypofractionated frameless stereotactic intensity-modulated radiotherapy with whole brain radiotherapy for the treatment of 1–3 brain metastases. *Neurol Sci* 34:647–653.

De Puysseleyr A, Van De Velde J, Speleers B, Vercauteren T, Goedgebeur A, Van Hoof T, Boterberg T, De Neve W, De Wagter C, Ost P (2014) Hair-sparing whole brain radiotherapy with volumetric arc therapy in patients treated for brain metastases: Dosimetric and clinical results of a phase II trial. *Radiat Oncol* 9:170.

Dhabaan A, Schreibmann E, Siddiqi A, Elder E, Fox T, Ogunleye T, Esiashvili N, Curran W, Crocker I, Shu HK (2012) Six degrees of freedom CBCT-based positioning for intracranial targets treated with frameless stereotactic radiosurgery. *J Appl Clin Med Phys* 13:3916.

Dincoglan F, Beyzadeoglu M, Sager O, Oysul K, Sirin S, Surenkok S, Gamsiz H, Uysal B, Demiral S, Dirican B (2012) Image-guided positioning in intracranial non-invasive stereotactic radiosurgery for the treatment of brain metastasis. *Tumori* 98:630–635.

Eaton BR, Gebhardt B, Prabhu R, Shu HK, Curran WJ, Jr, Crocker I (2013) Hypofractionated radiosurgery for intact or resected brain metastases: Defining the optimal dose and fractionation. *Radiat Oncol* 8:135.

Edwards AA, Keggin E, Plowman PN (2010) The developing role for intensity-modulated radiation therapy (IMRT) in the non-surgical treatment of brain metastases. *Br J Radiol* 83:133–136.

Ernst-Stecken A, Ganslandt O, Lambrecht U, Sauer R, Grabenbauer G (2006) Phase II trial of hypofractionated stereotactic radiotherapy for brain metastases: Results and toxicity. *Radiother Oncol* 81:18–24.

Fahrig A, Ganslandt O, Lambrecht U, Grabenbauer G, Kleinert G, Sauer R, Hamm K (2007) Hypofractionated stereotactic radiotherapy for brain metastases–results from three different dose concepts. *Strahlenther Onkol* 183:625–630.

Flickinger JC, Kondziolka D, Maitz AH, Lunsford LD (1998) Analysis of neurological sequelae from radiosurgery of arteriovenous malformations: How location affects outcome. *Int J Radiat Oncol Biol Phys* 40:273–278.

Fokas E, Henzel M, Surber G, Kleinert G, Hamm K, Engenhart-Cabillic R (2012) Stereotactic radiosurgery and fractionated stereotactic radiotherapy: Comparison of efficacy and toxicity in 260 patients with brain metastases. *J Neurooncol* 109:91–98.

Gondi V, Paulus R, Bruner DW, Meyers CA, Gore EM, Wolfson A, Werner-Wasik M, Sun AY, Choy H, Movsas B (2013) Decline in tested and self-reported cognitive functioning after prophylactic cranial irradiation for lung cancer: Pooled secondary analysis of radiation therapy oncology group randomized trials 0212 and 0214. *Int J Radiat Oncol Biol Phys* 86:656–664.

Gondi V, Pugh SL, Tome WA, Caine C, Corn B, Kanner A, Rowley H et al. (2014) Preservation of memory with conformal avoidance of the hippocampal neural stem-cell compartment during whole-brain radiotherapy for brain metastases (RTOG 0933): A phase II multi-institutional trial. *J Clin Oncol* 32:3810–3816.

Gondi V, Tolakanahalli R, Mehta MP, Tewatia D, Rowley H, Kuo JS, Khuntia D, Tome WA (2010a) Hippocampal-sparing whole-brain radiotherapy: A "how-to" technique using helical tomotherapy and linear accelerator-based intensity-modulated radiotherapy. *Int J Radiat Oncol Biol Phys* 78:1244–1252.

Gutierrez AN, Westerly DC, Tome WA, Jaradat HA, Mackie TR, Bentzen SM, Khuntia D, Mehta MP (2007) Whole brain radiotherapy with hippocampal avoidance and simultaneously integrated brain metastases boost: A planning study. *Int J Radiat Oncol Biol Phys* 69:589–597.

Hall MD, McGee JL, McGee MC, Hall KA, Neils DM, Klopfenstein JD, Elwood PW (2014) Cost-effectiveness of stereotactic radiosurgery with and without whole-brain radiotherapy for the treatment of newly diagnosed brain metastases. *J Neurosurg* 121(Suppl):84–90.

Hsu F, Carolan H, Nichol A, Cao F, Nuraney N, Lee R, Gete E, Wong F, Schmuland M, Heran M, Otto K (2010) Whole brain radiotherapy with hippocampal avoidance and simultaneous integrated boost for 1–3 brain metastases: A feasibility study using volumetric modulated arc therapy. *Int J Radiat Oncol Biol Phys* 76(5):1480–1485.

Hsu F, Nichol A, Ma R, Kouhestani P, Toyota B, McKenzie M (2015) Stereotactic radiosurgery for metastases in eloquent central brain locations. *Can J Neurol Sci* 42(5):333–337.

Jiang XS, Xiao JP, Zhang Y, Xu YJ, Li XP, Chen XJ, Huang XD, Yi JL, Gao L, Li YX (2012) Hypofractionated stereotactic radiotherapy for brain metastases larger than three centimeters. *Radiat Oncol* 7:36.

Kased N, Huang K, Nakamura JL, Sahgal A, Larson DA, McDermott MW, Sneed PK (2008) Gamma knife radiosurgery for brainstem metastases: The UCSF experience. *J Neurooncol* 86:195–205.

Kilburn JM, Ellis TL, Lovato JF, Urbanic JJ, Daniel Bourland J, Munley MT, Deguzman AF et al. (2014) Local control and toxicity outcomes in brainstem metastases treated with single fraction radiosurgery: Is there a volume threshold for toxicity? *J Neurooncol* 117:167–174.

Kim KH, Cho BC, Lee CG, Kim HR, Suh YG, Kim JW, Choi C, Baek JG, Cho J (2015) Hippocampus-sparing whole-brain radiotherapy and simultaneous integrated boost for multiple brain metastases from lung adenocarcinoma: Early response and dosimetric evaluation. *Technol Cancer Res Treat* 2015 Jan 18. pii: 1533034614566993. [Epub ahead of print].

Kocher M, Soffietti R, Abacioglu U, Villa S, Fauchon F, Baumert BG, Fariselli L et al. (2011) Adjuvant whole-brain radiotherapy versus observation after radiosurgery or surgical resection of one to three cerebral metastases: Results of the EORTC 22952-26001 study. *J Clin Oncol* 29:134–141.

Koyfman SA, Tendulkar RD, Chao ST, Vogelbaum MA, Barnett GH, Angelov L, Weil RJ, Neyman G, Reddy CA, Suh JH (2010) Stereotactic radiosurgery for single brainstem metastases: The Cleveland clinic experience. *Int J Radiat Oncol Biol Phys* 78:409–414.

Kwon AK, Dibiase SJ, Wang B, Hughes SL, Milcarek B, Zhu Y (2009) Hypofractionated stereotactic radiotherapy for the treatment of brain metastases. *Cancer* 115:890–898.

Lagerwaard FJ, van der Hoorn EA, Verbakel WF, Haasbeek CJ, Slotman BJ, Senan S (2009) Whole-brain radiotherapy with simultaneous integrated boost to multiple brain metastases using volumetric modulated arc therapy. *Int J Radiat Oncol Biol Phys* 75:253–259.

Leeman JE, Clump DA, Wegner RE, Heron DE, Burton SA, Mintz AH (2012) Prescription dose and fractionation predict improved survival after stereotactic radiotherapy for brainstem metastases. *Radiat Oncol* 7:107.

Li J, Bentzen SM, Renschler M, Mehta MP (2007) Regression after whole-brain radiation therapy for brain metastases correlates with survival and improved neurocognitive function. *J Clin Oncol* 25:1260–1266.

Lindvall P, Bergstrom P, Lofroth PO, Henriksson R, Bergenheim AT (2005) Hypofractionated conformal stereotactic radiotherapy alone or in combination with whole-brain radiotherapy in patients with cerebral metastases. *Int J Radiat Oncol Biol Phys* 61:1460–1466.

Martens B, Janssen S, Werner M, Fruhauf J, Christiansen H, Bremer M, Steinmann D (2012) Hypofractionated stereotactic radiotherapy of limited brain metastases: A single-centre individualized treatment approach. *BMC Cancer* 12:497-2407-12-497.

McDuff SG, Taich ZJ, Lawson JD, Sanghvi P, Wong ET, Barker FG II, Hochberg FH et al. (2013) Neurocognitive assessment following whole brain radiation therapy and radiosurgery for patients with cerebral metastases. *J Neurol Neurosurg Psychiatry* 84:1384–1391.

Mehta MP (2015) The controversy surrounding the use of whole-brain radiotherapy in brain metastases patients. *Neuro Oncol* 17:919–923.

Meyers CA, Smith JA, Bezjak A, Mehta MP, Liebmann J, Illidge T, Kunkler I et al. (2004) Neurocognitive function and progression in patients with brain metastases treated with whole-brain radiation and motexafin gadolinium: Results of a randomized phase III trial. *J Clin Oncol* 22:157–165.

Murai T, Ogino H, Manabe Y, Iwabuchi M, Okumura T, Matsushita Y, Tsuji Y, Suzuki H, Shibamoto Y (2014) Fractionated stereotactic radiotherapy using CyberKnife for the treatment of large brain metastases: A dose escalation study. *Clin Oncol (R Coll Radiol)* 26:151–158.

Narayana A, Chang J, Yenice K, Chan K, Lymberis S, Brennan C, Gutin PH (2007) Hypofractionated stereotactic radiotherapy using intensity-modulated radiotherapy in patients with one or two brain metastases. *Stereotact Funct Neurosurg* 85:82–87.

Nataf F, Schlienger M, Liu Z, Foulquier JN, Gres B, Orthuon A, Vannetzel JM et al. (2008) Radiosurgery with or without A 2-mm margin for 93 single brain metastases. *Int J Radiat Oncol Biol Phys* 70:766–772.

Nichol A, McKenzie M, Ma R, Hsu F, Cheung AK, Schellenberg D, Gondara L et al. (2015) Whole brain radiation therapy with simultaneous integrated boost using volumetric modulated arc therapy: A phase II study for 1 to 10 brain metastases. *Int J Radiat Oncol Biol Phys* 90:S25.

Oehlke O, Wucherpfennig D, Fels F, Frings L, Egger K, Weyerbrock A, Prokic V, Nieder C, Grosu AL (2015) Whole brain irradiation with hippocampal sparing and dose escalation on multiple brain metastases: Local tumour control and survival. *Strahlenther Onkol* 191(6):461–469.

Ogura K, Mizowaki T, Ogura M, Sakanaka K, Arakawa Y, Miyamoto S, Hiraoka M (2012) Outcomes of hypofractionated stereotactic radiotherapy for metastatic brain tumors with high risk factors. *J Neuro-Oncol* 109:425–432.

Patchell RA, Tibbs PA, Regine WF, Dempsey RJ, Mohiuddin M, Kryscio RJ, Markesbery WR, Foon KA, Young B (1998) Postoperative radiotherapy in the treatment of single metastases to the brain: A randomized trial. *JAMA* 280:1485–1489.

Peng JL, Liu C, Chen Y, Amdur RJ, Vanek K, Li JG (2011) Dosimetric consequences of rotational setup errors with direct simulation in a treatment planning system for fractionated stereotactic radiotherapy. *J Appl Clin Med Phys* 12:3422.

Rajakesari S, Arvold ND, Jimenez RB, Christianson LW, Horvath MC, Claus EB, Golby AJ et al. (2014) Local control after fractionated stereotactic radiation therapy for brain metastases. *J Neurooncol* 120:339–346.

Regine WF, Scott C, Murray K, Curran W (2001) Neurocognitive outcome in brain metastases patients treated with accelerated-fractionation vs. accelerated-hyperfractionated radiotherapy: An analysis from radiation therapy oncology group study 91-04. *Int J Radiat Oncol Biol Phys* 51:711–717.

Rodrigues G, Eppinga W, Lagerwaard F, de Haan P, Haasbeek C, Perera F, Slotman B, Yaremko B, Yartsev S, Bauman G (2012b) A pooled analysis of arc-based image-guided simultaneous integrated boost radiation therapy for oligometastatic brain metastases. *Radiother Oncol* 102:180–186.

Rodrigues G, Warnemr A, Zindler J, Slotman B, Lagerwaard F (2014) A clinical nomogram and recursive partitioning analysis to determine the risk of regional failure after radiosurgery alone for brain metastases. *Radiother Oncol* 111:52–58.

Rodrigues G, Yartsev S, Tay KY, Pond GR, Lagerwaard F, Bauman G (2012a) A phase II multi-institutional study assessing simultaneous in-field boost helical tomotherapy for 1–3 brain metastases. *Radiat Oncol* 7:42.

Rodrigues G, Yartsev S, Yaremko B, Perera F, Dar AR, Hammond A, Lock M et al. (2011) Phase I trial of simultaneous in-field boost with helical tomotherapy for patients with one to three brain metastases. *Int J Radiat Oncol Biol Phys* 80:1128–1133.

Roos DE, Wirth A, Burmeister BH, Spry NA, Drummond KJ, Beresford JA, McClure BE (2006) Whole brain irradiation following surgery or radiosurgery for solitary brain metastases: Mature results of a prematurely closed randomized Trans-Tasman radiation oncology group trial (TROG 98.05). *Radiother Oncol* 80:318–322.

Sahgal A (2015) Point/counterpoint: Stereotactic radiosurgery without whole-brain radiation for patients with a limited number of brain metastases: The current standard of care? *Neuro Oncol* 17:916–918.

Sahgal A, Aoyama H, Kocher M, Neupane B, Collette S, Tago M, Shaw P, Beyene J, Chang EL (2015) Phase 3 trials of stereotactic radiosurgery with or without whole-brain radiation therapy for 1 to 4 brain metastases: Individual patient data meta-analysis. *Int J Radiat Oncol Biol Phys* 91:710–717.

Samblas JM, Sallabanda K, Bustos JC, Gutierrez-Diaz JA, Peraza C, Beltran C, Samper PM (2009) Radiosurgery and whole brain therapy in the treatment of brainstem metastases. *Clin Transl Oncol* 11:677–680.

Scorsetti M, Facoetti A, Navarria P, Bignardi M, De Santis M, Ninone SA, Lattuada P, Urso G, Vigorito S, Mancosu P, Del Vecchio M (2009) Hypofractionated stereotactic radiotherapy and radiosurgery for the treatment of patients with radioresistant brain metastases. *Anticancer Res* 29:4259–4263.

Soffietti R, Kocher M, Abacioglu UM, Villa S, Fauchon F, Baumert BG, Fariselli L et al. (2013) A European organisation for research and treatment of cancer phase III trial of adjuvant whole-brain radiotherapy versus observation in patients with one to three brain metastases from solid tumors after surgical resection or radiosurgery: Quality-of-life results. *J Clin Oncol* 31:65–72.

Soon YY, Tham IW, Lim KH, Koh WY, Lu JJ (2014) Surgery or radiosurgery plus whole brain radiotherapy versus surgery or radiosurgery alone for brain metastases. *Cochrane Database Syst Rev* 3:CD009454.

Sperduto PW, Shanley R, Luo X, Andrews D, Werner-Wasik M, Valicenti R, Bahary JP, Souhami L, Won M, Mehta M (2014) Secondary analysis of RTOG 9508, a phase 3 randomized trial of whole-brain radiation therapy versus WBRT plus stereotactic radiosurgery in patients with 1–3 brain metastases; poststratified by the graded prognostic assessment (GPA). *Int J Radiat Oncol Biol Phys* 90:526–531.

Sterzing F, Welzel T, Sroka-Perez G, Schubert K, Debus J, Herfarth KK (2009) Reirradiation of multiple brain metastases with helical tomotherapy: A multifocal simultaneous integrated boost for eight or more lesions. *Strahlenther Onkol* 185:89–93.

Theelen A, Martens J, Bosmans G, Houben R, Jager JJ, Rutten I, Lambin P, Minken AW, Baumert BG (2012) Relocatable fixation systems in intracranial stereotactic radiotherapy. Accuracy of serial CT scans and patient acceptance in a randomized design. *Strahlenther Onkol* 188:84–90.

Theunissen EA, Zuur CL, Yurda ML, van der Baan S, Kornman AF, de Boer JP, Balm AJ, Rasch CR, Dreschler WA (2014) Cochlea sparing effects of intensity modulated radiation therapy in head and neck cancers patients: A long-term follow-up study. *J Otolaryngol Head Neck Surg* 43:30.

Thomas EM, Popple RA, Wu X, Clark GM, Markert JM, Guthrie BL, Yuan Y, Dobelbower MC, Spencer SA, Fiveash JB (2014) Comparison of plan quality and delivery time between volumetric arc therapy (RapidArc) and gamma knife radiosurgery for multiple cranial metastases. *Neurosurgery* 75:409–417; discussion 417–418.

Tomita N, Kodaira T, Tachibana H, Nakamura T, Nakahara R, Inokuchi H, Shibamoto Y (2008) Helical tomotherapy for brain metastases: Dosimetric evaluation of treatment plans and early clinical results. *Technol Cancer Res Treat* 7:417–424.

Tsao MN, Lloyd N, Wong RK, Chow E, Rakovitch E, Laperriere N, Xu W, Sahgal A (2012) Whole brain radiotherapy for the treatment of newly diagnosed multiple brain metastases. *Cochrane Database Syst Rev* 4:CD003869.

Valery CA, Boskos C, Boisserie G, Lamproglou I, Cornu P, Mazeron JJ, Simon JM (2011) Minimized doses for linear accelerator radiosurgery of brainstem metastasis. *Int J Radiat Oncol Biol Phys* 80(2):362–368.

Wang SZ, Li J, Miyamoto CT, Chen F, Zhou LF, Zhang HY, Yang G et al. (2009) A study of middle ear function in the treatment of nasopharyngeal carcinoma with IMRT technique. *Radiother Oncol* 93:530–533.

Weber DC, Caparrotti F, Laouiti M, Malek K (2011) Simultaneous in-field boost for patients with 1 to 4 brain metastasis/es treated with volumetric modulated arc therapy: A prospective study on quality-of-life. *Radiat Oncol* 6:79.

Wegner RE, Oysul K, Pollock BE, Sirin S, Kondziolka D, Niranjan A, Lunsford LD, Flickinger JC (2011) A modified radiosurgery-based arteriovenous malformation grading scale and its correlation with outcomes. *Int J Radiat Oncol Biol Phys* 79:1147–1150.

Wiggenraad R, Verbeek-de Kanter A, Kal HB, Taphoorn M, Vissers T, Struikmans H (2011) Dose-effect relation in stereotactic radiotherapy for brain metastases. A systematic review. *Radiother Oncol* 98:292–297.

Wolfson AH, Bae K, Komaki R, Meyers C, Movsas B, Le Pechoux C, Werner-Wasik M, Videtic GM, Garces YI, Choy H (2011) Primary analysis of a phase II randomized trial radiation therapy oncology group (RTOG) 0212: Impact of different total doses and schedules of prophylactic cranial irradiation on chronic neurotoxicity and quality of life for patients with limited-disease small-cell lung cancer. *Int J Radiat Oncol Biol Phys* 81:77–84.

Yuan J, Wang JZ, Lo S, Grecula JC, Ammirati M, Montebello JF, Zhang H, Gupta N, Yuh WT, Mayr NA (2008) Hypofractionation regimens for stereotactic radiotherapy for large brain tumors. *Int J Radiat Oncol Biol Phys* 72:390–397.

17 Image-guided hypofractionated radiation therapy for high-grade glioma

John Cuaron and Kathryn Beal

Contents

17.1 INTRODUCTION

Glioblastoma is the most common high-grade glioma and the most common primary brain tumor in adults. It is characterized by an extremely aggressive clinical course and poor prognosis. Even with optimal treatment of maximal resection followed by concurrent chemoradiation and adjuvant chemotherapy, patients typically experience recurrence and have a median survival of less than 2 years. Following surgery, radiation has been the mainstay of treatment since the 1970s following the positive result of the BTSG studies (Walker et al., 1978, 1980). Since that time, many studies of radiation therapy (RT) have been conducted with little avail in terms of improving outcomes. Specifically, studies of dose escalation, hyperfractionation, radiosurgery, and brachytherapy boosts have not yielded benefit. However, technological advances in the delivery of RT can allow for the use of image-guided hypofractionated stereotactic radiotherapy (IG-HSRT) in an attempt to improve outcomes in this aggressive disease. IG-HSRT may improve the therapeutic ratio with increased effectiveness, decreased side effects, and shorter treatment times for patients. This chapter will review (1) current indications and evidence supporting IG-HSRT for both newly diagnosed and recurrent malignant glioma, (2) technical approaches to the planning and delivery of IG-HSRT, (3) potential toxicities and their management, and (4) controversies and future directions for the use of IG-HSRT for malignant glioma.

17.2 INDICATIONS

17.2.1 NEWLY DIAGNOSED GLIOBLASTOMA

The current standard of care for the treatment of newly diagnosed glioblastoma was defined by the EORTC/NCIC trial (Stupp et al., 2005, 2009). In this study, 573 patients were randomized between postoperative RT alone and RT with concurrent and adjuvant temozolomide. Combined modality treatment significantly improved median overall survival from 12.1 to 14.6 months. Subsequent efforts to improve outcomes by investigating various radiotherapy techniques and fractionation schemes have largely been unsuccessful. The use of a stereotactic radiosurgery boost after chemoradiation also failed to show benefit in median survival or quality of life in the RTOG 9305 study (Loeffler et al., 1992). Recursive partitioning analysis of a hyperfractionation study from the RTOG also showed no benefit over standard therapy (Scott et al., 1998). Similarly, no improvement in overall survival was seen with the use of I-125 interstitial brachytherapy boosts at the time of surgery (Selker et al., 2002) nor with dose escalation utilizing conventional fractionation (Graf et al., 2005; Tsien et al. 2009; Watkins et al., 2009).

These data have called into question the efficacy of standard fractionation doses and have provided rationale of investigating larger doses per fraction. The use of hypofractionation in the delivery of radiotherapy has several theoretical advantages. First, hypofractionation offers the radiobiological advantages of delivering a higher dose per fraction, allowing for higher direct tumor cell kill and inhibiting cellular repopulation in glioblastoma cells (Kaaijk et al., 1997). Estimates of the alpha/beta ratio of glioblastoma range from 2 to 5, which argues for a higher dose per fraction than conventionally fractionated treatments. Finally, a hypofractionated approach is more convenient for patients by minimizing the amount of time that patients with a short life expectancy spend within the radiation oncology department and the time traveling to and from appointments each day.

Early studies of abbreviated RT and high doses per fraction were conducted with patients with an especially poor prognosis, including elderly patients and those with a poor performance status (Roa et al., 2004; Hingorani et al., 2012). Recent prospective efforts have taken advantage of intensity modulation and simultaneous integrated boost capabilities to study hypofractionation in less selected populations. Floyd et al. (2004) reported on 20 patients treated with 50 Gy delivered in 5 Gy daily fractions to the enhancing disease and simultaneous 30 Gy in 3 Gy daily fractions to the surrounding edema and showed low survival and significant rates of radiation necrosis requiring surgical excision. Investigators from the University of Colorado reported on a phase I dose escalation trial (Chen et al., 2011) and subsequent phase II trial delivering dose-painted 30–60 Gy in 10 fractions with concurrent temozolomide (Reddy et al., 2012). Median survival was comparable to rates achieved in the Stupp trial, although four out of six patients who underwent surgical excision for suspected recurrence were found to have >80% necrosis.

An encouraging median survival of 20 months with dose painting up to 68 Gy in 8 fractions was reported by Iuchi et al. (2014), although the authors did not report the rates of radiation necrosis requiring surgical intervention. Most recently, investigators at Memorial Sloan Kettering Cancer Center conducted a phase II trial investigating concurrent bevacizumab and temozolomide with dose-painted image-guided hypofractionated stereotactic radiotherapy (Omuro et al., 2014). With a median follow-up of 42 months among surviving patients, the median overall survival was 19 months, and the overall survival was 93% at 1 year. Notably, both the clinical and radiographic incidence of radionecrosis was lower in this study compared to the other studies investigating hypofractionation in the up-front setting. Studies investigating the use of hypofractionation with 10 or less fractions in newly diagnosed glioma are summarized in Table 17.1. Overall, these studies demonstrate IG-HSRT to be a safe, effective, and convenient therapy approach for newly diagnosed glioma that requires further study in the randomized trial setting.

17.2.2 RECURRENT GLIOBLASTOMA

Although adjuvant chemoradiation therapy was shown to improve overall survival in a randomized phase III trial for newly diagnosed malignant glioblastoma (Stupp et al., 2005, 2009), recurrence at the site of the original tumor is nearly inevitable. Treatment options include surgical resection, chemotherapy, and reirradiation. Because of the pattern of spread of recurrent tumors, maximal surgical excision is often not possible. Although brachytherapy and single-fraction stereotactic radiosurgery have been employed in the recurrent setting, they have been associated with significant toxicity. With limited treatment options, limited volume hypofractionated reirradiation is sometimes employed.

As with use in the up-front setting, IG-HSRT in the recurrent setting provides both radiobiological and technical advantages over conventionally fractionated schedules. Tumor that has recurred after conventional chemoradiation has declared itself to be radioresistant, and the use of higher doses per fraction is presumably more radiobiologically efficacious. Further, the use of intensity modulation and CT-based image guidance allows for the optimal sparing of normal structures and mitigation of toxicity

Table 17.1 Studies of hypofractionated stereotactic radiation therapy in newly diagnosed high-grade glioma

AUTHOR	PATIENTS	STUDY TYPE	MEDIAN DOSE	NUMBER OF FRACTIONS	BED (a/b = 3)	MEDIAN TUMOR VOLUME (RANGE)	CONCURRENT SYSTEMIC AGENT	TOXICITY REQUIRING SURGICAL INTERVENTION	MEDIAN OVERALL SURVIVAL (MONTHS)
Floyd et al. (2004)	20	Prospective phase I/II	50/30[a]	10	133.3/60	NR[b]	None	15%	7
Reddy et al. (2012)	24	Prospective phase II	60/30[a]	10	180/60	97.87 (53.9–145.09)/258.04 (126–452.49)[c]	TMZ	17%	16.6
Iuchi et al. (2014)	46	Prospective phase II	68/40/32[a]	8	260.7/106.7/74.7	80.9 (26.5–267.2)/160.7 (78.5–374.3)/0.6 (0–65.9)[d]	TMZ	NR[b]	20
Omuro et al. (2014)	40	Prospective phase II	36/24[a]	6	108/56	NR[b]	TMZ + bevacizumab	5%	19

[a] Dose painting.
[b] Not reported.
[c] PTV$_1$/PTV$_2$.
[d] PTV$_1$/PTV$_2$/PTV$_3$.

risk associated with re-irradiation. There is both retrospective and prospective data on the safety and efficacy of reirradiation with hypofractionated RT (using 10 fractions or less) for glioblastoma, summarized in Table 17.2. The interest and use of this treatment approach has been growing as evidenced by the number of studies published, especially within the past 10 years. An early study by Laing et al. (1993) used a range of 20–50 Gy delivered in 4–10 fractions and demonstrated a median overall survival of 9.8 months. Other studies have shown median overall survival rates after reirradiation ranging from 7.0 to 18.0 months. These rates compare favorably to series utilizing stereotactic radiotherapy with standard fraction sizes. Cho et al. (1999) reported a median survival of 12 months after the delivery of 37.5 Gy in 2.5 Gy fractions and, in the largest series of patients with recurrent glioma treated with fractionated stereotactic reirradiation to date, Combs et al. (2005) reported a median survival of 8 months for patients with GBM treated to a median dose of 36 Gy in 2 Gy fractions. Outcomes of IG-HSRT for recurrent glioma also compare favorable to the use of both cytotoxic and molecularly targeted chemotherapy. Series investigating treatment approaches with chemotherapy alone have reported median overall survival times ranging from 4.4 to 9.8 months (Bambury and Morris, 2014).

The rates of radiation necrosis requiring surgical intervention after IG-HSRT appear to be acceptably low (ranging from 0% to 15%) with the exception of the Voynov and Kohshi studies, which reported rates of 20% and 28%, respectively (Voynov et al., 2002; Kohshi et al., 2007). Various systemic agents, including cytotoxic chemotherapies and targeted agents, have been combined with reirradiation without evidence of excessive toxicity. Our institution is currently enrolling patients on a phase I dose escalation protocol using IG-HSRT in combination with bevacizumab for recurrent high-grade glioma, based on encouraging early results from a previously published pilot study (Gutin et al., 2009). However, follow-up is limited, and prognosis remains poor for patients with recurrent disease.

17.3 SIMULATION, TARGET DELINEATION, AND TREATMENT DELIVERY

At our institution, prior to simulation, patients undergo an MRI of the brain with contrast with thin cuts, ideally 1 mm in size. Patients are simulated in the supine position utilizing a three-point thermoplastic mask. More recently, immobilization is achieved with a custom head mold and open-face mask (CDR Systems, Calgary, Alberta, Canada). A CT scan utilizing 1 mm slice thickness through the entire brain and brain stem and with IV contrast is obtained, and images are fused to the MRI.

For patients with newly diagnosed glioblastoma, we employ a dose-painting technique. The initial gross tumor volume (GTV_1) is defined as any postoperative residual contrast-enhancing tumor. A 5 mm margin is added to create a CTV_1 and an additional 5 mm is added to create the planning target volume (PTV_1). GTV_2 is created by contouring the FLAIR abnormality as defined on the postoperative MRI. A 2 cm margin is added to create CTV_2 and an additional 5 mm is added to create PTV_2. For patients treated in the recurrent setting, the GTV is generated by contouring all enhancing disease on the T1 postcontrast sequence. If a gross total resection was performed, the GTV is defined as the postoperative cavity. Adjacent T2 FLAIR signal abnormality is also included in the GTV at the discretion of the treating physician. A PTV is directly generated by expanding the GTV by 5 mm. The eyes, optic nerves, optic chiasm, brain stem, and spinal cord are contoured as organs at risk.

Inverse treatment planning is performed with Eclipse software (Varian Medical Systems, Palo Alto, CA). For patients treated in the up-front setting, a plan is generated to deliver a total of 36 Gy to the PTV_1 and 24 Gy to PTV_2 in 5 fractions, utilizing intensity-modulated RT with sliding window technique. For recurrent glioma, the PTV is treated to 30 Gy in 5 fractions. Coverage is deemed acceptable if the prescription dose is delivered to at least 95% of the PTV. Dose constraints for normal structures include a max point dose of 23 Gy to the optic structures, max point dose of 30 Gy to the spinal cord, and D_{05} of 30 Gy to the brain stem.

Treatment is delivered every other day with a Varian TrueBeam Linear Accelerator with onboard imaging capability. Patients are first set up conventionally using room lasers and manual shifts. If the CDR head frame is used, an optical surface imaging program (AlignRT, VisionRT, London, UK) is used in conjunction with a treatment couch with 6 degrees of freedom to correct translational and rotational shifts.

Table 17.2 Studies of hypofractionated stereotactic radiation therapy in recurrent high-grade glioma

AUTHOR	PATIENTS	STUDY TYPE	MEDIAN DOSE	MEDIAN NUMBER OF FRACTIONS	BED (a/b = 3)	MEDIAN TUMOR VOLUME (cm^3)	CONCURRENT SYSTEMIC AGENT	TOXICITY REQUIRING SURGICAL INTERVENTION	MEDIAN OS AFTER SALVAGE RT
Laing et al. (1993)	22	Prospective phase I/II	40 (20–50)	8 (4–10)	106.7	25 (1–93)	None	NR	9.8 months
Glass et al. (1997)	20	Prospective phase I/II	42 (37.5–42)	7 (7–10)	126	14.3 (1.76–122)	Cisplatin	15%	12.7 months
Shepherd et al. (1997)	29	Retrospective	35 (20–50)	7 (4–10)	93.3	24 (3–93)	None	6%	10.7 months
Hudes et al. (1999)	19	Prospective phase I	30 (21–35)	10 (7–10)	60	12.66 (0.89–47.5)	None	0%	10.5 months
Lederman et al. (2000)	88	Prospective phase I/II	24 (18–36)	4	72	32.7 (1.5–150.3)	Paclitaxel	8%	7 months
Voynov et al. (2002)	10	Retrospective	30 (25–40)	4 (2–5)	105	34.69 (4.29–75.23)	Various	20%	10.1 months
Vordermark et al. (2005)	19	Retrospective	30 (20–30)	5 (2–6)	90	15 (4–70)[a]	None	0%	9.3 months
Grosu et al. (2005)	44	Prospective phase I/II	30	6	80	19	Temozolomide	7%	8 months
Kohshi et al. (2007)	25 (11 GBM, 14 AA)	Retrospective	22 (18–27)	8	42.2	8.7 (1.7–159.3)	None	28%	11 months (GBM)
Ernst-Strecken et al. (2007)	15	Prospective phase I/II	35	7	93.3	5.75 (0.77–21.94)	None	NR	12 months
Patel et al. (2009)	10	Retrospective	36	6	108	51.1 (16.1–123.3)	None	10%	7.4 months
Gutin et al. (2009)	25	Prospective phase I/II	30	5	90	34 (2–62)[a]	Bevacizumab	4%	12.5 months

(Continued)

Table 17.2 (Continued) Studies of hypofractionated stereotactic radiation therapy in recurrent high-grade glioma

AUTHOR	PATIENTS	STUDY TYPE	MEDIAN DOSE	MEDIAN NUMBER OF FRACTIONS	BED (a/b = 3)	MEDIAN TUMOR VOLUME (cm³)	CONCURRENT SYSTEMIC AGENT	TOXICITY REQUIRING SURGICAL INTERVENTION	MEDIAN OS AFTER SALVAGE RT
Henke et al. (2009)	31	Retrospective	20 (20–25)	5 (4–5)	46.7	55 (0.9–277)[a]	Various	6%	10.2 months
Fokas et al. (2009)	53	Retrospective	32.5 (20–60)	10 (5–30)	67.7	35.01 (3–204)[a]	TMZ, ACNU/ VM-26, or PCV	0%	9 months
Fogh et al. (2010)	147	Retrospective	35	10	75.8	22 (0.6–104)	Various	0%	11 months (GBM)
Kim et al. (2011)	8	Retrospective	25	5	66.7	69.5[a]	None	13%	7.6 months
Minniti et al. (2011)	54 (42 GBM, 12 AA)	Prospective phase I/II	25	5	66.7	13.1 (1–35.3)	Bevacizumab or fotemustine	5%	13.0 months, AA 9 months, GBM
McKenzie et al. (2013)	35	Retrospective	30 (8–30)	5 (1–5)	90	8.54 (0.4–46.56)[a]	Various	0%	8.6 months
Ciammella et al. (2013)	15	Retrospective	25	5	66.7	NR[b]	None	0%	9.5 months
Wuthrick et al. (2014)	11	Prospective phase I/II	35 (30–42)	10 (10–15)	75.8	16.75 (0.05–72.01)	Sunitinib	0%	11 months
Miwa et al. (2014)	21	Prospective phase I/II	25–35	5	66.7–116.7	27.4 (3.4–102.9)[a]	None	4.8%	11 months
Navarria et al. (2015)	25	Retrospective	25 (20–50)	5 (5–10)	66.7	35 (2.46–116.7)	Various	0%	18 months

[a] Planning target volume.
[b] Not reported.

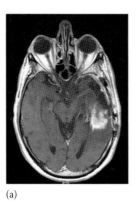
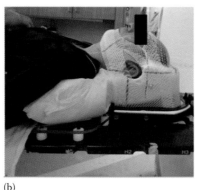
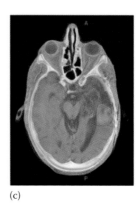

(a)   (b)   (c)

Figure 17.1 Fifty-eight-year-old male with WHO Grade IV glioblastoma who underwent resection and concurrent chemoradiation therapy. He experienced a recurrence 2 years later and was treated with partial resection of the tumor. He was recommended to undergo postoperative hypofractionated radiation therapy. An MRI of the brain (a) was used for treatment planning purposes. The patient was simulated in a CDR head frame (b). Images from the planning MRI were fused with CT images from the simulation in order to aid in target delineation (c). GTV— blue; PTV—red.

A cone-beam CT is obtained and the patient is matched to bony anatomy. The patient position is then confirmed using 2D kilovoltage imaging prior to treatment delivery with 6 MV photon beams. An example of the pretreatment MRI, CDR immobilization setup, and contours of a patient treated in the recurrent setting is shown in Figure 17.1.

The timing of concurrent chemotherapy is variable and should be instituted according to institutional practice and experience. At our institution, patients with recurrent high-grade glioma treated per protocol are treated with bevacizumab 10 mg/kg IV once every 2 weeks on days 1 and 15 of every cycle. Cycles are defined as 28 days. Patients are treated with one cycle of bevacizumab and evaluated with a new MRI for intratumoral hemorrhage and tumor response. This MRI is also used for treatment planning purposes. IG-HSRT is started on days 7–10 of cycle 2. Following RT, patients continue bevacizumab 10 mg/kg IV approximately every 2 weeks.

17.4 TOXICITY AND MANAGEMENT

Expected acute side effects from treatment to the brain include localized skin erythema, alopecia, fatigue, and mild headache. Skin effects can be managed with nonscented, noncolored moisturizers. Mild headache can be managed with over-the-counter medication. In the reirradiation setting, permanent hair loss may occur.

Unlikely acute side effects include headache requiring narcotic medication, nausea, vomiting, exacerbation of existing neurological symptoms, and the development of new focal neurological symptoms. The pathophysiology surrounding these effects is almost always due to an acute radiation induced inflammatory reaction. In such cases, antiedema medication such as a dexamethasone bolus followed by a taper should be considered. When used concurrently, the effects of bevacizumab may offset the signs and symptoms of acute inflammation and radionecrosis and decrease or replace the need for corticosteroids.

Possible but unlikely long-term side effects include changes in memory or cognitive function, white matter changes, and radiation necrosis. The risk of symptomatic radiation necrosis is estimated to be less than 10%. In certain cases of symptomatic radiation necrosis that is refractory to steroid medication, surgical excision should be considered.

In general, the risk of both acute and late toxicities increases with increasing treatment volume. In general, we limit IG-HSRT to patients with tumor volumes of <60 cc in the up-front setting and PTV size <40 cc in the recurrent setting.

17.5 CONTROVERSIES AND FUTURE DIRECTIONS

Despite these important data demonstrating IG-HSRT for newly diagnosed glioblastoma to be well tolerated for small volume disease, with evidence of treatment response and prolonged survival in some cases, the prognosis remains poor. While there have been promising results in a phase I trial combining hypofractionated RT with bevacizumab, recently published phase III trials of utilizing standard chemoradiation with bevacizumab failed to show an overall survival benefit (Chinot et al., 2014; Gilbert et al., 2014) and have curbed enthusiasm for the use of this expensive agent in the up-front setting. The use of hypofractionation in the up-front setting may have the added disadvantage of lessening the time that patients are treated concurrently with temozolomide, from 6 weeks with a standard regimen to only 1–2 weeks with a hypofractionated approach. Continuing investigations of novel systemic therapies and altered fractionation radiotherapy are still needed to help improve treatment results.

In the recurrent setting, the use of IG-HSRT for limited volume disease has a firm rationale in using high doses per fraction to treat radioresistant disease, image guidance to ensure accurate treatment delivery, and intensity modulation to help spare previously treated structures at risk. It appears to be well tolerated with low rates of radiation necrosis requiring surgical intervention, especially for small volume disease. However, controversies still exist in the management of recurrent patients. Further study is needed to clarify which patients should undergo surgical resection, definitive reirradiation, chemotherapy, or observation. Furthermore, the prognostic implications of biological characteristics of the tumor, including MGMT promoter status and IDH gene mutations, remain to be determined in patients with recurrent disease. Finally, the ability to accurately diagnose recurrent disease and distinguish it from pseudoprogression poses a significant challenge, especially with the use of antiangiogenic therapy. Efforts to investigate, validate, and standardize radiographic criteria of recurrent and progressive disease are urgently needed.

CHECKLIST: KEY POINTS FOR CLINICAL PRACTICE

✓	ACTIVITY	SOME CONSIDERATIONS
	Patient selection	*Is the patient appropriate for HSRT?* • Consider for patients that have a unifocal tumor or unifocal recurrence • KPS > 50 • No recent cytotoxic chemotherapy • Up-front setting: GTV < 60 cc • Recurrent setting: PTV < 40 cc
	SRS vs HSRT	*Is the lesion amenable to single-fraction SRS?* • In general, patients with HGG are treated with either HSRT or conventionally fractionated RT. The role of single-fraction SRS is limited in this disease
	Simulation	*Immobilization* • Supine position with three point thermoplastic mask or custom head mold and open-face mask *Imaging* • Both CT scan and MRI should be performed in treatment position • MRI ideally should be fine cut (1 mm slices), T1 sequence with IV contrast + T2 FLAIR sequence • Accurate image fusion should be verified by clinician and used for contouring targets

(Continued)

✓	ACTIVITY	SOME CONSIDERATIONS
	Treatment planning	*Contours* Newly diagnosed glioblastoma • Dose-painting technique • GTV1 should include postoperative residual contrast-enhancing tumor • A 5 mm margin is added to create a CTV_1 • Additional 5 mm is added to create the PTV_1 • GTV_2 is created by contouring the FLAIR abnormality as defined on the postoperative MRI • A 2 cm margin is added to create CTV_2 • Additional 5 mm is added to create PTV_2 • Organs at risk should be contoured on CT and verified via MRI Recurrent glioblastoma • GTV includes all enhancing disease on the T1 postcontrast sequence (or enhancing postoperative cavity if disease is resected) • Adjacent T2 FLAIR signal abnormality is also included in the GTV at the discretion of the treating physician • PTV is directly generated by expanding the GTV by 5 mm *Treatment planning* • Inverse treatment planning • Dose • Up-front setting: 36 Gy to the PTV_1 and 24 Gy to PTV_2 in 5 fractions, utilizing intensity-modulated RT with sliding window technique • Recurrent: PTV is treated to 30 Gy in 5 fractions • Coverage • D95% ≥ Rx • Dose constraints • Optic structures: max point dose (MPD) 23 Gy • Spinal cord: MPD 30 Gy • Brain stem: D_{05} of 30 Gy
	Treatment delivery	*Imaging* • Patients are first set up conventionally using room lasers and manual shifts • OSI program if the CDR head frame is used • Treatment couch with 6 degrees of freedom to correct translational and rotational shifts • Cone-beam CT to match to bony anatomy • Position confirmed using 2D kilovoltage imaging *Concurrent systemic therapy* • Variable • Should be instituted per protocol based on center practice and experience

REFERENCES

Bambury RM, Morris PG (2014) The search for novel therapeutic strategies in the treatment of recurrent glioblastoma multiforme. *Expert Rev Anticancer Ther* 14(8):955–964.

Chen C, Damek D, Gaspar LE, Waziri A, Lillehei K, Kleinschmidt-DeMasters BK, Robischon M, Stuhr K, Rusthoven KE, Kavanagh BD (2011) Phase I trial of hypofractionated intensity-modulated radiotherapy with temozolomide chemotherapy for patients with newly diagnosed glioblastoma multiforme. *Int J Radiat Oncol Biol Phys* 81(4):1066–1074.

Chinot OL, Wick W, Mason W, Henriksson R, Saran F, Nishikawa R, Carpentier AF et al. (2014) Bevacizumab plus radiotherapy-temozolomide for newly diagnosed glioblastoma. *N Engl J Med* 370(8):709–722.

Cho KH, Hall WA, Gerbi BJ, Higgins PD, McGuire WA, Clark HB (1999) Single dose versus fractionated stereotactic radiotherapy for recurrent high-grade gliomas. *Int J Radiat Oncol Biol Phys* 45(5):1133–1141.

Ciammella P, Podgornii A, Galeandro M, D'Abbiero N, Pisanello A, Botti A, Cagni E, Iori M, Iotti C (2013) Hypofractionated stereotactic radiation therapy for recurrent glioblastoma: Single institutional experience. *Radiat Oncol* 8:222.

Combs SE, Thilmann C, Edler L, Debus J, Schulz-Ertner D (2005) Efficacy of fractionated stereotactic reirradiation in recurrent gliomas: Long-term results in 172 patients treated in a single institution. *J Clin Oncol* 23(34):8863–8869.

Ernst-Stecken A, Jeske I, Hess A, Rodel F, Ganslandt O, Grabenbauer G, Sauer R, Brune K, Blumcke I (2007) Hypofractionated stereotactic radiotherapy to the rat hippocampus. Determination of dose response and tolerance. *Strahlenther Onkol* 183(8):440–446.

Floyd NS, Woo SY, Teh BS, Prado C, Mai WY, Trask T, Gildenberg PL et al. (2004) Hypofractionated intensity-modulated radiotherapy for primary glioblastoma multiforme. *Int J Radiat Oncol Biol Phys* 58(3):721–726.

Fogh, SE, Andrews DW, Glass J, Curran W, Glass C, Champ C, Evans JJ et al. (2010) Hypofractionated stereotactic radiation therapy: An effective therapy for recurrent high-grade gliomas. *J Clin Oncol* 28(18):3048–30453.

Fokas E, Wacker U, Gross MW, Henzel M, Encheva E, Engenhart-Cabillic R (2009) Hypofractionated stereotactic reirradiation of recurrent glioblastomas: A beneficial treatment option after high-dose radiotherapy? *Strahlenther Onkol* 185(4):235–240.

Gilbert MR, Dignam JJ, Armstrong TS, Wefel JS, Blumenthal DT, Vogelbaum MA, Colman H et al. (2014) A randomized trial of bevacizumab for newly diagnosed glioblastoma. *N Engl J Med* 370(8):699–708.

Glass J, Silverman CL, Axelrod R, Corn BW, Andrews DW (1997) Fractionated stereotactic radiotherapy with cis-platinum radiosensitization in the treatment of recurrent, progressive, or persistent malignant astrocytoma. *Am J Clin Oncol* 20(3):226–229.

Graf R, Hildebrandt B, Tilly W, Sreenivasa G, Ullrich R, Felix R, Wust P, Maier-Hauff K (2005) Dose-escalated conformal radiotherapy of glioblastomas—Results of a retrospective comparison applying radiation doses of 60 and 70 Gy. *Onkologie* 28(6–7):325–330.

Grosu AL, Weber WA, Franz M, Stark S, Piert M, Thamm R, Gumprecht H, Schwaiger M, Molls M, Nieder C (2005) Reirradiation of recurrent high-grade gliomas using amino acid PET (SPECT)/CT/MRI image fusion to determine gross tumor volume for stereotactic fractionated radiotherapy. *Int J Radiat Oncol Biol Phys* 63(2):511–519.

Gutin PH, Iwamoto FM, Beal K, Mohile NA, Karimi S, Hou BL, Lymberis S, Yamada Y, Chang J, Abrey LE (2009) Safety and efficacy of bevacizumab with hypofractionated stereotactic irradiation for recurrent malignant gliomas. *Int J Radiat Oncol Biol Phys* 75(1):156–163.

Henke G, Paulsen F, Steinbach JP, Ganswindt U, Isijanov H, Kortmann RD, Bamberg M, Belka C (2009) Hypofractionated reirradiation for recurrent malignant glioma. *Strahlenther Onkol* 185(2):113–119.

Hingorani M, Colley WP, Dixit S, Beavis AM (2012) Hypofractionated radiotherapy for glioblastoma: Strategy for poor-risk patients or hope for the future? *Br J Radiol* 85(1017):e770–e781.

Hudes RS, Corn BW, Werner-Wasik M, Andrews D, Rosenstock J, Thoron L, Downes B, Curran WJ, Jr. (1999) A phase I dose escalation study of hypofractionated stereotactic radiotherapy as salvage therapy for persistent or recurrent malignant glioma. *Int J Radiat Oncol Biol Phys* 43(2):293–298.

Iuchi T, Hatano K, Kodama T, Sakaida T, Yokoi S, Kawasaki K, Hasegawa Y, Hara R (2014) Phase 2 trial of hypofractionated high-dose intensity modulated radiation therapy with concurrent and adjuvant temozolomide for newly diagnosed glioblastoma. *Int J Radiat Oncol Biol Phys* 88(4):793–800.

Kaaijk P, Troost D, Sminia P, Hulshof MC, van der Kracht AH, Leenstra S, Bosch DA (1997) Hypofractionated radiation induces a decrease in cell proliferation but no histological damage to organotypic multicellular spheroids of human glioblastomas. *Eur J Cancer* 33(4):645–651.

Kim B, Soisson E, Duma C, Chen P, Hafer R, Cox C, Cubellis J, Minion A, Plunkett M, Mackintosh R (2011) Treatment of recurrent high grade gliomas with hypofractionated stereotactic image-guided helical tomotherapy. *Clin Neurol Neurosurg* 113(6):509–512.

Kohshi K, Yamamoto H, Nakahara A, Katoh T, Takagi M (2007) Fractionated stereotactic radiotherapy using gamma unit after hyperbaric oxygenation on recurrent high-grade gliomas. *J Neurooncol* 82(3):297–303.

Laing RW, Warrington AP, Graham J, Britton J, Hines F, Brada M (1993) Efficacy and toxicity of fractionated stereotactic radiotherapy in the treatment of recurrent gliomas (phase I/II study). *Radiother Oncol* 27(1):22–29.

Lederman G, Wronski M, Arbit E, Odaimi M, Wertheim S, Lombardi E, Wrzolek M (2000) Treatment of recurrent glioblastoma multiforme using fractionated stereotactic radiosurgery and concurrent paclitaxel. *Am J Clin Oncol* 23(2):155–159.

Loeffler JS, Alexander E, III, Shea WM, Wen PY, Fine HA, Kooy HM, Black PM (1992) Radiosurgery as part of the initial management of patients with malignant gliomas. *J Clin Oncol* 10(9):1379–1385.

McKenzie JT, Guarnaschelli JN, Vagal AS, Warnick RE, Breneman JC (2013) Hypofractionated stereotactic radiotherapy for unifocal and multifocal recurrence of malignant gliomas. *J Neurooncol* 113(3):403–409.

Minniti G, Armosini V, Salvati M, Lanzetta G, Caporello P, Mei M, Osti MF, Maurizi RE (2011) Fractionated stereotactic reirradiation and concurrent temozolomide in patients with recurrent glioblastoma. *J Neurooncol* 103(3):683–691.

Miwa K, Matsuo M, Ogawa S, Shinoda J, Yokoyama K, Yamada J, Yano H, Iwama T (2014) Re-irradiation of recurrent glioblastoma multiforme using 11C-methionine PET/CT/MRI image fusion for hypofractionated stereotactic radiotherapy by intensity modulated radiation therapy. *Radiat Oncol* 9:181.

Navarria P, Ascolese AM, Tomatis S, Reggiori G, Clerici E, Villa E, Maggi G et al. (2015) Hypofractionated stereotactic radiation therapy in recurrent high-grade glioma: A new challenge. *Cancer Res Treat*. doi: 10.4143/crt.2014.259.

Omuro A, Beal K, Gutin P, Karimi S, Correa DD, Kaley TJ, DeAngelis LM et al. (2014) Phase II study of bevacizumab, temozolomide, and hypofractionated stereotactic radiotherapy for newly diagnosed glioblastoma. *Clin Cancer Res* 20(19):5023–5031.

Patel M, Siddiqui F, Jin JY, Mikkelsen T, Rosenblum M, Movsas B, Ryu S (2009) Salvage reirradiation for recurrent glioblastoma with radiosurgery: Radiographic response and improved survival. *J Neurooncol* 92(2):185–191.

Reddy K, Damek D, Gaspar LE, Ney D, Waziri A, Lillehei K, Stuhr K, Kavanagh BD, Chen C (2012) Phase II trial of hypofractionated IMRT with temozolomide for patients with newly diagnosed glioblastoma multiforme. *Int J Radiat Oncol Biol Phys* 84(3):655–660.

Roa W, Brasher PM, Bauman G, Anthes M, Bruera E, Chan A, Fisher B et al. (2004) Abbreviated course of radiation therapy in older patients with glioblastoma multiforme: A prospective randomized clinical trial. *J Clin Oncol* 22(9):1583–1588.

Scott CB, Scarantino C, Urtasun R, Movsas B, Jones CU, Simpson JR, Fischbach AJ, Curran WJ Jr (1998) Validation and predictive power of Radiation Therapy Oncology Group (RTOG) recursive partitioning analysis classes for malignant glioma patients: A report using RTOG 90-06. *Int J Radiat Oncol Biol Phys* 40(1):51–55.

Selker RG, Shapiro WR, Burger P, Blackwood MS, Arena VC, Gilder JC, Malkin MG et al. (2002) The Brain Tumor Cooperative Group NIH Trial 87-01: A randomized comparison of surgery, external radiotherapy, and carmustine versus surgery, interstitial radiotherapy boost, external radiation therapy, and carmustine. *Neurosurgery* 51(2):343–355; discussion 355–357.

Shepherd SF, Laing RW, Cosgrove VP, Warrington AP, Hines F, Ashley SE, Brada M (1997) Hypofractionated stereotactic radiotherapy in the management of recurrent glioma. *Int J Radiat Oncol Biol Phys* 37(2):393–398.

Stupp R, Hegi ME, Mason WP, van den Bent MJ, Taphoorn MJ, Janzer RC, Ludwin SK et al. (2009) Effects of radiotherapy with concomitant and adjuvant temozolomide versus radiotherapy alone on survival in glioblastoma in a randomised phase III study: 5-year analysis of the EORTC-NCIC trial. *Lancet Oncol* 10(5):459–466.

Stupp R, Mason WP, van den Bent MJ, Weller M, Fisher B, Taphoorn MJ, Belanger K et al. (2005) Radiotherapy plus concomitant and adjuvant temozolomide for glioblastoma. *N Engl J Med* 352(10):987–996.

Tsien C, Moughan J, Michalski JM, Gilbert MR, Purdy J, Simpson J, Kresel JJ et al. (2009) Phase I three-dimensional conformal radiation dose escalation study in newly diagnosed glioblastoma: Radiation Therapy Oncology Group Trial 98-03. *Int J Radiat Oncol Biol Phys* 73(3):699–708.

Vordermark D, Kolbl O, Ruprecht K, Vince GH, Bratengeier K, Flentje M (2005) Hypofractionated stereotactic re-irradiation: Treatment option in recurrent malignant glioma. *BMC Cancer* 5:55.

Voynov G, Kaufman S, Hong T, Pinkerton A, Simon R, Dowsett R (2002) Treatment of recurrent malignant gliomas with stereotactic intensity modulated radiation therapy. *Am J Clin Oncol* 25(6):606–611.

Walker MD, Alexander E Jr, Hunt WE, MacCarty CS, Mahaley MS Jr, Mealey J Jr, Norrell HA et al. (1978) Evaluation of BCNU and/or radiotherapy in the treatment of anaplastic gliomas. A cooperative clinical trial. *J Neurosurg* 49(3):333–343.

Walker MD, Green SB, Byar DP, Alexander E Jr, Batzdorf U, Brooks WH, Hunt WE et al. (1980) Randomized comparisons of radiotherapy and nitrosoureas for the treatment of malignant glioma after surgery. *N Engl J Med* 303(23):1323–1329.

Watkins JM, Marshall DT, Patel S, Giglio P, Herrin AE, Garrett-Mayer E, Jenrette JM III (2009) High-dose radiotherapy to 78 Gy with or without temozolomide for high grade gliomas. *J Neurooncol* 93(3):343–348.

Wuthrick EJ, Curran WJ Jr, Camphausen K, Lin A, Glass J, Evans J, Andrews DW et al. (2014) A pilot study of hypofractionated stereotactic radiation therapy and sunitinib in previously irradiated patients with recurrent high-grade glioma. *Int J Radiat Oncol Biol Phys* 90(2):369–375.

Benign brain tumors

Or Cohen-Inbar and Jason P. Sheehan

Contents

18.1 INTRODUCTION

Since the emergence of modern neurosurgery, the focus of neurosurgeons has been the skillful complete surgical resection of the skull base lesion, with minimal morbidity and mortality, relying on a detailed anatomical knowledge and technical expertise. Despite tremendous advances in neurosurgical technique and equipment, postoperative morbidity continues to taint open complete removal of many cranial base tumors. The incidence of temporary and permanent cranial nerve deficits has been reported to be as high as 44% and 56%, respectively (Bricolo et al., 1992; De Jesús et al., 1996; Arnautovic et al., 2000;

Tuniz et al., 2009). Postoperative mortality rates as high as 9% (median, 3.6%) have been reported (Samii, 1992; DeMonte et al., 1994; Couldwell et al., 1996; Samii et al., 1996, 1997; George, 1997; Spallone et al., 1999; Natarajan et al., 2007). As a consequence, clinicians must choose between complete resection (with its substantial risk of morbidity) and subtotal resection followed by stereotactic radiosurgery (SRS). Progression rates after partial removal of a meningioma with no radiosurgery or radiation therapy have been reported to be as high as 70% (Condra et al., 1997). These are supported by reports of a 4 mm mean annual rate of tumor growth after subtotal resection among patients with partially resected petroclival meningiomas (Kirkpatrick et al., 2008). In contrast, Goldsmith et al. (1994) reported a 98% 5-year progression-free survival for patients with incompletely removed meningiomas treated with adjuvant conventionally fractionated external beam radiotherapy (Maire et al., 1995).

Different radiation delivery techniques can represent primary therapeutic alternatives and/or adjuvant treatment plans after open resection of benign cranial base tumors. With the emergence of radiosurgical methods more than four decades ago, radiosurgery has become the principal tool for this purpose. Following advances made in the technological aspects of radiosurgical delivery, many of the initial hurdles and limitations imposed on radiosurgery hold less true. Lesion size and location's initial technical restrictions are less constraining than they once were. First-generation, frame-based techniques initially allowed for only limited access to low-lying tumors (skull base and upper neck). Still, in specific circumstances of lesion location and size, the safe administration of an effective radiosurgical dose with minimal side effects remains a challenge. As the target lesion size increases, so does the area of normal brain and other healthy tissue that is irradiated, thereby increasing the risk of radiation-induced complications. Because of such limitations, radiosurgery has been traditionally used primarily to treat "smaller" lesions, defined as less than 3 cm in largest diameter (or smaller than 10–15 cm^3 in volume). When a larger cranial base neoplasm is confronted, the most commonly employed strategy is now cytoreductive surgery followed by single or hypofractionated SRS. Such an approach places a premium on neurological preservation.

18.1.1 CYTOREDUCTIVE SURGERY FOLLOWED BY STEREOTACTIC RADIOSURGERY

For meningiomas in critical locations, the success of SRS—confirmed by studies with excellent long-term results (Lee et al., 2002; Hasegawa et al., 2007; Kollova et al., 2007; Kondziolka et al., 2008; Haselsberger et al., 2009)—has led to a shift in the treatment paradigms from radical surgery to a combined surgical/radiosurgical approach as attempts of achieving gross total resection are associated with a high risk of surgical morbidity. In an attempt to develop a comprehensive treatment plan, utilizing the strengths and advantages of both surgery and radiosurgery, the term "cytoreductive surgery" was coined. Cytoreductive surgery followed by radiosurgery advocates a less aggressive surgical resection with the intent on functional preservation. Practically, it refers to the combination of a planned subtotal surgical debulking with radiosurgery to the smaller surgical remnant, which is ideally left adjacent to the most critical brain structures. Such a remnant can often be more safely addressed with radiosurgery than microsurgery. Today, many cranial benign neoplasms are surgically mitigated with preservation techniques. For many patients, combining surgery with radiosurgery offers the best and most durable tumor control while offering neurological preservation or improvement. Mathiesen et al. (2007) have coined the term "Simpson grade 4 gamma" for a combination of judicious (subtotal) surgery followed by SRS.

18.1.2 CONVENTIONAL RADIATION THERAPY VERSUS RADIOSURGERY

The development of radiosurgery has led to changes in the field of radiation therapy, driving practitioners to seek and adopt better delivery techniques and to refine targeting on the basis of image guidance. When any type of radiation treatment is being considered, one must find a way to mitigate the very real possibility that radiosensitive normal structures in the brain, such as the anterior visual pathways, may lie within the region of marginal prescribed dose. With the goal of protecting radiation-sensitive adjacent brain structures, the emergence of image-guided radiosurgery and the use of the nonisocentric beam delivery and inverse planning algorithms (as opposed to isocentric approaches) enables a significant measure of dose homogeneity. This homogeneity may be beneficial in these clinical situations as one is less likely to have hot spots adjacent to or even within critical normal tissues (Adler et al., 2008). Despite a more homogeneous dose distribution, single-fraction radiosurgery is less forgiving with respect to late adverse effects as compared to conventional radiation therapy. As a result, the practice of hypofractionated radiosurgery (keeping the dose per fraction ≥5 Gy) emerged with the intent to still deliver an ablative dose of radiation but in a few fractions (as opposed to 4–6 weeks as per conventional fractionated radiotherapy [CFRT]). It was proposed that the principles of hypofractionated protocols when adapted to radiosurgery may aid in yielding an additional radioprotective effect. Considering the advantages and limitations of each technique, hypofractionated (2–5 fractions) radiosurgery, as described in this chapter and the book overall, can play an important role in managing many benign cranial base tumors.

Generally speaking, compared with conventionally fractionated radiation, radiosurgical methods may have the advantage of sparing normal brain and critical organs, such as the pituitary gland and the cranial nerves. CFRT, while undoubtedly demonstrates better delivery and focus than in years past, uses a broader strategy in terms of targeting. Thus, radiation therapy cannot achieve either the conformity or selectivity that radiosurgery provides. The entire issue of biologically equivalent doses (BED), comparing the biological cumulative effects of radiation for the purpose of regimen comparison, is an imperfect solution. As a result, it is difficult to compare outcomes between radiosurgery and CFRT without bias. No randomized trial has ever been performed to determine which method is more efficacious or safe with respect to late effects. What is not disputable is that fractionation during the delivery of ionizing radiation has benefits with respect to late-responding normal tissues by allowing for radiation repair in between fractions.

18.1.3 SINGLE-SESSION VERSUS HYPOFRACTIONATED RADIOSURGERY

In contrast to fractionated radiotherapy, which is generally most effective in killing rapidly dividing cells, SRS (linear accelerator [LINAC] based, CyberKnife, and Gamma Knife based) is thought to inactivate target cells regardless of their mitotic activity or inherent radiosensitivity (Niranjan et al., 2004). With a 5-year tumor control rate that exceeds 90%, single-session radiosurgery is safe and effective for many parasellar lesions (Kondziolka et al., 1999; Stafford et al., 2001; Adler et al., 2008). Still, single-session radiosurgery has limitations, particularly with restricting dose to nearby eloquent structures. The dose gradient that can be achieved with all forms of single-session photon radiosurgery can be inadequate for optimal treatment of some perioptic lesions. In instances in which a segment of the optic nerve or chiasm is exposed to more than 8–12 Gy in a single fraction, studies demonstrate an elevated risk of visual injury (Tishler et al., 1993; Girkin, 1997; Leber et al., 1998). Consequently, when the distance between tumor and anterior visual pathways is less than 3 mm, radiosurgery may be contraindicated because of the difficulty in delivering an effective dose to the tumor while maintaining a tolerable dose to the optic apparatus. The tolerable dose to the optic apparatus, a traditional radiation sensitive structure, may be even lower for patients who have had prior radiation therapy or who have preexisting visual deficits. Although our knowledge and understanding of the precise threshold dose of radiation that results in optic nerve or chiasm damage is far from complete and continues to evolve, this basic principle is widely accepted. Furthermore, in some cases, it is difficult to reliably delineate an optic apparatus that is effaced or displaced by a tumor, even with the best of computerized imaging. In addition, when all other variables are kept constant, the risk of radiosurgical complications increases with greater treatment volume.

Hypofractionated radiosurgery serves to integrate the benefits of focused high-dose radiation and conformity typically associated with single-session radiosurgery, with the benefits of repair and repopulation of the normal tissues and the potential to enhance tumor cell kill by reoxygenation and redistribution to more sensitive phases of the cell cycle in between fractions (Nguyen et al., 2014). In particular, hypofractionated radiosurgery may help to overcome the limitations of single-session radiosurgery associated with larger volume targets and suboptimal dose falloff (Adler et al., 2008; Milker-Zabel et al., 2009; Tuniz et al., 2009; Sayer et al., 2011). As the number of LINAC-based radiosurgical modalities has expanded, hypofractionated radiosurgery has become more common (Adler et al., 2008; Killory et al., 2009). Reflecting this trend, the definition of radiosurgery has been written to encompass treatment delivery in one to five sessions (Barnett et al., 2007). Hypofractionated radiosurgery has been employed in an effort to reduce the risk of late complications when treating patients with large or inoperable benign cranial base tumors (Tanaka et al., 1996; Firlik et al., 1998; Sirin et al., 2006; Tuniz et al., 2009).

Many sellar and parasellar tumors at our institution are resected and/or treated with single-session SRS. Hypofractionated radiosurgery is considered for patients with sellar, parasellar tumors, and para-acoustic lesions when the patient refuses surgery, when surgery is not considered safe due to medical comorbidities, or when the tumor has a broad contact area with the optic apparatus, brain stem, or other cranial nerves. Such a broad contact area between the tumor and the optic apparatus would preclude limiting dose to a D_{max} of 12 Gy to the visual pathways (Nguyen et al., 2014). Hypofractionated radiosurgery with the Gamma Knife has been performed using different protocols in the past. Staged Gamma Knife radiosurgery (GKRS) has also been employed for large meningiomas (Pendl et al., 2000; Iwai et al., 2001; Haselsberger et al., 2009). GKRS with the frame left attached throughout the course of treatment has been attempted (Simonova et al., 1995); however, this is obviously clinically not desirable for the patient nor for the accuracy of therapy. In 1998, Sweeney et al. (1998) reported a noninvasive fixation system that relied on a vacuum-assisted dental impression (Sweeney et al., 2001) and represented a major advance in enabling hypofractionated GKRS.

18.1.4 TEMPORAL VERSUS SPATIAL FRACTIONATION

As the indications for radiosurgery have expanded over time and as the radiobiology of hypofractionated treatment regimens have become better understood, evidence has shown that in certain situations, various types of hypofractionated radiosurgery may be advantageous (Adler et al., 2008; Kim et al., 2008; Tuniz et al., 2009). One should distinguish a multisession hypofractionated treatment of the same individual lesion, from multiple single-fraction treatments of different lesions in patients with multiple lesions of distant recurrences. Multiple serial sessions of single-fraction radiosurgery are employed for most part in patients with brain metastases, where new lesions not seen at the time of treatment are seen on follow-up scans. A third variation to this definition, the multistaging approach (spatial fractionation), refers to the segmentation of a large lesion (or arteriovenous malformation [AVM]) to smaller adjacent volume treatment sessions. This has been employed in AVM radiosurgery for decades (Tanaka et al., 1996; Firlik et al., 1998; Sirin et al., 2006; Faisal et al., 2011; Nguyen et al., 2014).

As discussed later, treating intracranial pathologies in multiple sessions can be advantageous in certain situations. Hypofractionated treatment regimens can improve the therapeutic ratio and better spare nearby normal tissue (Park et al., 2008). As such, patients with multiple lesions may be treated with repeated single-fraction radiosurgeries to control the local progression of each lesion without the potentially deleterious effects of traditional fractionated whole-brain radiation therapy (Faisal et al., 2011). Alternatively, volume-staged radiosurgery for intracranial meningiomas exceeding 3 cm in diameter has shown tumor control rates of 90% with low morbidity (Haselsberger et al., 2009). Similar results (86% tumor control) were shown for ominous large petroclival and cavernous sinus meningiomas treated with two-staged GKRS (Iwai et al., 2001). Thus, an accumulating body of evidence suggests that staged radiosurgery for meningiomas represent a safe option that can be recommended for large meningiomas in critical locations, either after incomplete surgery or as primary treatment (Haselsberger et al., 2009).

18.1.5 DOSE SELECTION AND FRACTIONATION

When reviewing the literature on past radiosurgical experience in both volume-staged and hypofractionated radiosurgery, using either LINAC-based CyberKnife or Gamma Knife–based systems, a problem emerges. Although all series report using two to five sessions, these treatment methods and protocols are not perfectly comparable, as discussed later. As a consequence, there is very little precedent to guide us in choosing multifraction radiosurgical parameters for the treatment of different benign lesions. We did observe that hypofractionated multistage (i.e., temporal and spatial fractionation) treatments have been performed for years, using frame-based GKRS, to treat selected large cranial base meningiomas and brain AVMs (Tanaka et al., 1996; Firlik et al., 1998; Sirin et al., 2006).

Treatment schemes vary considerably (discussed in depth per etiology in subsequent sections). For example, Pendl et al. (2000) reported a 100% local control and 17% cranial nerve toxicity incidence in a series of 12 large cranial base meningiomas treated with volume-staged (spatial fractionation) radiosurgery performed over multiple months. Marginal doses ranged from 10 to 25 Gy prescribed to the 30%–50% isodose. The time interval between sessions varied from 1 to 8 months, and the mean follow-up time was 27.9 months. Iwai et al. (2001) reported achieving a local control in six of seven large petroclival and cavernous sinus meningiomas. These tumors were treated with volume-staged GKRS involving two sessions, with a treatment interval of 6 months and a marginal dose of 8–12 Gy. Therefore, most centers, being conservative in their initial dose selection and having established safety with the earliest-treated patients, began a gradual mild dose escalation over the past 14 years with the intent to maximize the likelihood of long-term tumor control safely. Whether or not a further increase in BED is warranted is uncertain.

Dose selection is influenced by a multitude of factors including tumor type, volume, proximity, and extent of irradiated optic nerve, as well as a previous history of radiation therapy. Although BED formulas were used at first (for temporally fractionated radiosurgery), one should remember that in most studies performed to date, the initial choice of number of sessions was to a large extent empirically based. Protocols were usually derived from earlier experiences with hypofractionated frame-based radiosurgery in patients with no other treatment options (Adler et al., 2008). With reference to tumors adjacent to or involving the visual pathways, this was an area first explored, as with hypofractionation, the radiation tolerance to the anterior optic apparatus was thought to be extendable to 15.3–17.4 Gy for 3 fractions and 23–25 Gy for 5 fractions (Benedict et al., 2010). This has been shown to be safe with now long-term outcomes. In our opinion, the choice of more protracted regimens should be reserved for patients with the longest involvement of the optic apparatus or where the optic nerve or chiasm are most displaced and, as such, cannot be clearly contoured.

18.1.5.1 Large dose per session: Rationale

The reason for using larger doses per session of radiation stems from basic radiobiological principles. Although there are no controlled comparison studies to demonstrate the benefits of larger fraction size in treating benign as opposed to malignant brain tumors, there is however a sound theoretical basis for such a conclusion (Brenner, 1994). Studies comparing SRS with CFRT in the management of benign tumors demonstrate high rates of tumor control with both modalities. Nevertheless, the larger dose per session that characterizes radiosurgery results in a higher biological equivalent dose and subsequently correlates with greater tumor shrinkage on follow-up imaging (Metellus, 2005). In particular, a dose of >5 Gy per fraction appears to induce distinct subcellular mechanisms and vascular change that differ from effects triggered by doses of 1.8–2 Gy per fraction (Garcia-Barros, 2003).

Fractionation schemes used for hypofractionated SRS varied widely since the mid-1990s and within different institutions, ranging from near "conventional" daily fractionations, to 6 fractions in 2 weeks, 2 fractions per week, 2–10 daily fractions, or the 2–5 daily fractions scheme used by most other institutes as well as ourselves.

18.1.6 RADIATION THERAPY NEAR THE ANTERIOR VISUAL PATHWAYS

Owing to steep falloff and multiple beams, radiosurgery has the capacity to minimize the irradiation of nearby critical structures and, thereby, restrict collateral damage. This ability to limit radiation damage to normal brain anatomy would seem intrinsically desirable even if some benefits defy easy identification and would be particularly beneficial in the treatment of perioptic lesions in which the radiation tolerance of the optic apparatus is so critical. The anterior visual pathways (namely, the optic nerve and optic chiasm) pose a particular challenge for inactivating "perioptic" tumors, having unique radiation sensitivity that precludes conventional radiosurgery when a lesion is within 2–3 mm of the anterior visual pathways (Tishler et al., 1993; Leber et al., 1995, 1998; George, 1997; Ove et al., 2000; Heilbrun et al., 2002; Mehta et al., 2002; Stafford et al., 2003). Thus, most tumors encroaching or abutting the optic chiasm and optic nerve are treated using either standard fractionated radiotherapy or adaptive hybrid surgery (i.e., a combination of microsurgical resection followed by radiosurgical ablation).

Our experience, supported by findings reported by other groups as well (Adler et al., 2008), utilizing both Gamma Knife–based and CyberKnife-based systems, suggests that the radiation tolerance of the optic apparatus and chiasm seems to be not only related to dose but also inversely proportional to the length of irradiated nerve (Mehta et al., 2002). This "volume effect" is an important phenomenon throughout single-fraction radiosurgery practice and also plays a key role in the relative safety of the hypofractionated radiosurgery approach.

Treatment of benign lesions with doses of radiation between 45 and 55 Gy using 1.8–2 Gy fractions successfully results in tumor growth control in most patients (Taylor et al., 1988; McCollough et al., 1991). Long-term (10 years) local control rates range from 68% to 89% for meningiomas and schwannomas (Barbaro et al., 1987; Taylor et al., 1988; Goldsmith et al., 1994) and 89% for pituitary adenoma (Salinger et al., 1992). During the course of conventionally fractionated focal radiation therapy, a lesion and the immediately surrounding normal brain is bathed with tumor-static radiation doses. Although there are numerous publications, establishing dose and fractionation regimens for conventional focal radiotherapy, when adhered to strictly an approximate 3% risk of optic neuropathy for pituitary tumors (Brada et al., 1993; Fisher et al., 1993; Hughes et al., 1993; Movsas et al., 1995; McCord et al., 1997; Paek et al., 2005) and less than 3% for cranial base meningiomas (Maire et al., 1995; Maguire et al., 1999; Moyer et al., 2000; Selch et al., 2004; Milker-Zabel et al., 2005), is expected. However, in terms of efficiency of such regimens, that is, tumor control and tumor shrinkage after radiotherapy for many perioptic lesions, reported efficacy is not quite as good as radiosurgery (discussed in length later) (Estrada et al., 1997; Cozzi et al., 2001; Metellus, 2005).

Patient setup and immobilization for conventional radiation therapy inherently results in spatial inaccuracies that are larger than for radiosurgery. As such, larger margins are typically used to compensate, and compounded by the different planning strategy utilized in conventional radiotherapy, greater doses to the surrounding brain parenchyma occur. When discussing perioptic- and parasellar-based lesions, these regions may include longer segments of the optic apparatus, diencephalic structures (pituitary, hypothalamus), and medial temporal lobe. Short-term effects of such inaccuracies and irradiation may go unnoticed and are usually minor, but when considering the long life expectancy of patients treated for benign lesions, these merit consideration. The longer-term consequences of such irradiation are largely undealt with in current literature, yet potentially deleterious. Pituitary failure occurring a decade or more after regional radiation therapy, being one such example, is a well-described phenomenon (Agha et al., 2005; Darzy, 2005). The risk of iatrogenic, radiation-induced hypopituitarism with conventional radiotherapy is reported to be in the range of 30%–70% (Nelson et al., 1989; McCollough et al., 1991; Tsang et al., 1994). Secondary malignancies and temporal lobe necrosis are other established late complications (Brada et al., 1992; Tsang et al., 1994; Breen et al., 1998; Hoshi et al., 2003; Kry, 2005; Sachs et al., 2005). There are other additional shortcomings to conventional radiotherapy. When conventionally fractionated treatment fails in controlling tumor size in a perioptic tumor, a second course of treatment to the recurrent lesions is a more risky approach. Another shortcoming to standard radiation, albeit minor, is the fact that a 6-week course of therapy may be inconvenient for many patients.

With respect to dose tolerance, Mayo et al. (2010) report that the risk of toxicity to the optic nerves and chiasm markedly increases at doses beyond 60 Gy when given in 1.8 Gy/fraction and at greater than 12 Gy for single-fraction SRS. However, the optic apparatus tolerance for hypofractionated SRS of 2–5 fractions is not well defined in terms of the maximum dose and dose per fraction (Taylor et al., 1988; Leber et al., 1995; George, 1997; Benedict et al., 2010). Results from recent studies using different radiosurgical modalities have thus far echoed the low visual complication rate associated with multisession radiosurgery for lesions that lie in close proximity to the optic system (Nguyen et al., 2014). In a retrospective review of 49 patients with tumors located within 2 mm of the optic system treated with CyberKnife in two to five sessions, Adler et al. (2008) observed visual deterioration in 6% ($n = 3$). Each of these patients was associated with tumor progression that eventually resulted in death. Consistent with this result, Kim et al. (2008) reported visual deterioration in 4.5% ($n = 1$), also accompanied by tumor progression, in their retrospective review of 22 patients treated using traditional GKRS in three to four sessions. Therefore, for many types of indications such as meningiomas, pituitary adenomas, and metastasis, it appears that a therapeutic margin dose with low complication rates can be achieved in the setting of multisession radiosurgery.

18.1.7 COMPARING TREATMENT REGIMENS AND RADIOBIOLOGICAL EFFECTS OF HYPOFRACTIONATED RADIOSURGERY

Using hypofractionated approaches, higher total doses as compared to single-fraction irradiation are required to produce the same dose effect due to the factors of DNA repair and cell repopulation in between fractions (Girkin, 1997). For example, a marginal dose of 19.7 Gy to a mean irradiation volume of 8.55 cc at the 50% isodose line over three to five treatment sessions, as we recently reported (Nguyen et al., 2014) for parasellar and perioptic lesions, translates to a mean BED (α/β = 2.5) of 60.9 Gy. This is comparable to 12 Gy when given in a single session (BED [α/β = 2.5] = 69.6 Gy) and to doses frequently delivered with CFRT (i.e., 45–60 Gy) for intracranial tumors. Kim et al. (2008) used similar dose plans with a median margin dose of 20 Gy in an average volume of 4.1 cc at isodose lines ranging from 46% to 50% in their multisession GKRS for benign perioptic lesions. In comparison, the group from Stanford University delivered an average margin dose of 20.3 Gy to a mean volume of 7.7 cc at isodose lines that ranged from 70% to 90% using CyberKnife (Adler et al., 2008).

18.1.7.1 Equivalence and isoeffective doses at large doses per fraction

What is the optimal method of comparing or equating different dose and fractionation schemes? As basic a question as it seems, no one consensus answer can be given. A common practice is to calculate the BED using the linear-quadratic (LQ) model (Little et al., 2005). Despite its popularity, stemming from its simplicity and hence practicability, the LQ model has shortcomings. Many have raised concerns about the accuracy of the LQ model to accurately predict the biological effects at large doses per fraction (Kirkpatrick et al., 2008). Still others argue that this model is reasonably valid up to 18 Gy per fraction (Brenner, 2008). The LQ model may be hampered if the value of α/β is dose range dependent and/or when extrapolating to fractions of high dose, because the LQ curve bends continuously on the log–linear plot. This does not coincide with what is observed experimentally in many clonogenic cell survival studies at high dose wherein radiation dose–response relationships more closely approximate a straight line (Park et al., 2008; Choi et al., 2011).

Alternative models have been proposed (Guerrero, 2004; Park et al., 2008), yet these are not without criticism and still require validation using in vitro and clinical studies (Astrahan, 2008; Kavanagh et al., 2008; McKenna, 2009). The universal survival curve (USC) (Park et al., 2008) constructed by hybridizing the LQ model and the multitarget model can be used to compare the dose-fractionation schemes of both CFRT and SRS. The USC is said to provide an empirically and a clinically well-justified rationale for SBRT while preserving the strengths of the LQ model for CFRT (Park et al., 2008). Another model suggests that BED calculations may better approximate reality by using a dose range-independent LQ-linear formulation, which better fits the experimental data across a wider range of dose. Still a third model (Guerrero, 2004) suggests extending the conventional LQ model with one additional parameter to more accurately describe the acute high-dose regime. The resulting modified LQ (MLQ) model essentially maps the lethal and potentially lethal (LPL) model at both high and low doses and dose rates, retaining its mechanistic nature. At the same time, in the low-dose regime, the conventional LQ model is recovered with the well-known α and β parameters. The MLQ is said to give a better fit to the data than the LQ (Guerrero, 2004). In short, which model best describes cell survival after radiosurgery remains controversial.

After grasping the scope of the relevant key issues, we turn our discussion to a review on current treatment approaches and published data on three common benign brain lesions. In Table 18.1, we summarize results segregated per pathology from noteworthy hypofractionated radiosurgery series and a typical checklist for hypofractionated SRS is provided in detail at the end of this chapter.

18.2 MENINGIOMAS

18.2.1 OPTIC NERVE SHEATH MENINGIOMAS

18.2.1.1 Introduction

Optic nerve sheath meningiomas (ONSMs) are rare tumors, representing 2% of all orbital tumors, 1%–2% of intracranial meningiomas, and one-third of optic nerve lesions (Leber et al., 1998; Marchetti et al., 2011). Middle-aged woman are the most often affected (Alper, 1981; Stafford et al., 1998). Primary ONSMs arise from the arachnoid cap cells forming the fibrous dural capsule of the optic nerve and usually grow circumferentially along the nerve. Secondary ONSMs usually arise from the sphenoid ridge or the tuberculum sellae and subsequently spread into the optic canal and orbit (Marchetti et al., 2011). The most frequent symptom at presentation is visual loss (reduction in visual acuity, visual fields, or both),

Table 18.1 Hypofractionated radiosurgery for benign brain lesions, literature review

INDICATION	AUTHORS	YEAR	N	SRS DEVICE	NUMBER OF FX	TOTAL MARGIN DOSE (Gy)	MEAN FOLLOW-UP (MONTHS)	TUMOR CONTROL (%)	COMPLICATIONS (%)
Meningiomas									
ONSM[a]	Marchetti et al. (2011)	2011	21	CK[b]	5	25	36	100	0
Perioptic	Adler et al. (2008)	2008	27	CK[b]	2–5	20.3	46	94	7
	Kim et al. (2008a)	2008	13	GK[c]	3–4	20	29	100	0
	Kim et al. (2008b)	2008	5	GK	3–4	20	20.9	80[d]	0
Skull base	Haselsberger et al. (2009)	2009	20	GK	2–5 (volume staged)	20–75	60.5 (median)	100 (10% OFR[e])	0
	Iwai et al. (2001)	2001	7	GK	2	18	39	86	0
	Pendl et al. (2000)	2000	12	GK	2–5	20–75	27.9 (median)	100	16.6
Pituitary adenoma									
Functioning	Adler et al. (2008)	2008	7	CK	2–5	20.3	49	94	14.3 (one patient)
	Kim et al. (2008a)	2008	1	GK	3–4	20	29	100	0
Nonfunctioning	Adler et al. (2008)	2008	12	CK	2–5	20.3	49	94	0
	Kim et al. (2008a)	2008	2	GK	3–4	20	29	100	0
Vestibular schwannomas	Williams(2002)	2002	150	LINAC	5	25 (90% of patients)	12	100	G-R[f] grade 1 or 2, 70% hearing preservation
	Meijer et al. (2003)	2000	80	LINAC	5	20–25	60	94	4% facial nerve, 34% loss of hearing
	Chang et al. (2005)	2005	61	CK	3	18–21	48	98	10% loss of hearing
	Hansasuta et al. (2011)	2011	383	CK	3	18	43	99%/96% at 3 and 5 years	3.8%–9.3% nonauditory, 24% loss of hearing

a Optic Nerve Sheath Meningiomas.
b CyberKnife.
c Gamma Knife.
d One patient suffered recurrence had a WHO-II meningioma.
e Out-of-field recurrence.
f Gardner–Robertson.

resulting from either direct compression of the optic nerve (and resultant optic atrophy) or from vascular rearrangement (the so-called optociliary shunts, which are usually a late and rare sign, resulting from the direct compressive optic neuropathy). This triad of visual disturbances, optic nerve atrophy, and optociliary shunts are pathognomonic of ONSM (Sibony et al., 1984). Traditional treatment options for ONSM included observation, surgery, radiotherapy, or radiosurgery. Conservative treatment, whose reasoning stems from the benign nature of the tumor, invariably leads to visual deterioration and complete blindness (Shields et al., 2004; Kim et al., 2005) and thus has fallen out of favor. Surgery best suits those instances where histological confirmation is required, and an immediate reduction in tumor burden in the frame of deteriorating vision or blindness is feasible. However, the complete removal of the tumor is in most instances impossible or carries too much risk for the optic nerve, central retinal artery, or ophthalmic artery damage (Dutton, 1992; Carrasco, 2004; Berman, 2006), including the contralateral side (Clark, 1989). Moreover, the postoperative course is often characterized by worsening of symptoms (Alper, 1981; Dutton, 1992; Andrews et al., 2002; Carrasco, 2004). For these reasons, surgery is often limited to patients who lack useful vision, since visual preservation following microsurgical resection is rare (Alper, 1981; Clark et al., 1989). Furthermore, a nonnegligible fact is that surgery may lead to very poor aesthetical results.

18.2.1.2 Radiosurgery for the treatment of ONSM

Unless the patient's vision is substantially compromised in the affected eyed, high-dose single-session SRS has seldom if ever been proposed as ONSM treatment due to the known dose tolerance of the optic nerve (Tishler et al., 1993; Leber et al., 1998; Stafford et al., 2003; Kwon et al., 2005). However, hypofractionated radiosurgery can be used to treat ONSM. Marchetti et al. (2011) reported their experience with 21 patients treated with radiosurgery using a frameless CyberKnife system, with 5 fractions of 5 Gy each prescribed to the 75%–85% isodose line. Mean follow-up time was 30 months. Ninety-five percent (n = 20) of patients responded well, with only one patient developing a transient steroid responsive optic neuropathy. Their results, comparable to most other series (Leber et al., 1998; Milker-Zabel et al., 2009), showed a 100% tumor control rate, with visual function improved in 35% patients or maintained stable in an addition 65%, with no permanent worsening (visual follow-up of 36 months). This fractionation schedule, according to the LQ model, provided an equivalent dose at least comparable with conventional fractionated regimens, which deliver 50.4–56 Gy in 1.8–2 Gy fractions (Marchetti et al., 2011).

18.2.2 PERIOPTIC MENINGIOMAS

As part of a larger report of treatment experience with perioptic lesions, Adler et al. (2008) report their experience with hypofractionated CyberKnife-based SRS for 27 perioptic meningiomas, located in the medical sphenoid wing (n = 3), cavernous sinus with posterior (n = 9), cavernous sinus with posterior orbital involvement (n = 6), orbital apex (n = 2), petroclival (n = 1), or tuberculum sella (n = 6). By definition, every tumor was within 2 mm of, at times displacing or even completely obscuring portions, the anterior visual pathways. For the entire series, SRS was delivered in two to five sessions to an average target volume of 7.7 cm³ using a total marginal dose of 20.3 Gy. Treatment dose was prescribed to a mean isodose line of 80%, normalized to an average maximum dose of 25.5 Gy. Individual SRS sessions were separated by 12 (n = 3) or 24 h (n = 46). Overall mean radiographic follow-up was 46 months. Tumor control was seen in 94% (n = 46) of cases. Two patients showed tumor progression close to or within the treatment field. These patients lost vision in the ipsilateral eye in the setting of tumor progression. In that same series, mean visual field follow-up was 49 months. Visual fields remained stable or improved in 25 out of 27 (93%) (Adler et al., 2008).

In a similar format report of perioptic lesions, Kim et al. (2008a) report their experience with hypofractionated GKRS for 13 perioptic meningiomas all within 1 mm of the optic apparatus. Radiation was delivered in 3–4 fractions with a median-cumulated marginal dose of 20 Gy. Mean follow-up time was 29 months, during which tumor control was achieved in 100% (n = 13). Of these, tumor volume decreased in 92% (n = 12) and remained stable in 8% (n = 1) at the last follow-up. Visual function was stable or improved in all patients. No other complications were observed (Kim et al., 2008a).

In their report focusing on hypofractionated GKRS for the treatment of orbital tumors, Kim et al. (2008b) report on an experience with five orbital meningiomas. Patients had preserved vision, and tumors

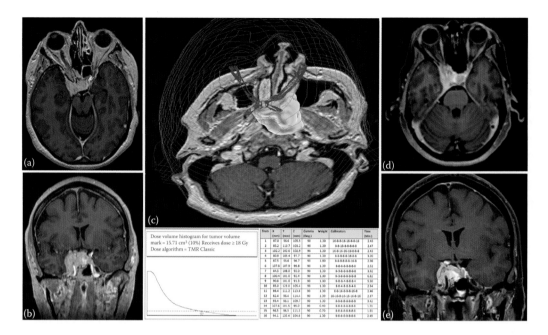

Figure 18.1 Sample patient. (a and b) Pre-SRS T1WI MRI. Axial (a) and coronal (b) views. A right parasellar meningioma is shown, occupying the right cavernous sinus and the sella turcica, abutting the right optic nerve and chiasm as well as the right oculomotor (CN-III) nerve. (c) Treatment plan. the patient was treated with hypofractionated Gamma-Knife radiosurgery (5 fractions of 5 Gy). The target volume was 15.71 cm³. The Dose histogram and isocenter description is given. (d and e) Follow-up T1WI+Contrast MRI scan taken 12 months post-SRS. A significant reduction in lesion size is shown, especially evident is the reduced compression of the anterior visual pathways. No visual deterioration was noted during follow-up at 1, 3, 5 years.

were located near the optic nerve. The clinicians opted to treat with 3 or 4 fractions GKRS. The mean tumor volume was 3695 mm³ and median cumulative marginal dose was 20 Gy (15–20 Gy). Mean follow-up time was 20.9 months; tumor control was confirmed in four of five patients. Of note, the patient who suffered tumor recurrence had a WHO-II rhabdoid meningioma, which required surgical resection. The remaining four patients showed no change (*n* = 1) or improved (*n* = 3) visual fields. No other complications were reported (Kim et al., 2008b). Refer to Figure 18.1 for a sample case.

18.2.3 VOLUME-STAGED HYPOFRACTIONATED SRS FOR MENINGIOMAS IN OTHER LOCATIONS

Haselsberger et al. (2009) reported their series of 20 patients, harboring large benign critically located meningiomas treated during the time period of 1992–2008 using volume-staged hypofractionated GKRS. Fourteen patients had at least one surgical partial "Simpson grade-IV gamma" resection; one patient had a prior tumor embolization. In five patients (25%), multistaged GKRS was the primary treatment. Tumor volume ranged from 13.6 to 79.8 cm³ (median 33.3), and treatment volumes ranged from 5.4 to 42.9 cm³ (median 19.0). The median time interval between the staged procedures was 6 months (range, 1–12 months). Of 41 treatments, the prescription dose at the tumor margin was 12 Gy for 33 treatments, 10 Gy for 1 treatment, 14 Gy for 4 treatments, 15 Gy for 1 treatment, and 25 Gy for 2 treatments (median 12 Gy to a marginal isodose of 45%). Tumor control was achieved in 90% (25% volume regression, 65% stable size). Two patients (10%) experienced out-of-field tumor progression, treated by an additional radiosurgery. In these two patients, tumor volume decreased in one patient and remained stable in the second one. Clinically, 9 patients (45%) improved, and 11 (55%) remained unchanged. The authors concluded staged radiosurgery for meningiomas represents a safe treatment modality for meningiomas in critical locations either after incomplete surgery or as primary treatment for patients with significant comorbidities (Haselsberger et al., 2009).

Iwai et al. (2001) reported their experience with two-staged GKRS for the treatment of seven patients harboring large petroclival and cavernous sinus meningiomas. Three of these patients had prior surgical debulking done; the other four were only followed. Tumor volume ranged from 34.5 to 101 cm³ (mean 53.5). Treatment volume ranged from 6.8 to 29.6 cm³ (mean 18.6). The treatment interval between sessions was 6 months, and the marginal tumor doses were 8–12 Gy (mean of 9 Gy). Mean follow-up time was 39 months. Six patients (86%) demonstrated tumor growth control. Tumor regression was observed in three patients (43%). Three patients (43%) showed clinical improvement. No patient suffered from symptomatic adverse radiation effects injury. The authors similarly concluded that "two staged GKRS… may be a very useful option for large petroclival and cavernous sinus meningiomas in selected patients" (Iwai et al., 2001).

Pendl et al. (2000) reported their experience treating 12 large meningiomas with volume-staged hypofractionated GKRS (3 clival, 3 parasellar, 2 sphenoid ridge, 1 tentorium, 1 falx, 1 olfactory groove). All patients had a previous surgical debulking. Tumor volume ranged from 19 to 90 cm³. Marginal dose ranged from 10 to 25 Gy prescribed to the 30%–50% isodose. The time interval between fractions varied from 1 to 8 months. This interval decision relied on a previous report by Kondziolka et al. (1998, 1999) implying that staged radiosurgery should be performed during the early postoperative interval (2–6 months). These authors reported that when radiosurgery was performed at the time of symptomatic tumor regrowth, unsatisfactory tumor control rates were observed. With follow-up ranging from 5 to 89 months, no deterioration or further tumor growth was observed. Neurological examination showed improvement in six patients (50%) due to amelioration of cranial nerve deficits. Four patients (33%) remained neurologically stable, whereas two patients complained of new cranial nerve deficits.

18.3 PITUITARY ADENOMA

18.3.1 INTRODUCTION

Pituitary adenomas are common among the general population and comprise about 10%–20% of all intracranial tumors (Vance, 2004; Melmed, 2006; Dekkers et al., 2008). While many anatomic and histological classifications of pituitary adenomas exist, they are classically divided by size and functional status. Microadenomas are defined as measuring less than 1 cm in size, and macroadenomas are defined as measuring at least 1 cm in size. Functional status refers to hormone hypersecretion from functioning lesions and lack of abnormal hormonal production from nonfunctioning lesions. Surgical resection achieves tumor control in only 50%–80% of cases (Hoybye, 2009). Thus, these lesions frequently require additional treatment such as radiosurgery. Radiosurgery provides an excellent solution for patients having residual tumor postoperatively, tumor progression, or recurrence. Another potential complication is radiation damage to the anterior optic pathways.

Current experience and published literature on hypofractionated radiosurgical approaches to pituitary adenomas is scant as well as inconsistent in terms of methodology and definition (reviewed below). Our current knowledge as to treatment protocols, effective dose plans per adenoma type, and prognosis (remission rates, hypopituitarism, and side effects) is taken from the single-session SRS literature. A short description of common features and treatment protocols of different pituitary pathologies follows. Inevitably, any hypofractionated protocol suggested for the different pituitary pathologies should take under account these points (Hoybye, 2009; Ding et al., 2014).

18.3.2 FUNCTIONING VERSUS NONFUNCTIONING PITUITARY ADENOMAS

While postradiosurgical tumor control rates for nonfunctioning adenomas are excellent and approach 90%, the rates of biochemical remission for functioning adenomas are lower than tumor control rates. The highest endocrine remission rates are achieved in patients with Cushing's disease and the lowest in those with prolactinomas. In a recent review of literature (Izawa et al., 2000; Melmed, 2006; Jagannathan et al., 2007; Hoybye, 2009; Starke et al., 2012; Sheehan et al., 2013; Ding et al., 2014), the major radiosurgical series since 2002 that detail outcome in nonfunctioning adenoma patients were reviewed. Single-session radiosurgery margin doses of 12–20 Gy, with a median of 16 Gy, were used for most patients with nonfunctioning adenomas. Tumor control rates ranged from 83% to 100% with a mean of 95.2%. Hypopituitarism following radiosurgery was observed in 0%–40% with a mean of 8.8%. At our institution, we reported similarly a 90% tumor control and a 30% occurrence of postradiosurgery hypopituitarism in a series of 140 patients with nonfunctioning pituitary adenomas (Starke et al., 2012). In a recent multicenter trial evaluating the role of GKRS for 512 patients with nonfunctioning pituitary adenomas, and a median follow-up of 36 months, an overall tumor control rate of 93% was reported. Hypopituitarism following GKRS was noted in 21% of patients (Sheehan et al., 2013). Positive prognosticators included patients older than 50 years, those with a tumor volume less than 5 cc, and those without prior radiation. These prognostic factors were integrated into a radiosurgical pituitary score (Sheehan et al., 2013).

18.3.3 HYPOFRACTIONATED SRS FOR PITUITARY ADENOMAS

As part of a general report of treatment experience with multisession CyberKnife-based SRS for perioptic lesions, Adler et al. (2008) report their experience with 19 cases of histologically confirmed residual and recurrent pituitary adenomas, 37% ($n = 7$) of which hormonally active (acromegaly in $n = 4$, Cushing's disease in $n = 2$ and prolactinoma $n = 1$). By definition, every tumor involving variable portions of the sella and adjacent cavernous sinus had a suprasellar component situated within 2 mm of, at times displacing or even completely obscuring portions of, the anterior visual pathways. SRS was delivered in two to five sessions to an average target volume of 7.7 cm^3 using a total marginal dose of 20.3 Gy. Overall mean radiographic follow-up was 46 months, demonstrating a 94% ($n = 46$) tumor size reduction or volume control ($n = 15$). Mean visual field follow-up was 49 months. One patient with Cushing's disease had an initial good radiographic and hormonal response to each of three radiosurgical sessions before subsequently developing further tumor recurrences. This patient also suffered visual loss in one eye. In this patient, radiation injury to the optic nerve was the presumed culprit (Adler et al., 2008).

In a similar format report of treatment experience with perioptic lesions, Kim et al. (2008a) report their experience with hypofractionated GKRS for three perioptic pituitary adenomas. Two patients had a nonfunctional adenoma. One patient had a growth hormone–secreting adenoma. Radiation was delivered in 3–4 fractions with a median cumulated marginal dose of 20 Gy. Mean follow-up time was 29 months, during which tumor control was achieved in all patients. Of these, tumor volume decreased in two patients and remained stable in one patient. Visual function was improved or not changed in all patients. Growth hormone levels were normalized at 23 months after hypofractionated GKRS, and it was accompanied by a decrease in tumor volume (Kim et al., 2008a).

18.4 VESTIBULAR SCHWANNOMAS

18.4.1 INTRODUCTION

A vestibular schwannoma (VS) is a benign Schwann cell-derived tumor arising from the vestibular portion of the eighth cranial nerve. The overall incidence is 1:100,000, with as much as 3,000 cases diagnosed per year in the United States. Likely because of wide availability of imaging, there is a rise in both incidence and earlier detection of VS (Propp et al., 2006). Common presenting symptoms include hearing loss, tinnitus, vertigo, imbalance, and gait disturbance. Trigeminal symptoms (pain mainly) and hemifacial spasms are less common. Tumor progression can lead to brain stem compression, cranial neuropathies, and hydrocephalus. Treatment options span from observation, microsurgical resection, conventionally fractionated radiotherapy, and SRS. Reports of excellent tumor control rates after single-session SRS (Rowe et al., 2003; Hasegawa et al., 2005a,b; Lunsford et al., 2005; Friedman et al., 2006; Hempel et al., 2006; Chopra et al., 2007; Fukuoka et al., 2009) have established SRS as a good treatment option. Schwannomas appearing in the frame of neurofibromatosis-II prove more resistant to treatment, carrying a higher failure and complications rate, as these do in other treatment modalities as well. The consequent treatment-related cranial nerve complications can include hearing loss, facial nerve palsy, and trigeminal dysfunction.

18.4.2 SINGLE-SESSION VERSUS HYPOFRACTIONATED SRS FOR VESTIBULAR SCHWANNOMAS

High rates of tumor control have been reported with the Gamma Knife–based (Rowe et al., 2003; Hasegawa et al., 2005a,b; Chopra et al., 2007; Fukuoka et al., 2009) and LINAC-based SRS (Friedman et al., 2006; Kalogeridi et al., 2009; Hsu et al., 2010) and conventionally fractionated stereotactic radiotherapy (Andrews et al., 2001; Sawamura et al., 2003; Chan et al., 2005; Combs et al., 2005; Kopp et al., 2011) and proton beam irradiation (Harsh et al., 2002; Hansasuta et al., 2011). Stereotactic radiosurgery has been proven to be an effective alternative to microsurgical resection for small to medium tumors, with rates of tumor control that range from 92% to 100%. The likelihood of hearing preservation in earlier clinical series was noted to be 51%–60% (Hirato, 1995; Hirato et al., 1996; Kondziolka et al., 1998; Iwai et al., 2003; Chang et al., 2005), but more recent studies in which more isocenters were used to enhance conformal treatment suggest that intermediate-term hearing preservation rates can be improved to between 71% and 73% (Flickinger et al., 1999; Niranjan et al., 1999; Spiegelmann et al., 1999; Chang et al., 2005). Given the high rates of tumor control, more attention is being paid to functional outcomes such as hearing, trigeminal nerve, and facial nerve preservation. In an effort to improve functional outcomes, the dose for single-fraction SRS has been reduced from previous doses of 16–20 Gy to the currently accepted dose of 11–13 Gy (Flickinger et al., 1996, 2001, 2004; Lunsford et al., 2005). Analogous to the dose deescalation trend observed with single-session SRS, hypofractionated fraction dose has also undergone adaptation, reduced in some protocols from 21 Gy in three sessions to 18 Gy in three sessions; the rate of tumor control with the lower dose of 18 Gy was shown to be equivalent to earlier experience using 21 Gy in three sessions (Poen et al., 1999).

Two key treatment parameters in SRS are dose and fractionation. Improved hearing preservation can be achieved with a lower radiation dose so long as the tumor is controlled. A recent literature review of 45 published articles (Yang, 2010) representing 4234 patients treated with GKRS showed an overall hearing preservation rate of 51%; the authors found dose to be a statistically significant factor associated with hearing preservation (hearing preservation rate of 60.5% at <13 Gy vs 50.4% at >13 Gy: $P = 0.001$) (Yang, 2010). The effect of the second treatment parameter, fractionation, on hearing preservation is less clear. The radiobiological rationale behind fractionation is to minimize radiation-induced normal tissue complications (Hall, 2006). Although the 70%–100% hearing preservation rate reported with fractionated radiotherapy or hypofractioned SRS (Andrews et al., 2001; Sawamura et al., 2003; Chan, 2005; Combs et al., 2005; Kopp et al., 2011) is promising, no prospective randomized study has evaluated the effects of fractionation (with either conventional fractionation schedules or two- to five-session SRS) on hearing preservation.

Still, one should bear in mind that tumor control is not a rationale for staging radiosurgery in this location. Existing single-session radiosurgery techniques and protocols consistently result in high rates of tumor control. The primary objective for staging radiosurgical treatment in VS is to minimize radiation-related cranial nerve complications. Despite this goal, it is important to note that by combining a lower dose and greater conformality, contemporary single-session SRS techniques for VS already have reduced the risk of radiation-induced facial nerve injury to less than 2% in most series.

18.4.3 HYPOFRACTIONATED SRS FOR VESTIBULAR SCHWANNOMA

The proponents of hypofractionated SRS for VS contend that a fractionated approach may confer a greater chance of neurological preservation and in particular hearing preservation. However, the literature does not permit a clear consensus as to the merits of such an approach (Murphy et al., 2011). Nevertheless, we summarize the experience in hypofractionated SRS for VS patients to date.

Initial reports used a wide array of protocols. Williams (2002) reported the outcome of 150 VS patients treated with either 5 fractions of 5 Gy for tumors smaller than 3.0 cm or 10 fractions of 3 Gy for tumors larger than 3.0 cm. Ninety percent of patients were treated using the 5-fraction 5 Gy regimen protocol. During 1 year of follow-up, no tumor increased in size, and there was no instance of facial nerve injury. At 1 year, Gardner–Robertson grade 1 or 2 hearing was maintained in 70% of patients when the tumor was smaller than 3.0 cm and treated with the 5-fraction 5 Gy regimen. Meijer et al. (2003) reported on their results of hypofractionated radiosurgery in 80 patients with VS using either five treatments of 4 Gy or five treatments of 5 Gy. The overall 5-year actuarial tumor control rates and facial nerve preservation were 94% and 96%, respectively. Hearing preservation rate at 5 years was reported as 66% (Meijer et al., 2000). Besides using different dose and fraction schemes, both of these clinical series used radiosurgical techniques that are less conformal and less accurate than that used in current practice and thus should be viewed with caution.

Chang et al. (2005) reported their experience using staged CyberKnife-based SRS treatment of unilateral VS in 61 patients using three sessions, 13% (n = 8) of which had prior surgery and the residual tissue was targeted. Mean pretreatment maximal tumor dimension was 18.5 mm (range, 5–32 mm). After radiosurgery, 29 (48%) of 61 tumors decreased in size, and 31 tumors (50%) were stable resulting in a tumor control rate of 98%. The mean follow-up period for hearing assessment was 48 months (range, 36–62 months). Of the 61 patients in this series, 13 patients had no measurable hearing (Gardner–Robertson grade 5) and were not tested with serial audiograms after radiosurgery. Of the remaining 48 patients, all had Gardner–Robertson grade 1–3 hearing before treatment. At last follow-up, 43 (90%) of these patients maintained Gardner–Robertson grade 1–3 hearing.

Hansasuta et al. (2011) reported their experience of tumor control and complication results of hypofractionated CyberKnife-based SRS treatment of VS. With over 1 year, 383 patients' follow-up data were analyzed. Facial nerve function, hearing, and tumor volume/mass effect were classified with the House–Brackmann (House, 1985), Gardner–Robertson (Gardner, 1988), and Koos scales (Koos et al., 1998), respectively. Multisession SRS was the primary treatment in 89% (n = 342). On presentation, 91% (n = 350) of patients had a House–Brackmann facial nerve function grade I, hemifacial spasm was present in 2% (n = 8), hearing loss in 92% (n = 353), and tinnitus was reported in 49% (n = 188). Ninety percent (n = 368) of patients were treated with 18 Gy in three sessions, targeting a median tumor volume of 1.1 cm^3 (range, 0.02–19.8 cm^3), and 9.6% (n = 22) of patients were neurofibromatosis-II patients or were treated prior to year 2000 (when treatment dose was reduced) (Hansasuta et al., 2011).

Of the 383 patients cohort, 10 tumors exhibited progressive growth requiring additional treatment (microsurgery [n = 9] or repeat SRS [n = 1]). Median follow-up was 3.6 years, and the 3- and 5-year Kaplan–Meier resection/repeat SRS-free tumor control rates were 99% and 96%, respectively. The NF2-associated tumors had worse tumor control compared with sporadic tumors (in accordance with other reports). The 3- and 5-year Kaplan–Meier tumor control rates for sporadic versus NF2-associated tumors were 99% and 96% compared with 93% and 84%, respectively (P = 0.03) (Hansasuta et al., 2011). During a median follow-up period of 3 years for hearing, a crude serviceable hearing preservation rate of 76% was found. Smaller tumor volume was associated with a higher hearing preservation rate after SRS (P = 0.001, with tumor volume as a continuous variable). Tumors <3 cm^3 had a serviceable hearing preservation rate of 80% compared with 59% for tumors >3 cm^3 (P = 0.009). These results are in line with some single-institution series results (68%–77%) (Rowe et al., 2003; Hasegawa et al., 2005b; Chung et al., 2008; Fukuoka et al., 2009). Nonauditory complications (trigeminal and facial nerve symptoms, hydrocephalus, brain stem compression) were associated with a larger tumor volume, a largest quartile tumors (3.4 cm^3) having a complication rate of 9.6% compared with 3.5% for the rest (P = 0.03). Similarly, the rate of nonauditory complications for Koos stage IV tumors was 9.3% compared with 3.8% for Koos stage I, II, and III tumors (P = 0.05).

18.5 FUTURE FOR HYPOFRACTIONATED SRS FOR BENIGN BRAIN NEOPLASMS

Because of the increase in radiation dose to normal brain that is associated with larger tumor volumes, the general guideline for single-session radiosurgery are lesions smaller than 10–15 cm^3 in volume (or 2.5–3 cm in diameter) (Harsh et al., 2002). However, hypofractionated radiosurgery offers a reasonable approach to many patients wherein the approach reduces tissue toxicity while maintaining a high local tumor control rates. Hypofractionated radiosurgery (temporal and spatial summation SRS) can take advantage of the beneficial effect that fractionation has on early- and late-responding tissues as seen in conventional radiotherapy. The integration of onboard imaging such as cone-beam CT and even MRI to SRS devices will only further expand the role and ease of implementation of hypofractionated SRS for intracranial pathologies.

CHECKLIST: KEY POINTS FOR CLINICAL PRACTICE

✓	ACTIVITY	SOME CONSIDERATIONS
	Patient selection	*Is the patient appropriate for HSRT?* • Consider for patients that have a unifocal tumor or unifocal recurrence • KPS ≥ 70 • No recent cytotoxic chemotherapy • Primary or recurrent setting: typically PTV < 40 cc
	SRS vs HSRT	*Is the lesion amenable to single-fraction SRS?* • In general, patients with benign intracranial tumors are treated with single-session SRS. Patients harboring larger tumors (>10 cc), those near radiation-sensitive critical structures, or those who have failed prior SRS are typical candidates for HSRT
	Simulation	*Immobilization* • Supine position with three-point thermoplastic mask, a custom head mold and open-face mask, or an open vacuum-assisted mouthpiece *Imaging* • Both CT scan and MRI should be performed in the treatment position • MRI ideally should be a thin-slice study (1 mm slices), including T1 sequence with IV contrast as well as T2 or FLAIR sequence • Further image sequences as per clinical preference and optimized for tumor histology (e.g., fat saturation for postresection pituitary adenomas) • Accurate image fusion should be verified by a physicist and the treating clinician and used for contouring targets
	Treatment planning	*Contours* Newly diagnosed benign intracranial tumor • Dose-painting technique • GTV$_1$ should include postoperative residual contrast tumor • No margin is added to create a clinical target volume (CTV$_1$) • Additional 1–2 mm is added to create the planning target volume (PTV$_1$) This is not always done in Gamma Knife–based systems • Organs at risk should be contoured on CT and verified via MRI Recurrent benign intracranial tumor • GTV includes all enhancing disease on the T1 postcontrast sequence (or enhancing postoperative cavity if disease is previously resected) • PTV is directly generated by expanding the GTV by 1–2 mm *Treatment planning* • Forward or inverse treatment planning depending upon the SBRT platform • Planning techniques vary by SBRT platform but can include noncoplanar IMRT or multiple VMAT arcs (LINAC), multiple isocenters (Gamma Knife), or nonisocentric beams (CyberKnife) • Dose • Varies depending upon the underlying tumor histology. Please refer to the chapter's text for further guidance

(Continued)

✓	ACTIVITY	SOME CONSIDERATIONS

	Treatment planning	• Coverage
		  • $D_{95\%} \geq$ Rx
		• Dose constraints
		  • Optic structures: max point dose (MPD) 23–25 Gy in up to 5 fractions
		  • Brain stem: D_{05} of 30 Gy in up to 5 fractions

	Treatment delivery	*Setup verification imaging (varies depending upon the SBRT platform)* • Patients are first set up conventionally using room lasers and manual shifts • OSI program if the CDR head frame is used • Cone-beam CT to match to bony anatomy • Treatment couch with 6 degrees of freedom to correct translational and rotational shifts • Position confirmed using 2D kilovoltage imaging *Manual setup verification (for SBRT platforms with no onboard imaging)* • Patient position measured via manual measurements or by matching field templates to light field *Treatment delivery* • Couch shifts are automated for some systems (Gamma Knife) and manual on others (some LINACs). Shifts on some platforms may require imaging for position verification

REFERENCES

Adler JR Jr, Gibbs IC, Puataweepong P, Chang SD (2008) Visual field preservation after multisession cyberknife radiosurgery for perioptic lesions [Reprint of *Neurosurgery*. 2006;59(2):244–254; discussion 244–254; PMID: 16883165]. *Neurosurgery* 62(Suppl 2):733–743.

Agha A, Sherlock M, Brennan S, O'Connor SA, O'Sullivan E, Rogers B, Faul C, Rawluk D, Tormey W, Thompson CJ (2005) Hypothalamic-pituitary dysfunction following irradiation of non-pituitary brain tumours in adults. *J Clin Endocrinol Metab* 90:6355–6360.

Alper MG (1981) Management of primary optic nerve meningiomas. Current status-therapy in controversy. *J Clin Neuroophthalmol* 1:101–117.

Andrews DW, Faroozan R, Yang BP, Hudes RS, Werner-Wasik M, Kim SM, Sergott RC et al. (2002) Fractionated stereotactic radiotherapy for the treatment of optic nerve sheath meningiomas: Preliminary observations of 33 optic nerves in 30 patients with historical comparison to observation with or without prior surgery. *Neurosurgery* 51:890–902; discussion 903–894.

Andrews DW, Suarez O, Goldman HW, Downes MB, Bednarz G, Corn BW, Werner-Wasik M, Rosenstock J, Curran WJ Jr (2001) Stereotactic radiosurgery and fractionated stereotactic radiotherapy for the treatment of acoustic schwannomas: Comparative observations of 125 patients treated at one institution. *Int J Radiat Oncol Biol Phys* 50(5):1265–1278.

Arnautovic KI, Al-Mefty O, Husain M (2000) Ventral foramen magnum meningiomas. *J Neurosurg* 92:71–80.

Astrahan M (2008) BED calculations for fractions of very high dose: In regard to Park et al. (*Int J Radiat Oncol Biol Phys* 2007;69:S623–S624). *Int J Radiat Oncol Biol Phys* 71(3):963; author reply 963–964.

Barbaro NM, Gutin PH, Wilson CB, Sheline GE, Boldrey EB, Wara WM (1987) Radiation therapy in the treatment of partially resected meningiomas. *Neurosurgery* 20:525–528.

Barnett GH, Linskey ME, Adler JR, Cozzens JW, Friedman WA, Heilbrun MP, Lunsford LD et al. (2007) American Association of Neurological surgeons; Congress of Neurological Surgeons Washington Committee Stereotactic Radiosurgery Task Force. Stereotactic radiosurgery—An organized neurosurgery-sanctioned definition. *J Neurosurg* 106(1):1–5.

Benedict SH, Yenice KM, Followill D, Galvin JM, Hinson W, Kavanagh B, Keall P et al. (2010) Stereotactic body radiation therapy: The report of AAPM Task Group 101. *Med Phys* 37(8):4078–4101.

Berman D, Miller NR (2006) New concepts in the management of optic nerve sheath meningiomas. *Ann Acad Med Singapore* 35:168–174.

Brada M, Ford D, Ashley S, Bliss JM, Crowley S, Mason M, Rajan B, Traish D (1992) Risk of second brain tumour after conservative surgery and radiotherapy for pituitary adenoma. *BMJ* 304:1343–1346.

Brada M, Rajan B, Traish D, Ashley S, Holmes-Sellors PJ, Nussey S, Uttley D (1993) The long-term efficacy of conservative surgery and radiotherapy in the control of pituitary adenomas. *Clin Endocrinol* 38:571–578.

Breen P, Flickinger JC, Kondziolka D, Martinez AJ (1998) Radiotherapy for non-functional pituitary adenoma: Analysis of long-term tumor control. *J Neurosurg* 89:933–938.

Brenner DJ (2008) The linear-quadratic model is an appropriate methodology for determining isoeffective doses at large doses per fraction. *Semin Radiat Oncol* 18(4):234–239.

Brenner DJ, Hall EJ (1994) Stereotactic radiotherapy of intracranial tumors: An ideal candidate for accelerated treatment. *Int J Radiat Oncol Biol Phys* 28:1039–1047.

Bricolo AP, Turazzi S, Talacchi A, Cristofori L (1992) Microsurgical removal of petroclival meningiomas: A report of 33 patients. *Neurosurgery* 31:813–828.

Cantore WA (2000) Neural orbital tumors. *Curr Opin Ophthalmol* 11:367–371.

Carrasco JR, Penne RB (2004) Optic nerve sheath meningiomas and advanced treatment options. *Curr Opin Ophthalmol* 15:406–410.

Chan AW, Black P, Ojemann RG, Barker FG II, Kooy HM, Lopes VV, McKenna MJ, Shrieve DC, Martuza RL, Loeffler JS (2005) Stereotactic radiotherapy for vestibular schwannomas: Favorable outcome with minimal toxicity. *Neurosurgery* 57(1):60–70.

Chang SD, Gibbs IC, Sakamoto GT, Lee E, Oyelese A, Adler JR Jr (2005) Staged stereotactic irradiation for acoustic neuroma. *Neurosurgery* 56(6):1254–1261.

Choi CY, Soltys SG, Gibbs IC, Harsh GR, Sakamoto GT, Patel DA, Lieberson RE, Chang SD, Adler JR (2011) Stereotactic radiosurgery of cranial nonvestibular schwannomas: Results of single- and multisession radiosurgery. *Neurosurgery* 68(5):1200–1208.

Chopra R, Kondziolka D, Niranjan A, Lunsford LD, Flickinger JC (2007) Long-term follow-up of acoustic schwannoma radiosurgery with marginal tumor doses of 12 to 13 Gy. *Int J Radiat Oncol Biol Phys* 68(3):845–851.

Chung WY, Shiau CY, Wu HM, Liu KD, Guo WY, Wang LW, Pan DH (2008) Staged radiosurgery for extra-large cerebral arteriovenous malformations: Method, implementation, and results. *J Neurosurg* 109(Suppl):65–72.

Clark WC, Theofilos CS, Fleming JC (1989) Primary optic nerve sheath meningiomas. Report of nine cases. *J Neurosurg* 70:37–40.

Combs SE, Volk S, Schulz-Ertner D, Huber PE, Thilmann C, Debus J (2005) Management of acoustic neuromas with fractionated stereotactic radiotherapy (FSRT): Long-term results in 106 patients treated in a single institution. *Int J Radiat Oncol Biol Phys* 63(1):75–81.

Condra KS, Buatti JM, Mendenhall WM, Friedman WA, Marcus RB Jr, Rhoton AL (1997) Benign meningiomas: Primary treatment selection affects survival. *Int J Radiat Oncol Biol Phys* 39:427–436.

Couldwell WT, Fukushima T, Giannotta SL, Weiss MH (1996) Petroclival meningiomas: Surgical experience in 109 cases. *J Neurosurg* 84:20–28.

Couldwell WT, Kan P, Liu JK, Apfelbaum RI (2006) Decompression of cavernous sinus meningioma for preservation and improvement of cranial nerve function. *J Neurosurg* 105:148–152.

Cozzi R, Barausse M, Asnaghi D, Dallabonzana D, Lodrini S, Attanasio R (2001) Failure of radiotherapy in acromegaly. *Eur J Endocrinol* 145:717–726.

Craig WM, Gogela LJ (1949) Intraorbital meningiomas; a clinicopathologic study. *Am J Ophthalmol* 32:1663–1680, illust.

Darzy KH, Shalet SM (2005) Hypopituitarism after cranial irradiation. *J Endocrinol Invest* 28:78–87.

De Jesús O, Sekhar LN, Parikh HK, Wright DC, Wagner DP (1996) Long-term follow-up of patients with meningiomas involving the cavernous sinus: Recurrence, progression, and quality of life. *Neurosurgery* 39:915–920.

Dekkers OM, Pereira AM, Romijn JA (2008) Treatment and follow-up of clinically nonfunctioning pituitary macroadenomas. *J Clin Endocrinol Metab* 93(10):3717–3726.

DeMonte F, Smith HK, Al-Mefty O (1994) Outcome of aggressive removal of cavernous sinus meningiomas. *J Neurosurg* 81:245–251.

Ding D, Starke RM, Sheehan JP (2014) Treatment paradigms for pituitary adenomas: Defining the roles of radiosurgery and radiation therapy. *J Neurooncol* 117(3):445–457.

Dodd RL, Ryu M, Kamnerdsupaphon P, Gibbs IC, Chang SD, Adler JR Jr (2006) Cyberknife radiosurgery treatment of benign intradural extramedullary spinal tumors. *Neurosurgery* 58:674–685.

Dutton JJ (1992) Optic nerve sheath meningiomas. *Surv Ophthalmol* 37:167–183.

Estrada J, Boronat M, Mielgo M, Magallón R, Millan I, Díez S, Lucas T, Barceló B (1997) The long-term outcome of pituitary irradiation after unsuccessful transsphenoidal surgery in Cushing's disease. *N Engl J Med* 336:172–177.

Firlik AD, Levy EI, Kondziolka D, Yonas H (1998) Staged volume radiosurgery followed by microsurgical resection: A novel treatment for giant cerebral arteriovenous malformations: Technical case report. *Neurosurgery* 43:1223–1228.

Fisher BJ, Gaspar LE, Noone B (1993) Radiation therapy of pituitary adenoma: Delayed sequelae. *Radiology* 187:843–846.

Flickinger JC, Kondziolka D, Lunsford LD (1999) Dose selection in stereotactic radiosurgery. *Neurosurg Clin North Am* 10:271–280.

Flickinger JC, Kondziolka D, Niranjan A, Lunsford LD (2001) Results of acoustic neuroma radiosurgery: An analysis of 5 years' experience using current methods. *J Neurosurg* 94(1):1–6.

Flickinger JC, Kondziolka D, Niranjan A, Maitz A, Voynov G, Lunsford LD (2004) Acoustic neuroma radiosurgery with marginal tumor doses of 12 to 13 Gy. *Int J Radiat Oncol Biol Phys* 60(1):225–230.

Flickinger JC, Kondziolka D, Pollock BE, Lunsford LD (1996) Evolution in technique for vestibular schwannoma radiosurgery and effect on outcome. *Int J Radiat Oncol Biol Phys* 36(2):275–280.

Friedman WA, Bradshaw P, Myers A, Bova FJ (2006) Linear accelerator radiosurgery for vestibular schwannomas. *J Neurosurg* 105(5):657–661.

Fukuoka S, Takanashi M, Hojyo A, Konishi M, Tanaka C, Nakamura H (2009) Gamma knife radiosurgery for vestibular schwannomas. *Prog Neurol Surg* 22:45–62.

Ganz JC (2011) *Gamma Knife Neurosurgery*. Heidelberg, Germany: Springer.

Garcia-Barros M, Paris F, Cordon-Cardo C, Lyden D, Rafii S, Haimovitz-Friedman A, Fuks Z, Kolesnick R (2003) Tumor response to radiotherapy regulated by endothelial cell apoptosis. *Science* 300:1155–1159.

Gardner G, Robertson JH (1988) Hearing preservation in unilateral acoustic neuroma surgery. *Ann Otol Rhinol Laryngol* 97(1):55–66.

George B, Lot G, Boissonnet H (1997) Meningioma of the foramen magnum: A series of 40 cases. *Surg Neurol* 47:371–379.

Girkin CA, Comey CH, Lunsford LD, Goodman ML, Kline LB (1997) Radiation optic neuropathy after stereotactic radiosurgery. *Ophthalmology* 104:1634–1643.

Goldsmith BJ, Wara WM, Wilson CB, Larson DA (1994) Postoperative irradiation for subtotally resected meningiomas. A retrospective analysis of 140 patients treated from 1967 to 1990. *J Neurosurg* 80:195–201.

Guerrero M, Li XA (2004) Extending the linear-quadratic model for large fraction doses pertinent to stereotactic radiotherapy. *Phys Med Biol* 49(20):4825–4835.

Hall E, Giaccia A (2006) *Radiobiology for the Radiologist*, 6th edn. Philadelphia, PA: Lippincott Williams & Wilkins.

Hansasuta A, Choi CY, Gibbs IC, Soltys SG, Tse VC, Lieberson RE, Hayden MG et al. (2011) Multisession stereotactic radiosurgery for vestibular schwannomas: Single-institution experience with 383 cases. *Neurosurgery* 69(6):1200–1209.

Harsh GR, Thornton AF, Chapman PH, Bussiere MR, Rabinov JD, Loeffler JS (2002) Proton beam stereotactic radiosurgery of vestibular schwannomas. *Int J Radiat Oncol Biol Phys* 54(1):35–44.

Hasegawa T, Fujitani S, Katsumata S, Kida Y, Yoshimoto M, Koike J (2005a) Stereotactic radiosurgery for vestibular schwannomas: Analysis of 317 patients followed more than 5 years. *Neurosurgery* 57(2):257–265.

Hasegawa T, Kida Y, Kobayashi T, Yoshimoto M, Mori Y, Yoshida J (2005b) Long-term outcomes in patients with vestibular schwannomas treated using gamma knife surgery: 10-year follow up. *J Neurosurg* 102(1):10–16.

Hasegawa T, Kida Y, Yoshimoto M, Koike J, Iizuka H, Ishii D. (2007) Long-term outcomes of gamma knife surgery for cavernous sinus meningioma. *J Neurosurg* 107:745–751.

Haselsberger K, Maier T, Dominikus K, Holl E, Kurschel S, Ofner-Kopeinig P, Unger F (2009) Staged gamma knife radiosurgery for large critically located benign meningiomas: Evaluation of a series comprising 20 patients. *J Neurol Neurosurg Psychiatry* 80(10):1172–1175.

Heilbrun MP, Mehta VK, Le QT, Chang SD, Adler JR Jr, Martin DP (2002) Staged image guided radiosurgery for lesions adjacent to the anterior visual pathways. *Acta Neurochir* 144:1101 (abstr).

Hempel JM, Hempel E, Wowra B, Schichor C, Muacevic A, Riederer A (2006) Functional outcome after gamma knife treatment in vestibular schwannoma. *Eur Arch Otorhinolaryngol* 263(8):714–718.

Higuchi Y, Serizawa T, Nagano O, Matsuda S, Ono J, Sato M, Iwadate Y, Saeki N (2009) Three-staged stereotactic radiotherapy without whole brain irradiation for large metastatic brain tumors. *Int J Radiat Oncol Biol Phys* 74:1543–1548.

Hirato M (1995) Gamma knife radiosurgery for acoustic schwannoma: Early effects and preservation of hearing. *Neurol Med Chir* 35:737–741.

Hirato M, Inoue H, Zama A, Ohye C, Shibazaki T, Andou Y (1996) Gamma Knife radiosurgery for acoustic schwannoma: Effects of low radiation dose and functional prognosis. *Stereotact Funct Neurosurg* 66(Suppl 1):134–141.

Hoshi M, Hayashi T, Kagami H, Murase I, Nakatsukasa M (2003) Late bilateral temporal lobe necrosis after conventional radiotherapy. *Neurol Med Chir* 43:213–216.

House JW, Brackmann DE (1985) Facial nerve grading system. *Otolaryngol Head Neck Surg* 93(2):146–147.

Hoybye C, Rahn T (2009) Adjuvant Gamma Knife radiosurgery in non-functioning pituitary adenomas; low risk of long-term complications in selected patients. *Pituitary* 12(3):211–216.

Hsu PW, Chang CN, Lee ST, Huang YC, Chen HC, Wang CC, Hsu YH, Tseng CK, Chen YL, Wei KC (2010) Outcomes of 75 patients over 12 years treated for acoustic neuromas with linear accelerator-based radiosurgery. *J Clin Neurosci* 17(5):556–560.

Hughes MN, Llamas KJ, Yelland ME, Tripcony LB (1993) Pituitary adenomas: Long-term results for radiotherapy alone and post-operative radiotherapy. *Int J Radiat Oncol Biol Phys* 27:1035–1043.

Iwai Y, Yamanaka K, Nakajima H (2001) Two-staged gamma knife radiosurgery for the treatment of large petroclival and cavernous sinus meningiomas. *Surg Neurol* 56:308–314.

Iwai Y, Yamanaka K, Shiotani M, Uyama T (2003) Radiosurgery for acoustic neuromas: Results of low-dose treatment. *Neurosurgery* 53:282–288.

Izawa M, Hayashi M, Nakaya K, Satoh H, Ochiai T, Hori T, Takakura K (2000) Gamma Knife radiosurgery for pituitary adenomas. *J Neurosurg* 93(Suppl 3):19–22.

Jagannathan J, Sheehan JP, Pouratian N, Laws ER, Steiner L, Vance ML (2007) Gamma Knife surgery for Cushing's disease. *J Neurosurg* 106(6):980–987.

Javalkar V, Pillai P, Vannemreddy P, Caldito G, Ampil F, Nanda A (2009) Gamma knife radiosurgery for arteriovenous malformations located in eloquent regions of the brain. *Neurol India* 57:617–621.

Jung HW, Yoo H, Paek SH, Choi KS (2000) Long-term outcome and growth rate of subtotally resected petroclival meningiomas: Experience with 38 cases. *Neurosurgery* 46:567–575.

Kalogeridi MA, Georgolopoulou P, Kouloulias V, Kouvaris J, Pissakas G (2009) Long-term results of LINAC-based stereotactic radiosurgery for acoustic neuroma: The Greek experience. *J Cancer Res Ther* 5(1):8–13.

Kavanagh BD, Newman F (2008) Toward a unified survival curve: In regard to Park et al. (*Int J Radiat Oncol Biol Phys* 2008;70:847–852) and Krueger et al. (*Int J Radiat Oncol Biol Phys* 2007;69:1262–1271). *Int J Radiat Oncol Biol Phys* 71(3):958–959.

Killory BD, Kresl JJ, Wait SD, Ponce FA, Porter R, White WL (2009) Hypofractionated CyberKnife radiosurgery for perichiasmatic pituitary adenomas: Early results. *Neurosurgery* 64:A19–A25.

Kim JW, Im YS, Nam DH, Park K, Kim JH, Lee JI (2008a) Preliminary report of multisession gamma knife radiosurgery for benign perioptic lesions: Visual outcome in 22 patients. *J Korean Neurosurg Soc* 44:67–71.

Kim JW, Rizzo JF, Lessell S (2005) Controversies in the management of optic nerve sheath meningiomas. *Int Ophthalmol Clin* 45:15–23.

Kim MS, Park K, Kim JH, Kim YD, Lee JI (2008b) Gamma knife radiosurgery for orbital tumors. *Clin Neurol Neurosurg* 110:1003–1007.

King CR, Lehmann J, Adler JR, Hai J (2003) CyberKnife radiotherapy for localized prostate cancer: Rationale and technical feasibility. *Technol Cancer Res Treat* 2:25–30.

Kirkpatrick JP, Meyer JJ, Marks LB (2008) The linear-quadratic model is inappropriate to model high dose per fraction effects in radiosurgery. *Semin Radiat Oncol* 18(4):240–243.

Kollova´ A, Liscák R, Novotný J Jr, Vladyka V, Simonová G, Janousková L (2007) Gamma knife surgery for benign meningioma. *J Neurosurg* 107:325–336.

Kondziolka D, Flickinger JC, Perez B (1998) Judicious resection and/or radiosurgery for parasagittal meningiomas: Outcomes from a multicenter review. *Neurosurgery* 43:405–414.

Kondziolka D, Levy EI, Niranjan A, Flickinger JC, Lunsford LD (1999) Long-term outcomes after meningioma radiosurgery: Physician and patient perspectives. *J Neurosurg* 91:44–50.

Kondziolka D, Lunsford LD, McLaughlin MR, Flickinger JC (1998) Long-term outcomes after radiosurgery for acoustic neuromas. *N Engl J Med* 339:1426–1433.

Kondziolka D, Mathieu D, Lunsford LD, Martin JJ, Madhok R, Niranjan A, Flickinger JC (2008) Radiosurgery as definitive management of intracranial meningiomas. *Neurosurgery* 62:53–60.

Koos WT, Day JD, Matula C, Levy DI (1998) Neurotopographic considerations in the microsurgical treatment of small acoustic neurinomas. *J Neurosurg* 88(3):506–512.

Kopp C, Fauser C, Müller A, Astner ST, Jacob V, Lumenta C, Meyer B, Tonn JC, Molls M, Grosu AL (2011) Stereotactic fractionated radiotherapy and LINAC radiosurgery in the treatment of vestibular schwannoma: Report about both stereotactic methods from a single institution. *Int J Radiat Oncol Biol Phys* 80(5):1485–1491.

Kry SF (2005) The calculated risk of fatal secondary malignancies from intensity-modulated radiation therapy. *Int J Radiat Oncol Biol Phys* 62:1195–1203.

Kwon Y, Bae JS, Kim JM, Lee DH, Kim SY, Ahn JS, Kim JH, Kim CJ, Kwun BD, Lee JK (2005) Visual changes after gamma knife surgery for optic nerve tumors. Report of three cases. *J Neurosurg* 102(Suppl):143–146.

Leber KA, Bergloff J, Langmann G, Mokry M, Schrottner O, Pendl G (1995) Radiation sensitivity of visual and oculomotor pathways. *Stereotact Funct Neurosurg* 1:233–238.

Leber KA, Bergloff J, Pendl G (1998) Dose-response tolerance of the visual pathways and cranial nerves of the cavernous sinus to stereotactic radiosurgery. *J Neurosurg* 88:43–50.

Lee JYK, Niranjan A, McInerney J, Kondziolka D, Flickinger JC, Lunsford LD (2002) Stereotactic radiosurgery providing long-term tumor control of cavernous sinus meningiomas. *J Neurosurg* 97:65–72.

Linskey ME, Martinez AJ, Kondziolka D, Flickinger JC, Maitz AH, Whiteside T, Lunsford LD (1993) The radiobiology of human acoustic schwannoma xenografts after stereotactic radiosurgery evaluated in the subrenal capsule of athymic mice. *J Neurosurg* 78:645–653.

Little KM, Friedman AH, Sampson JH, Wanibuchi M, Fukushima T (2005) Surgical management of petroclival meningiomas: Defining resection goals based on risk of neurological morbidity and tumor recurrence rates in 137 patients. *Neurosurgery* 56:546–559.

Liu L, Bassano DA, Prasad SC, Hahn SS, Chung CT (2003) The linear-quadratic model and fractionated stereotactic radiotherapy. *Int J Radiat Oncol Biol Phys* 57(3):827–832.

Lunsford LD, Niranjan A, Flickinger JC, Maitz A, Kondziolka D (2005) Radiosurgery of vestibular schwannomas: Summary of experience in 829 cases. *J Neurosurg* 102(Suppl):195–199.

MacNally SP, Rutherford SA, Ramsden RT, Evans DG, King AT (2008) Trigeminal schwannomas. *Br J Neurosurg* 22(6):729–738.

Maguire PD, Clough R, Friedman AH, Halperin EC (1999) Fractionated external-beam radiation therapy for meningiomas of the cavernous sinus. *Int J Radiat Oncol Biol Phys* 44:75–79.

Maire JP, Caudry M, Guérin J, Célérier D, San Galli F, Causse N, Trouette R, Dautheribes M (1995) Fractionated radiation therapy in the treatment of intracranial meningiomas: Local control, functional efficacy, and tolerance in 91 patients. *Int J Radiat Oncol Biol Phys* 33:315–321.

Marchetti M, Bianchi S, Milanesi I, Bergantin A, Bianchi L, Broggi G, Fariselli L (2011) Multisession radiosurgery for optic nerve sheath meningiomas—An effective option: Preliminary results of a single-center experience. *Neurosurgery* 69(5):1116–1122; discussion 1122–1123.

Mathiesen T, Gerlich A, Kihlström L, Svensson M, Bagger-Sjöbäck D (2007) Effects of using combined transpetrosal surgical approaches to treat petroclival meningiomas. *Neurosurgery* 60:982–992.

Mayo C, Martel MK, Marks LB, Flickinger J, Nam J, Kirkpatrick J (2010) Radiation dose-volume effects of optic nerves and chiasm. *Int J Radiat Oncol Biol Phys* 76:S28–S35.

McCollough WM, Marcus RB Jr, Rhoton AL Jr, Ballinger WE, Million RR (1991) Long-term follow-up of radiotherapy for pituitary adenoma: The absence of late recurrence after greater than or equal to 4500 cGy. *Int J Radiat Oncol Biol Phys* 21:607–614.

McCord MW, Buatti JM, Fennell EM, Mendenhall WM, Marcus RB Jr, Rhoton AL, Grant MB, Friedman WA (1997) Radiotherapy for pituitary adenoma: Long-term outcome and sequelae. *Int J Radiat Oncol Biol Phys* 39:437–444.

McKenna F, Ahmad S (2009) Toward a unified survival curve: In regard to Kavanagh and Newman (*Int J Radiat Oncol Biol Phys* 2008;71:958–959) and Park et al. (*Int J Radiat Oncol Biol Phys* 2008;70:847–852). *Int J Radiat Oncol Biol Phys* 73(2):640; author reply 640–641.

Mehta VK, Lee QT, Chang SD, Cherney S, Adler JR Jr (2002) Image guided stereotactic radiosurgery for lesions in proximity to the anterior visual pathways: A preliminary report. *Technol Cancer Res Treat* 1:173–180.

Meijer OW, Vandertop WP, Baayen JC, Slotman BJ (2003) Single-fraction versus fractionated linac-based stereotactic radiosurgery for vestibular schwannoma: A single-institution study. *Int J Radiat Oncol Biol Phys* 56:1390–1396.

Meijer OW, Wolbers JG, Baayen JC, Slotman BJ (2000) Fractionated stereotactic radiation therapy and single high-dose radiosurgery for acoustic neuroma: Early results of a prospective clinical study. *Int J Radiat Oncol Biol Phys* 46:45–49.

Melmed S (2006) Medical progress: Acromegaly. *N Engl J Med* 355(24):2558–2573.

Metellus P (2005) Evaluation of fractionated radiotherapy and gamma knife radiosurgery in cavernous sinus meningiomas: Treatment strategy. *Neurosurgery* 57:873–886.

Milker-Zabel S, Huber P, Schlegel W, Debus J, Zabel-du Bois A (2009) Fractionated stereotactic radiation therapy in the management of primary optic nerve sheath meningiomas. *J Neurooncol* 94:419–424.

Milker-Zabel S, Zabel A, Schulz-Ertner D, Schlegel W, Wannenmacher M, Debus J (2005) Fractionated stereotactic radiotherapy in patients with benign or atypical intracranial meningioma: Long-term experience and prognostic factors. *Int J Radiat Oncol Biol Phys* 61:809–816.

Minniti G, Traish D, Ashley S, Gonsalves A, Brada M (2005) Risk of second brain tumor after conservative surgery and radiotherapy for pituitary adenoma: Update after an additional 10 years. *J Clin Endocrinol Metab* 90:800–804.

Movsas B, Movsas TZ, Steinberg SM, Okunieff P (1995) Long-term visual changes following pituitary irradiation. *Int J Radiat Oncol Biol Phys* 33:599–605.

Moyer PD, Golnik KC, Breneman J (2000) Treatment of optic nerve sheath meningioma with three-dimensional conformal radiation. *Am J Ophthalmol* 129:694–696.

Murphy ES, Suh JH (March 15, 2011) Radiotherapy for vestibular schwannomas: A critical review. *Int J Radiat Oncol Biol Phys* 79(4):985–997.

Natarajan SK, Sekhar LN, Schessel D, Morita A (2007) Petroclival meningiomas: Multimodality treatment and outcomes at long-term follow-up. *Neurosurgery* 60:965–981.

Nelson PB, Goodman ML, Flickenger JC, Richardson DW, Robinson AG (1989) Endocrine function in patients with large pituitary tumors treated with operative decompression and radiation therapy. *Neurosurgery* 24:398–400.

Nguyen JH, Chen CJ, Lee CC, Yen CP, Xu Z, Schlesinger D, Sheehan JP (2014) Multisession gamma knife radiosurgery: A preliminary experience with a non-invasive, relocatable frame. *World Neurosurg* 82(6):1256–1263.

Niranjan A, Gobbel GT, Kondziolka D, Flickinger JC, Lunsford LD (2004) Experimental radiobiological investigations into radiosurgery: Present understanding and future directions. *Neurosurgery* 55:495–504.

Niranjan A, Lunsford LD, Flickinger JC, Maitz A, Kondziolka D (1999) Dose reduction improves hearing preservation rates after intracanalicular acoustic tumor radiosurgery. *Neurosurgery* 45:753–765.

Ove R, Kelman S, Amin PP, Chin LS (2000) Preservation of visual fields after peri-sellar gamma-knife radiosurgery. *Int J Cancer* 90:343–350.

Paek SH, Downes MB, Bednarz G, Keane WM, Werner-Wasik M, Curran WJ Jr, Andrews DW (2005) Integration of surgery with fractionated stereotactic radiotherapy for treatment of nonfunctioning pituitary macroadenomas. *Int J Radiat Oncol Biol Phys* 61:795–808.

Park C, Papiez L, Zhang S, Story M, Timmerman RD (2008) Universal survival curve and single fraction equivalent dose: Useful tools in understanding potency of ablative radiotherapy. *Int J Radiat Oncol Biol Phys* 70:847–852.

Pendl G, Eustacchio S, Unger F (2001) Radiosurgery as alternative treatment for skull base meningiomas. *J Clin Neurosci* 1:12–14.

Pendl G, Unger F, Papaefthymiou G, Eustacchio S (2000) Staged radiosurgical treatment for large benign cerebral lesions. *J Neurosurg* 93(Suppl 3):107–112.

Poen JC, Golby AJ, Forster KM, Martin DP, Chinn DM, Hancock SL, Adler JR Jr (1999) Fractionated stereotactic radiosurgery and preservation of hearing in patients with vestibular schwannoma: A preliminary report. *Neurosurgery* 45(6):1299–1305.

Pollock BE, Lunsford LD, Kondziolka D, Flickinger JC, Bissonette DJ, Kelsey SF, Jannetta PJ (1995) Outcome analysis of acoustic neuroma management: A comparison of microsurgery and stereotactic radiosurgery. *Neurosurgery* 36:215–224; discussion 224–219.

Pourel N, Auque J, Bracard S, Hoffstetter S, Luporsi E, Vignaud JM, Bey P (2001) Efficacy of external fractionated radiation therapy in the treatment of meningiomas: A 20-year experience. *Radiother Oncol* 61:65–70.

Propp JM, McCarthy BJ, Davis FG, Preston-Martin S (2006) Descriptive epidemiology of vestibular schwannomas. *Neuro Oncol* 8(1):1–11.

Rowe J, Grainger A, Walton L, Silcocks P, Radatz M, Kemeny A (2007) Risk of malignancy after gamma knife stereotactic radiosurgery. *Neurosurgery* 60:60–66.

Rowe JG, Radatz MW, Walton L, Hampshire A, Seaman S, Kemeny AA (2003) Gamma knife stereotactic radiosurgery for unilateral acoustic neuromas. *J Neurol Neurosurg Psychiatry* 74(11):1536–1542.

Ruschin M, Komljenovic PT, Ansell S, Ménard C, Bootsma G, Cho YB, Chung C, Jaffray D (2013) Cone beam computed tomography image guidance system for a dedicated intracranial radiosurgery treatment unit. *Int J Radiat Oncol Biol Phys* 85:243–250.

Sachs RK, Brenner DJ (2005) Solid tumor risks after high doses of ionizing radiation. *Proc Natl Acad Sci USA* 102:13040–13045.

Salinger DJ, Brady LW, Miyamoto CT (1992) Radiation therapy in the treatment of pituitary adenomas. *Am Clin Oncol* 15:467–473.

Samii M, Carvalho GA, Tatagiba M, Matthies C (1997) Surgical management of meningiomas originating in Meckel's cave. *Neurosurgery* 41:767–775.

Samii M, Klekamp J, Carvalho G (1996) Surgical results for meningiomas of the craniocervical junction. *Neurosurgery* 39:1086–1095.

Samii M, Tatagiba M (1992) Experience with 36 surgical cases of petroclival meningiomas. *Acta Neurochir* 118:27–32.

Sawamura Y, Shirato H, Sakamoto T, Aoyama H, Suzuki K, Onimaru R, Isu T, Fukuda S, Miyasaka K (2003) Management of vestibular schwannoma by fractionated stereotactic radiotherapy and associated cerebrospinal fluid malabsorption. *J Neurosurg* 99(4):685–692.

Sayer FT, Sherman JH, Yen CP, Schlesinger DJ, Kersh R, Sheehan JP (2011) Initial experience with the eXtend system: A relocatable frame system for multiple-session GK radiosurgery. *World Neurosurg* 75(5–6):665–672.

Selch MT, Ahn E, Laskari A, Lee SP, Agazaryan N, Solberg TD, Cabatan-Awang C, Frighetto L, Desalles AA (2004) Stereotactic radiotherapy for treatment of cavernous sinus meningiomas. *Int J Radiat Oncol Biol Phys* 59:101–111.

Sheehan JP, Starke RM, Mathieu D, Young B, Sneed PK, Chiang VL, Lee JY et al. (2013) Gamma Knife radiosurgery for the management of nonfunctioning pituitary adenomas: A multicenter study. *J Neurosurg* 119(2):446–456.

Shields JA, Shields CL, Scartozzi R (2004) Survey of 1264 patients with orbital tumors and simulating lesions: The 2002 Montgomery Lecture, part 1. *Ophthalmology* 111:997–1008.

Sibony PA, Krauss, HR, Kennerdell JS, Maroon JC, Slamovits TL (1984) Optic nerve sheath meningiomas. Clinical manifestations. *Ophthalmology* 91:1313–1326.

Simonova G, Novotny J, Novotny J Jr, Vladyka V, Liscak R (1995) Fractionated stereotactic radiotherapy with the Leksell Gamma Knife: Feasibility study. *Radiother Oncol* 37:108–116.

Sirin S, Kondziolka D, Niranjan A, Flickinger JC, Maitz AH, Lunsford LD (2006) Prospective staged volume radiosurgery for large arteriovenous malformations: Indications and outcomes in otherwise untreatable patients. *Neurosurgery* 58:17–27.

Spallone A, Makhmudov UB, Mukhamedjanov DJ, Tcherekajev VA (1999) Petroclival meningioma. An attempt to define the role of skull base approaches in their surgical management. *Surg Neurol* 51:412–420.

Spiegelmann R, Gofman J, Alezra D, Pfeffer R (1999) Radiosurgery for acoustic neurinomas (vestibular schwannomas). *Isr Med Assoc J* 1:8–13.

Stafford SL, Perry A, Suman VJ, Meyer FB, Scheithauer BW, Lohse CM, Shaw EG (1998) Primarily resected meningiomas: Outcome and prognostic factors in 581 Mayo Clinic patients, 1978 through 1988. *Mayo Clin Proc* 73:936–942.

Stafford SL, Pollock BE, Foote RL, Link MJ, Gorman DA, Schomberg PJ, Leavitt JA (2001) Meningioma radiosurgery: Tumor control, outcomes, and complications among 190 consecutive patients. *Neurosurgery* 49:1029–1037.

Stafford SL, Pollock BE, Leavitt JA, Foote RL, Brown PD, Link MJ, Gorman DA, Schomberg PJ (2003) A study on the radiation tolerance of the optic nerves and chiasm after stereotactic radiosurgery. *Int J Radiat Oncol Biol Phys* 55:1177–1181.

Starke RM, Williams BJ, Jane JA Jr, Sheehan JP (2012) Gamma Knife surgery for patients with nonfunctioning pituitary macroadenomas: Predictors of tumor control, neurological deficits, and hypopituitarism. *J Neurosurg* 117(1):129–135.

Stripp D, Maity A, Janss AJ, Belasco JB, Tochner ZA, Goldwein JW, Moshang T et al. (2004) Surgery with or without radiation therapy in the management of craniopharyngiomas in children and young adults. *Int J Radiat Oncol Biol Phys* 58:714–720.

Sweeney RA, Bale R, Auberger T, Vogele M, Foerster S, Nevinny-Stickel M, Lukas P (2001) A simple and non-invasive vacuum mouthpiece-based head fixation system for high precision radiotherapy. *Strahlenther Onkol* 177:43–47.

Sweeney R, Bale R, Vogele M, Nevinny-Stickel M, Bluhm A, Auer T, Hessenberger G, Lukas P (1998) Repositioning accuracy: Comparison of a noninvasive head holder with thermoplastic mask for fractionated radiotherapy and a case report. *Int J Radiat Oncol Biol Phys* 41:475–483.

Tanaka T, Kobayashi T, Kida Y (1996) Growth control of cranial base meningiomas by stereotactic radiosurgery with a gamma knife unit. *Neurol Med Chir* 36:7–10.

Taylor BW Jr, Marcus RB Jr, Friedman WA, Ballinger WE Jr, Million RR (1988) The meningioma controversy: Postoperative radiation therapy. *Int J Radiat Oncol Biol Phys* 15:299–304.

Tishler RB, Loeffler JS, Lunsford LD, Duma C, Alexander E III, Kooy HM, Flickinger JC (1993) Tolerance of cranial nerves of the cavernous sinus to radiosurgery. *Int J Radiat Oncol Biol Phys* 27:215–221.

Tsang RW, Brierley JD, Panzarella T, Gospodarowicz MK, Sutcliffe SB, Simpson WJ (1994) Radiation therapy for pituitary adenoma: Treatment outcome and prognostic factors. *Int J Radiat Oncol Biol Phys* 30:557–565.

Tuniz F, Soltys SG, Choi CY, Chang SD, Gibbs IC, Fischbein NJ, Adler JR Jr (2009) Multisession cyberknife stereotactic radiosurgery of large, benign cranial base tumors: Preliminary study. *Neurosurgery* 65:898–907; discussion 907.

Vance ML (2004) Treatment of patients with a pituitary adenoma: One clinician's experience. *Neurosurg Focus* 16(4):E1.

Williams JA (2002) Fractionated stereotactic radiotherapy for acoustic neuromas. *Acta Neurochir* 144:1249–1254.

Yang I, Sughrue ME, Han SJ, Aranda D, Pitts LH, Cheung SW, Parsa AT (2010) A comprehensive analysis of hearing preservation after radiosurgery for vestibular schwannoma. *J Neurosurg* 112(4):851–859.

19 Radiation necrosis

Kenneth Y. Usuki, Susannah Ellsworth,
Steven J. Chmura, and Michael T. Milano

Contents

19.1 INTRODUCTION

Single-fraction stereotactic radiosurgery (SRS) and 2- to 5-fraction stereotactic radiotherapy (hSRT) are widely used and are efficacious treatments for brain/spine metastases, meningiomas, arteriovenous malformations (AVMs), vestibular schwannomas (also referred to as acoustic neuromas), trigeminal neuralgia, and recurrent gliomas. Modern treatment planning, in conjunction with stereotactic techniques discussed elsewhere in this book, achieves a sharp dose gradient allowing an ablative dose of radiation to the planning target volume (PTV) while reducing significant dose delivery to adjacent healthy critical structures. While stereotactic techniques can be used in conjunction with conventional fractionation (1.8–2 Gy), hypofractionation (larger fractional dose) is more convenient to the patient and likely affords a different and increased radiobiologic effect on the target.

A general tenet of radiation oncology is that with increased fraction size, there is greater risk of late normal tissue complication. While penumbra is not a clinically significant issue for low-dose standard fractionated therapy, at high doses of SRS/hSRT, the penumbra dose becomes clinically significant and can be damaging to healthy tissue, possibly leading to radiation necrosis. With SRS/hSRT, stereotactic techniques are used to minimize the volume of normal tissue exposed to deleterious doses, mitigating but not eliminating early as well as late toxicity risk. The historically used linear quadratic (LQ) model may be less predictive of normal tissue effects after SRS/hSRT, likely due to different biologic mechanisms with high-dose per fraction schedules (Milano et al., 2011), although this is controversial (Shuryak et al., 2015). The linear quadratic model is derived from *in vitro* cell survival assays of cancer cell lines, and it has been shown to be clinically predictive at low-dose per fraction treatment. This model is not necessarily expected to predict *in vivo* toxicity with increased fraction sizes to normal tissues for which injury of different cell types and varied intracellular components may be of increased relevance (Glatstein, 2008). The doses needed to sterilize tumor cells *in vitro* are much higher than the doses to attain long-term control of metastasis of similar size using SRS.

Looking at cell survival curves of EMT6 cells, Miyakawa and colleagues concluded that the linear quadratic model is not applicable to single-fraction and hypofractionated irradiation (Miyakawa et al., 2014). In the cell line investigated, the linear quadratic model was considered applicable to 7- to 20-fraction irradiation or doses per fraction of 2.57 Gy or smaller. In a Hungarian study which analyzed 18 resected brain metastases (of 2020 that were treated with SRS), excised lesions had parenchymal changes, stromal alterations, and vasculopathies not commonly seen in normal fractionated radiation; these changes were correlated with the length of local control (Szeifert et al., 2006). Thus, the importance of clinical outcome data for assessing risk of late complications including parenchymal brain radiation necrosis after SRS/hSRT is paramount.

19.2 SYMPTOMS AND ENDPOINTS

Brain radiation necrosis is the most significant late complication after cranial SRS/hSRT, resulting from tissue damage/breakdown involving vascular endothelial injury and glial injury (Sheline et al., 1980; Schultheiss et al., 1995; Chao et al., 2013), with blood–brain barrier (BBB) disruption likely a key component (Li et al., 2004; Brown et al., 2005). This normal tissue parenchymal breakdown is generally associated with surrounding edema. Edema may also occur in the absence of normal tissue necrosis. Brain edema and/or necrosis can be symptomatic or asymptomatic.

There have been many studies describing the frequency of symptomatic and asymptomatic brain parenchymal necrosis with variation partially attributable to differences in endpoints used and length of follow-up. There has been a wide range of the reported frequency of necrosis from 2% to 24% (Chin et al., 2001; Petrovich et al., 2002; Lutterbach et al., 2003; Andrews et al., 2004; Minniti et al., 2011; Sneed et al., 2015). Symptoms of edema and/or necrosis include headache, nausea, seizure, ataxia, and localized neurologic deficits (from necrotic brain parenchymal changes, its associated edema, and/or normal brain compression). These focal symptoms depend upon the region(s) of brain affected and may be completely or partially reversible.

The Radiation Therapy Oncology Group/ European Organization for Research and Treatment (RTOG/ EORTC) Late Effects Normal Tissue Task Force (LENT)-Subjective, Objective, Management, Analytic (SOMA) scale published in 1995 graded central nervous system (CNS) injury into categories as follows: fully functional with minor neurologic findings (grade 1); neurologic findings requiring home care, nursing assistance, and/or medications (grade 2); neurologic symptoms requiring hospitalization (grade 3); and serious impairment that includes paralysis, coma, and medication-resistant seizures with hospitalization required (grade 4) (Cox et al., 1995). More detailed grading, grouped by symptomatic, objective, management, and analytic criteria, was also provided in the organ-specific RTOG/EORTC LENT SOMA reviews (Schultheiss et al., 1995). Limited perilesional necrosis (grade 2), focal necrosis with mass effect (grade 3), and pronounced mass effect requiring surgical intervention (grade 4) were described as MRI criteria. The common terminology criteria for adverse events versions 3 and 4 (Trotti et al., 2003) scored CNS necrosis as asymptomatic (grade 1), moderate symptoms requiring corticosteroids (grade 2), severe symptoms requiring medical intervention (grade 3), and life-threatening symptoms requiring urgent intervention (grade 4).

19.3 FACTORS AFFECTING NECROSIS RISK

Several studies have correlated risk of necrosis with treatment-related factors. In the RTOG 90-05 study of SRS dose escalation for retreated primary and metastatic malignant brain tumors, the maximal tolerated marginal dose was 24 Gy for ≤2.0 cm lesions (dose tolerance was not met as there was doubt to the clinical relevance of continued dose escalation), 18 Gy for 2.1–3.0 cm lesions, and 15 Gy for 3.1–4.0 cm lesions. In this study, tumor volume >8.2 cc and a ratio of maximum dose to prescription dose >2 were associated with unacceptable toxicity. In a multivariate analysis, maximum tumor diameter was one variable associated with a significantly increased risk of grade 3, 4, or 5 neurotoxicity. Tumors 21–40 mm were 7.3–16 times more likely to develop grade 3–5 neurotoxicity compared to tumors <20 mm. The overall actuarial incidence of radionecrosis was 5%, 8%, 9%, and 11% at 6, 12, 18, and 24 months following radiosurgery (Shaw et al., 2000).

The RTOG 95-08 study randomized 333 patients with one to three brain metastases to whole-brain radiation with or without SRS, using the recommended RTOG 90-05 phase I dose levels (Andrews et al., 2004). Acute grade 3 (severe neurologic symptoms requiring medication) to grade 4 (life-threatening neurologic symptoms) toxicities occurred in 3% of those receiving SRS versus none among those not receiving SRS; late grade 3–4 toxicities occurred in 6% versus 3%, respectively. These differences were not significant, and among patients with solitary brain metastases, the risk of toxicity did not significantly differ between the 3 dose/size levels. In a University of Kentucky study of 160 patients with 468 metastases treated with SRS (median follow-up of 7 months), the study found that peripheral doses <20 versus 20 Gy resulted in inferior tumor control, while doses >20 Gy versus equal to 20 Gy did not result in improved tumor control, and there was a trending association with greater grade 3–4 neurologic toxicity risk (5.9% vs 1.9%, $p = 0.078$) (Shehata et al., 2004).

A study from University of California, San Francisco found prior SRS to the same lesion to be an important factor with a 20% 1-year risk of symptomatic necrosis, 4% for prior whole-brain radiotherapy (WBRT), 8% for concurrent WBRT versus 3% for no prior treatment (Sneed et al., 2015). When they excluded lesions treated previously with SRS, the 1-year probabilities of symptomatic radiation necrosis were <1% for 0.3–0.6 cm, 1% for 0.7–1.0 cm, 3% for 1.1–1.5 cm, 10% for 1.6–2.0 cm, and 14% for 2.1–5.1 cm for target maximum diameter (Sneed et al., 2015).

The most studied treatment-related factor related to brain radiation necrosis is the volume of tissue irradiated at, or greater than, a specific dose. Many of these studies are summarized in Table 19.1; select studies are discussed below. Different studies have examined different definitions of tissue volumes (i.e., "tissue," normal tissue, normal brain tissue, treated tissue including target volume).

Flickinger and colleagues have published several studies looking at the relationship between volume and dose in regard to risk of radiation necrosis (Flickinger, 1989; Flickinger et al., 1990, 1991a,b; Flickinger and Steiner, 1990). In a study of neurodiagnostic imaging changes after Gamma Knife SRS for post-AVM treatment, the only factor that correlated with imaging changes indicative of radiation necrosis was treatment volume (mean 3.75 cm^3) (Flickinger et al., 1992). The AVM Study Group specifically found, using multivariate analysis, the effects of AVM location and volume of tissue receiving 12 Gy or more were significant in predicting permanent sequelae (Flickinger et al., 2000). In addition, AVM location impacted the risk of necrosis, with the lowest to increasing risk as follows: frontal, temporal, intraventricular, parietal, cerebellar, corpus callosum, occipital, medulla, thalamus, basal ganglia, and pons/midbrain. The authors created a statistical model to predict risk of permanent symptomatic sequelae using both location and 12 Gy volume (which did not exclude target volume). Marginal 12 Gy volume (target volume excluded) did not significantly improve the risk-prediction model for permanent sequelae.

A study from UCSF found the volume parameters that correlated with symptomatic radiation necrosis included target, prescription isodose, V12 and V10 (Sneed et al., 2015). Excluding lesions treated with repeat SRS, the 1-year probability of symptomatic radiation necrosis leveled off at 13%–14% for brain metastases maximum diameter >2.1 cm, target volume >1.2 cm^3, prescription isodose volume >1.8 cm^3, 12 Gy volume >3.3 cm^3, and 10 Gy volume >4.3 cm^3. Interestingly, capecitabine was the only systemic therapy within 1 month of SRS that appeared to increase symptomatic radiation necrosis risk.

A study from the University of Cincinnati looked at the relationship between dose and volume in a group of patients where 63% had received previous whole-brain irradiation with a mean prescribed SRS dose of 18 Gy (Blonigen et al., 2010). Symptomatic radiation necrosis was observed in 10% and asymptomatic radiation necrosis in 4% of lesions treated. Multivariate regression analysis showed V8–V16 to be the most predictive of symptomatic radiation necrosis. For V10 and V12, they showed that the threshold volumes for which radionecrosis significantly increased were between the 75th and 90th percentiles. These percentiles corresponded to a V10 between 6.4 and 14.5 cm^3 and a V12 distribution between 4.8 and 10.9 cm^3. The midpoint of each interval was 10.45 and 7.85 cm^3, respectively. The risk of radionecrosis for V10 volume <2.2 cm^3 was 4.7%, for 2.2–6.3 cm^3 was 11.9%, for 6.4–14.5 cm^3 was 34.6%, and for >14.5 cm^3 was 68.8%. The risk of radionecrosis was the same for the V12 volumes of <1.6, 1.6–4.7,

Table 19.1 Select studies analyzing brain dose–volume metrics for complications after stereotactic radiosurgery

STUDY	STUDY COHORT	FOLLOW-UP (MEDIAN)	ENDPOINT/ OUTCOME	ONSET	DOSE–VOLUME PARAMETER	RISK		p VALUE
U. Cologne (Voges et al., 1996)	135 tumors or AVMs	9–59 (28) months	Radiation-induced tissue reactions (*edema ± ring enhancement*)	4–35 months	Tissue V10[a]			<0.0001
					≤10 mL	0%		
					>10 mL	24%		
Case Western (Korytko et al., 2006)	198 tumors (not AVMs)	Not reported	Symptomatic necrosis (decline in neurologic function that is associated with imaging changes)	Not reported	Tissue V12			Significant
					0–5 mL	23%		
					5–10 mL	20%		
					10–15 mL	54%		
					>15 mL	57%		
			Asymptomatic necrosis	Not applicable	Tissue V12			Not significant
					<10 mL	19%		
					>10 mL	19%		
AVM Radiosurgery Study Group (Flickinger et al., 2000)	422 AVMs	9–140 (34) months *for those with complications*	Symptomatic complications/ necrosis	Not reported	Tissue V12	[b]		0.0001
		24–92 (45) months *for those without complications*						

(*Continued*)

Table 19.1 (Continued) Select studies analyzing brain dose–volume metrics for complications after stereotactic radiosurgery

STUDY	STUDY COHORT	FOLLOW-UP (MEDIAN)	ENDPOINT/ OUTCOME	ONSET	DOSE–VOLUME PARAMETER		RISK	p VALUE
U. Maryland (Chin et al., 2001)	243 tumors[c]	Until death or >15 months	Radiation necrosis: *based upon MRI + pathology or necrotic lesion that resolved*	2–14 (median 4) months	Tissue V10 median of 28.4 vs 7.8 mL		—	0.007
					Normal brain V10 median of 19.8 vs 7.1 mL		—	0.005
U. Florida (Friedman et al., 2003)	269 AVMs	Not reported	Permanent radiation-induced complications	Not reported	Tissue V12		—	0.047–0.080
			Transient radiation-induced complications	Not reported	Tissue V12		—	0.052–0.145
U. Cincinnati (Blonigen et al., 2010)	173 brain metastases	3.5–51 (14) months	Asymptomatic or symptomatic necrosis[d]	*Symptomatic:* 2–41 months	Tissue V10 and V12[d]			<0.0001
						<2.2 mL	<1.6 mL	5%
		>6 months unless developed necrosis		*Asymptomatic:* 3–19 months	2.2–6.3 mL	1.6–4.7 mL	12%
					6.4–14.5 mL	4.8–10.8 mL	35%
					>14.5 mL	>10.8 mL	69%
U. Maryland (Chin et al., 2001)	243 tumors[c]	Until death or >15 months	Radiation necrosis: *see previous*	2–14 (median 4) months	Tumor volume		—	0.04
					Median of 4.4 vs 1.5 mL			

(Continued)

Table 19.1 (*Continued*) Select studies analyzing brain dose–volume metrics for complications after stereotactic radiosurgery

STUDY	STUDY COHORT	FOLLOW-UP (MEDIAN)	ENDPOINT/ OUTCOME	ONSET	DOSE-VOLUME PARAMETER		RISK	*p* VALUE
U. Pittsburgh (Varlotto et al., 2003)	208 brain metastases	12–122 (18) months	Neurologic complications[e]	10–25 months	Treatment volume			0.009
						≤2 mL	2% at 1 year	
						≤2 mL	4% at 5 years	
						≤2 mL	3% at 1 year	
						≥2 mL	16% at 5 years	
UCSF (Miyawaki et al., 1999)	73 AVMs	3–93 months	Neurologic complications requiring steroids, anticonvulsants or surgery	3–62 months	Treatment volume			0.04
						<1.0 mL	0%	
						1.0–3.9 mL	15%	
						4.0–13.9 mL	14%	
						≥14 mL	27%	
			Necrosis requiring resection			<14 mL	0%	0.01
						≥14 mL	13%	
			MRI T2 abnormalities			<1.0 mL	13%	
						1.0–3.9 mL	31%	
						4.0–13.9 mL	50%	
						≥14 mL	69%	
UCSF (Miyawaki et al., 1999)	1181 AVMs or tumors	>2 months	Grade ≥ 3 neurologic complications	0.3–17.6 months	Prescription volume			0.009
						0.05–0.66 mL	0% at 1.5 years	
						0.67–3.0 mL	3% at 1.5 years	
						3.1–8.6 mL	7% at 1.5 years	
						8.7–95.1 mL	9% at 1.5 years	

(*Continued*)

Table 19.1 (Continued) Select studies analyzing brain dose–volume metrics for complications after stereotactic radiosurgery

STUDY	STUDY COHORT	FOLLOW-UP (MEDIAN)	ENDPOINT/ OUTCOME	ONSET	DOSE–VOLUME PARAMETER	RISK	p VALUE
Sant'Andrea Hosp. (Minniti et al., 2011)	310 brain metastases	2–42 (9.4) months	Neurologic complications	*Symptomatic:* median 11 months	Brain V10 and V12		
			RTOG grade 3–4			<3.3 mL	2.6%
			MR imaging	*Asymptomatic:* median 10 months		3.3–5.9 mL	11%
						6.0–10.9 mL	24%
						>10.9 mL	47%
						>15.4 mL	62%
					V10 and V12		p = 0.001
Gifu U. (Ohtakara et al., 2012)	131 brain metastases	7–45.9 (18.2) months	Clinical symptoms or MR imaging	*Symptomatic:* 2.2–24.2 (median 3.7)	Non-WBRT cases		
					V12 cutoff values		
				Asymptomatic: 2.5–8.4 (median 6.9)	*Symptomatic:* 8.87 m	—	0.006
					All: 8.62 mL	—	0.008
					V22 cutoff-values	—	
					Symptomatic: 2.62 mL	—	0.001
					All: 2.14 mL	—	<0.001
					WBRT cases	—	
					V12 cutoff values	—	

(Continued)

Table 19.1 (Continued) Select studies analyzing brain dose–volume metrics for complications after stereotactic radiosurgery

STUDY	STUDY COHORT	FOLLOW-UP (MEDIAN)	ENDPOINT/ OUTCOME	ONSET	DOSE–VOLUME PARAMETER	RISK	p VALUE
					Symptomatic: 8.39 mL	—	0.009
					All: 8.39 mL	—	<0.001
					V15 cutoff values	—	
					Symptomatic: 5.20 mL	—	0.006
					All: 2.14 mL	—	<0.001
					V18 cutoff values	—	
					Symptomatic: 1.72 mL	—	0.088
					*All:*1.72 mL	—	0.063

Source: Modified from Milano, M.T. et al., *Cancer Treat. Rev.,* 37, 567, 2011.

[a] Minimal and maximal target dose and target volume were not significant. The V10 of the tissue minus target (i.e., normal brain V10) was not significant.

[b] Of the patients, 85 developed symptomatic necrosis, and 38/85 were classified as having permanent symptomatic necrosis, with unchanged symptoms ≥2 years after SRS.

[c] Matched-pair analyses.

[d] Symptomatic necrosis defined as requiring steroids, hyperbaric oxygen, vitamin E, or pentoxifylline therapy, or with new neurologic complaints. Tissue was V8–18 significant for symptomatic necrosis and tissue V8–14 significant for asymptomatic necrosis. Symptomatic necrosis was observed in 10% and asymptomatic necrosis in 4% of lesions treated.

[e] Of 11 complications, mostly included necrosis (n = 4) and persistent or symptomatic edema (n = 4).

AVM, arteriovenous malformation; U, University.

Crude risk unless other specified.

"Tissue" implies target + normal brain.

4.8–10.8, and >10.8 cm^3, respectively. There were no cases of radionecrosis below the 25th percentile for either V10 (<0.68 cm^3) or V12 (<0.5 cm^3). The data demonstrate that the risk of radiation necrosis exists over a relatively wide range of dose/volume exposures, although risks are low at smaller volumes; risks are unlikely to be zero at therapeutic doses and likely impacted by other (nondosimetric) factors.

In a study from Sant'Andrea Hospital, patients received SRS as their primary and only brain metastasis treatment (Minniti et al., 2011). Brain radionecrosis occurred in 24% of treated lesions and was symptomatic in 10% and asymptomatic in 14%. V10 Gy and V12 Gy were the most predictive. For V10 Gy >12.6 cm^3 and V12 Gy >10.9 cm,^{3} the risk of radionecrosis was 47%. Lesions with V12 Gy >8.5 cm^3 carried a risk of radionecrosis >10%.

In another volume-based study, Japanese investigators from Gifu University analyzed whether a superficial location, which could lead to spillage of dose into the extraparenchymal tissue outside of the brain, might decrease the risk of brain radiation necrosis (Ohtakara et al., 2012). They evaluated 131 lesions, 43.5% of which received prior whole-brain radiotherapy. A three-tiered location grade was defined with grade 1 involving less than or equal to 5 mm depth from the brain surface, grade 2 located at greater than 5 mm, and grade 3 located in the brain stem, cerebellar peduncle, diencephalon, or basal ganglion. Symptomatic radiation necrosis and asymptomatic radiation necrosis were observed in 8.4% and 6.9% of cases, respectively. Multivariate analysis indicated that the significant factors for both types of necrosis were location grade, V12 Gy and V22 Gy. In all cases, V12 was the most significant dosimetric variable for radionecrosis. Looking at non-WBRT cases, V22 and location grade were the most significant. For WBRT cases, V15 and location grade had the strongest correlation. For the non-WBRT cases, the cutoff values of V22 Gy were 2.62 and 2.14 cm^3 for symptomatic necrosis and combined (symptomatic and asymptomatic), respectively. For the WBRT cases, the cutoff values of V15 Gy were 5.61 and 5.20 cm^3 for symptomatic necrosis and combined, respectively. In addition to the dose–volume data, location grade helped predict the risk of radiation necrosis.

In addition to the different dosimetric variables for treating superficial and deep brain lesions, different anatomic regions of the CNS may possess varying susceptibilities to injury. If this is the case, the difference may be related to tissue vascularity, glial cell population, or repair capacity. Permanent, late clinical manifestations of radiation injury likely impacted not only by location but also by tumor-related factors (histology, genetic morphology), patient-related factors (comorbid conditions, prior surgery, prior radiation, sex), and redundancy in brain function and plasticity (repair and recruitment of function from other regions of the brain). As previously mentioned, the University of Pittsburgh analysis also suggests that occipital, parietal, cerebellar, corpus callosum, and intraventricular AVMs are at greater risk of symptomatic necrosis versus frontal or temporal locations (Flickinger et al., 2000).

In a study from Case Western Reserve University, occipital and temporal lobe non-AVM tumors had a greater likelihood of symptomatic necrosis. In addition to increased fractional dose (Ganz et al., 1996; Kalapurakal et al., 1997) and increased tumor size (Kalapurakal et al., 1997; Kondziolka et al., 1998), several meningioma studies have implicated parasagittal/midline location as a significant variable with greater risks of symptomatic perilesional edema after SRS for meningiomas (Ganz et al., 1996; Kalapurakal et al., 1997; Chang et al., 2003; Patil et al., 2008). In the AVM studies from the University of Pittsburgh group, post-SRS imaging changes and symptomatic radiation necrosis (Flickinger et al., 1997, 1998, 2000) after SRS for AVMs were increased among patients with brain stem targets. These findings raise the question of whether or not the brain stem is more susceptible to radiation injury or whether radiation injury to this structure results in more symptomatic injury versus other areas of the CNS (Loeffler, 2008).

Several papers have reported toxicity outcomes after SRS for brain stem metastases, with a broad range of peripheral doses (9–30 Gy), with median peripheral doses of 15–20 Gy (Huang et al., 1999; Shuto et al., 2003; Fuentes et al., 2006; Yen et al., 2006; Hussain et al., 2007; Kased et al., 2008; Lorenzoni et al., 2009; Koyfman et al., 2010). Symptomatic brain stem radionecrosis is relatively uncommon after brain stem SRS with good reported local tumor control (76%–96%) (Hatiboglu et al., 2011; Sengöz et al., 2013), likely attributable to the smaller size of brain stem lesions at presentation with cone usage decreasing penumbra allowing lower peripheral brain stem doses while still maintaining clinically significant doses within the target. In addition, the prescribed dose used by practitioners tends to be lower due to the critical nature of the structure. Unfortunately, these patients suffer from poor survival (median 4–11 months) and may not have the opportunity to manifest late radiation toxicity. When the possibility of radiation toxicity occurs, it is often difficult to discern from symptomatic progression. Reported toxicities include hemiparesis, ataxia, cranial nerve deficits, headaches, nausea/vomiting, and seizures (Huang et al., 1999; Hussain et al., 2007; Kased et al., 2008; Hatiboglu et al., 2011).

A study from the University of Pittsburgh reported on adverse imaging findings and neurologic deficits in 38 patients with benign tumors who were followed 6–84 (median 41) months after SRS (Sharma et al., 2008). Not all patients with adverse imaging findings developed deficits arising within the brain stem long tracts or adjacent cranial nerve. Some patients developed neurologic deficits in the absence of adverse imaging findings. Interestingly, there was no correlation between marginal dose and adverse imaging findings or neurologic deficits. Interestingly, marginal doses >18 Gy were associated with less neurologic deficits than marginal doses of 15–17 Gy (16.6% vs 19.1%, not significant), which the authors attribute to differences in distribution of target types between different dose groups (i.e., the 15–17 Gy group represented mostly cavernomas and the >18 Gy represented mostly AVMs).

MD Anderson reported that 20% of their patients developed complications likely related to SRS as early as 1 month after radiosurgery and that multivariate analysis showed pre-SRS tumor volume and male sex were associated with a significantly shorter overall survival interval (Hatiboglu et al., 2011). Sengöz et al. (2013) found peritumoral changes were detected radiologically in 4% of the metastatic lesion sites treated with Gamma Knife radiosurgery, but none of the patients exhibited symptoms. Female gender, Karnofsky performance status (KPS) 70, mesencephalon tumor location, and response to treatment were associated with longer survival (Hatiboglu et al., 2011; Sengöz et al., 2013). Based upon the published studies, a brain stem maximum dose of 10–12 Gy is expected to result in a minimal (<1%–2%) risk of brain stem toxicity.

From the quantitative analyses of normal tissue effects in the clinic (QUANTEC) review, for doses of 12.5, 14.2, 16.0, and 17.5 Gy: partial volume irradiation to one-third of the brain stem results in normal tissue complication probabilities (NTCPs) of 1%, 13%, 61%, and 94%, respectively; "small partial volume" irradiation (1% used to approximate a brain stem maximum dose) results in NTCPs of 0.2%, 3.2%, 26%, and 68%, respectively (Mayo et al., 2010). For benign tumors compressing the brain stem, for which patients are already symptomatic or at risk of becoming symptomatic, marginal doses of 13 Gy appear to be well tolerated (Nakaya et al., 2010). The generally accepted dose constraint to the brain stem is considered to be on the order of 10–12 Gy, and efforts to minimize brain stem dose are recommended. In general, the brain stem maximum should be maintained below 12 Gy if feasible, though when therapeutic dose to the target is compromised with such a constraint, particularly when the risks of treatment failure outweighs the risks of treatment toxicity, higher maximal doses to the brain stem should be considered. Poor local control has been shown to decrease survival and higher doses improve local control. If tumor control is compromised with lower SRS doses, progression of the metastatic lesion or primary tumor in the brain stem could be fatal. Even when exceeding tolerance to small volumes, the risk of brain stem toxicity is low and is less likely to cause a poor outcome than a failure to achieve brain stem tumor control. In the treatment of trigeminal neuralgia, we know that negligible volumes of brain stem may even safely receive doses >20–50 Gy. When selecting the aggressiveness of the prescription and brain stem dose, factors such as histology, KPS, primary tumor/other metastatic disease control, and the presence of extracranial metastatic disease should be taken into consideration.

19.4 CORTICOSTEROID USE IN THE MANAGEMENT OF BRAIN NECROSIS

Radiation necrosis following SRS/hSRT tends to be a temporary problem, with most of the symptoms typically caused by brain edema and not from parenchymal damage (although this may occur as well). SRS/hSRT-induced necrosis can usually be managed conservatively with observation in asymptomatic or minimally symptomatic patients. The first-line treatment of symptomatic radiation necrosis is corticosteroids due to the drugs' rapid ability to reduce cerebral edema. Corticosteroids are often well tolerated, particularly when using moderate doses over a short period of time. In patients who are suffering from symptomatic edema, corticosteroids usually briskly alleviate or improve symptoms. Once symptoms are controlled, corticosteroids should be gradually tapered with the speed of the taper dependent on the length of steroid dependency. If steroids were given for less than 2 weeks, a taper is likely not necessary. One should try to avoid use of corticosteroids for an extended time period as their deleterious effects tend to be a function of length of use and the dose. Patients should be prescribed the lowest possible dose to alleviate their symptoms. Steroids may be difficult to tolerate in patients with a psychiatric history who may be more prone to steroid-induced psychosis and diabetics who may have a difficult time controlling blood sugar levels. Careful attention needs to be paid to these two groups. Diabetics may benefit from recommendations by their primary care physician or endocrinologist concerning any changes in diabetic therapies and monitoring that may help the patient tolerate the temporary steroid treatment. The possible side effects of corticosteroids are many and include anxiety, depression, osteonecrosis, psychosis, sleeplessness, aggressive behavior, infection, diabetes, GI irritation, increased appetite, osteoporosis, and facial swelling.

19.5 WARFARIN, PENTOXIFYLLINE, AND VITAMIN E

Vitamin E and pentoxifylline given in combination are reported to benefit other body sites afflicted with radiation damage. The University of Pittsburgh group performed a pilot study looking at oral pentoxifylline and vitamin E therapy given to 11 patients with suspected adverse radiation effects after SRS (Williamson et al., 2008). These patients were followed and their magnetic resonance imaging fluid-attenuated inversion recovery (MR FLAIR) volume changes were plotted over time. The change in edema volume varied from 59.6 mL in one patient (worse edema) to –324.2 mL (improvement). The average change in edema from pre- to posttreatment was –72.3 mL. One patient had more edema despite treatment; this patient was found to have tumor recurrence. Two patients discontinued pentoxifylline because of persistent nausea and abdominal discomfort (Williamson et al., 2008). Since radiation necrosis is usually a transient occurrence after SRS/hSRT, more extensive data are needed before any conclusion on the efficacy of vitamin E, and pentoxifylline can be drawn.

In a study from Brown University, the effect of anticoagulation with heparin and warfarin on radiation necrosis was analyzed in 11 patients with late radiation-induced nervous system injuries (eight with cerebral radionecrosis, one with a myelopathy, and two with plexopathies, all unresponsive to dexamethasone and prednisone) (Glantz et al., 1994). In five of the eight patients with cerebral radionecrosis, some recovery of function occurred. Anticoagulation was continued for 3–6 months. In one patient with cerebral radionecrosis, symptoms recurred after discontinuation of anticoagulation and disappeared again after reinstitution of treatment. The group hypothesized that anticoagulation may improve small vessel endothelial injury leading to clinical improvement.

19.6 BEVACIZUMAB IN THE TREATMENT AND PREVENTION OF CENTRAL NERVOUS SYSTEM RADIATION NECROSIS

Vascular endothelial growth factor (VEGF) is released in the setting of radiation injury to the BBB and is a key mediator of radiation-induced white matter toxicity and CNS radiation necrosis (Kim et al., 2004). Bevacizumab, a humanized monoclonal VEGF antibody, therefore represents a promising targeted agent for brain radionecrosis. It has been studied as a treatment for known brain radionecrosis and as an adjunct to radiation dose escalation or dose intensification with the goal of preventing the development of brain radionecrosis.

Despite considerable clinical interest in bevacizumab as a treatment for brain radionecrosis, most reports on its use in this setting have been case reports or retrospective series. Tye and colleagues conducted a pooled analysis of the 16 studies published through 2012 (including 71 cases) of brain radionecrosis treatment with bevacizumab (Tye et al., 2014). The overall radiographic response rate was 97%, and the clinical improvement rate (measured in terms of performance status) was 79%. The median decreases in T1 contrast-enhancing area and FLAIR signal abnormality were 63% and 59%, respectively, and dexamethasone dose decreased by a median of 6 mg.

Additionally, a single-institution study randomized 14 patients with known brain radionecrosis (diagnosed via MRI and/or biopsy) to bevacizumab versus placebo and permitted crossover in patients who had neurologic or radiographic progression (Levin et al., 2011). Ultimately, all the patients in the placebo arm received bevacizumab, rendering this essentially a trial of early versus delayed bevacizumab for brain radionecrosis. The study reported a radiographic response rate of 100% in both arms. Symptomatically, no differences were observed between the two groups, and four of the five patients who were on dexamethasone at the time of study enrollment were able to decrease their steroid doses. Observed adverse events included sagittal sinus thrombosis, pulmonary embolism, and pneumonia (in one patient each).

Studies of bevacizumab for the prevention of radionecrosis in brain tumor patients undergoing dose-escalated radiation therapy or reirradiation have produced mixed results. Memorial Sloan Kettering Cancer Center reported a prospective trial of bevacizumab in conjunction with hypofractionated stereotactic reirradiation for recurrent malignant glioma (Gutin et al., 2009). Patients without evidence of disease progression after one cycle of bevacizumab (10 mg/kg every 14 days of a 28-day cycle) went on to receive 30 Gy in 5 fractions. Bevacizumab was subsequently continued until disease progression or unacceptable toxicity occurred. The study excluded patients with tumors larger than 3.5 cm in diameter. No instances of radionecrosis occurred in this cohort, even though all but two of the patients received reirradiation to areas that had previously received 60 Gy. However, three patients (12%) stopped therapy due to toxicities that included brain hemorrhage, intestinal perforation, and wound dehiscence. A fourth patient developed GI bleeding 3 weeks after discontinuing study treatment due to tumor progression.

More recently, the University of Colorado reported a prospective clinical trial of an aggressively hypofractionated combination regimen in previously untreated glioblastoma (Ney et al., 2015). Dose painting was used to treat the surgical cavity or enhancing tumor (plus a 1 cm margin) to 60 Gy in 10 fractions; areas of T2 abnormality (also plus a 1 cm margin) received 30 Gy in 10 fractions. Patients received radiation Monday through Friday for 2 weeks without a scheduled break. Concurrent temozolomide was administered at standard doses (75 mg/m^2) during radiation, and patients received bevacizumab (10 mg/kg) on days 1 and 15. Adjuvant bevacizumab and temozolomide (200 mg/m^2) were restarted 4–6 weeks after radiation. This study was stopped early when interim analysis revealed that 50% of patients had developed radionecrosis (pathologically proven by surgical samples or autopsy in six individuals). Other potentially treatment-related toxicities included stroke ($n = 1$), grade 3 wound dehiscence ($n = 2$), and pulmonary embolism ($n = 2$). The unexpectedly high rate of radionecrosis in this study was thought to be due to a combination of large target volumes and a high total RT dose; the median 30 Gy volume was 342.6 cm^3 and the median 60 Gy volume was 131.1 cm^3.

For comparison, a previous prospective trial of concurrent hypofractionated RT/temozolomide/bevacizumab in newly diagnosed glioblastoma prescribed a total dose of 36 Gy in 6 fractions to enhancing tumor and 24 Gy in 6 fractions to the T2 abnormality (Omuro et al., 2014). Additionally, that study excluded patients with a tumor volume >60 cm^3 and reported a much lower rate of radionecrosis, with two cases of biopsy-proven radionecrosis (5%) observed among the 40 participants at a median follow-up of 42 months.

Overall, bevacizumab appears to be a promising treatment for brain radionecrosis. Although the available data are relatively sparse, studies published to date have reported impressively high radiographic and clinical response rates in patients with radiographic or pathologic evidence of radiation necrosis. The efficacy of bevacizumab as prophylaxis against radionecrosis for patients treated with intensified radiation regimens remains in question and requires further study. Considerable caution must be exercised in selecting patients for treatment with bevacizumab, which is associated with significant toxicity, particularly hemorrhage, thrombosis, and wound healing difficulties. The authors therefore recommend that patients with known brain radionecrosis either demonstrate resistance to steroid therapy or have intolerable steroid toxicity before being considered for treatment with bevacizumab. Contraindications to bevacizumab use include the following: coagulation disorders (bleeding diathesis or hypercoagulability); use of antiplatelet agents or anticoagulants; pregnancy/nursing; a history of significant cardiovascular disease (stroke, cerebral hemorrhage, heart failure, uncontrolled hypertension, recent myocardial infarction, peripheral vascular disease, or aortic aneurysm/dissection); hemoptysis; and a recent history of GI bleeding, perforation, abscess, or fistula (Hershman et al., 2013). In patients who have undergone surgery, bevacizumab treatment should be delayed for at least 28 days postoperatively to allow adequate wound healing (Genentech/Roche, 2015).

19.7 HYPERBARIC OXYGEN

Hyperbaric oxygen (HBO) is a treatment where patients enter a sealed chamber with 100% oxygen that is up to three times greater than the atmospheric pressure. This allows dissolving of oxygen into the plasma component, bypassing the need for hemoglobin saturation to deliver oxygen to tissue. This aids in delivering oxygen through the plasma to necrotic tissue or fibrotic tissue where the damaged vasculature may not allow effective hemoglobin passage, theoretically, reducing fibrosis by stimulating angiogenesis and allowing stem cells to be recruited to the irradiated tissue. The treatment schedule is usually 5 days a week, and a patient may be prescribed between 20 and 40 treatments with each session taking 120 minutes.

Data for HBO therapy are paltry and the reported effectiveness in CNS radionecrosis is mostly limited to case studies (Leber et al., 1998; Kohshi et al., 2003; Valadao et al., 2003; Pérez-Espejo et al., 2009). There is legitimate, though unproven, concern that HBO may stimulate malignant cells. In addition, there are side effects associated with HBO although severe life-threatening side effects are rare. Side effects may include seizure in 1%–2%, pulmonary symptoms in 15%–20%, and reversible myopia in 20% of patients treated (Leach et al., 1998).

HBO may have use as prophylaxis against radiation necrosis. A Japanese study looked at 78 patients presenting with 101 brain metastases treated with SRS (Ohguri et al., 2007). Of these, 32 patients with 47 brain metastases were treated with prophylactic HBO, which included all 21 patients who underwent subsequent or prior radiotherapy and 11 patients with common predictors of longer survival, such as inactive extracranial tumors and younger age. The other 46 patients with 54 brain metastases did not undergo HBO. The radiation-induced brain injuries were divided into two categories, white matter injury and radiation necrosis on the basis of imaging findings. Radiation-induced brain injury occurred in 5 lesions (white matter injury, 2; necrosis, 3) (11%) in the HBO group. In the non-HBO, 11 lesions occurred (white matter injury, 9; necrosis, 2) (20%). While the data are promising, more research is needed into the subject of HBO for prophylaxis of brain radiation injury. HBO is an option for corticosteroid refractory radiation necrosis after SRS, particularly if bevacizumab or surgery is not an option for treatment.

19.8 SURGICAL REMOVAL OF BRAIN NECROSIS

When corticosteroids and antiangiogenics fail to alleviate the pressure created by the radiation necrosis–induced edema with neurological deficits, surgical resection of a radiation necrotic lesion can relieve increased intracranial pressure immediately, and it can quickly improve or halt progression of a neurologic disability. The surgical removal of the necrosis may provide significant reversal of the edema-created neurological deficits, and many patients can be weaned from corticosteroids.

If bevacizumab has been given, some surgeons will delay surgery at least a month to avoid surgical morbidity and wound healing issues. Surgical resection also has the benefit of giving information as to whether radiographic changes and neurologic symptoms are from tumor progression, radiation necrosis, or both. Surgery is associated with tangible risks including neurologic deficit, infection, anesthesia risk, and the need for an inpatient hospital stay.

In a study from the University of Cincinnati, 11 patients who underwent surgical resection for radiation necrosis had an initial diagnosis of a malignant brain tumor: 8 patients with high-grade glioma and 3 patients with metastatic brain tumors. All patients underwent at least one radiation treatment, and many had multiple treatments. Three patients underwent whole-brain radiation therapy administered for metastatic disease, eight patients with malignant gliomas received 3D conformal radiation therapy. In addition, four patients received 12,000–15,000 cGy using I-125 seeds at the time of the first surgical resection. Seven of the patients were treated with SRS at the time of recurrence. The diagnosis of radiation necrosis was based primarily on MRI and clinical suspicion. Optimal resection was confirmed by intraoperative MRI (in nine patients) and achieved in all patients by the use of frameless stereotaxis, with no additional resection performed in any patient. In the nine patients taking steroids before treatment of necrosis, all had a substantial reduction in steroid dosage (pre- to postoperative dose 24–8 mg/day) after surgical treatment. Postoperatively, KPS improved in four patients, remained stable in four, and worsened in three. Complications from surgery included wound infection, asymptomatic carotid dissection, and pulmonary embolism. The authors concluded that morbidity including both surgical complications and neurologic deterioration was 54%, and given the success of medical therapies, they recommended that surgical treatment of radiation necrosis should be reserved for symptomatic patients in whom medical therapy has failed (McPherson and Warnick, 2004). Surgery should be considered in moderately to severely symptomatic patients in whom medical therapy fails and/or pathological examination could affect the treatment course.

19.9 FRACTIONATION

Lastly, what is the role of hypofractionated (25–50 Gy in 2–6 fractions) hSRT? Fractionation benefits hSRT normal tissue tolerance, bringing the isodose curves that the normal brain can safely tolerate (taking into account normal tissue repair ability in between fractions) closer to the target. A theoretical negative is that the increased and likely different biologic *in vivo* effect associated with high doses may be diminished with fractionation. This would depend on if the superior hypofractionated biologic effect has a threshold dose and/or increases with rising dose. Even if there is a gradient dose response in the hSRT dose range to get the improved biologic effect with higher dose per fraction, could fractionation, within hSRT dose ranges, with a higher cumulative dose compensate or be superior? Fractionation with hSRT doses may give some of the increased biologic effect of large doses per fraction and spare more normal brain tissue. hSRT may allow for equivalent, or near equivalent, or possibly superior (Shuryak et al., 2015) local control and less toxicity in large targets, multiple targets close by, or when the target is near a sensitive critical structure. Groups have treated brain metastases with hSRT and reported promising tumor control and toxicity, even for lesions larger than 4 cm or 3–5 mL (which are generally less amenable to single-fraction SRS due to greater risk of necrosis) and/or lesions involving or in close proximity to eloquent brain or optic nerve/chiasm (Manning et al., 2000; Aoyama et al., 2003; Ernst-Stecken et al., 2006; Fahrig et al., 2007; Narayana et al., 2007; Kwon et al., 2009; Feuvret et al., 2010; Kim et al., 2011; Marchetti et al., 2011). Retrospective studies have demonstrated comparable tumor control and neurologic toxicity of hSRT versus SRS, even though the hSRT group comprised patients with larger tumors (Feuvret et al., 2010; Kim et al., 2011) and/or adverse locations (Kim et al., 2011). While studies of hSRT have been undertaken in patients with brain metastases who are potentially eligible for single-fraction SRS (i.e., smaller lesions not involving or abutting eloquent structures) (Manning et al., 2000; Aoyama et al., 2003; Kwon et al., 2009), the role of hypofractionated hSRT versus single-fraction SRS in these patients is unclear and likely depends on many factors including histology and previous radiation (Linskey et al., 2010). Scant data exist for hypofractionated hSRT for meningiomas (Shrieve et al., 2004; Henzel et al., 2006; Gorman et al., 2008; Trippa et al., 2009). Due to the biologic mechanism of obliterating the nidus and AVM's low alpha/beta ratio, all favoring high single doses of radiation, the radiobiologic utility of hSRT for AVMs is controversial (Hall and Brenner, 1993; Wigg, 1999; Kocher et al., 2004; Qi et al., 2007; Vernimmen and Slabbert, 2010). Several studies have demonstrated the efficacy of hSRT for larger (>2.5–4 cm and/or >10–14 mL) AVMs where single-fraction ablative doses would be prohibitive (Lindvall et al., 2003; Chang et al., 2004; Silander et al., 2004; Veznedaroglu et al., 2004, 2008). Due to the previously mentioned radiobiologic advantages in using large single fractions, some studies have taken a volume-staged approach to large, even combining volume staging with surgery with good and possibly superior results to dose staging (Moosa et al., 2014; Abla et al., 2015).

Thus, size and location are valid reasons to consider fractionated hSRT. However, radiobiologically, hSRT may result in less effective tumor control and/or greater risk of normal tissue toxicity as compared to conventional fractionation (i.e., 1.8– 2 Gy), particularly with targets of low alpha/beta ratios (Shrieve et al., 2004; Vernimmen and Slabbert, 2010). In a study from the University of Erlangen, analyzing 51 patients (with 72 brain metastases) not considered candidates for single-fraction SRS (due to volume >3 mL or proximity/involvement of eloquent brain), a volume of normal brain ≥23 mL receiving 4 Gy/fraction (over 5 fractions) was associated with a significantly increased risk of brain necrosis (70% vs 14% for those with 4 Gy volume <23 mL, $p = 0.001$) (Ernst-Stecken et al., 2006). In another study from this same group, of 150 patients with 228 brain metastases, a PTV of >17 mL (corresponding to approximately 3+ cm in diameter) was associated with increased neurologic toxicity; furthermore, as the number of fractions increased (5 × 6–7 Gy vs 7 × 5 Gy vs 10 × 4 Gy), the risk of toxicity decreased (22%, 7%, and 0%, respectively) though the less protracted regimens were associated with greater tumor response (Fahrig et al., 2007).

19.10 CONCLUSION

Uncertainties remain about the risks for radiation necrosis and how best to utilize the varied brain dose–volume metrics. For example, what is the relevance of V10–12 to an individual lesion site versus a composite V10–12 for the entire SRS treatment? The number of lesions was not significant in one study (Varlotto et al., 2003). Based on our understanding of radiation-induced fibrosis and necrosis, it is reasonable to expect that a given volume >10–12 Gy is more likely to result in necrosis if that volume is not spread over multiple lesions but rather confined to one confluent location in the brain. Is the V10–12 Gy of the treatment volume versus normal tissue versus normal brain more clinically relevant?

There is a difference in how studies defined V12 and some studies did not specify a definition for V12. For studies of necrosis after treatment of AVMs, V12 tends to represent treated tissue receiving ≥12 Gy. Compared to benign and malignant tumors, the AVM nidus is more difficult to precisely define, as it is characterized by nidus and feeder vessels that interdigitate with normal brain. The size of the AVM may also be relevant in that with larger AVMs, it is possible that there is a greater impact of radiation in affecting regional blood flow and thus increasing necrosis risk. Flickinger and colleagues report that subtracting the target from V12 adds no additional benefit in correlating necrosis risk with V12. This finding may be partially attributable to the high doses used to treat AVMs, resulting in a significant component of the V12 volume being outside of the target, thus minimizing the predictive benefit of removing the target from V12. In some studies, the V12 is defined as "normal brain" receiving ≥12 Gy. This approach is logical for brain metastasis since normal brain is pushed out of the way by the target. In the treatment of brain metastases (and other tumors), the relative importance of necrosis within the target versus within the surrounding tissue is unclear, as both scenarios would likely result in surrounding vasogenic edema. Larger targets treated in excess of 12 Gy are likely to be at risk of tumor necrosis and surrounding transient edema, even if the V12 was somehow conformally kept entirely within the target volume. This edema caused by tumor necrosis is likely more temporary as the normal brain does not experience significant necrosis.

Interestingly, the aforementioned Japanese study (Ohtakara et al., 2012) evaluated the depth of the target lesion and concluded that adding depth grade to dose–volume metric indices is predictive of necrosis risk, as superficial metastasis treated utilizing normal tissue definition of V12 (normal brain + dura, skull, skin) would deposit some of the radiation necrosis correlative dose–volume into nonbrain tissue. In addition to dose being absorbed in nonbrain tissue for superficial lesions, the depth of the lesion also alters the relative size distribution of different radiation dose–volumes through the normal brain. As depth increases, more monitor units and geometric access (number/length of arcs and arc placement) become necessary to safely cover the target with a particular prescribed dose, which raises another question: what is the relevance of the dose–volume metrics of lower doses on necrosis risk? Perhaps brain V4–8 Gy correlates with late neurologic deficits, or perhaps the volume receiving even lower doses (perhaps V1–2 Gy) correlates with second malignancy risk. SRS-induced malignancy is reportedly rare (Loeffler et al., 2003; McIver and Pollock, 2003; Balasubramaniam et al., 2007; Rowe et al., 2007; Niranjan et al., 2009) as it is estimated to occur in fewer than 1 in 1,000–10,000 patients (McIver and Pollock, 2004; Murraciole and Regis, 2008; Niranjan et al., 2009), though it is unclear if this risk is actually higher, and will be better appreciated as more cases are ascertained (Loeffler et al., 2003), or if the risk is no different than malignant tumors in the general population (Rowe et al., 2007).

We concur with QUANTEC that "toxicity increases rapidly once the volume of the brain exposed to >12 Gy is >5–10 cm³" (Lawrence et al., 2010), and more stringent constraints should be considered in certain clinical contexts including eloquent brain regions and previous irradiation. There remains substantial variation between studies (reflected in Table 19.1), prohibiting more precise toxicity predictions. However, we can say that there are some factors likely predictive for the development of radionecrosis including tumor location, diameter, previous irradiation, V12, and male sex. It is clear that the technique of brain SRS/hSRT is a very heterogeneous treatment with different dosimetric planning techniques, physical delivery modalities, definitions of V12, definitions of target volume, definitions of toxicity, length of follow-up, depth of lesions, histology of lesions, and variability of individual brain volumes. More investigation is needed to understand the risks and indicators that predict for radiation necrosis in SRS and hSRT patients.

ACKNOWLEDGMENT

The authors thank Laura Finger for her editorial assistance.

REFERENCES

Abla AA, Rutledge WC, Seymour ZA, Guo D, Kim H, Gupta N, Sneed PK et al. (2015) A treatment paradigm for high-grade brain arteriovenous malformations: Volume-staged radiosurgical downgrading followed by microsurgical resection. *J Neurosurg* 122:419–432.

Andrews DW, Scott CB, Sperduto PW, Flanders AE, Gaspar LE, Schell MC, Werner-Wasik M et al. (2004) Whole brain radiation therapy with or without stereotactic radiosurgery boost for patients with one to three brain metastases: Phase III results of the RTOG 9508 randomized trial. *Lancet* 363:1665–1672.

Aoyama H, Shirato H, Onimaru R, Kagei K, Ikeda J, Ishii N, Sawamura Y, Miyasaka K (2003) Hypofractionated stereotactic radiotherapy alone without whole-brain irradiation for patients with solitary and oligo brain metastasis using noninvasive fixation of the skull. *Int J Radiat Oncol Biol Phys* 56:793–800.

Balasubramaniam A, Shannon P, Hodaie M, Laperriere N, Michaels H, Guha A (2007) Glioblastoma multiforme after stereotactic radiotherapy for acoustic neuroma: Case report and review of the literature. *Neuro Oncol* 9:447–453.

Blonigen BJ, Steinmetz RD, Levin L, Lamba MA, Warnick RE, Breneman JC (2010) Irradiated volume as a predictor of brain radionecrosis after linear accelerator stereotactic radiosurgery. *Int J Radiat Oncol Biol Phys* 77:996–1001.

Brown WR, Thore CR, Moody DM, Robbins ME, Wheeler KT (2005) Vascular damage after fractionated whole-brain irradiation in rats. *Radiat Res* 164:662–668.

Chang JH, Chang JW, Choi JY, Park YG, Chung SS (2003) Complications after gamma knife radiosurgery for benign meningiomas. *J Neurol Neurosurg Psychiatry* 74:226–230.

Chang TC, Shirato H, Aoyama H, Ushikoshi S, Kato N, Kuroda S, Ishikawa T, Houkin K, Iwasaki Y, Miyasaka K (2004) Stereotactic irradiation for intracranial arteriovenous malformation using stereotactic radiosurgery or hypofractionated stereotactic radiotherapy. *Int J Radiat Oncol Biol Phys* 60:861–870.

Chao ST, Ahluwalia MS, Barnett GH, Stevens GH, Murphy ES, Stockham AL, Shiue K, Suh JH (2013) Challenges with the diagnosis and treatment of cerebral radiation necrosis. *Int J Radiat Oncol Biol Phys* 87:449–457.

Chin LS, Ma L, DiBiase S (2001) Radiation necrosis following gamma knife surgery: A case-controlled comparison of treatment parameters and long-term clinical follow up. *J Neurosurg* 94:899–904.

Cox JD, Stetz J, Pajak TF (1995) Toxicity criteria of the Radiation Therapy Oncology Group (RTOG) and the European Organization for Research and Treatment of Cancer (EORTC). *Int J Radiat Oncol Biol Phys* 31:1341–1346.

Ernst-Stecken A, Ganslandt O, Lambrecht U, Sauer R, Grabenbauer G (2006) Phase II trial of hypofractionated stereotactic radiotherapy for brain metastases: Results and toxicity. *Radiother Oncol* 81:18–24.

Fahrig A, Ganslandt O, Lambrecht U, Grabenbauer G, Kleinert G, Sauer R, Hamm K (2007) Hypofractionated stereotactic radiotherapy for brain metastases—Results from three different dose concepts. *Strahlenther Onkol* 183:625–630.

Feuvret L, Vinchon S, Lamproglou I, Lang P, Assouline A, Hemery C, Boisserie G, Valery C, Mazeron J, Simon J (2010) Trifractionated stereotactic radiotherapy for large single brain metastases. *Int J Radiat Oncol Biol Phys* 78:S284.

Flickinger JC (1989) An integrated logistic formula for prediction of complications from radiosurgery. *Int J Radiat Oncol Biol Phys* 17:879–885.

Flickinger JC, Kondziolka D, Lunsford LD, Kassam A, Phuong LK, Liscak R, Pollock B (2000) Development of a model to predict permanent symptomatic postradiosurgery injury for arteriovenous malformation patients. Arteriovenous Malformation Radiosurgery Study Group. *Int J Radiat Oncol Biol Phys* 46:1143–1148.

Flickinger JC, Kondziolka D, Maitz AH, Lunsford LD (1998) Analysis of neurological sequelae from radiosurgery of arteriovenous malformations: How location affects outcome. *Int J Radiat Oncol Biol Phys* 40:273–278.

Flickinger JC, Kondziolka D, Pollock BE, Maitz AH, Lunsford LD (1997) Complications from arteriovenous malformation radiosurgery: Multivariate analysis and risk modeling. *Int J Radiat Oncol Biol Phys* 38:485–490.

Flickinger JC, Lunsford LD, Kondziolka D (1991a) Dose–volume considerations in radiosurgery. *Stereotact Funct Neurosurg* 57:99–105.

Flickinger JC, Lunsford LD, Kondziolka D, Maitz AH, Epstein AH, Simons SR, Wu A (1992) Radiosurgery and brain tolerance: An analysis of neurodiagnostic imaging changes after gamma knife radiosurgery for arteriovenous malformations. *Int J Radiat Oncol Biol Phys* 23:19–26.

Flickinger JC, Lunsford LD, Wu A, Kalend A (1991b) Predicted dose–volume isoeffect curves for stereotactic radiosurgery with the 60Co gamma unit. *Acta Oncol* 30:363–367.

Flickinger JC, Schell MC, Larson DA (1990) Estimation of complications for linear accelerator radiosurgery with the integrated logistic formula. *Int J Radiat Oncol Biol Phys* 19:143–148.

Flickinger JC, Steiner L (1990) Radiosurgery and the double logistic product formula. *Radiother Oncol* 17:229–237.

Friedman WA, Bova FJ, Bollampally S, Bradshaw P (2003) Analysis of factors predictive of success or complications in arteriovenous malformation radiosurgery. *Neurosurgery* 52:296–307. discussion -8.

Fuentes S, Delsanti C, Metellus P, Peragut JC, Grisoli F, Regis J (2006) Brainstem metastases: Management using gamma knife radiosurgery. *Neurosurgery.* 58:37–42.

Ganz JC, Schrottner O, Pendl G (1996) Radiation-induced edema after gamma knife treatment for meningiomas. *Stereotact Funct Neurosurg* 66(Suppl):129–133.

Genentech/Roche. Avastin (bevacizumab) prescribing information. Available at www.avastin.com [Accessed May 24, 2015].

Glantz MJ, Burger PC, Friedman AH, Radtke RA, Massey EW, Schold SC Jr (1994) Treatment of radiation-induced nervous system injury with heparin and warfarin. *Neurology* 44:2020.

Glatstein E (2008) Hypofractionation, long-term effects, and the alpha/beta ratio. *Int J Radiat Oncol Biol Phys* 72:11–12.

Gorman L, Ruben J, Myers R, Dally M (2008) Role of hypofractionated stereotactic radiotherapy in treatment of skull base meningiomas. *J Clin Neurosci* 15:856–862.

Gutin PH, Iwamoto FM, Beal K, Mohile NA, Karimi S, Hou BL, Lymberis S, Yamada Y, Chang J, Abrey LE (2009) Safety and efficacy of bevacizumab with hypofractionated stereotactic irradiation for recurrent malignant gliomas. *Int J Radiat Oncol Biol Phys* 75:156–163.

Hall EJ, Brenner DJ (1993) The radiobiology of radiosurgery: Rationale for different treatment regimes for AVMs and malignancies. *Int J Radiat Oncol Biol Phys* 25:381–385.

Hatiboglu MA, Chang EL, Suki D, Sawaya R, Wildrick DM, Weinberg JS (2011) Outcomes and prognostic factors for patients with brainstem metastases undergoing stereotactic radiosurgery. *Neurosurgery* 69:796–806.

Henzel M, Gross MW, Hamm K, Surber G, Kleinert G, Failing T, Strassmann G, Engenhart-Cabillic R (2006) Stereotactic radiotherapy of meningiomas: Symptomatology, acute and late toxicity. *Strahlenther Onkol* 182:382–388.

Hershman D, Wright JD, Lim E, Buono DL, Tsai WY, Neugut AI (2013) Contraindicated use of bevacizumab and toxicity in elderly patients with colorectal cancer. *J Clin Oncol* 31:3592–3599.

Huang CF, Kondziolka D, Flickinger JC, Lunsford LD (1999) Stereotactic radiosurgery for brainstem metastases. *J Neurosurg* 91:563–568.

Hussain A, Brown PD, Stafford SL, Pollock BE (2007) Stereotactic radiosurgery for brainstem metastases: Survival, tumor control, and patient outcomes. *Int J Radiat Oncol Biol Phys* 67:521–524.

Kalapurakal JA, Silverman CL, Akhtar N, Laske DW, Braitman LE, Boyko OB, Thomas PR (1997) Intracranial meningiomas: Factors that influence the development of cerebral edema after stereotactic radiosurgery and radiation therapy. *Radiology* 204:461–465.

Kased N, Huang K, Nakamura JL, Sahgal A, Larson DA, McDermott MW, Sneed PK (2008) Gamma knife radiosurgery for brainstem metastases: The UCSF experience. *J Neurooncol* 86:195–205.

Kim JH, Chung YG, Kim CY, Kim HK, Lee HK (2004) Upregulation of VEGF and FGF2 in normal rat brain after experimental intraoperative radiation therapy. *J Korean Med Sci* 19:879–886.

Kim YJ, Cho KH, Kim JY, Lim YK, Min HS, Lee SH, Kim HJ, Gwak HS, Yoo H, Lee SH (2011) Single-dose versus fractionated stereotactic radiotherapy for brain metastases. *Int J Radiat Oncol Biol Phys* 8:483–489.

Kocher M, Wilms M, Makoski HB, Hassler W, Maarouf M, Treuer H, Voges J, Sturm V, Müller RP (2004) Alpha/beta ratio for arteriovenous malformations estimated from obliteration rates after fractionated and single-dose irradiation. *Radiother Oncol* 71:109–114.

Kohshi K, Imada H, Nomoto S, Yamaguchi R, Abe H, Yamamoto H (2003) Successful treatment of radiation-induced brain necrosis by hyperbaric oxygen therapy. *J Neurol Sci* 209:115–117.

Kondziolka D, Flickinger JC, Perez B (1998) Judicious resection and/or radiosurgery for parasagittal meningiomas: Outcomes from a multicenter review. Gamma Knife Meningioma Study Group. *Neurosurgery* 43:405–413.

Koyfman SA, Tendulkar RD, Chao ST, Vogelbaum MA, Barnett GH, Angelov L, Weil RJ, Neyman G, Reddy CA, Suh JH (2010) Stereotactic radiosurgery for single brainstem metastases: The Cleveland Clinic experience. *Int J Radiat Oncol Biol Phys* 78:409–414.

Kwon AK, Dibiase SJ, Wang B, Hughes SL, Milcarek B, Zhu Y (2009) Hypofractionated stereotactic radiotherapy for the treatment of brain metastases. *Cancer* 115:890–898.

Lawrence YR, Li XA, el Naqa I, Hahn CA, Marks LB, Merchant TE, Dicker AP (2010) Radiation dose–volume effects in the brain. *Int J Radiat Oncol Biol Phys* 76:S20–S27.

Leach RM, Rees PJ, Wilmshurst P (1998) Hyperbaric oxygen therapy. *BMJ* 317:1140–1143.

Leber KA, Eder HG, Kovac H, Anegg U, Pendl G (1998) Treatment of cerebral radionecrosis by hyperbaric oxygen therapy. *Stereotact Funct Neurosurg* 70(Suppl 1):229–236.

Levin VA, Bidaut L, Hou P, Kumar AJ, Wefel JS, Bekele BN, Grewal J et al. (2011) Randomized double-blind placebo-controlled trial of bevacizumab therapy for radiation necrosis of the central nervous system. *Int J Radiat Oncol Biol Phys* 79: 1487–1495.

Li YQ, Chen P, Jain V, Reilly RM, Wong CS (2004) Early radiation-induced endothelial cell loss and blood-spinal cord barrier breakdown in the rat spinal cord. *Radiat Res* 161:143–152.

Lindvall P, Bergström P, Löfroth PO, Hariz MI, Henriksson R, Jonasson P, Bergenheim AT (2003) Hypofractionated conformal stereotactic radiotherapy for arteriovenous malformations. *Neurosurgery* 53:1036–1042.

Linskey ME, Andrews DW, Asher AL, Burri SH, Kondziolka D, Robinson PD, Ammirati M et al. (2010) The role of stereotactic radiosurgery in the management of patients with newly diagnosed brain metastases: A systematic review and evidence-based clinical practice guideline. *J Neurooncol* 96:45–68.

Loeffler JS (2008) Radiation tolerance limits of the brainstem. *Neurosurgery* 63:733.

Loeffler JS, Niemierko A, Chapman PH (2003) Second tumors after radiosurgery: Tip of the iceberg or a bump in the road? *Neurosurgery* 52:1436–1440.

Lorenzoni JG, Devriendt D, Massager N, Desmedt F, Simon S, Van Houtte P, Brotchi J, Levivier M (2009) Brain stem metastases treated with radiosurgery: Prognostic factors of survival and life expectancy estimation. *Surg Neurol* 71:188–195.

Lutterbach J, Cyron D, Henne K, Ostertag CB (2003) Radiosurgery followed by planned observation in patients with one to three brain metastases. *Neurosurgery* 52:1066–1074.

Manning MA, Cardinale RM, Benedict SH, Kavanagh BD, Zwicker RD, Amir C, Broaddus WC (2000) Hypofractionated stereotactic radiotherapy as an alternative to radiosurgery for the treatment of patients with brain metastases. *Int J Radiat Oncol Biol Phys* 47:603–608.

Marchetti M, Milanesi I, Falcone C, De Santis M, Fumagalli L, Brait L, Bianchi L, Fariselli L (2011) Hypofractionated stereotactic radiotherapy for oligometastases in the brain: A single-institution experience. *Neurol Sci* 32:393–399.

Mayo C, Yorke E, Merchant TE (2010) Radiation associated brainstem injury. *Int J Radiat Oncol Biol Phys* 76:S36–S41.

McIver JI, Pollock BE (2004) Radiation-induced tumor after stereotactic radiosurgery and whole brain radiotherapy: Case report and literature review. *J Neurooncol* 66:301–305.

McPherson CM, Warnick RE (2004) Results of contemporary surgical management of radiation necrosis using frameless stereotaxis and intraoperative magnetic resonance imaging. *J Neurooncol* 68:41–47.

Milano MT, Usuki KY, Walter KA, Clark D, Schell MC (2011) Stereotactic radiosurgery and hypofractionated stereotactic radiotherapy: Normal tissue dose constraints of the central nervous system. *Cancer Treat Rev* 37:567–578.

Minniti G, Clarke E, Lanzetta G, Osti MF, Trasimeni G, Bozzao A, Romano A, Enrici RM (2011) Stereotactic radiosurgery for brain metastases: Analysis of outcome and risk of brain radionecrosis. *Radiat Oncol* 6:48.

Miyakawa A, Shibamoto Y, Otsuka S, Iwata H (2014) Applicability of the linear–quadratic model to single and fractionated radiotherapy schedules: An experimental study. *J Radiat Res* 55:451–454.

Miyawaki L, Dowd C, Wara W, Goldsmith B, Albright N, Gutin P, et al (1999) Five year results of LINAC radiosurgery for arteriovenous malformations: Outcome for large AVMS. *Int J Radiat Oncol Biol Phys* 44:1089–1106.

Moosa S, Chen CJ, Ding D, Lee CC, Chivukula S, Starke RM, Yen CP, Xu Z, Sheehan JP (2014) Volume-staged versus dose-staged radiosurgery outcomes for large intracranial arteriovenous malformations. *Neurosurg Focus* 37:E18.

Muracciole X, Regis J (2008) Radiosurgery and carcinogenesis risk. *Prog Neurol Surg* 21:207–213.

Nakaya K, Niranjan A, Kondziolka D, Kano H, Khan AA, Nettel B, Koebbe C, Pirris S, Flickinger JC, Lunsford LD (2010) Gamma knife radiosurgery for benign tumors with symptoms from brainstem compression. *Int J Radiat Oncol Biol Phys* 77:988–995.

Narayana A, Chang J, Yenice K, Chan K, Lymberis S, Brennan C, Gutin PH (2007) Hypofractionated stereotactic radiotherapy using intensity-modulated radiotherapy in patients with one or two brain metastases. *Stereotact Funct Neurosurg* 85:82–87.

Ney DE, Carlson JA, Damek DM, Gaspar LE, Kavanagh BD, Kleinschmidt-DeMasters BK, Waziri AE, Lillehei KO, Reddy K, Chen C (2015) Phase II trial of hypofractionated intensity-modulated radiotherapy combined with temozolomide and bevacizumab for patients with newly diagnosed glioblastoma. *J Neurooncol* 122:135–143.

Niranjan A, Kondziolka D, Lunsford LD (2009) Neoplastic transformation after radiosurgery or radiotherapy: Risk and realities. *Otolaryngol Clin North Am* 42:717–729.

Ohguri T, Imada H, Kohshi K, Kakeda S, Ohnari N, Morioka T, Nakano K, Konda N, Korogi Y (2007) Effect of prophylactic hyperbaric oxygen treatment for radiation-induced brain injury after stereotactic radiosurgery of brain metastases. *Int J Radiat Oncol Biol Phys* 67:248–255.

Ohtakara K, Hayashi S, Nakayama N, Ohe N, Yano H, Iwama T, Hoshi H (2012) Significance of target location relative to the depth from the brain surface and high-dose irradiated volume in the development of brain radionecrosis after micromultileaf collimator-based stereotactic radiosurgery for brain metastases. *J Neurooncol* 108:201–209.

Omuro A, Beal K, Gutin P, Karimi S, Correa DD, Kaley TJ, DeAngelis LM et al. (2014) Phase II study of bevacizumab, temozolomide, and hypofractionated stereotactic radiotherapy for newly diagnosed glioblastoma. *Clin Cancer Res* 20:5023–5031.

Patil CG, Hoang S, Borchers DJ III, Sakamoto G, Soltys SG, Gibbs IC, Harsh GR IV, Chang SD, Adler JR Jr (2008) Predictors of peritumoral edema after stereotactic radiosurgery of supratentorial meningiomas. *Neurosurgery* 63:435–440.

Pérez-Espejo MA, García-Fernández R, Tobarra-González BM, Palma-Copete JD, González-López A, De la Fuente-Muñoz I, Salinas-Ramos J et al. (2009) Usefulness of hyperbaric oxygen in the treatment of radionecrosis and symptomatic brain edema after LINAC radiosurgery. *Neurocirugia (Astur)* 20:449–453.

Petrovich Z, Yu C, Giannotta SL, O'Day S, Apuzzo ML (2002) Survival and pattern of failure in brain metastases treated with stereotactic Gamma Knife radiosurgery. *J Neurosurg* 97:499–506.

Pollock B (2003) Second tumors after radiosurgery: Tip of the iceberg or a bump in the road? *Neurosurgery* 52:1436–1440.

Qi XS, Schultz CJ, Li XA (2007) Possible fractionated regimens for image-guided intensity-modulated radiation therapy of large arteriovenous malformations. *Phys Med Biol* 52:5667–5682.

Rowe J, Grainger A, Walton L, Silcocks P, Radatz M, Kemeny A (2007) Risk of malignancy after gamma knife stereotactic radiosurgery. *Neurosurgery* 60:60–65.

Schultheiss TE, Kun LE, Ang KK, Stephens LC (1995) Radiation response of the central nervous system. *Int J Radiat Oncol Biol Phys* 31:1093–1112.

Sengöz M, Kabalay IA, Tezcanlı E, Peker S, Pamir N (2013) Treatment of brainstem metastases with gamma-knife radiosurgery. *J Neurooncol* 113:33–38.

Sharma MS, Kondziolka D, Khan A, Kano H, Niranjan A, Flickinger JC, Lunsford LD (2008) Radiation tolerance limits of the brainstem. *Neurosurgery* 63:728–732.

Shaw E, Scott C, Souhami L, Dinapoli R, Kline R, Loeffler J, Farnan N (2000) Single dose radiosurgical treatment of recurrent previously irradiated primary brain tumors and brain metastases: Final report of RTOG protocol 90-05. *Int J Radiat Oncol Biol Phys* 47:291–298.

Shehata MK, Young B, Reid B, Patchell RA, St Clair W, Sims J, Sanders M, Meigooni A, Mohiuddin M, Regine WF (2004) Stereotactic radiosurgery of 468 brain metastases < or =2 cm: Implications for SRS dose and whole brain radiation therapy. *Int J Radiat Oncol Biol Phys* 59:87–93.

Sheline GE, Wara WM, Smith V (1980) Therapeutic irradiation and brain injury. *Int J Radiat Oncol Biol Phys* 6:1215–1228.

Shrieve DC, Hazard L, Boucher K, Jensen RL (2004) Dose fractionation in stereotactic radiotherapy for parasellar meningiomas: Radiobiological considerations of efficacy and optic nerve tolerance. *J Neurosurg* 101(Suppl 3):390–395.

Shuryak I, Carlson DJ, Brown JM, Brenner DJ (2015) High-dose and fractionation effects in stereotactic radiation therapy: Analysis of tumor control data from 2965 patients. *Radiotherapy Oncol* 8:339–348.

Shuto T, Fujino H, Asada H, Inomori S, Nagano H (2003) Gamma knife radiosurgery for metastatic tumours in the brain stem. *Acta Neurochir* 145:755–760.

Silander H, Pellettieri L, Enblad P, Montelius A, Grusell E, Vallhagen-Dahlgren C, Isacsson U (2004) Fractionated, stereotactic proton beam treatment of cerebral arteriovenous malformations. *Acta Neurol Scand* 109:85–90.

Sneed PK, Mendez J, Vemer-van den Hoek JG, Seymour ZA, Ma L, Molinaro AM, Fogh SE, Nakamura JL, McDermott MW (2015) Adverse radiation effect after stereotactic radiosurgery for brain metastases: Incidence, time course, and risk factors. *J Neurosurg* 15:1–14.

Szeifert GT, Atteberry DS, Kondziolka D, Levivier M, Lunsford LD (2006) Cerebral metastases pathology after radiosurgery: A multicenter study. *Cancer* 106:2672–2681.

Trippa F, Maranzano E, Costantini S, Giorni C (2009) Hypofractionated stereotactic radiotherapy for intracranial meningiomas: Preliminary results of a feasible trial. *J Neurosurg Sci* 53:7–11.

Trotti A, Colevas AD, Setser A, Rusch V, Jaques D, Budach V, Langer C et al. (2003) CTCAE v3.0: Development of a comprehensive grading system for the adverse effects of cancer treatment. *Semin Radiat Oncol* 13:176–181. NIH Publication No. 09-5410. http://evs.nci.nih.gov/ftp1/CTCAE/CTCAE_4.03_2010-06-14_QuickReference_8.5x11.pdf. 2009.

Tye K, Engelhard HH, Slavin KV, Nicholas MK, Chmura SJ, Kwok Y, Ho DS, Weichselbaum RR, Koshy M (2014) An analysis of radiation necrosis of the central nervous system treated with bevacizumab. *J Neurooncol* 117:321–327.

Valadão J, Pearl J, Verma S, Helms A, Whelan H (2003) Hyperbaric oxygen treatment for post-radiation central nervous system injury: A retrospective case series. *J Neurol Sci* 209:115–117.

Varlotto JM, Flickinger JC, Niranjan A, Bhatnagar AK, Kondziolka D, Lunsford LD (2003) Analysis of tumor control and toxicity in patients who have survived at least one year after radiosurgery for brain metastases. *Int J Radiat Oncol Biol Phys* 57:452–464.

Vernimmen FJ, Slabbert JP (2010) Assessment of the alpha/beta ratios for arteriovenous malformations, meningiomas, acoustic neuromas, and the optic chiasma. *Int J Radiat Biol* 86:486–498.

Veznedaroglu E, Andrews DW, Benitez RP, Downes MB, Werner-Wasik M, Rosenstock J, Curran WJ Jr, Rosenwasser RH (2004) Fractionated stereotactic radiotherapy for the treatment of large arteriovenous malformations with or without previous partial embolization. *Neurosurgery* 55:519–530.

Veznedaroglu E, Andrews DW, Benitez RP, Downes MB, Werner-Wasik M, Rosenstock J, Curran WJ Jr, Rosenwasser RH (2008) Fractionated stereotactic radiotherapy for the treatment of large arteriovenous malformations with or without previous partial embolization. *Neurosurgery* 62(Suppl 2):763–775.

Wigg DR (1999) Is there a role for fractionated radiotherapy in the treatment of arteriovenous malformations? *Acta Oncol* 38:979–986.

Williamson R, Kondziolka D, Kanaan H, Lunsford LD, Flickinger JC (2008) Adverse radiation effects after radiosurgery may benefit from oral vitamin E and pentoxifylline therapy: A pilot study. *Stereotact Funct Neurosurg* 86:359–366.

Yen CP, Sheehan J, Patterson G, Steiner L (2006) Gamma knife surgery for metastatic brainstem tumors. *J Neurosurg* 105:213–219.

20 Vertebral compression fracture post–spine SBRT

Isabelle Thibault, Samuel Bergeron Gravel,
Cari Whyne, David Mercier, and Arjun Sahgal

Contents

20.1 INTRODUCTION

Recently, we have learned that vertebral compression fracture (VCF) is a relatively common adverse side effect following spine stereotactic body radiation therapy (SBRT). In 2009, the Memorial Sloan Kettering Cancer Center (MSKCC) reported a prohibitive risk of post-SBRT VCF at 39% (Rose et al., 2009). Since then, several groups have reported their results including a major multi-institutional study led by the University of Toronto (UofT) which clarified the post-SBRT risk, in general, with a reported 1-year cumulative incidence of VCF of 12% and a median time to VCF of 2.5 months (Sahgal et al., 2013b). A 40% risk of VCF was still observed, but secondary to high-dose per fraction SBRT when delivering 24 Gy in a single fraction. It has also been suggested that radiation necrosis is a causative factor. The aim of this chapter is to focus on the reported literature to guide the reader how we define, assess, and manage SBRT-induced VCF.

20.2 DEFINITION

A VCF is defined as a collapse of the vertebral body (VB). Spine SBRT-induced VCF can occur as de novo, in a treated VB with no baseline VCF, or as progression of an existing VCF within the treated vertebral segment (fracture progression). Those series reporting on spine SBRT-induced VCF have typically excluded segments that concomitantly experience local tumor progression, arguing that tumor progression itself might destabilize the spinal segment and predispose to a pathologic VCF rather than an iatrogenic VCF. Similarly, SBRT-induced VCF is usually reported in those nonoperated patients, to avoid the preventive stabilization effect of an existing surgical procedure or instrumentation.

20.3 RADIOLOGIC ASSESSMENT OF VCF AND SURVEILLANCE IMAGING

20.3.1 VERTEBRAL BODY HEIGHT MEASUREMENTS

The initial step in the radiologic assessment of VCF is to determine the treated VB height, based on the endplates, according to the baseline treatment planning CT and MRI. Ideally, that baseline VB height is compared to prior imaging to determine whether or not there is any baseline VCF or progression of an existing one. If prior imaging is not available, then the height is compared to the average VB height of the vertebrae located immediately above and below the treated segment.

Following SBRT, the baseline VB height is compared to subsequent follow-up imaging studies, preferably using the same imaging modality. Both sagittal and coronal views should be used if available; however, the key is to be consistent in the method. A difference of ≤5% in VB height is considered insignificant and could represent a measurement error and, therefore, a radiographic score of VCF is based on >5% height loss.

20.3.2 IMAGING MODALITY

The preferred imaging modality post-SBRT for follow-up assessment is MRI (T1 nonenhanced sequences). As compared to CT, MRI has superior diagnostic accuracy (Buhmann Kirchhoff et al., 2009) for spinal metastases and allows visualization of any epidural and/or paraspinal soft tissue extension. Given that VCF tends to occur shortly after spine SBRT, on average 2.5 months post–spine SBRT (Sahgal et al., 2013b), the first follow-up MRI should be performed within the first 2–3 months following treatment completion. Most spine SBRT experts perform routine follow-up imaging at 2–3-month intervals for at least the first years and at 3–6-month intervals thereafter.

If VCF is observed, then further imaging is often required to characterize the VCF. These include plain film x-rays and most importantly a CT spine. CT is very useful in the assessment of the integrity of the bony anatomy with respect to detection of cortical and trabecular bone destruction. CT further characterizes the tumor as osteolytic, osteosclerotic, or mixed type lesions; this may have specific implications for treatment as lytic-based VCF may be treated with cement augmentation procedures as opposed to purely sclerotic-based VCF. Determination of the spinal alignment also has implications on surgical decision making and characterizing the patient's degree of spinal instability. For example, if translation beyond a physiologically expected level is present, then this is a high-risk situation often requiring surgical stabilization, and kyphotic deformity may also influence the type of intervention. Figure 20.1 presents a case of SBRT-induced VCF resulting in kyphotic deformity and mechanical pain.

20.3.3 CLASSIFICATION OF FRACTURE

Wedge and burst fractures are the two most common types of vertebral fractures (Denis, 1983). A wedge fracture is defined as failure of the anterior half of the VB, while the posterior part remains intact. As a result, this type of fracture is typically not associated with subluxation or neurological compromise and is relatively stable. It results from compression forces, mainly in flexion. On the contrary, burst fractures are characterized by failure under axial load of the entire VB and more often are characterized as "potentially unstable." These fractures are typically associated with a risk of retropulsion of a bony fragment into the

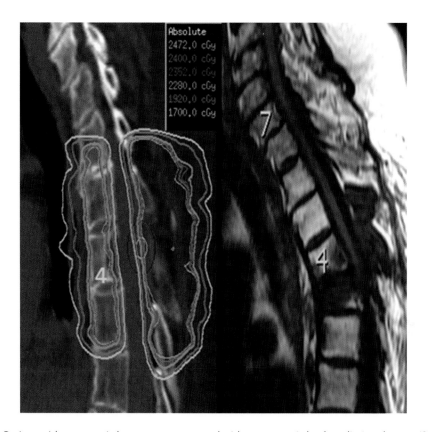

Figure 20.1 Patient with metastatic breast cancer treated with stereotactic body radiation therapy (SBRT) (24 Gy in 2 fractions) to T3–T4–T5. The sagittal dose distribution is shown on the left panel. Patient responded well to pain and 4 months later developed a sudden mechanical type of pain. Investigations showed fracture at T5 and resulting kyphotic deformity, and the sagittal T1-weighted magnetic resonance image is shown on the right panel, with numbered 7th cervical and 4th thoracic vertebral segments. The patient's pain settled and has been followed now 12 months following SBRT with no surgical intervention needed.

spinal canal which can lead to neurologic impairment. Wedge and burst fractures can be observed on the superior endplate, inferior endplate, or on both endplates. Although the shape of the fractured spinal segment provides anatomic details, it has only minor significance in the overall risk and management of spine instability. A comprehensive assessment of spinal instability is required in metastatic patients and will be further discussed in this chapter.

20.4 PATHOPHYSIOLOGY OF SBRT-INDUCED VCF

SBRT-induced VCF may occur as a relatively acute or late radiation effect on the vertebral bone. It is postulated that the high radiation doses inherent to SBRT induces radiation necrosis and/or fibrosis. This phenomenon was first observed after spine SBRT by Al-Omair et al. (2013), who identified radiation osteonecrosis and fibrosis in biopsy specimens of two cases with VCF suspected to be tumor progression. Osteoradionecrosis is a well-known late toxicity but rare after conventional radiotherapy and has been observed on the mandible (Marx and Johnson, 1987), hand (Walsh, 1897), and femoral head (Tai et al., 2000). It is characterized by osteolysis, altered collagen fibrils and loss of minerals, occurring most commonly in a hypoxic and avascular environment.

Radiation effect is only one component contributing to the pathophysiology of VCF (Sahgal et al., 2013a). The structural integrity and quality of the vertebral bone rely both on its material properties (tissue mineralization and collagen) and architecture. Strength, stiffness, and toughness is afforded by the hydroxyapatite content, arrangement of minerals within the collagen, and the collagen fiber network itself (Wang et al., 2001; Burr, 2002; Whyne, 2014). The presence of underlying metastatic foci, osteolytic or osteoblastic, may alter the trabecular network, collagen cross-linking, bone elasticity, and quality. Therefore, the interplay of tumor and treatment factors renders the pathophysiology complex, and the interested reader is directed to a recent review by Sahgal et al. (2013a).

20.5 INCIDENCE OF VCF

A summary of the published papers reporting on VCF after spine SBRT is presented in Table 20.1. The first major study identifying VCF as a serious toxicity following spine SBRT was performed by the MSKCC and reported a 39% risk of VCF following 18–24 Gy delivered in a single-fraction SBRT (Rose et al., 2009). Several centers have since reported their experience and some with similarly high complication rates. For example, Sung et al. reported a 36% risk of VCF following SBRT regimens ranging from 18 to 45 Gy in 1–5 fractions, corresponding to a single equivalent mean dose of 21 Gy (Sung and Chang, 2014).

In order to clarify the risk of VCF, a multi-institutional study was reported by Sahgal et al. pooling data from the MD Anderson Cancer Center (MDACC), Cleveland Clinic, and UofT. Based on 410 spinal segments treated, the crude rate of VCF was 14%, and the actuarial 1-year cumulative incidence was 12.4% (Sahgal et al., 2013b). The median and mean time to VCF was 2.46 and 6.33 months with two-thirds of VCFs developing within the first 4 months post-SBRT.

20.6 PREDICTIVE FACTORS OF VCF

20.6.1 EFFECT OF DOSE FRACTIONATION

High-dose per fraction SBRT exceeding 19 Gy has been identified as a major predictor of VCF. Of 410 treated spinal segments reported in the multi-institutional study by Sahgal et al. (2013b), those receiving ≥24 Gy/fraction (HR 5.25; 95% CI 2.29–12.01) or 20–23 Gy/fraction (HR 4.91; 95% CI 1.96–12.28) had higher VCF risk compared to those treated with SBRT of ≤19 Gy/fraction. SBRT dose per fraction of ≥20 Gy was also a significant predictor of VCF in the study of Cunha et al. (2012). The justification of SBRT regimens at ≥20 Gy is questionable given the increased incidence of VCF without any conclusive data to support increased efficacy. Table 20.2 summarizes those studies evaluating predictors of VCF post-SBRT.

20.6.2 BONE TUMOR TYPE

Osteolytic tumor type has also been shown to be a significant predictor of VCF. This factor was identified in all published studies evaluating risk factors for VCF, as shown in Table 20.2. For example, the multi-institutional Sahgal series reported 48 VCFs (18.8%) after SBRT among 256 lytic spinal metastases (HR 3.53; 95% CI 1.58–7.93). Lytic tumor is characterized by reduced bone mineralization, compromised inherent bone structure, and increased propensity for pressurization (Whyne, 2014). Thus, lytic tumor as a predictor of VCF makes biophysical sense as it is predisposed to fracture even prior to SBRT.

20.6.3 PREEXISTING FRACTURE

The presence of VCF at baseline has also been shown to predict for radiation-induced VCF, by both the MDACC series (Boehling et al., 2012) and the multi-institutional report (Sahgal et al., 2013b). Sahgal reported a crude risk of de novo VCF of 8.3% (27 of 327 segments) compared to 36.1% for those with a preexisting fracture (fracture progression in 30 of 83 segments with a baseline VCF). The question that these data pose is whether or not patients with preexisting VCF should be routinely treated with prophylactic stabilization surgery. With minimally invasive techniques consisting of cement augmentation or percutaneous instrumentation, some types of stabilization can be performed as an outpatient procedure prior to SBRT with significantly fewer adverse events as compared to those expected with traditional open surgery.

Gerszten et al. was one of the first to successfully combine kyphoplasty (Medtronic, Minnesota, MA) followed by SBRT for patients with painful pathologic VCF deemed eligible for SBRT (Gerszten et al., 2005). However, the risk of routine prophylaxis is that a significant proportion of patients would be overtreated, and there is still potential for significant adverse events despite the minimally invasive nature of the intervention. This highlights the importance of scoring systems, like the spinal instability neoplastic

Table 20.1 Comparison of studies reporting on vertebral compression fracture after spine stereotactic body radiation therapy

AUTHOR, SITE (YEAR)	NO. OF PTS.	NO. OF SEGMENTS	LYTIC SEGMENTS	BASELINE FRACTURED SEGMENTS[a]	SPINAL MALALIGNMENT	SBRT TOTAL DOSE/FX	INCIDENCE OF VCF (%)	MEDIAN TIME TO VCF (MONTHS)	SALVAGE INTERVENTION (%); TYPE
Rose et al., MSKCC (2009)	62	71	65%	28%	NA	18–24/1	39	25	3/27 (11%); 2 S, 1 CAP
Boehling et al., MDACC (2012)	93	123	58%	28%	NA	18–30/1–5	20	3	10/25 (40%); 10 CAP
Cunha et al., UofT (2012)	90	167	48%	17%	11%	8–35/1–5	11	2 (mean 3.3)	9/19 (47%); 3 S, 6 CAP
Balagamwala et al., CC (2012)	57	88	NA	30%	NA	8–16/1	14	NA	NA
Sahgal et al., Multi (2013a)	**252**	**410**	**62%**	**20%**	**8%**	**8–35/1–5**	**14**	**2.5 (mean 6.3)**	**24/57 (42%); 7 S, 17 CAP**
Thibault et al., UofT (2014)	37	61[b]	95%	21%	10%	18–30/1–5	16	1.6	4/10 (40%); 1 S, 3 CAP
Sung and Chang, Korea (2014)	72	72	NA	NA	11%	18–45/1–5	36	(mean: 1.5)	15/26 (58%); 5 S, 10 CAP
Guckenberger et al. (2014), Multi (2014)	301	387	72%	20%	NA	8–60/1–20	7.8	NA	NA

Note: The results of the multi-institutional study, by Sahgal et al., are reported in bold text.

a Indicates percentage of segments with a preexisting vertebral compression fracture at baseline, before spine stereotactic body radiation therapy.

b Of a cohort of 71 renal cell cancer spinal segments, 61 were analyzed for vertebral compression fracture risk as 10 were postoperative stereotactic body radiation therapy cases.

Abbreviations: VCF, vertebral compression fracture; pts, patients; fx, fractions; S, surgery; CAP, cement augmentation procedure; MSKCC, Memorial Sloan Kettering Cancer Center; MDACC, MD Anderson Cancer Center; UofT, University of Toronto; CC, Cleveland Clinic; Multi, Multi-institutions; NA, not available.

Table 20.2 Summary of the literature on risk factors for vertebral compression fracture after spine stereotactic body radiation therapy

AUTHOR, SITE (YEAR)	SIGNIFICANT PREDICTORS ON MVA
Rose et al., MSKCC (2009)	Lytic tumor (HR 3.8); 41%–60% vertebral body involved (HR 3.9)
Boehling et al., MDACC (2012)	Age >55 years (HR 5.7); baseline VCF [a] (HR 4.12); lytic tumor (HR 2.8)
Cunha et al., UofT (2012)	Lytic tumors (HR 12.2); malalignment (HR 11.1); ≥20 Gy/fraction (HR 6.8); lung histology (HR 4.3); liver histology (HR 34)
Sahgal et al., Multi (2013)	**Baseline VCF[a] (HR 8.9 if <50% VCF, HR 6.9 if ≥50% VCF); lytic tumor (HR 3.5); ≥20 Gy/fraction (HR 4.9 if 20–23 Gy, HR 5.3 if ≥24 Gy); spinal malalignment (HR 3.0)**
Sung and Chang, Korea (2014)	Vertebral body osteolysis rate of ≥60%

[a] Indicates preexisting vertebral compression fracture at baseline, prior to spine stereotactic body radiation therapy.
Abbreviations: VCF, vertebral compression fracture; MVA, multivariate analysis; HR, hazard ratio; Multi, multi-institutions.

score (SINS) (Fisher et al., 2010), which may help identify those patients a priori who are mechanically unstable and at risk of VCF. At this time, treatment with SBRT and intervention only upon VCF development is the standard of care, until robust patient selection methods are determined and validated specific to radiation-induced VCF risk. An additional benefit to this approach is to treat an undisturbed target for SBRT planning, and recently there has been reported the potential for tumor extravasation following cement injection (Cruz et al., 2014). These aspects will be discussed further in the chapter.

20.6.4 SPINAL MALALIGNMENT

Radiographic spinal malalignment, as either kyphotic/scoliotic deformity or subluxation/translation, is often the result of an unstable baseline VCF. The presence of a baseline spinal malalignment prior to SBRT was identified as an independent predictor of VCF by the large UofT series (Cunha et al., 2012) and confirmed by a multi-institutional report (Sahgal et al., 2013b). Evaluation of spinal alignment and stability is one key aspect as to why multidisciplinary management of these patients is essential as the nuances of this assessment are better performed by a spinal surgeon.

20.6.5 TUMOR HISTOLOGY

Cunha et al. identified lung and liver cancer histologies at increased risk of radiation-induced VCF (Cunha et al., 2012); however, these findings have not been reproduced in other studies (Table 20.2). Similarly, the VCF risks of 16% and 14% in renal cell cancer spinal metastases, reported by Thibault et al. and Balagamwala et al., respectively, are similar to those reported in the literature for various histologies, as shown in Table 20.1 (Balagamwala et al., 2012; Thibault et al., 2014). Therefore, at this time, there are insufficient data to identify those tumor types at greatest risk for post-SBRT VCF.

20.6.6 OTHER POTENTIAL FACTORS

Patient age, gender, obesity, presence of osteoporosis, use of bisphosphonate, use of narcotics, paraspinal tumor extension, and single versus multiple tumors treated within a single clinical target volume have been investigated but found not to impact on post-SBRT VCF risk (Rose et al., 2009; Boehling et al., 2012; Cunha et al., 2012; Sahgal et al., 2013b; Sung and Chang, 2014). However, the available data are still limited; as further pooled analyses and large single institution series are reported, we will gain a deeper understanding of risk factors to stratify patients into high-, mid-, and low-risk groups.

20.7 IDENTIFYING SPINAL INSTABILITY

Spine instability has been defined by the Spine Oncology Study Group (SOSG) as the "loss of spinal integrity as a result of a neoplastic process that is associated with movement-related pain, symptomatic or progressive deformity and/or neural compromise under physiological loads" (Fisher et al., 2010). Of note,

Table 20.3 **Spinal instability neoplastic score classification**

6 SINS FACTOR	DESCRIPTION	POINTS
Location of tumor	Junction: occiput–C2, C7–T2, T11–L1, L5–S1	3
	Mobile spine: C3–C6, L2–L4	2
	Semirigid T-spine: T3–T10	1
	Rigid sacrum: S2–S5	0
Pain	Mechanical pain[a]	3
	Occasional and nonmechanical	1
	None	0
Bone tumor type	Osteolytic	2
	Mixed	1
	Osteoblastic	0
Spinal malalignment	Subluxation or translation	4
	Kyphosis or scoliosis de novo deformity	2
	None	0
Vertebral body height collapse	≥50%	3
	<50%	2
	No compression fracture but >50% body involved	1
	None of the above	0
Tumoral involvement of posterior elements	Bilateral	3
	Unilateral	1
	None	0
Classification: Total SINS ranges from 0 to 18 points. According to SINS, stable segments have a score of 0–6, indeterminate/potentially unstable a score of 7–12, and unstable segments a score of 13–18.		

[a] Movement-related or axial loading pain and/or pain improvement with recumbency.
Abbreviation: SINS, spinal instability neoplastic score.

pain with movement or axial loading of the spine and/or pain relieved with recumbence is referring to mechanical pain. As mechanical pain is a key symptom to assess; some clinicians prefer to use the term mechanical instability rather than spinal instability (Laufer et al., 2013).

The SOSG developed the SINS, based on expert consensus, to aid clinicians in determining which spinal tumor is stable, potentially unstable, or unstable. This classification system has been validated as a reliable tool, with respect to intraobserver and interobserver reliability by surgeons, radiation oncologists, and radiologists (Fourney et al., 2011; Fisher et al., 2014a,b). The classification system is based on the assessment of six factors: location of the metastases, type of pain, bone lesion type, radiologic spinal alignment, VB collapse, and posterolateral involvement of spinal elements. Table 20.3 details the SINS classification.

Although the SINS criteria were originally developed to identify potentially unstable patients with spinal metastases, they were recently investigated for their predictive capacity with respect to developing SBRT-induced VCF. Of the six SINS criteria, three were found predictive: baseline VCF, osteolytic tumor, and malalignment (Sahgal et al., 2013b). The authors conclude that SINS is an important tool to identify patients at increased risk for SBRT-induced VCF together with SBRT dose per fraction.

20.8 GUIDELINES

With respect to VCF risk, patient selection is essential. There is greater attention to determining which patients should undergo a stabilization procedure prior to SBRT and which patients should be closely monitored following SBRT. It is our recommendation that patients presenting at baseline with a spinal subluxation or translation should be referred for consultation with a spine surgeon prior to radiation. Patients with a baseline VCF who also have mechanical pain and/or spinal malalignment should also be referred to a spine surgeon prior to SBRT. What is more controversial, as to prophylactic stabilization, are those patients pain-free with a baseline VCF and those with significant osteolytic tumor with no fracture. In these patients, close monitoring within the first 6 months following SBRT should be maintained as the risk of VCF is greatest during this time period, and the risk may be as high as 35%. Similarly, monitoring in the short term is needed for single-fraction SBRT of ≥20 Gy as the risk of VCF exceeds 20%.

20.9 SURGICAL MANAGEMENT OF SBRT-INDUCED VCF

Approximately 40% (range, 11%–58%) of SBRT-induced VCFs will be treated with a surgical salvage stabilization procedure (Table 20.1). Most commonly, vertebral cement augmentation procedures (e.g., kyphoplasty or vertebroplasty) have been performed; however, approximately 25% will be treated with an invasive instrumented surgical procedure (Table 20.1) (Rose et al., 2009; Boehling et al., 2012; Cunha et al., 2012; Sahgal et al., 2013b; Sung and Chang, 2014; Thibault et al., 2014). Intervention is needed most commonly due to the development of mechanical instability.

Traditionally, spine stabilization surgery was an open invasive surgery requiring a long incision, muscle dissection of paraspinal tissues, and instrumentation of levels above and below the injured segment. Epidural decompression or vertebrectomy could be performed within the same intervention due to the wide exposure. Such open surgery is associated with longer rehabilitation time, greater morbidity, and increased delay before initiating other oncologic treatments compared to noninvasive surgical intervention. As a result, surgery was offered cautiously and only to the most fit patients.

Since then, both surgical techniques and the decision-making process for surgery have evolved. The NOMS decision framework was introduced in 2013, integrating neurologic, oncologic, mechanical, and systemic considerations into the treatment decision making for spine metastases (Laufer et al., 2013). Mechanical instability, referring to severe mechanical pain and instability according to SOSG, was stipulated as an independent surgical indication, regardless of tumor epidural extension or radiosensitivity (Laufer et al., 2013). In patients with mechanical instability, a surgical stabilization intervention was recommended by the authors, based on radiation being unable to restore a spinal malalignment and steroids often failing to palliate mechanical pain. Most recently, the LMNOP system was proposed by the University of Saskatchewan, integrating SINS into a multifactorial decision-based and individualized approach for the general management of spinal metastases (Ivanishvili and Fourney, 2014). In potentially unstable metastases (SINS total score of 7–12), the authors' general guidance is to first consider a vertebral cement augmentation procedure if no cord compression is present, while pedicle screw fixation or more invasive procedures are usually reserved for unstable metastases (SINS of 13–18).

Percutaneous spinal interventions, such as vertebral cement augmentation (balloon kyphoplasty or vertebroplasty) and other minimally invasive spinal surgery (MISS) procedures, are now increasingly used and allow spinal procedures to be applied to a wider spectrum of spinal oncology patients. MISS refers to the application of instrumentation percutaneously and/or allows for epidural decompression via a tubular retraction system (Massicotte et al., 2012). MISS with limited decompression around the spinal cord is called separation surgery and is an emerging practice prior to spine SBRT.

Regarding cement augmentation procedures, vertebroplasty consists of a high-pressure cement injection (usually of polymethylmethacrylate), into the fractured vertebra under fluoroscopic guidance. In contrast, kyphoplasty utilizes balloon inflation within the fractured VB followed by a lower-pressure cement injection. Both procedures are associated with a risk of cement leakage, which can be of major consequence if cement leaks into the spinal canal causing cord compression or into the vasculature causing cement emboli. It has been reported that kyphoplasty is associated with fewer symptomatic cement leakage complications, although the risk is low at under 5% (Lee et al., 2009; Berenson et al., 2011). A multicenter randomized control trial evaluating kyphoplasty as compared to nonsurgical interventions has been reported in patients with cancer and a pathologic VCF. In those randomized to kyphoplasty, a significant relief of pain at the primary (1 month post-kyphoplasty) endpoint (mean score change from 17.6 to 9.1) was observed in addition to improved functional outcomes (Berenson et al., 2011). Although these data are specific to patients presenting with a pathologic VCF, it might be reasonable to expect similar outcomes after kyphoplasty for radiation-induced VCF.

Tumor extravasation following cement augmentation was recently reported as a possible iatrogenic complication (Cruz et al., 2014). Observed venous tumor extravasation and anterior subligamentous spread observed in two cases were postulated to be a direct result of the increased intravertebral pressure during balloon inflation and cement injection (Cruz et al., 2014). Importantly, the consequence of this complication impaired the feasibility of subsequent spine SBRT due to the difficulty in adequately delineating the target volume. Because of the risk of cement leakage and tumor extravasation, although rare, some spine SBRT experts prefer to treat with spine SBRT first followed by a cement augmentation procedure. The intent is to then perform the cement augmentation procedure at 6–8 weeks post-SBRT as a planned intervention.

20.10 NONSURGICAL MANAGEMENT OF SBRT-INDUCED VCF

For symptomatic patients not eligible for a stabilization surgical procedure, and/or declining a surgical intervention, pharmacologic pain management strategies consist of narcotic analgesics, nonsteroidal anti-inflammatory drugs, corticosteroids, membrane stabilizers, bisphosphonates, and tricyclic antidepressants. Treatment consideration may also include spinal orthosis (e.g., brace or cervical collar application).

20.11 FUTURE DIRECTIONS

VCF following spine SBRT has been identified as a significant adverse effect. In fact, the risk can approach nearly 40% of treated patients depending on the choice of dose. Predictors of VCF following spine SBRT have been identified over the last few years such that radiation oncologists can start to rely on these predictors to help in daily clinical practice. In particular, by identifying those with a high SINS, baseline fracture, significant osteolytic tumor burden, and spinal malalignment, appropriate referral to spine surgeons for consideration of a stabilization procedure prior to SBRT can be made. Further research to determine decision-making trees, recursive portioning analyses to stratify patients into risk groups, and a better understanding of the pathophysiology of SBRT-induced VCF will further our ability to choose the most appropriate patients for spine SBRT.

CHECKLIST: KEY POINTS FOR CLINICAL PRACTICE

- High-dose per fraction SBRT (≥20 Gy), osteolytic tumor, a preexisting VCF at baseline, and spinal malalignment are significant predictors of VCF following spine SBRT.
- VCF tends to occur shortly after spine SBRT, at a median time of 2.5 months. The first follow-up spine MRI should be performed within 2–3 months of spine SBRT.
- The pathophysiology of VCF is complex. Osteoradionecrosis has been reported in post-SBRT VCF giving rise to a potential mechanism.
- The SINS aims to identify patients who are unstable and has been shown to be a useful tool to help clinicians identify those at increased risk of VCF after spine SBRT.
- Patients with spinal malalignment, symptomatic baseline VCF, and/or mechanical pain should be evaluated by a spine surgeon prior to spine SBRT so as to consider a stabilization procedure.
- Surgical management of symptomatic SBRT-induced VCF includes cement augmentation procedures (kyphoplasty, vertebroplasty), minimally invasive surgeries, or invasive stabilization approaches with fixation.
- When a surgical procedure is required following spine SBRT, it is recommended to perform a biopsy in order to diagnose tumor progression versus necrosis as the cause of the VCF.

REFERENCES

Al-Omair A, Smith R, Kiehl TR, Lao L, Yu E, Massicotte EM, Keith J, Fehlings MG, Sahgal A (2013) Radiation-induced vertebral compression fracture following spine stereotactic radiosurgery: Clinicopathological correlation. *J Neurosurg Spine* 18:430–435.
Balagamwala EH, Angelov L, Koyfman SA, Suh JH, Reddy CA, Djemil T, Hunter GK, Xia P, Chao ST (2012) Single-fraction stereotactic body radiotherapy for spinal metastases from renal cell carcinoma. *J Neurosurg Spine* 17:556–564.
Berenson J, Pflugmacher R, Jarzem P, Zonder J, Schechtman K, Tillman JB, Bastian L, Ashraf T, Vrionis F (2011) Balloon kyphoplasty versus non-surgical fracture management for treatment of painful vertebral body compression fractures in patients with cancer: A multicentre, randomised controlled trial. *Lancet Oncol* 12:225–235.

Boehling NS, Grosshans DR, Allen PK, McAleer MF, Burton AW, Azeem S, Rhines LD, Chang EL (2012) Vertebral compression fracture risk after stereotactic body radiotherapy for spinal metastases. *J Neurosurg Spine* 16:379–386.

Buhmann Kirchhoff S, Becker C, Duerr HR, Reiser M, Baur-Melnyk A (2009) Detection of osseous metastases of the spine: Comparison of high resolution multi-detector-CT with MRI. *Eur J Radiol* 69:567–573.

Burr DB (2002) The contribution of the organic matrix to bone's material properties. *Bone* 31:8–11.

Cruz JP, Sahgal A, Whyne C, Fehlings MG, Smith R (2014) Tumor extravasation following a cement augmentation procedure for vertebral compression fracture in metastatic spinal disease. *J Neurosurg Spine* 21:372–377.

Cunha MV, Al-Omair A, Atenafu EG, Masucci GL, Letourneau D, Korol R, Yu E et al. (2012) Vertebral compression fracture (VCF) after spine stereotactic body radiation therapy (SBRT): Analysis of predictive factors. *Int J Radiat Oncol Biol Phys* 84:e343–e349.

Denis F (1983) The three column spine and its significance in the classification of acute thoracolumbar spinal injuries. *Spine (Phila PA 1976)* 8:817–831.

Fisher CG, DiPaola CP, Ryken TC, Bilsky MH, Shaffrey CI, Berven SH, Harrop JS et al. (2010) A novel classification system for spinal instability in neoplastic disease: An evidence-based approach and expert consensus from the Spine Oncology Study Group. *Spine (Phila PA 1976)* 35:E1221–E1229.

Fisher CG, Schouten R, Versteeg AL, Boriani S, Varga PP, Rhines LD, Kawahara N et al. (2014b) Reliability of the spinal instability neoplastic score (SINS) among radiation oncologists: An assessment of instability secondary to spinal metastases. *Radiat Oncol* 9:69.

Fisher CG, Versteeg AL, Schouten R, Boriani S, Varga PP, Rhines LD, Heran MK et al. (2014a) Reliability of the spinal instability neoplastic scale among radiologists: An assessment of instability secondary to spinal metastases. *Am J Roentgenol* 203:869–874.

Fourney DR, Frangou EM, Ryken TC, Dipaola CP, Shaffrey CI, Berven SH, Bilsky MH et al. (2011) Spinal instability neoplastic score: An analysis of reliability and validity from the spine oncology study group. *J Clin Oncol* 29:3072–3077.

Gerszten PC, Germanwala A, Burton SA, Welch WC, Ozhasoglu C, Vogel WJ (2005) Combination kyphoplasty and spinal radiosurgery: A new treatment paradigm for pathological fractures. *J Neurosurg Spine* 3:296–301.

Guckenberger M, Mantel F, Gerszten PC, Flickinger JC, Sahgal A, Letourneau D, Grills IS et al. (2014) Safety and efficacy of stereotactic body radiotherapy as primary treatment for vertebral metastases: A multi-institutional analysis. *Radiat Oncol* 9:226.

Ivanishvili Z, Fourney DR (2014) Incorporating the spine instability neoplastic score into a treatment strategy for spinal metastasis: LMNOP. *Global Spine J* 4:129–136.

Laufer I, Rubin DG, Lis E, Cox BW, Stubblefield MD, Yamada Y, Bilsky MH (2013) The NOMS framework: Approach to the treatment of spinal metastatic tumors. *Oncologist* 18:744–751.

Lee MJ, Dumonski M, Cahill P, Stanley T, Park D, Singh K (2009) Percutaneous treatment of vertebral compression fractures: A meta-analysis of complications. *Spine (Phila PA 1976)* 34:1228–1232.

Marx RE, Johnson RP (1987) Studies in the radiobiology of osteoradionecrosis and their clinical significance. *Oral Surg Oral Med Oral Pathol* 64:379–390.

Massicotte E, Foote M, Reddy R, Sahgal A (2012) Minimal access spine surgery (MASS) for decompression and stabilization performed as an out-patient procedure for metastatic spinal tumours followed by spine stereotactic body radiotherapy (SBRT): First report of technique and preliminary outcomes. *Technol Cancer Res Treat* 11:15–25.

Rose PS, Laufer I, Boland PJ, Hanover A, Bilsky MH, Yamada J, Lis E (2009) Risk of fracture after single fraction image-guided intensity-modulated radiation therapy to spinal metastases. *J Clin Oncol* 27:5075–5079.

Sahgal A, Atenafu EG, Chao S, Al-Omair A, Boehling N, Balagamwala EH, Cunha M et al. (2013b) Vertebral compression fracture after spine stereotactic body radiotherapy: A multi-institutional analysis with a focus on radiation dose and the spinal instability neoplastic score. *J Clin Oncol* 31:3426–3431.

Sahgal A, Whyne CM, Ma L, Larson DA, Fehlings MG (2013a) Vertebral compression fracture after stereotactic body radiotherapy for spinal metastases. *Lancet Oncol* 14:e310–e320.

Sung SH, Chang UK (2014) Evaluation of risk factors for vertebral compression fracture after stereotactic radiosurgery in spinal tumor patients. *Korean J Spine* 11:103–108.

Tai P, Hammond A, Dyk JV, Stitt L, Tonita J, Coad T, Radwan J (2000) Pelvic fractures following irradiation of endometrial and vaginal cancers-a case series and review of literature. *Radiother Oncol* 56:23–28.

Thibault I, Al-Omair A, Masucci GL, Masson-Cote L, Lochray F, Korol R, Cheng L et al. (2014) Spine stereotactic body radiotherapy for renal cell cancer spinal metastases: Analysis of outcomes and risk of vertebral compression fracture. *J Neurosurg Spine* 21:711–718.

Walsh D (1897) Deep Tissue traumatism from roentgen ray exposure. *Br Med J* 2:272–273.

Wang X, Bank RA, TeKoppele JM, Agrawal CM (2001) The role of collagen in determining bone mechanical properties. *J Orthop Res* 19:1021–1026.

Whyne CM (2014) Biomechanics of metastatic disease in the vertebral column. *Neurol Res* 36:493–501.

21 Spinal cord dose limits for stereotactic body radiotherapy

Ahmed Hashmi, Hiroshi Tanaka, Shun Wong,
Hany Soliman, Sten Myrehaug, Chia-Lin Tseng,
Simon S. Lo, David Larson, Arjun Sahgal, and Lijun Ma

Contents

21.1 INTRODUCTION

Spine stereotactic body radiotherapy (SBRT), also known as spine stereotactic radiosurgery (SRS), is an emerging treatment option for patients with spinal bone metastases with or without a soft tissue component, and is rapidly being adopted in the clinic albeit with limited high-quality evidence (Sahgal et al., 2008, 2011; Husain et al., 2013; Thibault et al., 2014). Done correctly, it delivers high dose to spinal metastases while minimizing dose to the spinal cord. Initially, the technique was developed for spinal metastases in the reirradiation setting (Masucci et al., 2011). More recently, it has become a treatment for de novo and postoperative spinal metastases and even for benign tumors of the spinal cord (Sahgal et al., 2007, 2011; Al-Omair et al., 2013; Thibault et al., 2014).

The spinal cord, which is typically located in close proximity to the vertebral tumor target volume, is the most critical organ at risk (OAR) to spare. The spinal cord has been classically described as an organ with a serial functional architecture, and as a result, damage to small volumes within the structure can have a major impact on neurologic function. However, recent proton-based rat spinal cord irradiation experiments suggested that the spinal cord might also have a component of parallel architecture when irradiating with inhomogeneous dose distributions, such as those inherent to spine SBRT (Bijl et al., 2002, 2003). This implies that small volumes of spinal cord may tolerate higher doses than expected to small volumes, as long as the dose exposure is minimized. However, recent experiments with pigs treated with photon-based SBRT did not support these findings, and the data would suggest that the tolerance is similar to that expected with homogeneous irradiation (Medin et al., 2010). Ultimately, pig anatomy/physiology is closer to that of humans than rats, and therefore there is no convincing data to date supporting that higher than accepted dose exposure to small volumes of cord is absolutely safe. There is undoubtedly a range of tolerance among humans and animals; however, as this OAR is most critical, then to apply a dose constraint that does not cause myelopathy (or at very low probability) as opposed to assuming some degree of risk is prudent.

Radiation-induced myelopathy is one of the most feared complications associated with SBRT given that patients can be rendered permanently paralyzed and fatal if it occurs within the cervical spinal cord (Sahgal et al., 2012, 2013; Wong et al., 2015). Specific to SBRT, the issue of spinal cord tolerance was of paramount significance as there was little experience in exposing the spinal cord to short-course high-dose-per-fraction radiation. Furthermore, radiobiology would support greater sensitivity to normal tissue late effects as the dose per fraction increases. Early adopters either believed that the spinal cord tolerance was no greater than what would be traditionally expected and applied conservative limits to small volumes, such as the point maximum, of spinal cord. However, others believed in selective volume-based dose escalation, allowing higher than expected point maximum cord doses arguing that the point maximum volume is clinically insignificant as long as larger volumes such as 0.1 or 1 cc volumes are exposed to subthreshold doses.

This chapter summarizes clinical spinal cord tolerance based on reported SBRT-induced radiation myelopathy (RM) cases, in both those previously irradiated and unirradiated (de novo).

21.2 DEFINITION AND HISTOPATHOLOGY OF RADIATION MYELOPATHY

RM is a late effect of radiotherapy delivered to the spinal cord, with clinical effects ranging from minor sensory and/or motor deficits to complete paraplegia/quadraplegia and loss of autonomic functioning. It is a diagnosis of exclusion, based on neurologic signs and symptoms consistent with damage to the irradiated segment of the spinal, without evidence of recurrent or progressive tumor affecting the spinal cord. Demyelination and necrosis of the spinal cord, typically confined to white matter, are the main histologic features of radiation-induced myelopathy, although they are not pathognomonic of radiation injury (Wong et al., 2015). Apart from white matter changes, varying degrees of vascular damage and glial reaction can often be seen. Injury of the microvasculature including disruption of the blood–spinal cord barrier (BSCB) has been implicated in the pathogenesis of RM, although vascular changes may be absent or inconspicuous histologically (Wong et al., 2015).

The underlying biologic mechanism of RM remains unclear. The prevalent model suggests that mitotic death of endothelial cells results in BSCB disruption. This leads to vasogenic edema, hypoxia, and an inflammatory cascade resulting in demyelination and necrosis. The pathobiology of RM was reviewed recently and further discussions can be found in the reference (Wong et al., 2013).

21.3 IMAGING CHARACTERISTICS

MRI is currently the most commonly used imaging tool in the diagnostic assessment of RM. As this is a rare complication, the literature consists of mainly case reports generally without histopathologic correlation (Sahgal et al., 2012, 2013). Characteristic MRI changes in the cord include areas of low signals on T1-weighted images, high signals on T2, and focal contrast enhancement (Wong et al., 2015). In the rat spinal cord, high signal intensity on T2 has been shown to correlate histopathologically with edema and confluent necrosis and enhancement postcontrast with BSCB disruption (Wong et al., 2013). Advanced quantitative MRI techniques such as apparent diffusion coefficients, magnetization transfer, and diffusion tensor imaging may provide additional information regarding structural changes after radiation, particularly in white matter after radiation.

21.4 CONTOURING THE SPINAL CORD OAR

Contouring the spinal cord in the context of SBRT is challenging, but critical for safe practice. On the one hand, if safe spinal cord dose limits (e.g., 12.4 Gy in a single fraction) are applied to a cord contour that is too generous than what is "true," one risks underdosing the epidural space, and it has been shown that progression within the epidural space is the most common pattern of failure (Sahgal et al., 2011). On the other hand, if the safe dose limit is applied to an inaccurately small cord volume than what is "true," then one risks delivering a true spinal cord dose beyond tolerance. This is why multimodal imaging is required for proper spinal cord contouring and, typically, involves thin-slice T1 and T2 axial volumetric MR images fused to the treatment planning CT and/or a CT myelogram (Sahgal et al., 2011). This way the closest contour to what is "true" may be achieved. Additionally, one must bear in mind that simple window leveling by itself can alter what is contoured as the "true" cord and highlights the challenge of spine SBRT.

Beyond contouring issues, other significant considerations include inter- and intrafraction patient motion (Ma et al., 2009; Hyde et al., 2012), motion of the spinal cord itself (Tseng et al., 2015), inaccuracy in the fusion of MR images to the treatment planning CT, and dose calculation uncertainties inherent to the treatment planning system. Although each on their own can be relatively minor, the cumulative impact of all these sources of uncertainty needs to be considered when determining what dose limit is to be applied to the cord and to what contour.

Spinal cord motion was recently assessed by Tseng et al. (2015). In 65 patients with spinal metastases planned for spine SBRT, spinal cord motion was visualized with dynamic axial and sagittal MR reconstructions. Dynamic MRI sequences were taken over a 137 second time interval. Motion of the cord consisted of two components: first, oscillatory motion, which is associated with physiological cardiorespiratory motion and cerebrospinal fluid pulsations, and second, bulk motion, which arises from random gross patient shifts and drifts. A maximum oscillatory cord motion of 0.39 mm in the anterior–posterior (AP), 0.44 mm in the lateral (LR), and 0.77 mm in the superior–inferior (SI) direction was observed, with median values of 0.16 mm in the AP, 0.17 mm in the lateral (LR), and 0.44 mm in the SI direction. The maximal bulk motion was 2.21, 2.87, and 3.90 mm AP, LR, and SI direction, and the median motion was 0.51, 0.59, and 0.66 mm AP, LR, and SI direction. This emphasizes the importance of adding a margin beyond the contoured cord on the static imaging sequences. Furthermore, support the need for rigid and reproducible immobilization, as even over a short period of 137 seconds, significant shifts were observed.

Ultimately, given the uncertainties both with respect to organ motion (Tseng et al., 2015) and intra- and interfraction patient motion (Hyde et al., 2012), several groups apply their estimated safe dose limit not to the contoured cord itself but to the cord plus a geometric margin around the cord (Foote et al., 2011). The resultant volume is known as the planning OAR volume (PRV). A margin of 1.5 mm is considered a reasonable PRV and is typically equivalent to the thecal sac contour. The thecal sac is a structure considered to be a reasonable surrogate for the true spinal cord and the approach used in the landmark spinal cord tolerance papers specific to spinal cord tolerance for spine SBRT (Sahgal et al., 2012, 2013).

21.5 SBRT-SPECIFIC SPINAL CORD TOLERANCE: RADIATION NAÏVE PATIENTS

RM has been considered a rare adverse event since the seminal work by Wong et al. (1994), as most practitioners do not surpass the recommended 45–50 Gy cord dose tolerance limit when given in 2 Gy/day fractions, At this exposure, the risk of long-term spinal cord injury has been estimated to be between 0.03% and 0.2%, and myelopathy has been considered a rare complication over the last decade.

More recently, however, RM associated with spinal SBRT has been reported (Sahgal et al., 2012, 2013). Only recently, has the much needed high-quality evidence been developed specific to the high dose per fraction and inhomogeneous radiation inherent to spine SBRT (Sahgal et al., 2010, 2012, 2013).

For de novo patients, Sahgal and colleagues performed a multi-institutional study where the dose–volume histogram (DVH) data for nine cases of RM from spinal SBRT were compared to a cohort of 66 spine SBRT patients with no RM (Sahgal et al., 2013). Notably, one patient in this series had been treated with SBRT as a boost 6 weeks after 30 Gy in 10 fractions delivered conventionally (SBRT thecal sac point maximum dose was 15 Gy). This case was included in the series, as within a 6-week time period, one would not expect sufficient repair to classify it as a re-RM case. The median follow-up intervals among patients with and without RM were 23 months (range 8–40 months) and 15 months (range 4–64 months), respectively. The median time to development of RM was 12 months (range 3–15 months).

Among the nine cases of myelopathy, it was surprising that in two patients relatively low absolute single-fraction point maximum thecal sac doses caused myelopathy, at 10.6 and 13.1 Gy. In five single-fraction cases, myelopathy was not unexpected, given single-fraction thecal sac exposures of 14.8, 15, 15.7, 16.2, and

16.5 Gy. The remaining two cases resulting in myelopathy were secondary to hypofractionated treatments with thecal sac point maximum doses of 25.6 Gy given in 2 fractions and 30.9 Gy given in 3 fractions. At these dose exposures, it is no surprise that myelopathy was observed.

Based on these nine cases of myelopathy, and the thecal sac point maximum doses at which toxicity was observed, we can conclude that when traditional dose limits within the point maximum volume of the cord are clearly exceeded, then toxicity can result. Moreover, although there are some patients that can tolerate high cord doses beyond what we would consider tolerant (Daly et al., 2011), others will still develop myelopathy despite slightly greater dose exposure to what we know based on both conventionally delivered single-fraction exposure and spine SBRT to be safe at 10 Gy (Macbeth et al., 1996).

To enable comparison of various fractionation schemes, all doses in Sahgals' report were converted to an equivalent dose in 2 Gy fractions (EQD2), also known as the normalized 2 Gy equivalent biologically effective dose (nBED) (Sahgal et al., 2012, 2013). The alpha/beta ratio of the spinal cord was assumed to be 2. The 1%, 2%, and 5% probabilities of RM based on linear regression analysis corresponded to an nBED of 25.7, 33.8, and 44.7 $Gy_{2/2}$ for the thecal sac maximum point volume (Pmax), respectively (Sahgal et al., 2013). Table 21.1 summarizes the estimated absolute dose thresholds specific to the thecal sac Pmax for 1–5 fractions resulting in a 5% or less risk of RM.

Furthermore, the study attempted to determine a volume effect by comparing the thecal sac nBED in the controls to the RM cases, based on the dose exposure to a series of thecal sac volumes irradiated. The median and mean nBED for the point maximum volume up to 1 cc, in 0.1 cc increments, and the 2 cc volume were compared. Statistically significant differences were observed between the doses in the non-RM and RM cohorts at the Pmax and up to the 0.8 cc volumes. This result was interpreted as indirect evidence to support the recommendation that small volumes of cord irradiated should be an important consideration. Furthermore, given that the most significant difference was observed at the point maximum volume comparison, applying the constraint on the Pmax thecal sac volume was recommended.

With respect to whether or not there is partial volume tolerance, when comparing the mean nBED at each volume in the no-RM and RM cohorts, interesting observations were made. First, in the no-RM cohort, the greatest mean nBED (or EQD2) observed was at the Pmax volume at ~39 $Gy_{2/2}$, which is conservative for cord tolerance. Subsequent volumes were exposed to even lower doses. Therefore, we cannot derive any partial volume tolerance inference, as the tolerance was not exceeded even at the smallest volumes in that cohort. In the RM cohort, the greatest mean nBED (or EQD2) observed was at the Pmax volume at ~70 $Gy_{2/2}$, which does exceed traditional limits considering conventional fractionated radiotherapy. However, even by the 0.1 cc volume, the nBED was within what we would consider low risk for RM at <60 $Gy_{2/2}$. Therefore, we can simply state that dose to small volumes of spinal cord such as the Pmax should be respected, and we cannot infer that larger volumes, such as 0.1 or 0.2 cc, can tolerate greater dose exposure than what is considered safe.

One of the issues that has been a source of research and modeling is how to account for the inhomogeneity of the SBRT dose distribution (Ma et al., 2007). This concept of accounting for the gradient of dose exposure within the organ irradiated gave rise to the generalized biologically effective

Table 21.1 Predicted maximum point dose thecal sac volume absolute doses for 1–5 fractional spine stereotactic body radiotherapy that results in less than 1%–5% probability of radiation myelopathy

COMPLICATION PROBABILITY (%)	1-FRACTION Pmax LIMIT (Gy)	2-FRACTION Pmax LIMIT (Gy)	3-FRACTION Pmax LIMIT (Gy)	4-FRACTION Pmax LIMIT (Gy)	5-FRACTION Pmax LIMIT (Gy)
1	9.2	12.5	14.8	16.7	18.2
3	11.5	15.7	18.8	21.2	23.1
5	12.4	17.0	20.3	23.0	25.3

Source: Sahgal, A. et al., *Int. J. Radiat. Oncol. Biol. Phys.*, 85, 341, 2013.
Pmax, maximum point dose.

dose (gBED) equation, which was specifically developed for spine SBRT (Ma et al., 2007). From the initial publication by Sahgal et al. (2010), where five patients with RM were compared to a cohort of 19 controls, the same analysis was performed but with the gBED (Ma et al., 2007). Although the doses at which safe dose limits were recommended did not change, a significant difference in the mean point max and 0.1 cc volumes were observed (as compared to only the Pmax volume in that publication) between the no-RM and RM cohorts. This implies that there may be a better method to understand the dosimetric data by accounting for the full-dose distribution into the calculation, but this remains investigational.

21.6 SBRT SPINAL CORD TOLERANCE: REIRRADIATION SETTING

As spinal SBRT is increasingly established as the treatment of choice in the reirradiation of spinal metastases, the need for dose limits to the previously irradiated spinal cord became an urgent area of research. Sahgal and colleagues reported a DVH analysis of five cases of RM following reirradiation spinal SBRT and compared the DVH data with 16 retreatment spinal SBRT controls treated at University of California, San Francisco (Sahgal et al., 2012).

The thecal sac cumulative nBED was calculated by adding the nBED of the first course of conventional radiotherapy to the Pmax nBED from the retreatment SBRT course, assuming the alpha/beta ratio of the spinal cord was 2 Gy (Sahgal et al., 2012). The authors concluded that the cumulative nBED to the thecal sac Pmax should not exceed 70 $Gy_{2/2}$ based on the provision that the SBRT thecal sac retreatment Pmax nBED does not exceed 25 $Gy_{2/2}$, that the thecal sac SBRT Pmax nBED/cumulative Pmax nBED ratio does not exceed 0.5, and a minimum time interval to retreatment of 5 months is respected (Sahgal et al., 2012). Based on these principles, a set of guidelines were reported and summarized in Table 21.2. These doses have been strictly followed at the University of Toronto with several hundred cases and no RM observed (Dr. Arjun Sahgal, personal communication). Importantly, these guidelines are applicable to prior nBED radiation exposure ranging from 30 to 50 $Gy_{2/2}$. This is equivalent to prior conventional EBRT regimens such as 20 Gy/5, 30 Gy/10, 40 Gy/20, and 50 Gy/25. Of note, there was no recommendation for single-fraction retreatment spine SBRT, beyond an initial exposure of 42 $Gy_{2/2}$ due to the lack of data.

The tolerance levels in both Tables 21.1 and 21.2 represent what is thought to be safe. There is a range of tolerance for any OAR, such that some patient may indeed tolerate far greater spinal cord doses than what is recommended. Until we have predictive tests to determine who those patients are, caution is warranted when surpassing these evidence-based limits.

Table 21.2 Recommended reirradiation spine stereotactic body radiotherapy doses to the thecal sac maximum point volume following common initial conventional radiotherapy regimens

PRIOR CONVENTIONAL RADIOTHERAPY FRACTIONATIONS	PRIOR CONVENTIONAL RADIOTHERAPY nBED ($Gy_{2/2}$)	1-fx SBRT (Gy)	2-fx SBRT (Gy)	3-fx SBRT (Gy)	4-fx SBRT (Gy)	5-fx SBRT (Gy)
20 Gy/5 fx	30	9	12.2	14.5	16.2	18
30 Gy/10 fx	37.5	9	12.2	14.5	16.2	18
40 Gy/20 fx	40	N/A	12.2	14.5	16.2	18
45 Gy/25 fx	43	N/A	12.2	14.5	16.2	18
50 Gy/25 fx	50	N/A	11	12.5	14	15.5

Source: Sahgal, A. et al., Int. J. Radiat. Oncol. Biol. Phys., 82, 107, 2012.
Abbreviations: N/A, not applicable; fx, fraction; nBED, normalized biologically effective dose; SBRT, stereotactic body radiotherapy.

21.7 LIMITATIONS OF THE LQ MODEL IN ASSESSING SPINAL CORD TOLERANCE

The guidelines by Sahgal et al. (2012, 2013) were based on linear-quadratic (LQ) modeling to convert the various dose–fractionation regimens into the nBED (EQD2), as it represents the most commonly used and simplest model to apply in the clinic with the least number of assumptions. However, the LQ model has recently been questioned (Park et al., 2008) as to its ability to accurately estimate the BED in the ablative dose range (>15 Gy/fraction), as used in SBRT. Just as the gBED was developed, so have other models to better account for the effects of extreme hypofractionated dose-per-fraction SBRT on the target tumor tissue. Wang and colleagues proposed a generalized LQ (gLQ) model, which provides a natural extension of the LQ model across the entire dose range (Wang et al., 2010). This model has been independently validated for tumor response in animal and in vitro studies by the group from Thomas Jefferson University (Ohri et al., 2012). Nevertheless, it had not been used to model toxicities in normal tissues until Huang et al. recently reanalyzed the published data from the study on spinal cord tolerance for reirradiation with SBRT by Sahgal et al., using the gLQ model (Huang et al., 2013). It was also determined that when the cumulative Pmax nBED to the thecal sac was ≤70 $Gy_{2/2}$, the incidence of RM was zero. Given the limited clinical data available, the gLQ model must be approached with caution and clinical validation is necessary.

21.8 CONCLUSION

Spine SBRT is increasingly adopted into mainstream radiation oncology practice. Robust evidence-based guidelines with respect to treatment planning, contouring, and OAR dose limits are of paramount importance. The data summarized in this chapter represent a detailed analysis of spinal cord tolerance for spine SBRT and can guide radiation oncologists with respect to safe practice.

CHECKLIST: KEY POINTS FOR CLINICAL PRACTICE

- The radiobiological architecture of the spinal cord at this time is to be considered as an organ functioned in series
- Dose to the point maximum spinal cord PRV or thecal sac volume should be respected
- Point max thecal sac dose limits for de novo spinal SBRT driven by evidence-based guidelines from Sahgal et al. (2013) consist of 12.4 Gy/1 fraction, 17 Gy/2 fractions, 20.3 Gy/3 fractions, 23 Gy/4 fractions, and 25.3 Gy/5 fractions
- Point max thecal sac dose limits for typical reirradiation spinal SBRT driven by evidence-based guidelines from Sahgal et al. (2012) consist of 9 Gy/1 fraction, 12.2 Gy/2 fractions, 14.5 Gy/3 fractions, 16.2 Gy/4 fractions, and 18 Gy/5 fractions. Note this varies according to the amount of prior exposure

REFERENCES

Al-Omair A, Masucci L, Masson-Cote L, Campbell M, Atenafu EG, Parent A, Letourneau D et al. (2013) Surgical resection of epidural disease improves local control following postoperative spine stereotactic body radiotherapy. *Neuro-Oncology* 15:1413–1419.

Bijl HP, van Luijk P, Coppes RP, Schippers JM, Konings AW, van der Kogel AJ (2002) Dose-volume effects in the rat cervical spinal cord after proton irradiation. *Int J Radiat Oncol Biol Phys* 52:205–211.

Bijl HP, van Luijk P, Coppes RP, Schippers JM, Konings AW, van der Kogel AJ (2003) Unexpected changes of rat cervical spinal cord tolerance caused by inhomogeneous dose distributions. *Int J Radiat Oncol Biol Phys* 57:274–281.

Daly ME, Choi CY, Gibbs IC, Adler JR, Jr., Chang SD, Lieberson RE, Soltys SG (2011) Tolerance of the spinal cord to stereotactic radiosurgery: Insights from hemangioblastomas. *Int J Radiat Oncol Biol Phys* 80:213–220.

Foote M, Letourneau D, Hyde D, Massicotte E, Rampersaud R, Fehlings M, Fisher C et al. (2011) Technique for stereotactic body radiotherapy for spinal metastases. *J Clin Neurosci* 18:276–279.

Huang Z, Mayr NA, Yuh WT, Wang JZ, Lo SS (2013) Reirradiation with stereotactic body radiotherapy: Analysis of human spinal cord tolerance using the generalized linear-quadratic model. *Future Oncol* 9:879–887.

Husain ZA, Thibault I, Letourneau D, Ma L, Keller H, Suh J, Chiang V et al. (2013) Stereotactic body radiotherapy: A new paradigm in the management of spinal metastases. *CNS Oncol* 2:259–270.

Hyde D, Lochray F, Korol R, Davidson M, Wong CS, Ma L, Sahgal A (2012) Spine Stereotactic body radiotherapy utilizing cone-beam CT image-guidance with a robotic couch: Intrafraction motion analysis accounting for all six degrees of freedom. *Int J Radiat Oncol Biol Phys* 82:e555–e562.

Ma L, Sahgal A, Hossain S, Chuang C, Descovich M, Huang K Gottschalk A, Larson DA (2009) Nonrandom intrafraction target motions and general strategy for correction of spine stereotactic body radiotherapy. *Int J Radiat Oncol Biol Phys* 75:1261–1265.

Ma L, Sahgal A, Larson D, Chuang C, Verhey L (2007) A generalized biologically effective dose model and its application for radiosurgery and hypofractionated body radiotherapy. *Med Phys* 34:2403.

Macbeth FR, Wheldon TE, Girling DJ, Stephens RJ, Machin D, Bleehen NM, Lamont A et al. (1996) Radiation myelopathy: Estimates of risk in 1048 patients in three randomized trials of palliative radiotherapy for non-small cell lung cancer. The Medical Research Council Lung Cancer Working Party. *Clin Oncol (R Coll Radiol)* 8:176–181.

Masucci GL, Yu E, Ma L, Chang EL, Letourneau D, Lo S, Leung E et al. (2011) Stereotactic body radiotherapy is an effective treatment in reirradiating spinal metastases: Current status and practical considerations for safe practice. *Exp Rev Anticancer Ther* 11:1923–1933.

Medin PM, Foster RD, van der Kogel AJ, Sayre JW, McBride WH, Solberg TD (2010) Spinal cord tolerance to single-fraction partial-volume irradiation: A swine model. *Int J Radiat Oncol Biol Phys* 79:226–232.

Ohri N, Dicker AP, Lawrence YR (2012) Can drugs enhance hypofractionated radiotherapy? A novel method of modeling radiosensitization using in vitro data. *Int J Radiat Oncol Biol Phys* 83:385–393.

Park C, Papiez L, Zhang S, Story M, Timmerman RD (2008) Universal survival curve and single fraction equivalent dose: Useful tools in understanding potency of ablative radiotherapy. *Int J Radiat Oncol Biol Phys* 70:847–852.

Sahgal A, Bilsky M, Chang EL, Ma L, Yamada Y, Rhines LD, Létourneau D et al. (2011) Stereotactic body radiotherapy for spinal metastases: Current status, with a focus on its application in the postoperative patient. *J Neurosurg Spine* 14:151–166.

Sahgal A, Chou D, Ames C, Ma L, Lamborn K, Huang K, Chuang C et al. (2007) Image-guided robotic stereotactic body radiotherapy for benign spinal tumors: The University of California San Francisco preliminary experience. *Technol Cancer Res Treat* 6:595–604.

Sahgal A, Larson DA, Chang EL (2008) Stereotactic body radiosurgery for spinal metastases: A critical review. *Int J Radiat Oncol Biol Phys* 71:652–665.

Sahgal A, Ma L, Gibbs I, Gerszten PC, Ryu S, Soltys S, Weinberg V et al. (2010) Spinal cord tolerance for stereotactic body radiotherapy. *Int J Radiat Oncol Biol Phys* 77:548–553.

Sahgal A, Ma L, Weinberg V, Gibbs IC, Chao S, Chang UK, Werner-Wasik M et al. (2012) Reirradiation human spinal cord tolerance for stereotactic body radiotherapy. *Int J Radiat Oncol Biol Phys* 82:107–116.

Sahgal A, Weinberg V, Ma L, Chang E, Chao S, Muacevic A, Gorgulho A et al. (2013) Probabilities of radiation myelopathy specific to stereotactic body radiation therapy to guide safe practice. *Int J Radiat Oncol Biol Phys* 85:341–347.

Thibault I, Al-Omair A, Masucci GL, Masson-Cote L, Lochray F, Korol R, Cheng L et al. (2014) Spine stereotactic body radiotherapy for renal cell cancer spinal metastases: Analysis of outcomes and risk of vertebral compression fracture. *J Neurosurg Spine* 21:711–718.

Tseng CL, Sussman MS, Atenafu EG, Letourneau D, Ma L, Soliman H, Thibault I et al. (2015) Magnetic resonance imaging assessment of spinal cord and cauda equina motion in supine patients with spinal metastases planned for spine stereotactic body radiation therapy. *Int J Radiat Oncol Biol Phys* 91:995–1002.

Wang JZ, Huang Z, Lo SS, Yuh WT, Mayr NA (2010) A generalized linear-quadratic model for radiosurgery, stereotactic body radiation therapy, and high-dose rate brachytherapy. *Sci Transl Med* 2:39ra48.

Wong CS, Fehlings MG, Sahgal A (2015) Pathobiology of radiation myelopathy and strategies to mitigate injury. *Spinal Cord* 53(8):574–580.

Wong CS, Van Dyk J, Milosevic M, Laperriere NJ (1994) Radiation myelopathy following single courses of radiotherapy and retreatment. *Int J Radiat Oncol Biol Phys* 30:575–581.

Wong K, Zeng L, Zhang L, Bedard G, Wong E, Tsao M, Barnes E et al. (2013) Minimal clinically important differences in the brief pain inventory in patients with bone metastases. *Support Care Cancer* 21:1893–1899.

22 Summary of image-guided hypofractionated stereotactic radiotherapy: Serious late toxicities and strategies to mitigate risk

Simon S. Lo, Kristin J. Redmond, Nina A. Mayr, William T. Yuh,
Zhibin Huang, Eric L. Chang, Bin S. Teh, and Arjun Sahgal

Contents

22.1 INTRODUCTION

Serious late complications are one of the most feared aspects of image-guided hypofractionated stereotactic radiotherapy (IG-HSRT). Albeit their relatively rare occurrence with modern technology and the experience gained in these procedures throughout these years, they have been observed and reported. Fortunately, many of these serious complications can be prevented by using appropriate strategies. This chapter will summarize the serious late complications associated with intracranial and spinal IG-HSRT and the strategies to mitigate the risk. In this chapter, the term "spinal HSRT" is used interchangeably with "spinal SBRT." Details pertaining to each disease condition have been discussed in previous chapters.

22.2 SERIOUS LATE TOXICITIES FROM INTRACRANIAL IG-HSRT AND STRATEGIES TO MITIGATE RISK

This section will cover the most important complications, including optic neuropathy and other cranial nerve injuries, vascular injury, brain stem injury, and radiation necrosis.

22.2.1 OPTIC NEUROPATHY

Optic neuropathy is one of the most feared complications from stereotactic radiosurgery (SRS) or HSRT. Overall, the incidence of optic neuropathy after SRS is low (Stafford et al., 2003; Leavitt et al., 2013; Pollock et al., 2014). The most commonly used constraint for the optic apparatus is 8 Gy (Tishler et al., 1993), but recent data from Mayo Clinic showed that the risk of developing a clinically significant radiation-induced optic neuropathy was 1.1% for patients receiving a single maximum point dose of 12 Gy or less to the optic apparatus (Stafford et al., 2003). Overall, radiation-induced optic neuropathy developed in less than 2% of patients, despite that 73% received >8 Gy to a short segment of the optic apparatus. In a follow-up study, colleagues from Mayo Clinic found that the risk of radiation-induced optic neuropathy was 0%, 0%, 0%, and 10% when the maximum radiation doses received by the anterior visual pathway were ≤8, 8.1–10.0, 10.1–12.0, and >12 Gy, respectively (Leavitt et al., 2013). The overall risk of radiation-induced optic neuropathy in patients receiving >8 Gy to the anterior visual pathway was 1.0% (Leavitt et al., 2013). In another series from Mayo Clinic where 133 patients (266 sides) with pituitary adenomas were treated with SRS, no optic neuropathy was observed at a median follow-up of 32 months when the maximum point dose to the optic apparatus did not exceed 12 Gy (Pollock et al., 2014).

Data on the optic nerve tolerance for other hypofractionated regimens other than a single fraction are limited. In a study from Japan, a dose of 24 Gy or less in 6 fractions appeared to be safe for the optic apparatus (Kanesaka et al., 2011). In another study from Japan, grade 2 visual disorder was observed in one patient who received 20.8 and 20.7 Gy in 3 fractions to the optic nerve and chiasm, respectively (Iwata et al., 2011). Iwata et al. reported a 1% risk of optic neuropathy using either 17–21 Gy in 3 fractions or 22–25 Gy in 5 fractions for HSRT for pituitary adenomas. In a study from Italy where 25 patients with perioptic meningioma were treated with CyberKnife-based HSRT using regimens including 18 Gy in 2 fractions, 18–21 Gy in 3 fractions, 20–22 Gy in 4 fractions, and 23–25 Gy in 5 fractions, none of the patients developed visual deterioration with a median follow-up of 57.5 months (Conti et al., 2015). The optic pathway constraints used were 10 Gy in 2 fractions, 15 Gy in 3 fractions, 20 Gy in 4 fractions, and 25 Gy in 5 fractions. Using the same constraints for 2–5 fractions, a further group of 39 patients was treated prospectively with HSRT regimens including 18 Gy in 2 fractions, 18–21 Gy in 3 fractions, 20–22 Gy in 4 fractions, 25 Gy in 5 fractions, 27.5 Gy in 6 fractions, 30 Gy in 9 fractions, 34 Gy in 10 fractions, and 40 Gy in 15 fractions. With a median follow-up of 15 months, no visual toxicities were observed (Conti et al., 2015). In a study from Barrow Neurological Institute, 20 patients with pituitary adenomas were treated with CyberKnife-based HSRT with the optic pathway constraint set at 25 Gy in 5 fractions. With a median follow-up of 26.6 months, none of the patients developed visual deficits (Killory et al., 2009). The authors concluded that the tolerance of optic pathway was 25 Gy in 5 fractions (Killory et al., 2009). In a study from University of Virginia, 15 patients with meningioma, pituitary adenoma, or pilocytic astrocytoma were treated with Gamma Knife–based HSRT using a relocatable system (Nguyen et al., 2014). The doses to optic pathway were tracked. The maximum doses delivered to the optic pathway were 3.6–14.4 Gy in 3 fractions, 2.8–22.8 Gy in 4 fractions, and 5–24.5 Gy in 5 fractions. With a median follow-up of 13.8 months (range, 4–44.3 months), no visual toxicities were observed (Nguyen et al., 2014).

22.2.2 III, IV, V, AND VI CRANIAL NERVE PALSIES

The situations in which the III, IV, and VI cranial nerves will be exposed to significant doses of radiation are when cavernous sinus meningiomas and pituitary adenomas are treated with SRS or HSRT. Data in the literature on SRS for skull base meningiomas seem to suggest that the III, IV, and VI nerves are able to tolerate a high single dose of radiation (Witt, 2003). Witt from Indiana University reviewed the data on 1255 patients who underwent SRS for pituitary adenoma with the marginal dose delivered ranging from 14 to 34 Gy in 1 fraction. Given the proximity of the sella to the cavernous sinus and the high doses used to treat pituitary adenomas, especially secretory tumors, the cranial nerves in the cavernous sinus can potentially be exposed to very high radiation doses. However, the overall incidence of permanent III, IV, or VI neuropathy was 0.4% (Witt, 2003). In a study from Israel, 102 patients with cavernous sinus meningiomas were treated with linear accelerator (LINAC)-based SRS to a dose of 12–17.5 Gy in 1 fraction; the incidence of VI nerve palsy was <2% (Spiegelmann et al., 2010). For the V nerve, the incidence of neuropathy was 0.2% based on the analysis of patients with pituitary adenoma treated with SRS, which can potentially expose the V nerve to a high single dose of radiation (Witt, 2003). In the study from Israel mentioned earlier, the incidence of V nerve injury was <2% (Spiegelmann et al., 2010). Other clinical scenarios where the V nerve can be exposed to a substantial level of radiation are when SRS is used for the treatment of vestibular schwannoma and trigeminal neuralgia. In the early series of SRS for vestibular schwannoma from University of Pittsburgh Medical Center (UPMC) where a higher dose of 14–20 Gy was used, the incidence of V nerve deficit was 16% (Kondziolka et al., 1998). Other series using a dose of 12–13 Gy, including a later series from UPMC, reported much lower V nerve–deficit rates of 2%–8% (Petit et al., 2001; Flickinger et al., 2004; Murphy and Suh, 2011). Based on the data from retrospective series, the risks of III, IV, V, and VI cranial nerve palsies are in general very low with SRS.

Much less data are available for HSRT. Wang and colleagues treated 14 patients with cavernous sinus hemangiomas with HSRT to a dose of 21 Gy in 3 fractions in a phase II trial. Among the six patients with cranial nerve deficits, they either showed complete recovery or improvement of function (Wang et al., 2012). Other patients did not experience cranial nerve palsies. In a study from Johns Hopkins University where vestibular schwannomas were treated with HSRT to 25 Gy in 5 fractions, the incidence of V nerve deficit was 7% (Song and Williams, 1999). In another study from Japan, HSRT for vestibular schwannoma delivering 18–21 Gy in 3 fractions or 25 Gy in 5 fractions resulted in a zero incidence of V nerve deficit at a median follow-up of 80 months (Morimoto et al., 2013). In the Georgetown University series of HSRT for vestibular schwannoma delivering 25 Gy in 5 fractions, the rate of V nerve injury was 5.5% (Karam et al., 2013).

22.2.3 VII AND VIII CRANIAL NERVE PALSIES

Most of the data on toxicities of VII and VIII cranial nerve palsies came from SRS series for vestibular schwannoma. In an early series reported by Kondziolka et al. where a dose of 14–20 Gy in 1 fraction was used for Gamma Knife SRS, the hearing retention rate was only 47% and the risk of VII nerve injury was 15% (Kondziolka et al., 1998). In most recent studies where a dose of 12–13 Gy in 1 fraction was used, the incidence of retention of serviceable hearing ranged from 44% to 88% with Gamma Knife– or LINAC-based SRS, while local control was not jeopardized (Petit et al., 2001; Flickinger et al., 2004; Murphy and Suh, 2011). The risk of VII nerve deficit was also observed to be <5% (Petit et al., 2001; Flickinger et al., 2004; Murphy and Suh, 2011). Interestingly, the hearing preservation rates were lower at ~30% when proton-based SRS was used even if a low-dose regimen of 12 Cobalt Gray equivalent (CGE) was used (Murphy and Suh, 2011).

Less data are available with VII and VIII nerve deficits from HSRT. Song from Johns Hopkins University published his experience with the use of HSRT for vestibular schwannoma using a regimen of 25 Gy in 5 fractions. With relatively short follow-up times of 6–44 months, the hearing preservation rate was 75% and the rate of VII nerve deficits was 0% (Song and Williams, 1999). In a study from Japan where 25 patients with 26 vestibular schwannomas were treated with CyberKnife-based HSRT, using regimens of 18–21 Gy in 3 fractions or 25 Gy in 5 fractions, the overall VII and VIII nerve preservation rates were 92% and 50%, respectively (Morimoto et al., 2013). In a study from Taiwan, treatment with 18 Gy in 3 fractions was associated with a serviceable hearing retention rate of 81.5% at a mean follow-up of 61.1 months (Tsai et al., 2013). Using the same regimen of 18 Gy in 3 fractions, colleagues from UPMC reported a serviceable hearing retention rate of 53.5% (Vivas et al., 2014). Colleagues from Georgetown University used mainly the regimen of 25 Gy in 5 fractions and observed a hearing preservation rate of 73% at 5 years and a zero incidence of VII nerve palsy (Karam et al., 2013). A Dutch study comparing SRS and HSRT for vestibular schwannoma showed that with regimens of 20–25 Gy in 5 fractions, the rates of preservation of VII and VIII nerve function were 97% and 61%, respectively (Meijer et al., 2003).

22.2.4 IX, X, XI, AND XII CRANIAL NERVE PALSIES

The toxicity data on IX, X, XI, and XII cranial nerves are mainly extracted from studies on SRS for glomus tumors, which are frequently located close to the internal jugular vein. There have been two meta-analyses examining the outcomes of SRS for glomus tumors (Guss et al., 2011; Ivan et al., 2011). In the meta-analysis from University of California, San Francisco (UCSF), among 339 patients treated with SRS alone, the incidence of IX, X, XI, and XII nerve deficits was 9.7%, 9.7%, 12%, and 8.7%, respectively (Ivan et al., 2011). However, no dosimetric data were available. The meta-analysis from Johns Hopkins University did not report any toxicity data (Guss et al., 2011). The prescribed dose ranged from 12 to 20.4 Gy (median, 15.1 Gy) in 1 fraction.

Data are even more limited on IX, X, XI, and XII nerve toxicities caused by HSRT. In a study from University of Texas Southwestern Medical Center, 31 patients with skull base glomus tumors were treated with HSRT to a dose of 25 Gy in 5 fractions, and no grade 3 or worse toxicities were observed with a median follow-up of 24 months (Chun et al., 2014).

22.2.5 VASCULAR INJURY

The circumstances in which major vessels such as internal carotid arteries will be exposed to high doses of radiation are when cavernous sinus meningiomas and pituitary adenomas are treated with SRS or HSRT. Witt from Indiana University did a systematic review on SRS for pituitary adenomas and found that among the 1255 patients evaluated in the studies, there were only three cases of internal carotid artery occlusion or stenosis (Witt, 2003). In one case, the dose to the internal carotid artery was estimated to be less than 20 Gy in 1 fraction. It was recommended that <50% of the circumference should get the prescribed dose, and the internal carotid artery dose should be kept below 30 Gy in 1 fraction (Witt, 2003).

Pertaining to HSRT, among the studies on pituitary adenoma or cavernous sinus tumors, no vascular injury was observed when regimens of 21 Gy in 3 fractions or 25 Gy in 5 fractions were used (Killory et al., 2009; Wang et al., 2012; Nguyen et al., 2014).

22.2.6 BRAIN STEM INJURY

While the brain stem tolerance to conventional radiotherapy is quite well established, there is much less data on the brain stem tolerance to SRS and HSRT. Mayo and colleagues published a comprehensive review on radiation-induced brain stem injury as part of the Quantitative Analyses of Normal Tissue Effects in the Clinic (QUANTEC) project in 2010 (Mayo et al., 2010b). Based on literature review, the authors have concluded that a maximum brain stem dose of 12.5 Gy in 1 fraction is associated with a <5% risk of brain stem injury (Mayo et al., 2010b). There have been studies on SRS delivering doses up to 15–20 Gy to the brain stem without very low incidence of complication (Mayo et al., 2010b). This phenomenon may be explained in patients with brain stem metastases treated with SRS, where the survival is expected to be too short to develop complications.

Data on brain stem injury from HSRT mainly derive from studies of HSRT for skull base tumors, mainly vestibular schwannoma. An early seminal study on HSRT for benign and malignant tumors at various locations from McGill University, Canada, utilizing a regimen of 42 Gy in 6 fractions observed late and serious brain stem complications in 4 of 77 patients treated (Clark et al., 1998). Based on the nonconformal isodose distribution shown in the publication, it was expected that a relatively large volume of the brain stem would be included in the prescribed isodose line (Clark et al., 1998). While the delivery of HSRT using dynamic rotation technique was regarded as state of the art at that time, with the quantum leap of technology, a highly conformal isodose distribution around the tumor volume is expected. As a result, the data presented in this Canadian study may not apply to modern HSRT.

In the phase II study of HSRT for cavernous sinus hemangiomas by Wang and colleagues, no brain stem injury was observed with a brain stem dose of 19.8 Gy (range, 12.4–22.8 Gy) in 3 fractions (Wang et al., 2012). In a Japanese study, no brain stem injury was observed when the maximum brain stem dose was limited to 35 Gy in 5 fractions or 27 Gy in 3 fractions (Morimoto et al., 2013). Karam et al. from Georgetown University reported zero incidence of brain stem injury when 25 Gy in 5 fractions or 21 Gy in 3 fractions was used as the HSRT regimen for vestibular schwannoma (Karam et al., 2013). Dosimetric details were not available. In the study from Johns Hopkins where a dose of 25 Gy in 5 fractions was used, no brain stem injury was observed (Song and Williams, 1999).

22.2.7 RADIATION NECROSIS

Both intracranial SRS and HSRT deliver ablative doses of radiation to the target volume as well as the brain parenchyma immediately adjacent to it, and hence predisposing the treated volume to a risk of radiation necrosis. Colleagues from UPMC first developed a model for predicting symptomatic radiation necrosis from SRS for arteriovenous malformation (AVM). They have determined that the AVM location and the volume of tissue receiving ≥12 Gy (12 Gy volume) were the best predictors and have used those parameters to construct a significant postradiosurgery injury expression (SPIE) score, which was directly proportional to the risk of symptomatic necrosis (Flickinger et al., 2000). Locations of AVM in ascending order of risk and SPIE score (0–10) were frontal lobe, temporal lobe, intraventricular area, parietal lobe, cerebellum, corpus callosum, occipital lobe, medulla, thalamus, basal ganglia, and pons/midbrain (Flickinger et al., 2000). The 12 Gy volume has since become the standard parameter to predict symptomatic necrosis for SRS for AVM. However, it was unclear whether this would apply to intracranial tumors. Korytko et al. from Case Western Reserve University did a retrospective review on their 129 patients with 198 non-AVM tumors treated with Gamma Knife-based SRS and found that, as in AVM, the 12 Gy volume correlated with risk of symptomatic necrosis (Figure 22.1). The risk was 23%, 20%, 54%, and 57% when the 12 Gy volume was 0–5, 5–10, 10–15, and >15 cc, respectively. The 12 Gy volume was not predictive of the risk of asymptomatic necrosis (Korytko et al., 2006; Figure 22.1).

The data on radiation necrosis from intracranial HSRT are even more limited. Studies with dosimetric correlation are lacking. In a study from Sweden where 56 patients with AVMs were treated with HSRT to 30–32.5 Gy in 5 fractions or 35 Gy in 5 fractions, all patients who developed symptomatic radiation necrosis received 35 Gy in 5 fractions. The target volumes ranged from 1.5 to 29 cc (Lindvall et al., 2010). Inoue et al. from Japan have attempted to estimate the risk of symptomatic necrosis in 78 patients with 85 large brain metastases treated with 5-fraction HSRT. The surrounding brain volumes encompassed by the 28.8 Gy isodose line (single-dose equivalent to 14 Gy or V14) were measured to evaluate the risk of radiation necrosis (Inoue et al., 2014). The risk of brain necrosis increased in patients long-term survival when V14 ≥ 7.0 cc and none of the patients with a V14 of <7.0 cc experienced brain necrosis requiring surgical intervention (Inoue et al., 2014).

22.2.8 STRATEGIES TO MITIGATE RISKS OF COMPLICATIONS FROM INTRACRANIAL IG-HSRT

22.2.8.1 Cranial nerve injury

For patients undergoing intracranial SRS or HSRT for targets close to the optic apparatus, it is crucial to have the optic nerves, optic chiasm, and optic tracks contoured to track the doses and to ascertain that the tolerance is not exceeded. It is prudent to limit the volume of the optic apparatus receiving the tolerance dose. If a Gamma Knife unit is used for SRS, several maneuvers can be used to decrease the dose to the optic apparatus. Witt (2003) has described that by orienting the stereotactic frame parallel to the long axis of the optic nerves and chiasm, the anteroposterior axis of the peripheral isodose curves will then be parallel to the optic apparatus in the sagittal plane, and the extremely steep falloff of radiation in the craniocaudal direction can be exploited. The use of multiple 4 mm shots can also steepen the dose gradient, potentially decreasing the dose delivered to the optic apparatus. The automatic beam-shaping feature of the Perfexion model can also facilitate sparing of the optic apparatus. If a LINAC-based or robotic system is used, the use of inverse treatment planning can steer the dose away from the optic structures, potentially decreasing the risk of complications.

Based on the data in the literature, there is adequate data to suggest that a maximum point dose of 12 Gy in 1 fraction is safe for the optic apparatus (Stafford et al., 2003; Leavitt et al., 2013; Pollock et al., 2014). The QUANTEC has estimated that the tolerance of the optic apparatus is 8–12 Gy in 1 fraction but has not made recommendations for 2–5 fractions (Mayo et al., 2010a). There is a fair amount of data from retrospective studies from multiple radiosurgery centers suggesting a maximum point dose of 25 Gy in 5 fractions should

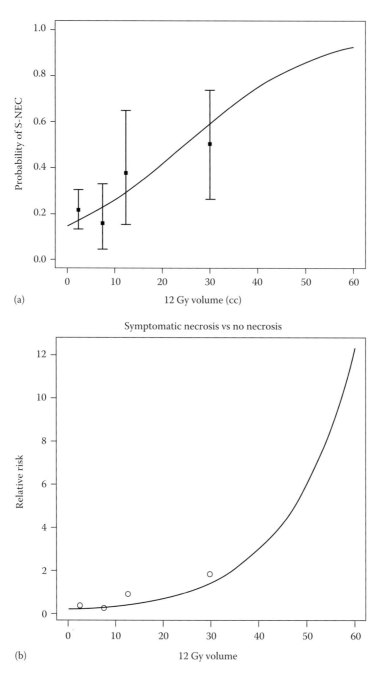

Figure 22.1 (a) Percentage of symptomatic radiation necrosis (S-NEC) versus 12 Gy volume. Logistic regression of the probability of S-NEC versus 12 Gy volume. The model was fitted with 12 Gy volume as the sole covariate. The error bars are Clopper–Pearson 95% confidence interval for the proportion of patients with S-NEC. (b) Relative risk of S-NEC versus 12 Gy volume. (Reprinted from *Int. J. Radiat. Oncol. Biol. Phys.*, 64 (2), Korytko, T., Radivoyevitch, T., Colussi, V., Wessels, B.W., Pillai, K., Maciunas, R.J., and Einstein, D.B., 12 Gy gamma knife radiosurgical volume is a predictor for radiation necrosis in non-AVM intracranial tumors, pp. 419–424, Copyright 2006, with permission from Elsevier.)

be safe for the optic apparatus (Killory et al., 2009; Iwata et al., 2011; Nguyen et al., 2014; Conti et al., 2015). Much less data are available for 2–4 fractions. Although it is tempting for one to extrapolate toxicity data from conventional fractionation, the derived tolerance is not rigorously tested in a clinical setting.

Studies of SRS and HSRT for cavernous sinus tumors and pituitary adenomas seem to suggest that the III, IV, V, and VI nerves are quite resistant to ablative doses of radiation (Witt, 2003; Spiegelmann et al., 2010). However, dosimetric analysis is lacking in those studies, and it is unclear exactly how much radiation is being delivered to those cranial nerves, which can be difficult to visualize even on magnetic resonance imaging (MRI). When the cavernous sinus is invaded by a meningioma or a macroadenoma, the III, IV, V, and VI nerves are likely to be included in the tumor volume to be treated and are likely receiving at least the prescribed dose. Based on the doses used in various studies, it appears that for SRS, the usual doses of 14–34 Gy in 1 fraction used for pituitary adenoma and 12–18 Gy in 1 fraction used for cavernous sinus meningioma are expected to result in a low risk of III, IV, and VI nerve damage (Witt, 2003; Spiegelmann et al., 2010). One note of caution is that although a single dose of up to 34 Gy has been used for SRS for pituitary adenomas, this regimen is only used in secretory microadenomas, and the doses delivered to the III, IV, V, and VI nerves are likely to be much lower. For HSRT, the use of regimens of 21 Gy in 3 fractions and 25 Gy in 5 fractions appears to be safe for the III, IV, V, and VI nerves, although the exact tolerances of those nerves to HSRT are unknown (Killory et al., 2009; Wang et al., 2012; Karam et al., 2013; Morimoto et al., 2013). Given the lack of information regarding the exact tolerance of the III, IV, V, and VI nerves to SRS and HSRT, it is imperative that every effort should be made to minimize high-dose spillage by creating a highly conformal isodose distribution around the periphery of the tumor. Significant hot spots are likely to be present inside the prescribed isodose line, usually 50% for Gamma Knife–based SRS and ranging from 70% to 85% for LINAC-based SRS/HSRT. By limiting the high-dose spillage, the III, IV, V, and VI nerves, which if involved by the tumor are expected to be at the periphery of the tumor, can be spared of an excessive dose of radiation from the hotspots.

Similar principles apply to the VII, VIII, IX, X, XI, and XII nerves. The use of highly conformal isodose plans can potentially spare cranial nerves (Figure 22.2). This can be accomplished in different ways depending on the device being used. For Gamma Knife, the use of smaller shots can create a highly conformal plan with a steep dose falloff beyond the tumor. For other LINAC-based treatment planning systems, the use of intensity-modulated radiotherapy (IMRT) or volumetric modulated arc therapy (VMAT) planning can achieve the same goals. The VII and VIII nerves are typically encountered in the setting of SRS or HSRT for vestibular schwannoma. For SRS, data in the literature suggest that a peripheral dose of 12–13 Gy in 1 fraction will result in <5% risk of VII nerve palsy and serviceable hearing rates of 44%–88% (Murphy and Suh, 2011). For vestibular schwannoma HSRT, regimens of 18 Gy in 3 fractions and 25 Gy in 5 fractions appear to be associated with good hearing preservation and low risk of VII nerve injury (Song and Williams, 1999; Meijer et al., 2003; Karam et al., 2013; Morimoto et al., 2013; Vivas et al., 2014). For the IX, X, XI, and XII nerves, the use of a dose of 15 Gy in 1 fraction for SRS or

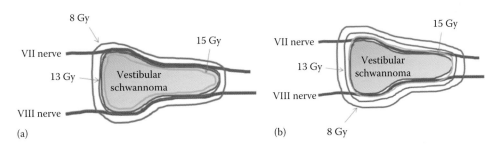

Figure 22.2 A dose of 13 Gy (red isodose line) in 1 fraction is prescribed to the vestibular tumor. The upper diagram (a) shows a highly conformal plan with the 15 Gy line located inside the tumor and the VII and VIII nerves that are displaced by the tumor to the periphery are receiving a dose close to the prescribed dose of 13 Gy. By contrast, the lower diagram (b) shows a nonconformal plan with significant high-dose spillage, resulting in the encompassment of the VII and VIII nerves within the 15 Gy line, potentially increasing the risk of cranial nerve injury.

25 Gy in 5 fractions for HSRT for glomus tumors appears to be associated with low to acceptable risk of injury (Guss et al., 2011; Ivan et al., 2011; Chun et al., 2014).

22.2.8.2 Vascular injury, brain stem injury, and radiation necrosis

Detailed dosimetric data on vascular injury caused by SRS or HSRT are very scarce. Based on the experience in the treatment of pituitary adenomas with SRS, the risk of internal carotid artery stenosis is very low at the dose range used (Witt, 2003). However, it is prudent to avoid having >50% of the circumference receiving the prescribed dose and to limit the maximum dose to 30 Gy in 1 fraction as suggested in the limited literature (Witt, 2003).

According to the literature and QUANTEC, the maximum point dose to the brain stem should be limited to 12.5 Gy in 1 fraction for SRS (Mayo et al., 2010b). For HSRT for tumors close to the brain stem, regimens of 21 Gy in 3 fractions and 25 Gy in 5 fractions appear to be safe for the brain stem (Song and Williams, 1999; Meijer et al., 2003; Karam et al., 2013; Morimoto et al., 2013; Vivas et al., 2014). When planning SRS or HSRT for tumors close to the brain stem, it is crucial to contour the brain stem to ascertain that the maximum brain stem dose does not exceed the tolerance. If SRS is Gamma Knife based, the isodose line can be shaped around the brain stem by manipulating the small shots as well as blocking one or more of the eight sectors for each shots. For LINAC-based SRS or HSRT, inverse planning of IMRT or VMAT can achieve the same purpose.

The risk of symptomatic radiation necrosis is a function of target location, volume treated, and radiation dose delivered. Data from UPMC has provided guidelines for estimation of risk of radiation necrosis from SRS for AVM, based on the SPIE and 12 Gy volume (Flickinger et al., 2000). This has become one of the most important strategies to decrease the risk of symptomatic necrosis after SRS for AVM. The applicability of 12 Gy volume as a parameter for risk estimation for non-AVM has been validated by the group from Case Western Reserve University. It appears that 12 Gy volume is a reasonable parameter to be used for SRS (Korytko et al., 2006). For HSRT, when a 5-fraction regimen is used, limited data seem to suggest that the volume encompassed by 28.8 Gy is a reasonable parameter to use to estimate the risk of symptomatic necrosis, but this is subject to further validation (Inoue et al., 2014).

22.3 SERIOUS LATE TOXICITIES FROM SPINAL IG-HSRT AND STRATEGIES TO MITIGATE RISK

This section will cover radiation myelopathy (RM), vertebral compression fracture (VCF), radiation plexopathy, neuropathy, pain flare, and esophageal injury.

22.3.1 RADIATION MYELOPATHY

Rare occurrence of RM has been observed after spinal SBRT in both radiation-naïve and radiation-reirradiated patients. Sahgal et al. (2013b) from Sunnybrook Health Science Centre at the University of Toronto have recently reported on nine cases of RM after spinal SBRT in radiation-naïve patients and provided an unprecedented detailed dose–volume histogram (DVH) analysis. The DVH data for the nine RM patients were compared to those of 66 controls. In the study, the thecal sac was contoured as a surrogate for spinal cord, taking into account potential sources of error such as physiologic spinal cord motion, intrafraction patient motion, variation in spinal cord delineation, potential errors in MRI and computed tomography (CT) image fusion, the treatment planning calculation algorithm, the image-guidance system, treatment couch motions, gantry rotation precision, and micro-multileaf collimator leaf position calibration. Based on the data analysis, to limit the risk of RM to below 5%, the thecal sac doses limited to 12.4 Gy in 1 fraction, 17 Gy in 2 fractions, 20.3 Gy in 3 fractions, 23 Gy in 4 fractions, and 25.3 Gy in 5 fractions were recommended (Sahgal et al., 2013b). Readers are advised to refer to Chapter 21 of this book for further details.

Sahgal et al. (2012) also performed analysis on five previously irradiated patients who developed RM after reirradiation with SBRT for spinal tumors, and the dosimetric parameters were compared with 16 retreatment spinal SBRT controls treated at the UCSF. Given the different fractionation regimens used, normalized 2 Gy equivalent biologically effective dose (nBED) was used to facilitate comparison.

The thecal sac cumulative nBED was then calculated by adding the nBED of the first course of radiotherapy to the point maximum nBED from the retreatment SBRT course, using an α/β ratio of two for the spinal cord. They concluded that the cumulative nBED to the thecal sac point maximum should not exceed 70 $Gy_{2/2}$ based on the conditions that the SBRT thecal sac retreatment point maximum nBED did not exceed 25 $Gy_{2/2}$, the thecal sac SBRT point maximum nBED/cumulative point maximum nBED ratio did not exceed 0.5, and the minimum time interval to retreatment was at least 5 months (Sahgal et al., 2012). Huang et al. (2013) reanalyzed these data using generalized linear-quadratic model and concluded that the thecal sac point maximum nBED should not exceed 70 $Gy_{2/2}$. Readers are advised to refer to Chapter 21 of this book for further details.

22.3.2 VERTEBRAL COMPRESSION FRACTURE

Vertebrae involved by metastases, especially lytic lesions, are predisposed to pathologic fractures due to the replacement of healthy bone with tumor. Radiotherapy, frequently used as a treatment for bone metastasis, can increase the risk of VCF, but the risk is deemed to be low in general with conventional palliative radiotherapy. However, HSRT or SBRT delivers ablative doses of radiation to the clinical target volume (CTV), which typically encompasses the entire vertebral body. Up until the recent years, there were no detailed data in the literature pertaining to VCF caused by spinal SBRT.

Colleagues from Memorial Sloan Kettering Cancer Center (MSKCC) first reported their observation of VCF after SBRT for spinal metastases to 18–24 Gy in 1 fraction, with a majority of patients receiving 24 Gy in 1 fraction (Rose et al., 2009). A progressive VCF rate of 39% was observed at a median time of 25 months. The location of spinal metastasis (above T10 vs T10 or below), the type of the spinal metastasis (lytic vs sclerotic and mixed), and the percentage of vertebral body involvement were identified as predictors of VCF (Rose et al., 2009). MD Anderson Cancer Center (MDACC) and University of Toronto also reported their series on VCF after spinal SBRT. However, they have reported a much lower rate of VCF occurring at 2–3.3 months after SBRT (Boehling et al., 2012; Cunha et al., 2012). In the MDACC study, the incidence of new or progressive VCF after SBRT in 93 patients with 123 spinal metastases was 20%. Approximately two-thirds of the patients in the MDACC series received either 27 Gy in 3 fractions or 20–30 Gy in 5 fractions. They have determined that factors predicting VCF were age >55 years, a preexisting fracture, and baseline pain (Boehling et al., 2012). Similarly, the University of Toronto study of 90 patients with 167 spinal metastases treated also included patients treated with 2–5 fractions in addition to those treated with 1 fraction. The presence of kyphosis/scoliosis, lytic appearance, primary lung and hepatocellular carcinoma, and a dose per fraction ≥20 Gy was identified as risk factors for VCF (Cunha et al., 2012). The crude rate of VCF was 11% and the 1-year fracture-free probability was 87.3%.

Sahgal et al. have pooled data of 252 patients with 410 spinal segments treated with SBRT at University of Toronto, MDACC, and Cleveland Clinic Foundation (CCF) and attempted to identify the risk factors of VCF. The Spinal Instability Neoplastic Scoring (SINS) system was also applied to determine predictive value (Sahgal et al., 2013a). There were 57 fractures (57 of 410, 14%), with 47% (27 of 57) new fractures, and 53% (30 of 57) fracture progression, and the median time to VCF was 2.46 months with 65% of VCFs occurring within the first 4 months. The 1- and 2-year cumulative incidences of fracture were 12.4% and 13.5%, respectively (Sahgal et al., 2013a). On multivariable analysis, dose per fraction (greatest risk for ≥24 vs 20–23 vs ≤19 Gy) and three of the six original SINS criteria, namely, baseline VCF, lytic tumor, and spinal deformity, were identified as significant predictors of VCF (Sahgal et al., 2013a).

22.3.3 RADIATION PLEXOPATHY/NEUROPATHY

The spinal nerves and nerve plexuses can potentially be injured by ablative doses of radiation delivered to the vertebrae via SBRT. Fortunately, radiation radiculopathy or plexopathy is uncommonly encountered after SBRT for spinal tumors. In a phase I/II trial of single-dose SBRT for radiation-naïve spinal metastases conducted at MDACC, 10 of 61 patients treated developed mild (grade 1 or 2) numbness and tingling and one developed grade 3 radiculopathy at L5 after 16–24 Gy (Garg et al., 2012). In a study from Beth Israel Deaconess Hospital, four cases of persistent or new radiculopathy were observed among their 60 patients treated with SBRT for recurrent epidural spinal metastases. However, it is uncertain whether the complications were caused by tumor progression, radiation injury of the spinal nerves, or a combination of

Table 22.1 Normal tissue constraints used by Radiation Therapy Oncology Group (RTOG) trials

NEURAL STRUCTURE	1 FRACTION	3 FRACTIONS	4 FRACTIONS	5 FRACTIONS
Brachial plexus	RTOG 0631 and 0915: 17.5 Gy (<0.03 cc or maximum)/14 Gy (<3 cc)	RTOG 0236 and 0618: 24 Gy (maximum) RTOG 1021: 24 Gy (maximum)/20.4 Gy (<3 cc)	RTOG 0915: 27.2 Gy (maximum)/ 23.6 Gy (<3 cc)	RTOG 0813: 32 Gy (maximum)/ 30 Gy (<3 cc)
Cauda equina	RTOG 0631: 16 Gy (<0.03 cc)/14 Gy (<5 cc)	Not available	Not available	Not available
Sacral plexus	RTOG 0631: 18 Gy (<0.03 cc)/14.4 Gy (<5 cc)	Not available	Not available	Not available

Source: Data extracted from www.rtog.org.
Disclaimer: These dose constraints are intended to be used in RTOG trials and have not been thoroughly tested clinically. The authors do not assume responsibility for the use of these dose limits.

both (Mahadevan et al., 2011) since all those patients had evidence of radiological progression of disease. In a study from MDACC, two cases of grade 3 lumbar plexopathy on reirradiation were observed with SBRT for recurrent spinal metastases in 59 patients (Garg et al., 2011).

Colleagues from Indiana University attempted to determine the tolerance of the brachial plexus to SBRT based on 36 patients with 37 apical primary lung cancer treated to a dose of 30–72 Gy in 3–4 fractions (Forquer et al., 2009). There were seven cases of grade 2–4 brachial plexopathy observed, and the cutoff dose was determined to be 26 Gy in 3–4 fractions. This result corroborates the constraint of 24 Gy in 3 fractions used in Radiation Therapy Oncology Group (RTOG) trials (Forquer et al., 2009). The 2-year brachial plexopathy rates were 46% and 8% when the maximum brachial plexus dose was >26 and ≤26 Gy, respectively (Forquer et al., 2009). However, in this study, the subclavian/axillary vessels served as a surrogate for brachial plexus, instead of the full brachial plexus contour. The constraints for nerves used by RTOG are listed in Table 22.1.

22.3.4 PAIN FLARE

Pain flare is a common side effect of SBRT for spinal metastases. Although it is in practical terms not a late complication, given the common occurrence, it is covered in this chapter. In the recent years, high-quality data have emerged from high-volume centers such as Sunnybrook Health Science Centre, University of Toronto, and the MDACC, Texas (Chiang et al., 2013; Pan et al., 2014; Khan et al., 2015). Colleagues from University of Toronto conducted a prospective clinical trial to determine the incidence of pain flare after spine SBRT in steroid-naïve patients and identify predictive factors. A total of 41 patients were enrolled and 18 patients were treated with 20–24 Gy in a single fraction, whereas 23 patients were treated with 24–35 Gy in 2–5 fractions. Pain flare was observed in 68.3% of patients, most commonly on day 1 after spinal SBRT. Significant predictive factors for pain flare included a higher Karnofsky performance status and tumor location in the cervical or lumbar region (Chiang et al., 2013). In those patients treated with dexamethasone, a significant decrease in pain scores over time were subsequently observed (Chiang et al., 2013). In a subsequent prospective observational study from University of Toronto, 47 patients were prophylactically treated with dexamethasone during spinal SBRT. The first cohort of 24 patients was treated with 4 mg of dexamethasone, and the second cohort of 23 patients with 8 mg of dexamethasone (Khan et al., 2015). The Brief Pain Inventory was used to score pain and functional interference each day during SBRT and for 10 days afterward. The total incidence of pain flare was 19%, and the incidence in the 4 and 8 mg cohorts was 25% and 13% (not significantly different), respectively (Khan et al., 2015).

The 4 mg cohort had better profile in walking ability as well as relationships with others, compared to the 8 mg cohort. Compared to the earlier steroid-naïve cohort, the use of dexamethasone was associated with lower worst pain scores and improved general activity interference outcome (Khan et al., 2015). In a secondary analysis of the phase I/II trials of SBRT for spinal tumors from MDACC, the incidence of pain flare was 23% and the median time to onset was 5 days. The only independent factor associated with pain flare was the number of fractions with a single-fraction regimen resulting in the highest rate (Pan et al., 2014). In that study, rigorous daily pain evaluation was not conducted, and therefore, it is difficult to compare the MDACC data to the discussed University of Toronto study.

22.3.5 ESOPHAGEAL TOXICITY

The esophagus is located adjacent to the thoracic spinal column and is, therefore, susceptible to toxicity from SBRT to spinal tumors, especially when delivered in a single fraction. The most robust data on esophageal toxicity from single-fraction spinal SBRT came from MSKCC. Two hundred and four spinal metastases abutting the esophagus in 182 patients were treated with single-dose SBRT to 24 Gy. The esophageal toxicity was scored using National Cancer Institute Common Toxicity Criteria for Adverse Events (version 4.0). The rates of acute and late esophageal toxicities were 15% and 12%, respectively (Cox et al., 2012). Grade 3 or higher acute or late toxicities occurred in 14 (6.8%) patients. For grade 3 or higher esophageal toxicities, the median splits were determined, and those for D2.5 cm^3 (the minimum dose to the 2.5 cc receiving the highest dose), V12 (the volume receiving at least 12 Gy), V15, V20, and V22 were 14 Gy, 3.78, 1.87, 0.11, and 0, respectively (Cox et al., 2012). The authors recommended the maximum point dose to be kept <22 Gy. Most notably, among those seven patients who developed grade 4 or higher toxicities, they had either radiation recall reactions after chemotherapy with doxorubicin or gemcitabine, or iatrogenic esophageal manipulation such as biopsy, dilatation, and stent placement (Cox et al., 2012).

Stephans et al. from CCF analyzed 52 patients treated with SBRT for liver or lung tumors who had a planning target volume (PTV) within 2 cm of the esophagus, trying to determine the esophageal tolerance to SBRT. The radiation dose given was 37.5–60 Gy in 3–10 fractions (median 50 Gy in 5 fractions). Two patients developed esophageal fistula and their maximum esophageal point doses were 51.5 and 52 Gy, and 1 cc doses were 48.1 and 50 Gy, respectively (Stephans et al., 2014). Interestingly, they both received adjuvant antiangiogenic agents within 2 months of completing SBRT.

22.3.6 STRATEGIES TO MITIGATE RISKS OF COMPLICATIONS FROM SPINAL IG-HSRT

To minimize the risks of serious complications from spinal SBRT or IG-HSRT, all relevant organs at risk (OARs) including the spinal cord, cauda equina, nerve plexuses/roots, and esophagus must be contoured, and the dose constraint for each OAR must be respected (Sahgal et al., 2008; Lo et al., 2010; Foote et al., 2011). Fusion of the spinal MRI (axial T1 and T2 sequences) with the treatment planning CT should be performed to facilitate the accurate contouring of neural structures such as the spinal cord and cauda equina. The accuracy of the fusion must be verified before the image sets are used for delineation of the neural structures. Alternatively, a CT myelogram can be used for delineation of the spinal cord in patients with contraindications to MRI or when there are significant metallic artifacts as in postoperative cases. It is imperative that the window leveling is correct; otherwise, the cord contour could be drawn incorrectly, resulting in an inaccurate determination of cord dose, which can potentially be detrimental.

Another important aspect to consider to avoid inaccurate determination of cord dose is the selection of an appropriate treatment planning algorithm, especially in the thoracic region. Okoye and colleagues from University Hospital Seidman Cancer Center, Case Western Reserve University, reviewed CyberKnife-based SBRT treatment plans in 37 patients with thoracic spinal tumors generated using Ray Tracing and Monte Carlo algorithms. They found discrepancies in the coverage of the PTV as well as the actual doses delivered to various OARs including the spinal cord (Okoye et al., 2015). In 14% of lesions, the actual dose delivered to the spinal cord determined with Monte Carlo calculation was found to be ≥5% (Okoye et al., 2015). These data underscore the importance of an optimal treatment planning algorithm especially in regions where there are tissues with substantially different electronic densities, such as the lungs. Interested readers are encouraged to visit http://rpc.mdanderson.org/rpc/Services/Anthropomorphic_%20Phantoms/TPS%20-%20algorithm%20list%20updated.pdf.

Given the fact that the spinal cord is in very close proximity to the spinal CTV, and the steep dose gradient between the spinal cord and spinal CTV, even a slight deviation in positioning can result in significant overdosing of the spinal cord, resulting in catastrophic complications such as RM (Wang et al., 2008). Therefore, robust immobilization is crucial. Colleagues from the University of Toronto have demonstrated that the BodyFIX (Elekta, Stockholm, Sweden) near-rigid body immobilization system is more robust in minimizing intrafraction motions than a simple vac-loc system, and it can limit the set-up error to 2 mm (Li et al., 2012). Therefore, a dual vacuum system is recommended for immobilization for spinal SBRT unless a CyberKnife system is used, as it can robotically track the spine in a near-real-time fashion.

Despite the most robust immobilization system and advanced technology, intrafraction patient motion can occur especially when the treatment time is anticipated to be long. Colleagues from University of Toronto showed that there could be an intrafraction motion of 1.2 mm and 1° even with near-rigid body immobilization with BodyFIX, image guidance with kilovoltage cone-beam CT, and a robotic couch capable of adjusting shifts with 6 degrees of freedom for their LINAC-based SBRT treatments (Hyde et al., 2012). In order to attain this level of precision, they recommended an intrafraction repeat cone-beam CT every 20 minutes to check for any positional deviation. With the availability of newer technologies such as VMAT and the high-dose rate flattening filter-free feature, treatment time can potentially be reduced dramatically, and in that situation, intrafraction cone-beam CT will not be unnecessary. Physiologic spinal cord motion can also potentially create uncertainties in the estimation of true spinal cord dose from SBRT. Data from multiple institutions have showed that the spinal cord motion is typically <1 mm (Cai et al., 2007; Tseng et al., 2015). Although some spinal SBRT physicians from very experienced centers use the true spinal cord as the avoidance structure for inverse planning, it is prudent to create a planning-at-risk volume (PRV), which typically ranges from 1.5 to 2.0 mm. Alternatively, the thecal sac can be contoured as a surrogate for spinal cord. This practice serves to decrease the risk of RM resulting from potential errors that can lead to overdosing of the spinal cord (Sahgal et al., 2012, 2013b).

Based on the analysis of radiation-naïve and radiation-reirradiated patients who developed RM after spinal SBRT, Sahgal et al. (2012, 2013b) have made recommendations on spinal cord constraints, and these have been discussed in detail in an earlier section of this chapter and an earlier chapter. When the recommendations are strictly followed, no RM has been observed at the University of Toronto in the past 7 years and approximately 1000 cases were treated (personal communication with one of the coauthors AS). Alternatively, in the setting of reirradiation with prior conventional radiotherapy dose ≤45 Gy (1.8–2.0 Gy per fraction), authors of this chapter (including SSL and ELC) have not observed RM when spinal cord dose constraints of 10 Gy in 5 fractions or 9 Gy in 3 fractions are used.

Based on the studies available, certain risk factors have been identified to predispose patients to VCF after spinal SBRT. It is reasonable to avoid using a single fraction of ≥20 Gy, especially in patients with risk factors such as baseline VCF, lytic tumor, and spinal deformity, all of which are parameters of SINS (Cunha et al., 2012; Sahgal et al., 2013a). Prophylactic kyphoplasty or vertebroplasty can be considered for patients with preexisting VCF before SBRT to reduce the risk of further fracture, relieve the mechanical pain, and render SBRT more tolerable.

In order to spare the nerve plexuses and nerves from damage from SBRT, these structures must be carefully and accurately contoured apart from respecting their tolerance to ablative radiation. MRI, especially T2 sequence, can be fused with treatment planning CT to facilitate accurate contouring of these structures. The sparing of these neural structures is particularly important at levels where the nerve roots or nerve plexuses are responsible for motor function of the extremities. Damage of nerve function at those levels will result in loss of vital extremity function. RTOG has included a brachial plexus contouring atlas on their website, and interested readers are encouraged to access the document (http://www.rtog.org/CoreLab/ContouringAtlases/BrachialPlexus-ContouringAtlas.aspx).

Pain flare is a relatively common phenomenon after spinal SBRT. Investigators from Sunnybrook Health Science Centre, University of Toronto, have reported in their prospective trial that approximately 2/3 of patients developed pain flare during or after SBRT, and the symptom can be effectively controlled with steroid therapy (Chiang et al., 2013). In another study, prophylactic dexamethasone with rapid taper has resulted in a much lower pain flare rate (Khan et al., 2015). There was no difference in pain flare rates between 4 and 8 mg per dose, but patients had better subsequent ability to walk and social relationships with others in the 4 mg cohort. It appears that dexamethasone 4 mg starting day 1, on days of SBRT, and 4 days post-SBRT, as used in the study, should be adequate for prophylaxis against pain flare. In the United States, a 6-day course with a Medrol (methylprednisolone) Dose Pack is very popular, and it can effectively control pain flare in most cases.

There can be considerable radiation dose exposure to the esophagus when the spinal segment to be treated with SBRT is located in the thoracic region. Therefore, it is imperative that the esophagus is contoured as an OAR and the dose constraints have to be respected in order to avoid serious complications. Data from MSKCC showed that the maximum single dose tolerated was volume dependent, and it was discussed in a previous section of this chapter (Cox et al., 2012). They also identified risks factors for serious post-SBRT esophageal toxicities, namely, post-SBRT doxorubicin- or gemcitabine-based chemotherapy or surgical manipulation of the esophagus (Cox et al., 2012). These treatments or procedures should be avoided after substantial radiation dose exposure from thoracic spinal SBRT. The use of a multiple session regimen (2–5 fractions) may also decrease the risk of serious esophageal toxicities from SBRT if the dose constraint cannot be met in 1 fraction. Data from CCF have showed that antiangiogenic agents should be avoided if the esophagus has been exposed to ablative doses of radiation from SBRT (Stephans et al., 2014). However, the study was based on lung tumors and, therefore, the dose used was beyond the range used for spinal tumors.

22.4 CONCLUSION

Serious complications have been observed after HSRT for intracranial and spinal tumors. In most cases, there are risk factors which can be identified. With the accumulation of experience with intracranial and spinal HSRT, normal tissue constraints of various OARs are better understood. With the sophistication of radiation technology and advancement of clinical expertise, it is possible to deliver very potent doses of radiation to brain and spinal tumors without causing excessive toxicities.

REFERENCES

Boehling NS, Grosshans DR, Allen PK, McAleer MF, Burton AW, Azeem S, Rhines LD, Chang EL (2012) Vertebral compression fracture risk after stereotactic body radiotherapy for spinal metastases. *J Neurosurg Spine* 16:379–386.
Cai J, Sheng K, Sheehan JP, Benedict SH, Larner JM, Read PW (2007) Evaluation of thoracic spinal cord motion using dynamic MRI. *Radiother Oncol* 84:279–282.
Chiang A, Zeng L, Zhang L, Lochray F, Korol R, Loblaw A, Chow E, Sahgal A (2013) Pain flare is a common adverse event in steroid-naive patients after spine stereotactic body radiation therapy: A prospective clinical trial. *Int J Radiat Oncol Biol Phys* 86:638–642.
Chun SG, Nedzi LA, Choe KS, Abdulrahman RE, Chen SA, Yordy JS, Timmerman RD, Kutz JW, Isaacson B (2014) A retrospective analysis of tumor volumetric responses to five-fraction stereotactic radiotherapy for paragangliomas of the head and neck (glomus tumors). *Stereotact Funct Neurosurg* 92:153–159.
Clark BG, Souhami L, Pla C, Al-Amro AS, Bahary JP, Villemure JG, Caron JL, Olivier A, Podgorsak EB (1998) The integral biologically effective dose to predict brain stem toxicity of hypofractionated stereotactic radiotherapy. *Int J Radiat Oncol Biol Phys* 40:667–675.
Conti A, Pontoriero A, Midili F, Iati G, Siragusa C, Tomasello C, La Torre D, Cardali SM, Pergolizzi S, De Renzis C (2015) CyberKnife multisession stereotactic radiosurgery and hypofractionated stereotactic radiotherapy for perioptic meningiomas: Intermediate-term results and radiobiological considerations. *Springerplus* 4:37.
Cox BW, Jackson A, Hunt M, Bilsky M, Yamada Y (2012) Esophageal toxicity from high-dose, single-fraction paraspinal stereotactic radiosurgery. *Int J Radiat Oncol Biol Phys* 83:e661–e667.
Cunha MV, Al-Omair A, Atenafu EG, Masucci GL, Letourneau D, Korol R, Yu E et al. (2012) Vertebral compression fracture (VCF) after spine stereotactic body radiation therapy (SBRT): Analysis of predictive factors. *Int J Radiat Oncol Biol Phys* 84:e343–e349.
Flickinger JC, Kondziolka D, Lunsford LD, Kassam A, Phuong LK, Liscak R, Pollock B (2000) Development of a model to predict permanent symptomatic postradiosurgery injury for arteriovenous malformation patients. Arteriovenous Malformation Radiosurgery Study Group. *Int J Radiat Oncol Biol Phys* 46:1143–1148.
Flickinger JC, Kondziolka D, Niranjan A, Maitz A, Voynov G, Lunsford LD (2004) Acoustic neuroma radiosurgery with marginal tumor doses of 12 to 13 Gy. *Int J Radiat Oncol Biol Phys* 60:225–230.

Foote M, Letourneau D, Hyde D, Massicotte E, Rampersaud R, Fehlings M, Fisher C et al. (2011) Technique for stereotactic body radiotherapy for spinal metastases. *J Clin Neurosci* 18:276–279.

Forquer JA, Fakiris AJ, Timmerman RD, Lo SS, Perkins SM, McGarry RC, Johnstone PA (2009) Brachial plexopathy from stereotactic body radiotherapy in early-stage NSCLC: Dose-limiting toxicity in apical tumor sites. *Radiother Oncol* 93:408–413.

Garg AK, Shiu AS, Yang J, Wang XS, Allen P, Brown BW, Grossman P et al. (2012) Phase 1/2 trial of single-session stereotactic body radiotherapy for previously unirradiated spinal metastases. *Cancer* 118:5069–5077.

Garg AK, Wang XS, Shiu AS, Allen P, Yang J, McAleer MF, Azeem S, Rhines LD, Chang EL (2011) Prospective evaluation of spinal reirradiation by using stereotactic body radiation therapy: The University of Texas MD Anderson Cancer Center experience. *Cancer* 117:3509–3516.

Guss ZD, Batra S, Limb CJ, Li G, Sughrue ME, Redmond K, Rigamonti D et al. (2011) Radiosurgery of glomus jugulare tumors: A meta-analysis. *Int J Radiat Oncol Biol Phys* 81:e497–e502.

Huang Z, Mayr NA, Yuh WT, Wang JZ, Lo SS (2013) Reirradiation with stereotactic body radiotherapy: Analysis of human spinal cord tolerance using the generalized linear-quadratic model. *Future Oncol* 9:879–887.

Hyde D, Lochray F, Korol R, Davidson M, Wong CS, Ma L, Sahgal A (2012) Spine stereotactic body radiotherapy utilizing cone-beam CT image-guidance with a robotic couch: Intrafraction motion analysis accounting for all six degrees of freedom. *Int J Radiat Oncol Biol Phys* 82:e555–e562.

Inoue HK, Sato H, Seto K, Torikai K, Suzuki Y, Saitoh J, Noda SE, Nakano T (2014) Five-fraction CyberKnife radiotherapy for large brain metastases in critical areas: Impact on the surrounding brain volumes circumscribed with a single dose equivalent of 14 Gy (V14) to avoid radiation necrosis. *J Radiat Res* 55:334–342.

Ivan ME, Sughrue ME, Clark AJ, Kane AJ, Aranda D, Barani IJ, Parsa AT (2011) A meta-analysis of tumor control rates and treatment-related morbidity for patients with glomus jugulare tumors. *J Neurosurg* 114:1299–1305.

Iwata H, Sato K, Tatewaki K, Yokota N, Inoue M, Baba Y, Shibamoto Y (2011) Hypofractionated stereotactic radiotherapy with CyberKnife for nonfunctioning pituitary adenoma: High local control with low toxicity. *Neuro Oncol* 13(8):916–922.

Kanesaka N, Mikami R, Nakayama H, Nogi S, Tajima Y, Nakajima N, Wada J et al. (2011) Preliminary results of fractionated stereotactic radiotherapy after cyst drainage for craniopharyngioma in adults. *Int J Radiat Oncol Biol Phys* 82(4):1356–1360.

Karam SD, Tai A, Strohl A, Steehler MK, Rashid A, Gagnon G, Harter KW et al. (2013) Frameless fractionated stereotactic radiosurgery for vestibular schwannomas: A single-institution experience. *Front Oncol* 3:121.

Khan L, Chiang A, Zhang L, Thibault I, Bedard G, Wong E, Loblaw A et al. (2015) Prophylactic dexamethasone effectively reduces the incidence of pain flare following spine stereotactic body radiotherapy (SBRT): A prospective observational study. *Support Care Cancer* 23(10):2937–2943.

Killory BD, Kresl JJ, Wait SD, Ponce FA, Porter R, White WL (2009) Hypofractionated CyberKnife radiosurgery for perichiasmatic pituitary adenomas: Early results. *Neurosurgery* 64:A19–A25.

Kondziolka D, Lunsford LD, McLaughlin MR, Flickinger JC (1998) Long-term outcomes after radiosurgery for acoustic neuromas. *N Engl J Med* 339:1426–1433.

Korytko T, Radivoyevitch T, Colussi V, Wessels BW, Pillai K, Maciunas RJ, Einstein DB (2006) 12 Gy gamma knife radiosurgical volume is a predictor for radiation necrosis in non-AVM intracranial tumors. *Int J Radiat Oncol Biol Phys* 64:419–424.

Leavitt JA, Stafford SL, Link MJ, Pollock BE (2013) Long-term evaluation of radiation-induced optic neuropathy after single-fraction stereotactic radiosurgery. *Int J Radiat Oncol Biol Phys* 87:524–527.

Li W, Sahgal A, Foote M, Millar BA, Jaffray DA, Letourneau D (2012) Impact of immobilization on intrafraction motion for spine stereotactic body radiotherapy using cone beam computed tomography. *Int J Radiat Oncol Biol Phys* 84:520–526.

Lindvall P, Bergstrom P, Blomquist M, Bergenheim AT (2010) Radiation schedules in relation to obliteration and complications in hypofractionated conformal stereotactic radiotherapy of arteriovenous malformations. *Stereotact Funct Neurosurg* 88:24–28.

Lo SS, Sahgal A, Wang JZ, Mayr NA, Sloan A, Mendel E, Chang EL (2010) Stereotactic body radiation therapy for spinal metastases. *Discov Med* 9:289–296.

Mahadevan A, Floyd S, Wong E, Jeyapalan S, Groff M, Kasper E (2011) Stereotactic body radiotherapy reirradiation for recurrent epidural spinal metastases. *Int J Radiat Oncol Biol Phys* 81:1500–1505.

Mayo C, Martel MK, Marks LB, Flickinger J, Nam J, Kirkpatrick J (2010a) Radiation dose-volume effects of optic nerves and chiasm. *Int J Radiat Oncol Biol Phys* 76:S28–S35.

Mayo C, Yorke E, Merchant TE (2010b) Radiation associated brainstem injury. *Int J Radiat Oncol Biol Phys* 76:S36–S41.

Meijer OW, Vandertop WP, Baayen JC, Slotman BJ (2003) Single-fraction vs. fractionated linac-based stereotactic radiosurgery for vestibular schwannoma: A single-institution study. *Int J Radiat Oncol Biol Phys* 56:1390–1396.

Morimoto M, Yoshioka Y, Kotsuma T, Adachi K, Shiomi H, Suzuki O, Seo Y et al. (2013) Hypofractionated stereotactic radiation therapy in three to five fractions for vestibular schwannoma. *Jpn J Clin Oncol* 43:805–812.

Murphy ES, Suh JH (2011) Radiotherapy for vestibular schwannomas: A critical review. *Int J Radiat Oncol Biol Phys* 79:985–997.

Nguyen JH, Chen CJ, Lee CC, Yen CP, Xu Z, Schlesinger D, Sheehan JP (2014) Multisession gamma knife radiosurgery: A preliminary experience with a noninvasive, relocatable frame. *World Neurosurg* 82:1256–1263.

Okoye CC, Patel RB, Hasan S, Podder T, Khouri A, Fabien J, Zhang Y et al. (2015) Comparison of ray tracing and monte carlo calculation algorithms for thoracic spine lesions treated with cyberknife-based stereotactic body radiation therapy. *Technol Cancer Res Treat* 2015 Jan 28. pii: 1533034614568026. [Epub ahead of print].

Pan HY, Allen PK, Wang XS, Chang EL, Rhines LD, Tatsui CE, Amini B, Wang XA, Tannir NM, Brown PD, Ghia AJ (2014) Incidence and predictive factors of pain flare after spine stereotactic body radiation therapy: Secondary analysis of phase 1/2 trials. *Int J Radiat Oncol Biol Phys* 90:870–876.

Petit JH, Hudes RS, Chen TT, Eisenberg HM, Simard JM, Chin LS (2001) Reduced-dose radiosurgery for vestibular schwannomas. *Neurosurgery* 49:1299–1306; discussion 1306–1297.

Pollock BE, Link MJ, Leavitt JA, Stafford SL (2014) Dose-volume analysis of radiation-induced optic neuropathy after single-fraction stereotactic radiosurgery. *Neurosurgery* 75:456–460; discussion 460.

Rose PS, Laufer I, Boland PJ, Hanover A, Bilsky MH, Yamada J, Lis E (2009) Risk of fracture after single fraction image-guided intensity-modulated radiation therapy to spinal metastases. *J Clin Oncol* 27:5075–5079.

Sahgal A, Atenafu EG, Chao S, Al-Omair A, Boehling N, Balagamwala EH, Cunha M et al. (2013a) Vertebral compression fracture after spine stereotactic body radiotherapy: A multi-institutional analysis with a focus on radiation dose and the spinal instability neoplastic score. *J Clin Oncol* 31:3426–3431.

Sahgal A, Larson DA, Chang EL (2008) Stereotactic body radiosurgery for spinal metastases: A critical review. *Int J Radiat Oncol Biol Phys* 71:652–665.

Sahgal A, Ma L, Weinberg V, Gibbs IC, Chao S, Chang UK, Werner-Wasik M et al. (2012) Reirradiation human spinal cord tolerance for stereotactic body radiotherapy. *Int J Radiat Oncol Biol Phys* 82:107–116.

Sahgal A, Weinberg V, Ma L, Chang E, Chao S, Muacevic A, Gorgulho A et al. (2013b) Probabilities of radiation myelopathy specific to stereotactic body radiation therapy to guide safe practice. *Int J Radiat Oncol Biol Phys* 85:341–347.

Song DY, Williams JA (1999) Fractionated stereotactic radiosurgery for treatment of acoustic neuromas. *Stereotact Funct Neurosurg* 73:45–49.

Spiegelmann R, Cohen ZR, Nissim O, Alezra D, Pfeffer R (2010) Cavernous sinus meningiomas: A large LINAC radiosurgery series. *J Neurooncol* 98:195–202.

Stafford SL, Pollock BE, Leavitt JA, Foote RL, Brown PD, Link MJ, Gorman DA, Schomberg PJ (2003) A study on the radiation tolerance of the optic nerves and chiasm after stereotactic radiosurgery. *Int J Radiat Oncol Biol Phys* 55:1177–1181.

Stephans KL, Djemil T, Diaconu C, Reddy CA, Xia P, Woody NM, Greskovich J, Makkar V, Videtic GM (2014) Esophageal dose tolerance to hypofractionated stereotactic body radiation therapy: Risk factors for late toxicity. *Int J Radiat Oncol Biol Phys* 90:197–202.

Tishler RB, Loeffler JS, Lunsford LD, Duma C, Alexander E III, Kooy HM, Flickinger JC (1993) Tolerance of cranial nerves of the cavernous sinus to radiosurgery. *Int J Radiat Oncol Biol Phys* 27:215–221.

Tsai JT, Lin JW, Lin CM, Chen YH, Ma HI, Jen YM, Ju DT (2013) Clinical evaluation of CyberKnife in the treatment of vestibular schwannomas. *Biomed Res Int* 2013:297093.

Tseng CL, Sussman MS, Atenafu EG, Letourneau D, Ma L, Soliman H, Thibault I et al. (2015) Magnetic resonance imaging assessment of spinal cord and cauda equina motion in supine patients with spinal metastases planned for spine stereotactic body radiation therapy. *Int J Radiat Oncol Biol Phys* 91:995–1002.

Vivas EX, Wegner R, Conley G, Torok J, Heron DE, Kabolizadeh P, Burton S, Ozhasoglu C, Quinn A, Hirsch BE (2014) Treatment outcomes in patients treated with CyberKnife radiosurgery for vestibular schwannoma. *Otol Neurotol* 35:162–170.

Wang H, Shiu A, Wang C, O'Daniel J, Mahajan A, Woo S, Liengsawangwong P, Mohan R, Chang EL (2008) Dosimetric effect of translational and rotational errors for patients undergoing image-guided stereotactic body radiotherapy for spinal metastases. *Int J Radiat Oncol Biol Phys* 71:1261–1271.

Wang X, Liu X, Mei G, Dai J, Pan L, Wang E (2012) Phase II study to assess the efficacy of hypofractionated stereotactic radiotherapy in patients with large cavernous sinus hemangiomas. *Int J Radiat Oncol Biol Phys* 83:e223–e230.

Witt TC (2003) Stereotactic radiosurgery for pituitary tumors. *Neurosurg Focus* 14:e10.

Index